T0229081

NORTH-H
PERSONAL LIBRARY

SOLITONS AND INSTANTONS

AN INTRODUCTION TO SOLITONS AND INSTANTONS IN QUANTUM FIELD THEORY

SOLITONS AND INSTANTONS

An Introduction to Solitons and Instantons in Quantum Field Theory

R. RAJARAMAN

Centre for Theoretical Studies,
Indian Institute of Science,
Bangalore 560012, India

NHPC

ELSEVIER
Amsterdam - Lausanne - New York - Oxford - Shannon - Tokyo

ELSEVIER SCIENCE B.V.
Sara Burgerhartstraat 25
P.O. Box 211, 1000 AE Amsterdam, The Netherlands

Hardbound edition 1982, Second impression 1984.
Paperback edition 1987, Second impression 1989, Third impression 1996, Fourth impression 2003

Library of Congress Cataloging in Publication Data
Rajaraman, R.
  Solitions and instantions.

  Bibliography: p.
  Includes index.
  1.Quantum field theory, 2.Solitons.
3.Instantons I.Title.
QC174.45.R34   530.1'43   81-22510
                            AACR2

ISBN: 0-444-87047-4

Transferred to digital printing 2005.

CONTENTS

ACKNOWLEDGEMENTS

Whatever understanding I have of this field has been gained largely from three people, through a combination of their written and spoken word. These are Professors Sidney Coleman, Roger Dashen and Roman Jackiw. I am very grateful to them.

In addition, I have benefited from discussions on this subject with numerous physicists from around the world. I thank them all: although it is not practicable to list all their names, some on whom I have imposed more than on others must be mentioned. These include Professors I. Affleck, N. Christ, B. Hasslacher, A. Neveu, R. Shankar, J. Strathdee, E. C. G. Sudarshan, C. Thorn, E. J. Weinberg and E. Witten. I am also indebted to Professors H. R. Krishnamurthy, R. Nityananda and S. Wadia for reading portions of the manuscript and for their valuable comments. Mrs. M. Raj Lakshmi's assistance in correcting the galley proofs is acknowledged with thanks. The blame for surviving errors, if any, falls entirely on me and not on those mentioned above.

I thank the Indian Institute of Science for permission to write this book, the Center for Continuing Education at the Institute for a grant towards preparing the typed manuscript, and Mr. M. S. Nagaraja for his efficient secretarial assistance. The generosity of the Institute for Advanced Study at Princeton, Harvard University and the Massachussetts Institute of Technology in offering visiting appointments which enabled me to interact with leading experts in this field is appreciated.

Last, but not least, I am grateful to my wife Indira and my parents for their encouragement.

CHAPTER 1

PREFACE CUM INTRODUCTION

We are deviating a little from normal practice in combining the preface with the introductory chapter. This is because we felt that a discussion of the scope of the book – the choice of topics included, the level of presentation and other qualifications and apologies – which normally appears separately in a Preface, would be more meaningful alongside a brief overview and assessment of this recently developed field.

This book centres on obtaining information about relativistic quantum field theories starting from classical solutions of the corresponding non-linear field equations. Apart from a few important but preliminary papers in earlier decades, the systematic development of this subject took place only in the mid-seventies. The first step in this development was the availability of solutions to an increasing number of classical relativistic non-linear wave equations. Initially limited to two-dimensional theories of single scalar fields, classical solutions (exact, or in good approximation) also became available soon thereafter for some coupled field equations involving scalar, vector and Dirac fields in higher dimensions.

These classical solutions correspond in some cases to field equations in the Minkowskian metric and in others to Euclidean equations. The use to which they are put in quantum theory differs accordingly. In both cases, an important characteristic of the solutions we will be concerned with is that they are localised. In particular, the Minkowskian solutions will have finite energy with a localised, non-dispersive energy density. Generally, they will travel undistorted in shape, with some uniform velocity. In the bulk of the literature such solutions are called "solitons", and we shall use the same term. (Strictly speaking they should be called solitary waves. The definition of solitons calls for more stringent requirements, not satisfied by many of our solutions. The finer distinction between these terms is explained in the next chapter.) In that they are non-dispersive localised packets of energy moving uniformly, solitons resemble extended particles, even though they are solutions of non-linear *wave* equations. Now, elementary particles in nature are also localised packets of energy, and are furthermore believed to be described by some relativistic field theory. Of course, the field theories describing elementary particles are *quantum* theories, whereas the solitons are, to start with, solutions of *classical* field equations. This provided the motivation for the next stage in the development of this subject, where the correspondence between classical soliton solutions of any given field theory and extended-particle states of the quantised version of that theory was sought. On intuitive grounds one would expect some such correspondence, but a systematic formalism for establishing it is a non-trivial matter. Procedures for establishing this correspondence, which may be termed 'quantisation of solitons', were developed using a variety of techniques by several independent sets of theorists during the years 1974–75. These methods amount to a generalisation of the well-known semiclassical expansion in non-relativistic quantum mechanics to relativistic quantum field theory. Through such a generalisation, it was shown that one could associate not only a quantum soliton-particle state with a classical soliton solution, but a whole series of excited states as well, by quantising fluctuations about the soliton. Properties of the quantum soliton particle such as its mass or form factors could be obtained in a systematic semiclassical expansion starting from corresponding properties of the classical soliton. Important technical difficulties like the ultraviolet renormalisation of quantum fluctuations and the zero-mode problems associated with continuous symmetries were investigated and treated.

As an offshoot of these developments was born the subject of instanton physics. Instantons are localised finite-action classical solutions of the *Euclidean* version of the field equations of any given model. Such solutions were obtained in exact analytical form for a variety of models including quantum chromodynamics (QCD). At the classical level, instantons are not very different from *static* soliton solutions of time-independent field equations. In fact, in most cases, instantons of a model in D dimensions are essentially the same as the static solitons of the same model in $D+1$ dimensions. However, in their impact on the corresponding quantum field theory, instantons are very different from solitons. Whereas the latter lead to extended-particle states, the former lead to tunnelling effects that can significantly affect the structure of the vacuum state. The vacuum state is the cornerstone of a quantum field theory. Any alterations in it will affect many other properties of the theory. For instance, in the two-dimensional abelian Higgs model, vacuum tunnelling due to instantons leads in turn to the confinement of charges – an effect that would not be revealed by standard perturbation theory. In QCD, where quark-confinement mechanisms have been eagerly sought, early hopes that instantons may lead to confinement have not materialised. But even in that model, instantons do yield significant contributions to the inter-quark forces.

Two features shared by both solitons and instantons make their quantum effects particularly interesting:

(i) Most of the soliton and instanton solutions are non-perturbative. That is, they cannot be obtained by starting from solutions of the corresponding linear part of the field equations and treating the non-linear terms perturbatively. Given that these classical solutions are themselves non-perturbative, the quantum effects obtained from them also turn out to be non-perturbative. These are not the first, or only, non-perturbative results available in field theory, but they do represent an important advance in the study of quantum fields.

(ii) Typically, soliton and instanton solutions are characterised by some topological index, related to their behaviour at spatial infinity. For solitons, this index turns out to be a conserved quantity which, in the quantised theory, becomes a conserved quantum number characterising the soliton state. Such a topological quantum number is quite different in origin from the familiar Noether charges associated with continuous symmetries of the Lagrangian. In the case of instantons, the existence of a non-zero topological index leads to the generation of a family of vacuum states, characterised by a vacuum angle θ. The extent of homotopy analysis involved in describing these phenomena is trivial by standards of mathematics; but the fact that such considerations can play a crucial role in quantum field theory is a new development.

Most of the results in quantum theory based on solitons or instantons are obtained by the semiclassical WKB method. The phrase 'semiclassical' carries a connotation, in the minds of some people, that the results are not quite quantum-theoretical. In response it should be pointed out that the results obtained using solitons or instantons are no less quantum-mechanical than the results of standard perturbation theory. In fact, standard perturbation theory can be viewed as a special case of the semiclassical method, where one quantises fluctuations around trivial classical solutions, whereas in soliton or instanton physics, one does the same thing around non-trivial, non-perturbative classical solutions. This point should be clear from chapters 5 and 6.

Although these methods yield qualitatively new results in quantum field theory, the basic physical concepts are quite easy to understand. Many of them are just generalisations to field theory of ideas that we are already familiar with in non-relativistic quantum mechanics. This is a gratifying feature of this subject. Of course detailed calculations or rigorous investigations of soliton/instanton effects can be complicated but the basic results can be appreciated without going into too many technicalities.

Having said this much in praise of our subject, we must also stress its limitations. Two major limitations stand out:

(i) The semiclassical method, whether it utilises solitons or instantons, demands a weak-coupling condition whereby an appropriate non-linear coupling parameter has to be small. This is required even though the final results are non-perturbative in that parameter, as explained in various sections of the book. In some cases, semiclassical WKB results turn out to give the correct answer, or a good numerical approximation to it, even when the coupling is large; but this cannot be taken to be a general feature of the method.

(ii) The method obviously requires that we know some non-trivial classical solutions in the first place. This means that for any given model, localised non-singular classical solutions must (a) exist and (b) be available either in exact analytical form or at least in fair approximation, along with their stability characteristics.

For some models of interest either one or both of the requirements mentioned above cannot be met. In particular, these requirements somewhat limit the usefulness of solitons and instantons in describing real hadron particles occurring in nature, whose physics is believed to be governed by quantum chromodynamics. Consider solitons first: QCD is an SU (3) gauge theory, in which the gauge bosons couple to fermionic quarks. There are different ways of setting up a semiclassical expansion in the presence of fermions (chapter 9). The simplest way is to start with classical solutions for which the Fermi field vanishes. For QCD, this amounts to finding solutions of just the self-coupled gauge-field system. Unfortunately it has been proved that pure gauge-field theory in $(3+1)$ dimensions can have no soliton solutions. Another way is to integrate over the fermionic degrees of freedom and obtain a more complicated effective action for the gauge fields. No exact soliton solutions corresponding to this effective action have been found either. The only significant use of solitons in describing hadrons has come from yet another variation in the semiclassical procedure (chapter 9). In this one begins with 'classical' solutions of the coupled boson–fermion field equations, where the term 'classical' for the spin-$\frac{1}{2}$ field in this case means replacing it by the c-number Dirac wavefunction. Even here, exact classical solutions are not available for QCD in $(3+1)$ dimensions, but approximate solutions have led to some success in describing hadrons. Such calculations, often called 'bag' models, are exemplified by the SLAC bag (Bardeen et al. 1975).

We next consider instantons: classical instanton solutions *are* available for the self-coupled Yang–Mills system (and by extension, for QCD) in exact analytical form. But when the quantum effects of these instantons are calculated the weak-coupling condition runs into difficulty. Instantons of QCD come in all sizes, i.e. they carry a scale parameter which can take any value from 0 to ∞. Computation of the net effect of instantons requires integrating over their size. Now, the effective gauge coupling constant, in the renormalised sense, varies with the distance scale. Indications are that it is large at large distance scales. Consequently a treatment of large instantons would involve a large coupling constant, violating the weak-coupling requirement of the semiclassical method. Despite this difficulty, valiant efforts have been made to extract as much information about QCD as possible, using instantons, but the results are somewhat clouded by the intrusion of strong coupling.

However, it should be emphasised that these difficulties do not imply that the semiclassical method is incorrect within the domain of its stated assumptions, but rather that those assumptions are not fully satisfied by QCD. For a variety of other field-theoretical models, the required classical solutions do exist and the weak-coupling condition can also be sustained. In such cases, the semiclassical method does yield important non-perturbative results. Our book describes numerous examples of such models. They are not unnatural or pathological examples, specially designed for this purpose; many of them had been studied as instructive models from other points of view, prior to these developments. Through these examples, the quantum effects of solitons and instantons have enlarged our understanding of the complexity and richness of structure of quantum field systems. Even in the case of QCD, a great deal of useful information has emerged, particularly regarding the homotopy classification of gauge transformations and vacuum states. It is unlikely that this information will be without value, despite the strong-coupling problems that plague instanton calculations in this model.

As investigations into relativistic quantum field theory, these matters hold their major interest for particle theorists. However the effect of solitons has also been studied in the statistical mechanics of continuum systems, with the mathematical similarity between statistical mechanics and (Euclidean) quantum field theory being put to further use.

The developments sketched above began in the early seventies, and for a few years there was prolific activity in this field around the world. As has happened before in theoretical particle physics, a large number of workers pounced on this newly emergent area and squeezed out, within a few years, most of the basic consequences as well as limitations of this approach. During the past year or so, the number of fresh papers on the semiclassical method has dwindled, and most of these papers deal with intricate technical matters. It seemed to us that, although it is a risky enterprise to write a book on a recently developed field, the time had come to attempt one on this subject. As a precaution, however, we have devoted the bulk of the book to the more basic and consolidated features of the subject. Subsequent investigations, when they are of a more elaborate, complicated or controversial nature, have been summarised in short paragraphs at the end of the appropriate sections.

The book is organised as follows:

The next three chapters (chapters 2, 3 and 4) are devoted to classical soliton and instanton solutions. Non-linear partial differential equations form a major topic in applied mathematics. We do not discuss the general methods and results dealing with this subject which have been covered in many review articles and books. While we do give the derivation of most of the classical solutions we present, particularly the ones obtained recently that deal with gauge theories, CP_N models etc., we do not go into more sophisticated techniques like the inverse scattering method or the Backlund transformations. It must be remembered that the main theme of this book is the study of the *quantum* effects of these solutions. The classical solutions themselves should be taken as given even though, for the sake of completeness, we have devoted three chapters to them. In these chapters, emphasis has been placed on those features of the classical solutions that play a role in the quantum effects described later. This is followed in chapters 5 to 8 by methods for quantising solitons to obtain quantum soliton particles and the associated sectors of states. First we describe the quantisation of static solitons, using intuitive arguments. This is later formalised using functional-integral techniques, and extended to quantise periodic solutions. Since many textbooks on field theory do not discuss functional-integral methods, we devote some space to introducing this technique before using it. Other methods for obtaining semiclassical results, including the canonical Hamiltonian approach, are also discussed. Technical matters like the renormalisation of ultraviolet divergences and the treatment of zero-modes are explained. These procedures are illustrated using two-dimensional scalar field theories as examples. In chapter 7, we have also included recent exact results on the *S*-matrices of a class of two-dimensional theories. This topic falls somewhat outside the semiclassical methodology characterising the rest of the book, but since it offers exact and elegantly derived results about soliton-bearing models, we could not resist the temptation to include it. In chapter 9, the functional-integral and semiclassical methods are extended to cover Fermi fields. A capsule introduction to Grassmann algebra is also given to provide the necessary mathematical machinery. Finally, chapters 10 and 11 deal with quantum effects of instantons. The basic vacuum-tunnelling phenomenon is discussed in chapter 10, first for a simple quantum mechanical system, then for an abelian gauge theory, and finally for the Yang–Mills system. The infrared problem caused by large instantons is brought out in a separate section. Chapter 11 discusses the effect of massless fermions on the vacuum-tunnelling phenomenon, and its relationship to the U (1) anomaly. The impact of instantons on the forces between external charges and on the Borel summability of perturbation expansions is also described. We conclude in section 11.5 with an open-ended summary of other developments, with references.

This is, by design, an elementary-level text, intended for research students but not for experts in the field. Senior physicists working in other areas who wish to understand the basic ideas in this field may also find the book useful. No claim is made of mathematical rigour, or of completeness in terms of covering each and every investigation that might have overlapped with our subject matter. About 350 references have been provided, but no doubt several others have been inadvertently left out. Our main emphasis has been on giving a pedagogical discussion of the basic ideas and techniques used in this field at as simple a level as possible. The discussion is carried out largely through illustrative examples. Wherever possible, we begin with examples from non-relativistic quantum mechanics. Although the physics of these examples is acknowledged as well known, we nevertheless discuss them at some length from a viewpoint tailored to our needs. The purpose is to point out that many concepts underlying the semiclassical method in field theory may be found in miniature within the familiar context of elementary quantum mechanics. This is followed, in the case of both solitons and instantons, by a detailed discussion of their effects in quantum field theory, once again through prototype examples.

By the time these developments took place, quantum field theory had already been under study for several decades. Needless to say, research work on our subject has used earlier results and techniques wherever necessary. Ideally, therefore, the reader needs to be well versed in the standard results of perturbative quantum field theory, gauge theory and renormalisation, as a prerequisite. However, this is not crucial for understanding most of our presentation, which is by-and-large self-contained. A strong background in quantum mechanics and some broad familiarity with field-theoretical methods will be sufficient for the most part. Of course at a few places in the book, especially near the end, more sophisticated inputs are unavoidable. Examples include the use of some general principles of quantum gauge theory, the running coupling constant, asymptotic freedom, current anomalies, the principles of S-matrix theory etc. These inputs are explicitly stated but not derived. They form major topics in their own right independently of the semiclassical method, and an attempt to introduce them in detail would have led to serious digression.

Finally, we have been handicapped by the absence of any other textbook on this subject. To the best of our knowledge at the time of writing, no textbook carrying a comprehensive treatment of our subject is available. We have, however, benefitted greatly from several review articles which have treated one or more aspects of this field. For a review of the classical solutions, see for example Scott et al. (1973), Makhankov (1978), Goddard and Olive (1978) and Actor (1979); on the quantisation of solitons see Rajaraman (1975), Jackiw (1977), Coleman (1977a) and Faddeev and Korepin (1978); on instanton physics see Coleman (1979), Marciano and Pagels (1978), Olive (1979), Sciuto (1979) and Jackiw (1980). In particular, the influence of Coleman's superb reviews pervades our whole presentation. The Faddeev–Korepin review on quantum solitons is also specially recommended, as its approach differs quite a lot from ours and it will make valuable complementary reading.

References to other good reviews, not dealing with the semiclassical theme but of relevance to our discussion, will be provided at appropriate places in the text.

CHAPTER 2

CLASSICAL SOLITONS AND SOLITARY WAVES

2.1. Introduction

Our principal theme, as emphasised in chapter 1, is the semiclassical method in second-quantised relativistic field theory. That is, given some or all solutions of a set of classical relativistic non-linear field equations, we will discuss methods by which such solutions could be used to study the vacuum, the one-particle states, and other features of the corresponding *quantum* field theory. The process of obtaining these classical solutions in the first place is not our main concern. Several impressive and complex techniques have been developed to study and solve non-linear wave equations. Such techniques, of which the inverse scattering method is one example, together merit a book in their own right. Excellent surveys of some of these methods are already available, as for instance in the book by Whitham (1974) and the review article by Scott et al. (1973). For us, it would be a digression to detail all these techniques here. Furthermore, knowledge of the methods of derivation of the classical solutions is not necessarily needed for understanding their subsequent use in quantum theory. We start with the classical solutions already in hand, so to speak.

Having made this disclaimer however, we will, for the sake of completeness, devote this chapter and the next two solely to classical solutions. Since the main beneficiary of relativistic quantum field theory is particle physics, we will concentrate on those classical solutions that resemble as closely as possible classical extended particles. In particular, it would be nice to find solutions that look like pulses of energy travelling without dissipation and with uniform velocity. It would be nicer still if the equation permitted several such pulses to collide and emerge after collision with their original shapes intact, at least asymptotically in time. Despite the complexity of non-linear partial differential equations, some solutions do exist with such properties. Such solutions are often referred to by the names solitary waves and solitons. Our first task will be to clarify the meaning of these two terms and distinguish them from one another. We will follow this with some examples of each kind. The sine–Gordon soliton, the 'kink' of the ϕ^4 theory, the 'tHooft–Polyakov monopole, the instanton and other objects prominent in recent literature in this area will be described. Despite the disclaimer in the previous paragraph, their derivation will also be given in the few cases where it is brief. We will emphasise those features of the solutions that are of relevance to subsequent quantisation.

While the examples given correspond to relativistic field equations and are of particular importance to particle physics, they are interesting in their own right as solutions to fairly difficult non-linear partial differential equations. Some of our discussion and terminology will be applicable to non-relativistic systems as well. It is therefore hoped that the contents of these early chapters would also hold some interest for physicists who study things other than elementary particles.

We will give a word about notation before proceeding to the next section. The bulk of our presentation will be in the standard covariant tensor notation. Space–time coordinates will be represented by the vector x^μ ($\mu = 0,1,2,3,; x^0 \equiv ct, x^1 = x, x^2 = y, x^3 = z$). Indices can be lowered or raised by the Minkowskian metric tensor $g_{\mu\nu} = g^{\mu\nu}$, where $g_{00} = -g_{11} = -g_{22} = -g_{33} = 1$ and $g_{\mu\nu} = 0$ for $\mu \neq \nu$. Repeated indices are to be summed over. ∂_μ refers to space–time derivatives $\partial/\partial x^\mu$. Sometimes we will deal with systems in lower dimensions; for instance, most systems in this chapter will be in two $(1+1)$ dimensions – one space and one time – in which case the range of the index μ will only be $\mu = 0, 1$. Whereas in subsequent chapters, use of this compact covariant notation will become essential, in this chapter we can afford to write space–time dependence in more explicit forms as well, intermingled with the covariant notation.

2.2. Solitary waves and solitons

The names solitary waves and solitons refer to certain special solutions of non-linear wave equations. In order to fully appreciate such solutions, we recall some properties of the simplest of all relativistic wave equations, namely,

$$\Box\phi = \partial^\mu\partial_\mu\phi = \left(\frac{1}{c^2}\frac{\partial^2}{\partial t^2} - \frac{\partial^2}{\partial x^2}\right)\phi(x,t) = 0 \tag{2.1}$$

where $\phi(x, t)$ is a real scalar field in $(1+1)$ dimensions, and c is the velocity of light. The properties of this equation are only too well known. It is both linear and dispersionless. As a result its solutions have two features of relevance to our discussion:

(i) Any real well-behaved function of the form $f(x \pm ct)$ is a solution of (2.1). In particular if we choose a localised function f, we can construct a localised wave packet that will travel with uniform velocity $\pm c$, and *no distortion in shape.* This is obviously related to the fact that plane waves $\cos(kx \pm \omega t)$ and $\sin(kx \pm \omega t)$ with $\omega = kc$ form a complete set of solutions to (2.1). Any well-behaved localised function, say $f(x-ct)$, can be written as

$$f(x-ct) = \int \mathrm{d}k\,[a_1(k)\cos(kx-\omega t) + a_2(k)\sin(kx-\omega t)]. \tag{2.2}$$

The fact that the wave packet $f(x-ct)$ travels undistorted with velocity c is related to the fact that all its plane-wave components have the same wave velocity $\omega/k = c$.

(ii) Since the wave equation is linear, given two localised wave packet solutions $f_1(x-ct)$ and $f_2(x+ct)$, their sum $f_3(x, t) = f_1(x-ct) + f_2(x+ct)$ is also a solution. At large negative time $(t \to -\infty)$ $f_3(x, t)$ consists of the two packets widely separated and approaching each other essentially undistorted. At finite t, they collide. But after collision they will asymptotically (as $t \to +\infty$) separate into the same two packets retaining their original shapes and velocities. For the system (2.1), this property holds for more than two packets as well.

These two features, namely (i) the shape and velocity retention of a single packet and (ii) the asymptotic shape and velocity retention of several packets even after collision, clearly hold for (2.1) because that particularly simple system is both linear and dispersionless. However, typical wave equations in many branches of physics are much more complicated: they can contain non-linear terms, dispersive terms, and several coupled wave-fields with space-dimensionality equal to 1, 2 or 3. We are concerned with the question: can such equations, despite their complexity compared with (2.1), nevertheless yield at least some solutions which enjoy the attractive feature (i) and perhaps even (ii)?

Note that the addition of even the simplest kinds of terms to (2.1) tends to destroy these nice features, even in $(1+1)$ dimensions. Consider for instance the Klein–Gordon equation in two dimensions,

$$(\Box + m^2c^2)\,\phi(x, t) \equiv \left(\frac{1}{c^2}\frac{\partial^2}{\partial t^2} - \frac{\partial^2}{\partial x^2} + m^2c^2\right)\phi(x, t)$$
$$= 0. \tag{2.3}$$

This equation is still linear and plane waves $\cos(kx \pm \omega t)$ and $\sin(kx \pm \omega t)$ still form a complete set of solutions. But now $\omega^2 = k^2c^2 + m^2c^4$. Thus different wavelengths travel at different velocities $\omega(k)/k$, and the equation is dispersive. Any localised wave packet having, at $t = 0$, the form

$$\int dk\, [a_1(k)\cos kx + a_2(k)\sin kx]$$

will spread as time goes on. Thus, feature (i) is lost, and so is feature (ii). If even a single packet cannot retain its shape, there is no question of several packets retaining their shapes after collision. Similarly, consider adding a simple non-linear term to (2.1) as in

$$\left(\frac{1}{c^2}\frac{\partial^2}{\partial t^2} - \frac{\partial^2}{\partial x^2}\right)\phi(x, t) + \phi^3(x, t) = 0. \tag{2.4}$$

Not all solutions of this equation are known, but one can persuade oneself through numerical or approximate calculations that an arbitrary wave packet will spread.

It is however possible that for some equations where *both* dispersive and non-linear terms are present, their effects might balance each other in such a way that some special solutions do essentially enjoy feature (i). This can happen in one, two or three space dimensions, and such solutions are, roughly speaking, called solitary waves. In a small subset of these cases, feature (ii) is also exhibited. Those solutions are called solitons.

To get a more precise definition of these two terms, we need to quantify the requirements that have so far been qualitatively described as features (i) and (ii). These features were extracted by us from the behaviour of wave packets in linear dispersionless systems like (2.1). We need to generalise these requirements so as to encompass as many cases as possible where solutions enjoy the essential spirit of features (i) and (ii) if not in the particular way that wave packets of (2.1) do. The current literature unfortunately does not provide a universally accepted definition of solitons and solitary waves. Different authors offer slightly different variations. We shall now give a working definition that is a compromise between these variations.

Our definition will be in terms of the energy density rather than the wave-fields themselves, since the former is more significant for systems of our interest. This means that we are restricting ourselves to those field equations (for any set of coupled fields $\phi_1(x, t)$, $\phi_2(x, t)$ . . .) that have an associated energy density $\varepsilon(x, t)$ which is some function of the fields $\phi_i(x, t)$. Its space-integral is the conserved total energy functional $E[\phi_i]$. A large class of equations, including field equations in particle physics, satisfy this. Since physical systems have an energy bounded from below we can also, without loss of generality, set the minimal value reached by E as equal to zero. Given this framework, we shall use the adjective 'localised' for those solutions to the field equation whose energy density $\varepsilon(\boldsymbol{x}, t)$ at any finite time t is localised in space, i.e. it is finite in some finite region of space and falls to zero at spatial infinity sufficiently fast as to be integrable. Note that for those systems where $E[\phi_i] = 0$ if and only if $\phi_i(x, t) = 0$, a localised solution as defined above also has the fields themselves localised in space. For instance, equation (2.4) has an associated conserved energy given by

$$E[\phi] = \int_{-\infty}^{\infty} \mathrm{d}x \left[\frac{1}{2c^2}\left(\frac{\partial \phi}{\partial t}\right)^2 + \frac{1}{2}\left(\frac{\partial \phi}{\partial x}\right)^2 + \frac{1}{4}\phi^4 \right] \tag{2.5}$$

which is minimised by $\phi(x, t) = 0$. Localised solutions of this system, if any, would asymptotically go to $\phi(x, t) = 0$ as $x \to \pm\infty$, at any given t. The derivatives $(\partial\phi/\partial x)$ and $(\partial\phi/\partial t)$ must also vanish in this limit. By contrast, the equation

$$\frac{1}{c^2}\frac{\partial^2 \phi}{\partial t^2} - \frac{\partial^2 \phi}{\partial x^2} - \phi + \phi^3 = 0 \tag{2.6}$$

has an associated energy

$$E[\phi] = \int_{-\infty}^{\infty} \mathrm{d}x \left[\frac{1}{2c^2}\left(\frac{\partial \phi}{\partial t}\right)^2 + \frac{1}{2}\left(\frac{\partial \phi}{\partial x}\right)^2 + \frac{1}{4}(\phi^2 - 1)^2 \right]. \tag{2.7}$$

Here $E[\phi]$ is minimised by $\phi(x, t) = \pm 1$. This is the simplest example of 'spontaneous symmetry breaking' in the quantised version of field systems. Now, a localised solution must approach $\phi = \pm 1$ as $x \to \pm\infty$ at any instant.

Given localisation in this sense of energy density, we define a *solitary wave* as that localised non-singular solution of any non-linear field equation (or coupled equations, when several fields are involved) whose energy density, as well as being localised, has a space–time dependence of the form

$$\varepsilon(\boldsymbol{x}, t) = \varepsilon(\boldsymbol{x} - \boldsymbol{u}t) \tag{2.8}$$

where $\boldsymbol{u}$ is some velocity vector.

In other words, the energy density should move undistorted with constant velocity. Among systems that do have an associated energy density this definition permits a larger class of solutions than the one given by Scott et al. (1973) who require that the fields themselves have such a 'travelling wave' space–time dependence. For instance, interesting examples like the charged solitary waves to be discussed in chapter 8 are encompassed in (2.8), but would be excluded by the definition given by Scott et al. On the other hand their definition, being directly in terms of the fields, could in principle be applied to systems which do not have an associated conserved energy. We will in any case not be dealing with such systems. At the same time, our definition is more restrictive than that of Coleman (1977a). He uses the name 'lumps' for what are essentially non-singular solutions with finite localised energy, whether or not they obey (2.8). An example like the doublet solution of the sine–Gordon system (section 2.5) would unfortunately be excluded by (2.8), but would be included by Coleman as a 'quivering lump'. We pick the definition (2.8) not because it is intrinsically superior to others, but because it is optimal for our purposes. It is simple, fairly comprehensive and also enables us to define solitons in a reasonably simple way.

Note that (2.8) defines solitary waves in one or more space dimensions. Further, any static (time-independent) localised solution is automatically a solitary wave, with the velocity $\boldsymbol{u} = 0$. Many of the solitary waves we will discuss will be obtained as static solutions. However, for systems with relativistic (or for that matter, Galilean) invariance, once a static solution is known, moving solutions are trivially obtained by boosting, i.e. transforming to a moving coordinate frame.

We now turn to solitons: these are solitary waves with an added requirement given below, which is a somewhat generalised and precisely stated version of feature (ii). Consider some (possibly coupled) non-linear equation(s). Let them have a solitary wave solution whose energy density is some localised function $\varepsilon_0(\boldsymbol{x}-\boldsymbol{u}t)$. Consider any other solution of this system which in the far past consists of N such solitary waves, with arbitrary initial velocities and positions. Then the energy density $\varepsilon(\boldsymbol{x}, t)$ of this solution will have the form

$$\varepsilon(\boldsymbol{x}, t) \rightarrow \sum_{i=1}^{N} \varepsilon_0(\boldsymbol{x}-\boldsymbol{a}_i-\boldsymbol{u}_i t), \qquad \text{as } t \rightarrow -\infty. \tag{2.9}$$

Given this configuration at $t = -\infty$, it will then evolve in time as governed by the non-linear equations. Suppose this evolution is such that

$$\varepsilon(\boldsymbol{x}, t) \rightarrow \sum_{i=1}^{N} \varepsilon_0(\boldsymbol{x}-\boldsymbol{a}_i-\boldsymbol{u}_i t+\boldsymbol{\delta}_i), \qquad \text{as } t \rightarrow +\infty \tag{2.10}$$

where $\boldsymbol{\delta}_i$ are some constant vectors. Then such a solitary wave is called a *soliton*. In other words, solitons are those solitary waves whose energy density profiles are asymptotically (as $t \to \infty$) restored to their original shapes and velocities. The vectors $\boldsymbol{\delta}_i$ represent the possibility that the solitons may suffer a bodily displacement compared with their pre-collision trajectories. This displacement should be the sole residual effect of collisions if they are to be solitons. Obviously this is a remarkable property for solutions of a non-linear field equation to have.

While all solitons are solitary waves, the converse is clearly not true. The added requirement (2.10) on solitons is very stringent. Consequently, they exist for far fewer equations than do solitary waves and are also much harder to find. In order to find a solitary-wave solution to a given non-linear equation, we only need to look for one localised solution satisfying (2.8). This is often hard enough to do, but several equations have yielded solitary waves by now. In contrast, to ensure that a solution is a soliton, we must find not merely that solution, but infinitely many time-dependent solutions consisting of arbitrary numbers of solitons, and check that (2.9) and (2.10) are satisfied. Thus, it is very hard to tell, given a non-linear wave equation, whether it even permits soliton solutions, let alone evaluate them explicitly. Of course a large body of powerful techniques has been developed for solving soliton-bearing equations and studying their properties. These include the inverse scattering method, Backlund transformations, the use of conserved quantities etc. While these techniques offer elegant ways of solving and understanding such systems, they are not as yet very helpful towards identifying new equations carrying solitons, or in deciding whether a given equation has this property. Not surprisingly then, very few soliton-bearing equations have been found. The review by Scott et al. (1973) offers a handful of examples, all in one space-dimension. Of these, only the sine–Gordon equation is relativistic. We will discuss this equation in section 2.5 as an example of such systems.

The bulk of the localised solutions discussed in the physics literature seem only to be solitary waves. It cannot be ruled out that some of them may also be solitons, but in the absence of information on all time-dependent solutions for those systems, this is hard to prove or disprove conclusively. One can only say that since the requirement (2.10) is very stringent, in all likelihood most of them are only solitary waves. Fortunately the techniques for quantising classical solutions do not rely heavily on their being solitons: solitary waves will do.

Having clearly differentiated solitons from solitary waves – the former being a very restricted subset of the latter – we should mention that in much of the literature this distinction is allowed to be blurred. What are merely solitary waves are often called solitons, presumably because the latter name is more crisp and appealing. While names do not matter much in physics, it is important to distinguish, at least in classical field theory, between solutions that obey only (2.8) and those that also obey (2.9)–(2.10). In later chapters we too will succumb to the populist trend of using the name solitons for solitary waves. But at least in this chapter, the distinction will be maintained.

2.3. Some solitary waves in two dimensions

We shall now present some examples of solitary waves, beginning with the simplest. As mentioned earlier, any localised static (time-independent) solution is a solitary wave. We shall therefore concentrate in this section on static solutions in the simplest context where they occur, namely scalar fields in two (one space + one time) dimensions. Examples in higher dimensions and with more complicated fields will be discussed in the next chapter. Consider first a single scalar field $\phi(x,t)$ whose dynamics is governed by the Lorentz-invariant Lagrangian density

$$\mathscr{L}(x,t) = \tfrac{1}{2}(\dot{\phi})^2 - \tfrac{1}{2}(\phi')^2 - U(\phi) \tag{2.11}$$

where henceforth a dot or a prime represents differentiation with respect to time or the space variable x, respectively, and the velocity of light c is set equal to one. The 'potential' $U(\phi)$ is any positive semi-definite function of ϕ, reaching a minimum value of zero for some value or values of ϕ. When the variational action principle

$$\delta\left(\int \mathrm{d}t \int_{-\infty}^{\infty} \mathrm{d}x\, \mathscr{L}(x,t)\right) = 0 \tag{2.12}$$

is applied to this Lagrangian, one obtains the wave equation

$$\Box\phi \equiv \ddot{\phi} - \phi'' = -\frac{\partial U}{\partial \phi}(x,t) \tag{2.13}$$

This is our wave equation whose non-linear terms depend on the choice of $U(\phi)$. The equation conserves, as t varies, the total energy functional E given by

$$E[\phi] = \int_{-\infty}^{\infty} \mathrm{d}x\left[\tfrac{1}{2}(\dot{\phi})^2 + \tfrac{1}{2}(\phi')^2 + U(\phi)\right]. \tag{2.14}$$

Let the absolute minima of $U(\phi)$, which are also its zeros, occur at M points,

$M \geqslant 1$. That is, let

$$U(\phi) = 0 \qquad \text{for} \qquad \phi = g^{(i)}; \quad i = 1, \ldots, M. \tag{2.15}$$

Then the energy functional is also minimised when the field $\phi(x, t)$ is constant in space–time and takes any one of these values. That is,

$$E[\phi] = 0$$

if and only if

$$\phi(x, t) = g^{(i)}; \qquad i = 1, \ldots, M. \tag{2.16}$$

Now, we are interested in static solutions, for which (2.13) reduces to

$$\phi''(x) \equiv \frac{\partial^2 \phi}{\partial x^2} = +\frac{\partial U}{\partial \phi}(x). \tag{2.17}$$

Further, a solitary wave must have finite energy and a localised energy density. In view of (2.15) its field must approach, as $x \to \pm\infty$, one of the values $g^{(i)}$. If the $U(\phi)$ had a unique minimum at $\phi = g$, then our static solution must have $\phi(x) \to g$ as $x \to \pm\infty$. If there are several degenerate minima ($M > 1$ in (2.15)), then $\phi(x)$ must tend to any one of the g^i as $x \to -\infty$, and either the same or any other of the g^i as $x \to +\infty$.

Subject to these boundary conditions, we solve (2.17) for $\phi(x)$. Since this is an ordinary second-order differential equation, it can easily be solved by quadrature for any $U(\phi)$. Before doing that, it will be useful to notice that (2.17) has a mechanical analogue. Such mechanical analogues to static solutions have been pointed out by several people (see for instance Coleman 1977a, Christ and Lee 1975, and Friedberg and Lee 1977). If we think of the variable x as 'time' and ϕ as the coordinate of a unit-mass point-particle, then (2.17) is just Newton's second law for this particle's motion in a potential given by $[-U(\phi)]$. The solution $\phi(x)$ represents the motion of this analogue particle. The total 'energy' of this motion, conserved as x, the 'time', varies is given by

$$W \equiv \tfrac{1}{2}(\mathrm{d}\phi/\mathrm{d}x)^2 - U(\phi) \tag{2.18}$$

The boundary conditions discussed earlier demand that as $x \to \pm\infty$, $U(\phi) \to 0$ and $(\partial\phi/\partial x) \to 0$. Hence $W = 0$. The energy W of the analogue particle is not to be confused with the energy E, given in (2.14), of the original field system. For a static solution $\phi(x)$, E is given by

$$E = \int_{-\infty}^{\infty} \left[\frac{1}{2}\left(\frac{\mathrm{d}\phi}{\mathrm{d}x}\right)^2 + U(\phi)\right] \mathrm{d}x \tag{2.19}$$

and clearly represents the total 'action' of the analogue particle's motion. Our static solution therefore corresponds to some finite-action, zero-energy trajectory of the particle. Finally, upon multiplying (2.17) by ϕ' and integrating once, we have

$$\int \phi' \phi'' \, \mathrm{d}x = \int \frac{\mathrm{d}U}{\mathrm{d}\phi} \phi' \, \mathrm{d}x$$

or

$$\tfrac{1}{2}(\phi')^2 = U(\phi). \tag{2.20}$$

Since both ϕ' and $U(\phi)$ vanish at $x = -\infty$, the integration constant is zero. Equation (2.20) is just a virial theorem for the analogue-particle.

Armed with this mechanical analogy, we consider first a $U(\phi)$ which has a unique minimum, at $\phi = \phi_1$ where $U(\phi_1) = 0$. The analogue particle sees a potential $[-U(\phi)]$ as in fig. 1(a), with a *maximum* at $\phi = \phi_1$ and a negative value for all other ϕ. The boundary conditions specified demand a zero-energy trajectory beginning and ending at $\phi = \phi_1$ in the far past and far future ($x = \pm\infty$). One look at fig. 1(a) is enough to tell us that no such non-trivial motion is possible!

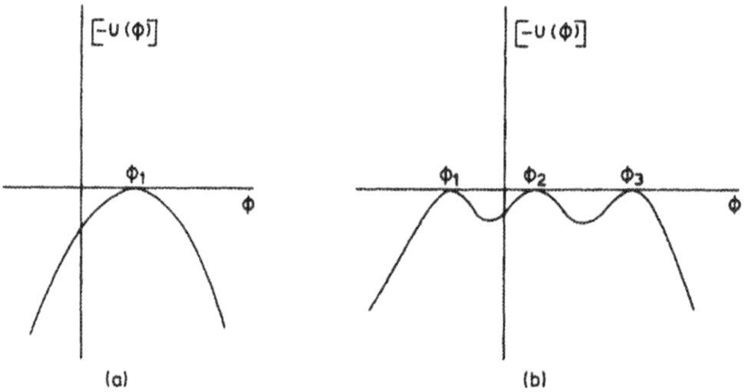

Fig. 1. (a) The potential $[-U(\phi)]$ of the analogue particle when $U(\phi)$ has a unique minimum at ϕ_1. There are no non-trivial static solutions here. (b) The case where $U(\phi)$ has three discrete degenerate minima. Here four non-trivial solutions are possible.

Once the particle takes off from $\phi = \phi_1$ in either direction, it will not return. Its kinetic energy will never be zero again since its zero total energy W will always be larger than its potential energy $[-U(\phi)]$. Consequently the particle can never stop and turn back towards ϕ_1. In terms of the static field solution $\phi(x)$, this means that once we fix the boundary condition as $\phi = \phi_1$ and $\mathrm{d}\phi/\mathrm{d}x = 0$ at $x = -\infty$, the same condition at $x = +\infty$ will not be satisfied by a non-trivial non-singular solution. Therefore, without explicitly solving (2.17) and independent of the details of $U(\phi)$, we see that if $U(\phi)$ has a unique absolute minimum, there can be no static solitary wave. Of course, the trivial solution $\phi(x) = \phi_1$ for all x, is permitted.

Next, let $U(\phi)$ have two or more degenerate minima, where it vanishes. Figure 1(b), where once again $[-U(\phi)]$, the analogue-particle's potential, is plotted, corresponds to an example where $U(\phi)$ has three minima at ϕ_1, ϕ_2 and ϕ_3. The boundary conditions now state that the particle must leave any one of these points at $x = -\infty$ and end up at $x = \infty$ at any one of them. This is now possible. It can take off from the top of the hill ϕ_1 at $x = -\infty$, and roll up to the top of the hill ϕ_2 asymptotically as $x \to +\infty$. Or, it can begin at ϕ_2 and end up at ϕ_3. Or, it can undergo the reverse of these two motions. These are the only four non-trivial possibilities for this example. It cannot, for instance, leave ϕ_1, go up to ϕ_2 and either return back to ϕ_1 or go on to ϕ_3. To see this, note that at ϕ_2 both $U(\phi)$ and $dU/d\phi$ vanish. Consequently, from (2.20) and (2.17), both the 'velocity' (ϕ') and the 'acceleration' (ϕ'') vanish there. Further,

$$\phi''' = \frac{d}{dx}\left(\frac{dU(\phi)}{d\phi}\right) = \frac{d^2U}{d\phi^2}\phi' = 0$$

$$\phi'''' = \frac{d^2U}{d\phi^2}\phi'' + \frac{d^3U}{d\phi^3}(\phi')^2 = 0 \quad \text{etc.} \tag{2.21}$$

Thus, all derivatives $d^n\phi/dx^n$ vanish at ϕ_2. The particle, having left ϕ_1 can barely make it to ϕ_2 as $x \to \infty$, where all derivatives of its motion vanish. It cannot return to ϕ_1 or proceed to ϕ_3.

Thus, the mechanical analogy helps us conclude that (i) when $U(\phi)$ has a unique absolute minimum, there can be no static solitary wave, and (ii) when $U(\phi)$ has n discrete degenerate minima, we can have $2(n-1)$ types of solutions, connecting any two neighbouring minima, as x varies from $-\infty$ to $+\infty$. It is of course understood that trivial space–time-independent solutions can exist in addition.

Apart from these general considerations, one can also explicitly solve (2.17) by quadrature. We have, from (2.20),

$$d\phi/dx = \pm[2U(\phi)]^{1/2}. \tag{2.22}$$

Upon integration

$$x - x_0 = \pm \int_{\phi(x_0)}^{\phi(x)} \frac{d\bar{\phi}}{[2U(\bar{\phi})]^{1/2}} \tag{2.23}$$

where x_0, the integration constant, is any arbitrary point in space where the field has value $\phi(x_0)$. Our earlier discussion tells us that as $x \to \pm\infty$, $\phi(x)$ must approach any two neighbouring minima of $U(\phi)$ and at in-between values of x, $\phi(x)$ lies between these two minima. Consequently, $U(\phi)$ will vanish only as $x \to \pm\infty$, and be positive for finite x. The integrand in (2.23) will therefore be non-singular except at the end points if $x \to \infty$ or $x_0 \to -\infty$. The solution $\phi(x)$ can be obtained in principle explicitly, given an x_0 and a $\phi(x_0)$, by integrating (2.23) and inverting it. In practice, it may be possible to do this analytically only for some $U(\phi)$. Note that varying x_0, keeping $\phi(x_0)$ fixed, merely shifts the same solution in x-space. This is just a reflection of the translational invariance of the parent equation (2.13).

As an illustration of this method, let us consider the 'kink' solution of the ϕ^4 theory (Dashen et al. 1974b, Goldstone and Jackiw 1975, Polyakov 1974). The Lagrangian density has the form (2.11) with

$$U(\phi) = \tfrac{1}{4}\lambda(\phi^2 - m^2/\lambda)^2 \tag{2.24}$$

where λ and m^2 are positive constants. The equation of motion,

$$\ddot{\phi} - \phi'' = m^2\phi - \lambda\phi^3 \tag{2.25}$$

is essentially the same as (2.6) except for constants. Here $U(\phi)$ vanishes at two degenerate minima $\phi = \pm m/\sqrt{\lambda}$. Consequently localised solutions must tend to $\pm m/\sqrt{\lambda}$ as $x \to \pm\infty$. In particular, static solutions can be of two types, as per earlier arguments. They can begin from $\phi = -m/\sqrt{\lambda}$ at $x = -\infty$ and end up with $\phi = +m/\sqrt{\lambda}$ at $x = \infty$, or vice versa. Specifically, the static equation

$$\phi'' = \mathrm{d}U/\mathrm{d}\phi = \lambda\phi^3 - m^2\phi \tag{2.26}$$

can be solved using (2.23) to give

$$x - x_0 = \pm \int_{\phi(x_0)}^{\phi(x)} \frac{\mathrm{d}\bar{\phi}}{\sqrt{\lambda/2}\,(\bar{\phi}^2 - m^2/\lambda)}. \tag{2.27}$$

This is a simple and well known integral. Upon choosing $\phi(x_0) = 0$, integrating over $\bar{\phi}$ and inverting, we have

$$\phi(x) = \pm(m/\sqrt{\lambda})\tanh[(m/\sqrt{2})(x - x_0)]. \tag{2.28}$$

The solution with the plus sign plotted in fig. 2(a) will be called the 'kink' and the one with the minus sign the 'antikink'. The effect of translational

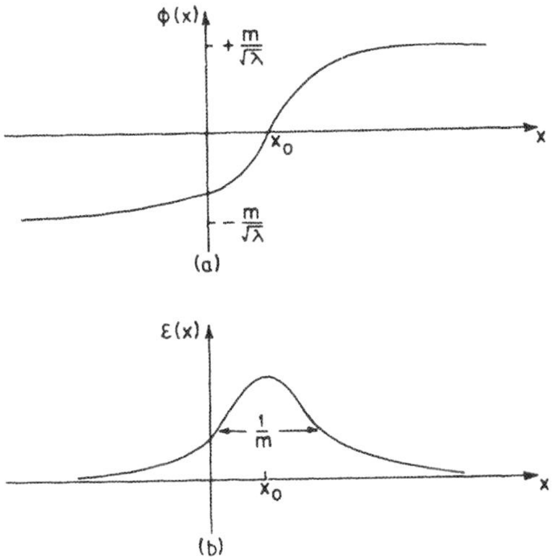

Fig. 2. (a) A schematic plot of the static kink solution (2.28). (b) The energy density of the kink. It is localised, with a width characterised by $1/m$.

invariance is explicitly seen, since a change in x_0 merely shifts the solution in space. The other symmetries of the Lagrangian, under $x \leftrightarrow -x$ and separately under $\phi \leftrightarrow -\phi$ are reflected in the relations which take on a particularly simple form when x_0 is chosen equal to zero:

$$\phi_{\text{kink}}(x) = -\phi_{\text{antikink}}(x) = \phi_{\text{antikink}}(-x). \tag{2.29}$$

The energy density of the kink solution,

$$\varepsilon(x) = \tfrac{1}{2}(\phi')^2 + U(\phi) = 2U(\phi), \text{ using (2.22),}$$

$$= (m^4/2\lambda)\,\text{sech}^4\,[m(x-x_0)/\sqrt{2}] \tag{2.30}$$

is plotted in fig. 2(b) and is clearly localised near x_0. The total kink energy, sometimes called the classical kinkmass M_{cl} is given by

$$M_{\text{cl}} = \int_{-\infty}^{\infty} dx\, \varepsilon(x) = \frac{2\sqrt{2}}{3}\frac{m^3}{\lambda} \tag{2.31}$$

and is finite. The kink is therefore a legitimate solitary wave. So is the antikink. However they are not solitons. Ordinarily, in order to be sure that some solitary waves are not solitons i.e. that they do not survive collisions, one needs information on time-dependent solutions involving several such waves. But in the present example, a two-kink configuration cannot even exist with finite energy let alone scatter in some desired manner. The first kink must begin at $x = -\infty$ with $\phi = -m/\sqrt{\lambda}$ and tend to $\phi = +m/\sqrt{\lambda}$ on the right. If this were to be followed by a second kink, the latter would tend to $\phi = 2m/\sqrt{\lambda}$ as $x \to +\infty$. This would lead to a constant non-zero energy density as $x \to +\infty$ and hence to infinite total energy. A kink can of course be followed by an antikink, bringing the field ϕ back to $-m/\sqrt{\lambda}$. Here again, numerical calculations indicate that a kink and an antikink approaching one another do not retain their shapes ofter collisions. The kink is therefore a solitary wave but not a soliton. It resembles a 'lump' of matter in the sense that it is a static, self-supporting localised packet of energy. The resemblence to an extended particle goes farther: because the system is Lorentz invariant, given the static solution (2.28), one can Lorentz-transform it to obtain a moving kink solution. Remembering that ϕ is a scalar field, we need only to transform the coordinate variables in (2.28). This gives

$$\phi_u(x, t) = \frac{m}{\sqrt{\lambda}} \tanh\left[\frac{m}{\sqrt{2}}\left(\frac{(x - x_0) - ut}{\sqrt{1 - u^2}}\right)\right] \tag{2.32}$$

where $1 > u > -1$ is the velocity. This is a solution of the field equation (2.25) as can be verified by substitution. Whereas the spatial width of the static kink, in the sense of its energy density (2.30), is characterised by $1/m$, the corresponding width of the moving kink in (2.32) is $\sqrt{1 - u^2}/m$, as would happen from Lorentz contraction for a lump of matter. Further, the energy of the time-dependent solution (2.32) as per (2.14) is

$$\begin{aligned} E[\phi_u] &= \int_{-\infty}^{\infty} dx \left[\tfrac{1}{2}(\dot{\phi}_u)^2 + \tfrac{1}{2}(\phi_u')^2 + U(\phi_u)\right] \\ &= \int_{-\infty}^{\infty} dx \left[\left(\frac{m^4}{4\lambda}\frac{u^2}{1 - u^2} + \frac{m^4}{4\lambda}\frac{1}{1 - u^2} + \frac{m^4}{4\lambda}\right)\right. \\ &\qquad \left.\operatorname{sech}^4\left(\frac{m}{\sqrt{2}}\frac{x - x_0 - ut}{\sqrt{1 - u^2}}\right)\right] \end{aligned}$$

$$= \int_{-\infty}^{\infty} \mathrm{d}x \frac{m^4}{2\lambda(1-u^2)} \operatorname{sech}^4\left(\frac{m}{\sqrt{2}} \frac{x-x_0-ut}{\sqrt{1-u^2}}\right)$$

$$= \frac{2\sqrt{2}}{3} \frac{m^3}{\lambda} \frac{1}{\sqrt{1-u^2}} = \frac{M_{\mathrm{cl}}}{\sqrt{1-u^2}} \tag{2.33}$$

where M_{cl} is the static kink energy in (2.31). The relationship of (2.33) to (2.31) is again the same as the Einstein mass–energy equation for a particle. Therefore it should not be surprising that in the quantum version of this theory, the kink solution leads to a particle-state. Another important feature of $\phi_{\mathrm{kink}}(x)$ is that it is singular as the non-linearity parameter λ goes to zero. Thus it cannot be obtained by mere perturbation expansion starting from the linear equation. Since ϕ_{kink} is non-perturbative, so are many consequences which flow from it in the quantised theory. A final feature of ϕ_{kink}, its 'topological' index, will be discussed a little later.

It is evident that a similar analysis can be done for any $U(\phi)$ with degenerate minima. Another example is $U(\phi) = \frac{1}{2}\phi^2(\phi^2-1)^2$, which has three degenerate minima at $\phi = 0, \pm 1$. One would therefore expect four types of static solutions with boundary conditions

$$\phi = [-1, 0], [0, 1], [1, 0] \text{ and } [0, -1]$$
$$\text{at } x = [-\infty, \infty]$$

respectively. Explicit static solutions to (2.17) for this case are

$$\phi(x) = \pm(1 + \mathrm{e}^{\pm 2x})^{-1/2} \tag{2.34}$$

All these solutions have localised energy density and finite total energy. Examples of $U(\phi)$ with a unique minimum can lead to no static solitary waves, as argued earlier. Thus the widely-studied field theory with $U(\phi) = \frac{1}{2}m^2\phi^2 + \frac{1}{4}\lambda\phi^4$ which differs from (2.24) in the sign of the ϕ^2 term, has no localised static solution, other than of course the trivial solution $\phi(x) = 0$.

Let us move on to the next level of complexity, i.e. static solutions to systems of coupled scalar fields in two space–time dimensions. This already brings us to the stage where no general methods are available for obtaining all localised static solutions, given the field equations. However, some solutions, but by no means all, can be obtained for a class of such Lagrangians using a little trial and error (Rajaraman 1979).

Up to a certain point, the discussion proceeds along the same lines as before, suitably generalised to a set of coupled scalar fields $\phi_i(x, t)$,

$i = 1, \ldots, N$. Consider the Lagrangian density,

$$\mathscr{L}(x, t) = \sum_{i=1}^{N} \tfrac{1}{2}[(\dot{\phi}_i)^2 - (\phi_i')^2] - U(\{\phi_i\}) \tag{2.35}$$

where $U(\{\phi_i\})$ is some function of all the ϕ_i, having a minimal value of zero. The field equations are

$$\Box\phi_i = -\partial U/\partial\phi_i; \qquad i = 1, \ldots, N. \tag{2.36}$$

A static solution must obey

$$\phi_i'' = \partial U/\partial\phi_i. \tag{2.37}$$

Again, this equation has a mechnical analogue. Now the analogue particle moves, as the 'time' x varies, in N dimensions with coordinates ϕ_i, and a potential $[-U(\phi_i)]$. The requirement that the solution have finite energy leads to the boundary conditions $\phi_i' = 0$ for all i and $U(\phi_i) = 0$, as $x \to \pm\infty$. For the analogue-particle, this again corresponds to a finite-action, zero-energy trajectory in N-dimensional space, beginning and ending at some minimum of $U(\phi_i)$. However there are two differences as compared with the single-field case:

(i) There, no solution was possible when $U(\phi)$ had a unique minimum. The argument was that once the particle leaves the maximum of $[-U(\phi)]$ it always has a negative-definite potential energy, and a non-zero kinetic energy and hence cannot stop and turn back towards its starting point. For coupled fields, if $U(\phi_i)$ has a unique minimum it is still true that at any other point in the N-dimensional ϕ_i-space, the particle cannot reach zero velocity. But without doing so, it may nevertheless swing around along a closed curve and return to the starting point. Correspondingly a non-trivial solitary wave is possible. This will be illustrated below with an example.

(ii) Unlike (2.17) which could be integrated by quadrature, there is no correspondingly simple way of integrating the coupled differential equations (2.37). Here the analogy to mechanics does not help. Given a general potential and end-point boundary conditions, it is not easy to integrate Newton's law of motion in two or more dimensions. This may seem surprising in view of the long history of mechanics and the several classic texts on orbit theory. Unfortunately, the bulk of such literature is devoted to 'central potentials'. These would correspond here to systems that are invariant under the rotation of the ϕ_i's into one another. Such symmetry under rotations in field space is called 'internal' symmetry in particle-physics terminology. When internal symmetry is present, we can try to exploit it in analogy with the use of polar coordinates in the mechanics of central potentials. We will deal with internal symmetry and 'charged' solitary waves separately in a later chapter. In the absence of such symmetry, even if $U(\phi_i)$ is a fairly simple polynomial function, it is not possible in general to derive all solutions of (2.37). However, we can find some solutions for individual cases by adopting the following strategy.

We divide the task into two parts. First we find the 'orbit' of the analogue particle in N-dimensional space, subject to the boundary conditions. This orbit is a one-dimensional curve and is specified by $N-1$ relations among the N coordinates ϕ_i. Once the orbit is known, motion along the orbit, i.e. the x-dependence of the $\phi_i(x)$ can be obtained by quadrature, since it amounts to motion along a given one-dimensional curve. Of course all orbits permitted by a given potential cannot be easily derived. However, for many typical forms of $U(\phi_i)$, simple trial orbits seem to fit the bill. This is nowhere nearly as desirable as deriving all the orbits from the potential. But it does yield some explicit solutions in many cases.

We illustrate the procedure using the case of two scalar fields ϕ_1 and ϕ_2 coupled by a given $U(\phi_1, \phi_2)$. The analogue particle then moves in the $\{\phi_1, \phi_2\}$ plane. Let $U(\phi_1, \phi_2)$ have, say, two degenerate minima at points P and Q in the $\{\phi_1, \phi_2\}$ plane as in fig. 3. (Generalisations to cases with more degenerate minima, or more coupled fields is straightforward.) We look for orbits permitted to the analogue-particle by the potential $[-U(\phi_1, \phi_2)]$ that it feels.

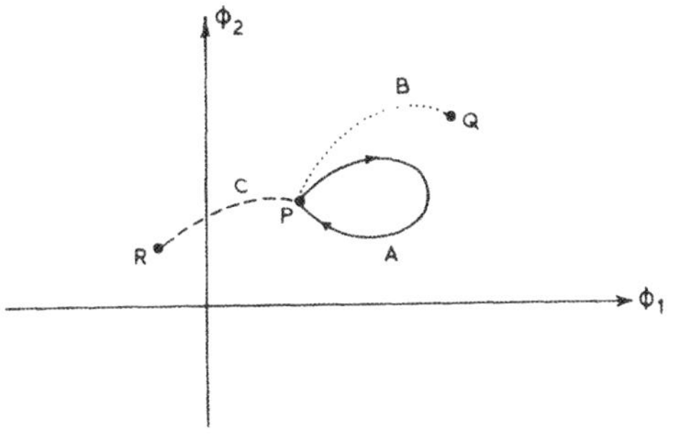

Fig. 3. Typical orbits corresponding to static localised solutions depicted here for an example with only two coupled fields ϕ_1 and ϕ_2 and two degenerate minima at P and Q of the potential $U(\phi_1, \phi_2)$. Type B orbits correspond, in the language of section 2.4 to topological solutions and type A to non-topological ones. An orbit of type C cannot exist corresponding to static solutions, but will play a role in 'charged' solutions to be discussed in chapter 8.

Let $g(\phi_1, \phi_2) = 0$ (2.38)

be the equation of an orbit. Then, we can relate the orbit directly to $U(\phi_1, \phi_2)$ by eliminating x. We have, from (2.38)

$$\frac{\mathrm{d}g}{\mathrm{d}x} = 0 = \frac{\partial g}{\partial \phi_1}\phi_1' + \frac{\partial g}{\partial \phi_2}\phi_2'.$$

Hence

$$\left(\frac{\partial g}{\partial \phi_1}\right)^2 (\phi_1')^2 = \left(\frac{\partial g}{\partial \phi_2}\right)^2 (\phi_2')^2. \tag{2.39}$$

But, on integrating (2.37),

$$\tfrac{1}{2}(\phi_i')^2 = \int \frac{\partial U(\phi_1, \phi_2)}{\partial \phi_i}\mathrm{d}\phi_i + A_i; \qquad i = 1, 2 \tag{2.40}$$

where A_i are integration constants. Inserting the two equations implied in (2.40) into (2.39), we have

$$\left(\frac{\partial g}{\partial \phi_1}\right)^2 \left(\int \frac{\partial U(\phi_1, \phi_2)}{\partial \phi_1}\mathrm{d}\phi_1 + A_1\right)$$
$$= \left(\frac{\partial g}{\partial \phi_2}\right)^2 \left(\int \frac{\partial U(\phi_1, \phi_2)}{\partial \phi_2}\partial\phi_2 + A_2\right). \tag{2.41}$$

This is the orbit equation which relates $g(\phi_1, \phi_2)$ directly to $U(\phi_1, \phi_2)$ without x explicitly appearing in it. However, note that the integrals have to be evaluated along the orbit, i.e. ϕ_1 and ϕ_2 are related by (2.38) in the integrands. Thus (2.41) is an integro-differential equation for the orbit in terms of the potential. It can also be recast in an alternate differential-equation form, but there is no general method for solving it in either form. This reflects the difficulties in orbit theory mentioned earlier. However, we can pick judicious trial functions for the orbit (2.38), insert them into (2.41) and see if the orbit parameters can be adjusted to satisfy (2.41).

The choice of trial orbits is simplified by our earlier arguments. Periodic orbits or orbits that flow to infinity, although important in genuine mechanics problems, are of no interest to us since they have infinite total action. Finiteness of the solitary wave's energy corresponds to finite-action, zero-energy orbits which begin at the minima P or Q and end at P or Q. Trial orbits of this kind can be of two types (fig. 3). Type A orbits begin and end at the same minimum, after looping around in the plane in a closed curve, while type B orbits connect different minima. A type C path, where the particle leaves from, say, P, goes to some arbitrary point R and retraces its path back to P, would have finite action, but is not permitted by the dynamics. At the turning point R, the particle must have zero kinetic energy and hence also $U = 0$ since it has zero total energy. But by hypothesis, the only zeros of U are at P and Q. Type C orbits therefore do not play a role in connection with static solitary waves, although they do come in for certain time-dependent 'charged' solutions which we will discuss in Chapter 8. Here, we need try only types A and B (If the potential had possessed a unique minimum, only type A would occur.)

The process of trying orbits is best illustrated by example. Consider a simple typical potential for coupled fields,

$$U(\phi_1, \phi_2) = \tfrac{1}{4}(\phi_1^2 - 1)^2 + \tfrac{1}{2}m^2(\phi_2)^2 + \tfrac{1}{4}\lambda(\phi_2)^4 + \tfrac{1}{2}d\phi_2^2(\phi_1^2 - 1) \quad (2.42)$$

where λ, d and $m^2 > 0$. More parameters could be introduced into this potential while keeping the same polynomial form, but they can always be removed by rescaling the fields and space–time coordinates. This is a very typical model, often discussed in the literature. It consists of two ϕ^4-systems, one with a negative ϕ^2 term as in (2.24) and one with a positive ϕ^2 term, coupled through the last term in (2.42). This $U(\phi_1, \phi_2)$ has two absolute minima, at $\{1, 0\}$ and $\{-1, 0\}$, in the $\{\phi_1, \phi_2\}$ plane. It vanishes only at these two points and is positive elsewhere if $\lambda > (d - m^2)^2$. Thus orbits of type A and type B may exist. Since the potential has a simple polynomial form, it is natural to try orbits which are also simple polynomials. For a type B orbit, which must go from one of these minima to the other, consider the family of trial orbits

$$g(\phi_1, \phi_2) = (\phi_2)^n + \alpha((\phi_1)^2 - 1) = 0 \quad (2.43)$$

where α and n are free to be determined. We insert this into (2.41) and integrate along the orbit starting from $(-1, 0)$; then we have, at any point $[\phi_1, \phi_2]$ on the orbit,

$$(2\alpha\phi_1)^2\left(\int_{-1}^{\phi_1} \mathrm{d}\tilde{\phi}_1(\tilde{\phi}_1^3 - \tilde{\phi}_1 + \mathrm{d}\tilde{\phi}_2^2\tilde{\phi}_1)\right)$$
$$= (n\phi_2^{n-1})^2\left(\int_0^{\phi_2} \mathrm{d}\tilde{\phi}_2(m^2\tilde{\phi}_2 + \lambda\tilde{\phi}_2^3 + \mathrm{d}\tilde{\phi}_2(\tilde{\phi}_1^2 - 1))\right) \quad (2.44)$$

where, since the integrals begin at $(-1, 0)$, where ϕ_1' and ϕ_2' vanish by boundary condition requirements, the integration constants A_1 and A_2 in (2.40) vanish. We use the orbit (2.43) to eliminate ϕ_1 in favour of ϕ_2 both inside and outside the integrals, and complete the integration to get

$$4\alpha^2\left(1-\frac{\phi_2^n}{\alpha}\right)\left(\frac{\phi_2^{2n}}{4\alpha^2}-\frac{nd}{2\alpha(n+2)}\phi_2^{n+2}\right)$$
$$=\tfrac{1}{2}m^2n^2\phi_2^{2n}+\tfrac{1}{4}\lambda n^2\phi_2^{2n+2}-\frac{dn^2\phi_2^{3n}}{\alpha(n+2)}. \tag{2.45}$$

Now it is only a matter of matching coefficients of powers of ϕ_2 on the two sides of this equation. There are two solutions:

(i) $n=2$, with

$$\alpha=\frac{(1-2m^2)}{d} \qquad \text{and} \qquad \lambda=\frac{d(d-2m^2)}{1-2m^2} \tag{2.46}$$

obtained respectively by matching coefficients of ϕ_2^4 and ϕ_2^6. The resulting orbit

$$\phi_2^2=[(1-2m^2)/d](1-\phi_1^2),$$

is a semi-ellipse connecting the two minima of U. The second relation in (2.46) is a constraint on the potential's parameters which needs to be satisfied, apart from the positivity requirement on U. Inserting this orbit into the field equations (2.37), we have

$$\phi_1''=\phi_1^3-\phi_1+d\phi_2^2\phi_1$$
$$=2m^2(\phi_1^3-\phi_1). \tag{2.47}$$

This is readily integrated giving (see also Montonen 1976),

$$\phi_1(x)=\tanh[m(x-x_0)]$$

and

$$\phi_2(x)=\pm[(1-2m^2)/d]^{1/2}\operatorname{sech}[m(x-x_0)]. \tag{2.48}$$

We thus have an explicit exact static solution with finite energy to the coupled field equations (2.37) for the potential (2.42). Another solution to (2.45) is

(ii) $n=1$ with

$$m^2=2, \qquad \lambda=\tfrac{8}{3}d \qquad \text{and} \qquad \alpha^2=(d-3)/2d. \tag{2.49}$$

The corresponding orbit is a parabolic arc:

$$\phi_2=[(d-3)/2d]^{1/2}(1-\phi_1^2),$$

connecting the two minima of U. Inserting this orbit into (2.37) we get

$$\begin{aligned}\phi_2'' &= m^2\phi_2 + \lambda\phi_2^3 + d\phi_2(\phi_1^2 - 1)\\ &= 2\phi_2 - d[2d/(d-3)]^{1/2}\phi_2^2 + \tfrac{8}{3}d\phi_2^3.\end{aligned} \tag{2.50}$$

This has the standard form

$$\phi_2'' = a\phi_2 - \tfrac{3}{2}b\phi_2^2 + 2c\phi_2^3 \tag{2.51}$$

for suitable constants a, b and c, and is readily integrated to yield

$$\phi_2(x) = \frac{2a}{(b^2 - 4ac)^{1/2}\cosh(\sqrt{a}\,x) + b}$$

and

$$\phi_1^2(x) = 1 - [2d/(d-3)]^{1/2}\phi_2(x). \tag{2.52}$$

This gives us another exact solitary wave to the system with the potential (2.42), but valid in a different range of λ, m^2 and d than the solution (2.48). Both (2.48) and (2.52) are type B solutions having qualitatively a bell-shaped ϕ_2 and a kink-shaped ϕ_1 as in fig. 4(a). For the same system, one can also try a type A solution, where the orbit begins and ends at, say, $(-1, 0)$. The simplest choice is the ellipse:

$$g(\phi_1, \phi_2) = \phi_2^2 - \beta(\phi_1 + 1)(\alpha - \phi_1) = 0. \tag{2.53}$$

This again works, in the sense that it satisfies (2.41) provided

$$\alpha = \frac{1 + 4m^2}{3}, \qquad \beta = \frac{2 - 4m^2}{1 + 4m^2}, \qquad d = \frac{3 + 12m^2}{4 + 4m^2}$$

and

$$\lambda = \frac{(5 - 4m^2)(1 + 4m^2)}{(8 + 8m^2)(1 - 2m^2)}. \tag{2.54}$$

These relations are easily derived following an identical procedure to that used in equations (2.44)–(2.45) for the type B orbit. The explicit solutions are again obtained by using the orbit equation (2.53) to eliminate one of the fields in the coupled equations (2.37). This gives

$$\begin{aligned}\phi_1'' = 4m^2(1 + \phi_1) &- \left(\frac{8(1 + m^2)^2 - 9}{2(1 + m^2)}\right)(1 + \phi_1^2)^2\\ &+ \left(\frac{8m^2 - 1}{2(1 + m^2)}\right)(1 + \phi_1)^3.\end{aligned} \tag{2.55}$$

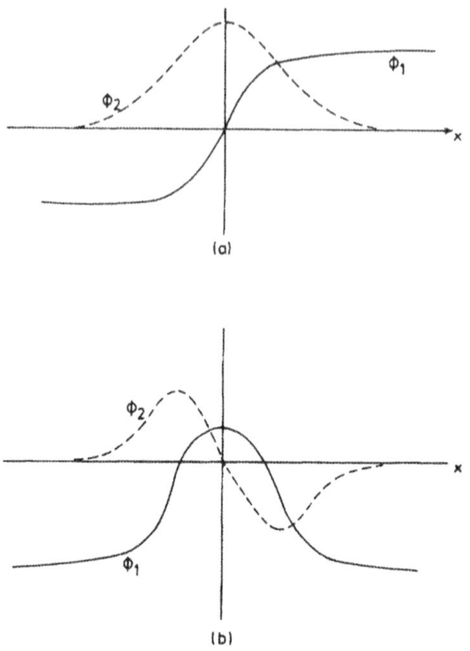

Fig. 4. (a) A sketch of the static solution (2.48). The solution (2.52) also has the same rough shape, but is a different function in detail. Both are of type B. (b) A sketch of the type A solution (2.56). This is a static non-topological solution. Such solutions do not exist when only a single scalar field is involved.

This again has the standard form (2.51) and is integrated to yield

$$1+\phi_1(x)=\frac{2a}{(b^2-4ac)^{1/2}\cosh(\sqrt{a}\,x)+b} \tag{2.56}$$

where once again a, b and c can be read off by comparing (2.55) with (2.51). The other field $\phi_2(x)$ is then obtained using the orbit equation (2.53). This solution, sketched in fig. 4(b), is typical of Type A and is interesting in that it is 'non-topological', i.e. it goes to the same limit as $x \to \pm\infty$, as distinct from earlier solutions in this chapter. But, while (2.56) is undoubtedly an exact solution, its stability against small disturbances is a different matter. Since it carries no non-zero topological index that could prevent it from dissipating into ripples (small-amplitude plane waves), it may well be unstable (see Subbaswamy and Trullinger 1980).

The exact solitary wave solutions in (2.48), (2.52) and (2.56) all correspond to the potential (2.42), but for different ranges of the parameters m^2, d and λ. This potential was just a typical example used to illustrate the method, which can be applied with profit to many other potentials (see for instance Ruck 1980).

The main weaknesses of this method are: (i) the need to try out orbits rather than derive them, and (ii) the constraints between the potential's parameters, such as in (2.46), required for the solution to work. The latter limitation arose because we tried particularly simple orbits. Orbits with more adjustable parameters could possibly loosen such constraints or even eliminate them. Finally, the generalisation of this method to more than two coupled fields is straightforward, although the algebra will be lengthy. One will have $(N-1)$ equations in the place of (2.41), relating the orbit to the potential.

2.4. Topological indices

It is often possible to make a topological classification of solutions of a given system of equations. Some early physics papers to which this concept can be traced are by Skyrme (1958), Finkelstein and Misner (1959) and Finkelstein (1966). A more recent and thorough discussion is given in Coleman's lectures (1977a). The central idea was already present in our discussion of the last section. We need only to extract and formalise it and extend it to more complicated systems. Specifically, one can define a topological index which is conserved in time. Like other conserved quantities it plays the important role of a 'quantum number' for particle states in the corresponding quantum field theory. It has, however, quite a different origin from that of the other familiar conserved quantities and quantum numbers.

Let us recall our discussion of a single scalar field $\phi(x,t)$ in two dimensions. Let the potential $U(\phi)$ in (2.11) have a discrete (though not necessarily finite) number of degenerate absolute minima, where it vanishes. The examples of the preceding section all satisfy this. So does the sine–Gordon system in the next section, which has a discrete infinity of degenerate minima. Now, we are interested in non-singular finite-energy solutions, of which solitary waves and solitons are special cases. Therefore the field, whether static or time-dependent, must tend at any instant t, to a minimum of $U(\phi)$ at every point on spatial infinity, in order that the energy E in (2.14) be finite. In one space-dimension, spatial infinity consists of two points, $x = \pm\infty$. Consider

$x = +\infty$, for instance. Let, at some given instant t_0,

$$\lim_{x \to \infty} \phi(x, t_0) \equiv \phi(\infty, t_0) = \phi_1 \tag{2.57}$$

where ϕ_1 has to be one of the minima of $U(\phi)$. Then, as time develops (either forward or backward, starting from t_0), the field $\phi(x, t)$ will change continuously with t at every x as governed by the differential equation (2.13). In particular, $\phi(\infty, t)$ will be some continuous function of t. On the other hand, since the energy of that solution is conserved and remains finite, $\phi(\infty, t)$ must always be one of the minima of $U(\phi)$, which are a discrete set. It cannot jump from ϕ_1 to another of the discrete minima if it is to vary continuously with t. Therefore $\phi(\infty, t)$ must remain stationary at ϕ_1. The same arguments apply to $x = -\infty$, where $\phi(-\infty, t) = \phi_2$, must also be time-independent and a minimum of $U(\phi)$, but not necessarily the same as ϕ_1 in the case of degenerate minima.

We can therefore divide the space of all finite-energy non-singular solutions into sectors, characterised by two indices, namely, the time-independent values of $\phi(x = \infty)$ and $\phi(x = -\infty)$. These sectors are topologically unconnected, in the sense that fields from one sector cannot be distorted continuously into another without violating the requirement of finite energy. In particular, since time evolution is an example of continuous distortion, a field configuration from any one sector stays within that sector as time evolves. Of course, when $U(\phi)$ has a unique minimum, there is only one permissible value for both $\phi(\infty)$ and $\phi(-\infty)$ and therefore only one sector of solutions exists.

As illustration consider the system (2.24) in the preceding section. The potential has two degenerate minima, at $\phi = (\pm m/\sqrt{\lambda})$. Consequently, all finite-energy non-singular solutions of this system, whether static or time-dependent, fall into four topological sectors. These are characterised by the pairs of indices $(-m/\sqrt{\lambda}, m/\sqrt{\lambda})$, $(m/\sqrt{\lambda}, -m/\sqrt{\lambda})$, $(-m/\sqrt{\lambda}, -m/\sqrt{\lambda})$ and $(m/\sqrt{\lambda}, m/\sqrt{\lambda})$ respectively, which represent the values of $(\phi(x = -\infty), \phi(x = +\infty))$. Thus, the kink, the antikink, and the trivial constant solutions $\phi(x) = \mp(m/\sqrt{\lambda})$, are members of the four sectors respectively. When a kink from the far left and an antikink from the far right approach one another, the field configuration belongs to the $(-m/\sqrt{\lambda}, -m/\sqrt{\lambda})$ sector. Even though we may not be able to calculate easily what happens after they collide, we can be sure that the resulting field configuration will always stay in the $(-m/\sqrt{\lambda}, -m/\sqrt{\lambda})$ sector.

A quantity sometimes called the 'topological charge' is often used in the literature. It can be defined here as

$$Q = (\sqrt{\lambda}/m)[\phi(x = \infty) - \phi(x = -\infty)] \tag{2.58}$$

with an associated conserved current,

$$k^\mu = (\sqrt{\lambda}/m)\varepsilon^{\mu\nu}\partial_\nu\phi, \tag{2.59}$$

where covariant summation notation

$$\mu,\nu = 0,1; \qquad g_{\mu\nu} = \begin{pmatrix} 1 & 0 \\ 0 & -1 \end{pmatrix}$$

has been used and $\varepsilon_{\mu\nu}$ is the antisymmetric tensor. Clearly

$$\partial_\mu k^\mu = 0 \qquad \text{and} \qquad Q = \int_{-\infty}^{\infty} \mathrm{d}x k_0. \tag{2.60}$$

This Q is just the difference between the two indices $(\sqrt{\lambda}/m)\phi(\infty)$ and $(\sqrt{\lambda}/m)\phi(-\infty)$. We mention it here because it is the analogue of topological indices in more complicated systems – such as gauge theories in four dimensions. Remember that to the extent that the field ϕ itself is a measurable physical quantity, we need both $\phi(\infty)$ and $\phi(-\infty)$ to identify a topological sector, and not Q alone. However, for cases where physical quantities depend only on differences in ϕ and not in the absolute value of ϕ, as happens with many applications of the sine–Gordon system, then Q becomes the sole relevant topological index. The adjective 'topological' is sometimes bestowed on solitary waves which have $Q \neq 0$. Waves with $Q = 0$ are 'non-topological'. Thus, the kink and the anti-kink of the system (2.24) are topological solutions, while the trivial solutions $\phi(x) = \pm(m/\sqrt{\lambda})$ are non-topological. One of our conclusions in the last section, couched in this terminology, says that for a single scalar field in two dimensions, non-trivial static solutions are necessarily topological.

Note that these topological indices are different from the more familiar conserved quantities like energy, momentum, charge etc. The latter, as is very well known in classical and quantum field theory, can be traced to the existence of continuous symmetries of the Lagrangian, such as under time translation, space translation, 'internal groups' and so on. By contrast, the topological indices are boundary conditions, conserved because of finiteness of energy. Indeed, in many cases, these indices are closely related to a certain kind of breaking of some symmetry. That is, suppose the Lagrangian and $U(\phi)$ are invariant under some symmetry transformation acting on $\phi(x)$. If $U(\phi)$ had a unique minimum at some $\phi = \phi_0$, then ϕ_0

itself must remain invariant under that transformation. But in order to get non-trivial topological sectors, we need to have two or more degenerate minima. In that case while the full set of minima is invariant under the transformation, each individual minimum need not be so. For instance, the system (2.24) which permits four topological sectors, has a $U(\phi)$ invariant under $\phi \leftrightarrow -\phi$. But its two minima, $\phi = -m/\sqrt{\lambda}$ and $\phi = +m/\sqrt{\lambda}$ are not separately invariant. Rather, they are transformed into one another. This fact has great importance in the quantum theory as well as the statistical mechanics of the field system and is called 'spontaneous symmetry breaking'. We will return to the phenomenon later, but at this stage we merely observe the relation of non-trivial topological sectors to the existence of several degenerate minima of the potential, which in turn is connected (often, but not always) to spontaneous symmetry breaking.

Generalisation of this idea to N coupled scalar fields $\phi_i(x, t)$ in two dimensions is straightforward. Let the degenerate minima of the potential $U(\phi_1, \ldots, \phi_N)$ be some discrete points $P^{(1)}, \ldots, P^{(m)}$ in the N-dimensional space of fields. The fields must approach the coordinates of any one of these points, say $P^{(r)}$, as $x \to \infty$, and another of these points $P^{(s)}$ as $x \to -\infty$. This pair of points $P^{(r)}$ and $P^{(s)}$ characterise the particular topological sector of finite-energy solutions. If $P^{(r)} = P^{(s)}$, the sector and solutions in it are 'non-topological'. If $P^{(r)} \neq P^{(s)}$, they are topological. We showed in the last section that, unlike the case of a single scalar field, static non-topological solutions *can* exist for coupled fields in two dimensions. These are the type A solutions, of which the one sketched in fig. 4(b) is an example.

As applied to scalar fields in two dimensions, the notion of topological sectors and indices seems too simple to merit so much fuss. However, when we go to more complicated systems in chapter 3, like gauge theories in four dimensions, similar ideas will hold. It will prove useful to have first encountered them here in a simpler context.

2.5. Solitons of the sine–Gordon system

The static solutions discussed in section 2.3 are, to the best of our knowledge, only solitary waves and not solitons. We will now discuss a system which yields solitons in the strict sense defined earlier. It consists of a single scalar field $\phi(x, t)$ in $(1+1)$ dimensions, governed by the Lagrangian density

$$\mathscr{L}(x, t) = \tfrac{1}{2}(\partial_\mu \phi)(\partial^\mu \phi) + (m^4/\lambda)\{\cos[(\sqrt{\lambda}/m)\phi] - 1\}. \qquad (2.61)$$

This system has been used in the study of a wide range of phenomena, including propagation of crystal dislocations, of splay waves in membranes, of magnetic flux in Josephson lines, Bloch wall motion in magnetic crystals, and two-dimensional models of elementary particles. Detailed references to these applications can be found in the reviews by Scott et al. (1973) and Barone et al. (1971). One can expand this Lagrangian density in powers of the coupling constant λ to get

$$\mathscr{L}(x, t) = \tfrac{1}{2}(\partial_\mu \phi)(\partial^\mu \phi) - \tfrac{1}{2}m^2\phi^2 + \frac{\lambda}{4!}\phi^4 - \frac{\lambda^2\phi^6}{m^2 6!} + \ldots \tag{2.62}$$

As $\lambda \to 0$, this is just the free Klein–Gordon system, and the $O(\lambda)$ term is the familiar (but here, attractive) quartic coupling. We will however work directly with the full Lagrangian (2.61) and look for classical solutions which are non-perturbative in λ. The field equation arising from (2.61) is the sine–Gordon equation:

$$\Box\phi + (m^3/\sqrt{\lambda})\sin[(\sqrt{\lambda}/m)\phi] = 0. \tag{2.63}$$

The name of the system is clearly inspired by the presence of the sine term compared with the Klein–Gordon equation – a piece of nomenclature that has attracted a certain amount of light-hearted controversy. Coleman (1977a) who calls the name 'silly', in turn quotes D. Finkelstein: 'I am sorry that I ever called it the sine–Gordon equation. It was a private joke between me and Julio Rubinstein and I never used it in print. By the time he used it as the title of a paper, he had earned his Ph.D. and was beyond the reach of justice'.

As historical evidence, the quote may stir more controversy than it will settle, particularly because Rubinstein (1970) in his paper passes on the parentage of the name to M. Kruskal. But in a field where the lighter vein is hard to come by, the quote earns the space it occupies!

Returning to the equation, we simplify it by changing variables to

$$\bar{x} = mx, \qquad \bar{t} = mt \qquad \text{and} \qquad \bar{\phi} = (\sqrt{\lambda}/m)\phi.$$

The Lagrangian density (2.61) becomes

$$\bar{\mathscr{L}}(\bar{x}, \bar{t}) = (m^4/\lambda)[\tfrac{1}{2}(\bar{\partial}_\mu\bar{\phi})(\bar{\partial}^\mu\bar{\phi}) + (\cos\bar{\phi} - 1)]. \tag{2.64}$$

The equation of motion is

$$\frac{\partial^2\bar{\phi}}{\partial\bar{t}^2} - \frac{\partial^2\bar{\phi}}{\partial\bar{x}^2} + \sin\bar{\phi}(\bar{x}, \bar{t}) = 0 \tag{2.65}$$

and the conserved energy integral is

$$E = \frac{m^3}{\lambda} \int d\bar{x} \left[\frac{1}{2}\left(\frac{\partial \bar{\phi}}{\partial \bar{t}}\right)^2 + \frac{1}{2}\left(\frac{\partial \bar{\phi}}{\partial \bar{x}}\right)^2 + (1 - \cos \bar{\phi}) \right]. \tag{2.66}$$

This Lagrangian and the field equation enjoy the discrete symmetries

$$\bar{\phi}(\bar{x}, \bar{t}) \to -\bar{\phi}(\bar{x}, \bar{t})$$

and

$$\bar{\phi}(\bar{x}, \bar{t}) \to \bar{\phi}(\bar{x}, \bar{t}) + 2N\pi; \qquad N = \ldots -2, -1, 0, 1, 2, \ldots \tag{2.67}$$

Consistent with these symmetries, the energy E vanishes at the absolute minima of

$$U(\bar{\phi}) = 1 - \cos \bar{\phi} \tag{2.68}$$

which are,

$$\bar{\phi}(\bar{x}, \bar{t}) = 2N\pi. \tag{2.69}$$

As per our earlier analysis, all finite-energy configurations, whether static or time-dependent, can be divided into an infinite number of topological sectors, each characterised by a conserved pair of integer indices (N_1, N_2) corresponding to the asymptotic values $2N_1\pi$ and $2N_2\pi$ that the field must approach as $\bar{x}$ tends to $-\infty$ and $+\infty$ respectively. If on physical grounds we decide that only $\bar{\phi}$ modulo 2π is meaningful, as will happen in applications where $\bar{\phi}$ is an angle variable, then only the topological charge

$$Q \equiv N_1 - N_2 = \frac{1}{2\pi} \int_{-\infty}^{\infty} d\bar{x} \, \frac{\partial \bar{\phi}}{\partial \bar{x}} \tag{2.70}$$

matters. By $\bar{\phi}$ modulo 2π, we mean that at any one space–time point $\bar{\phi}(\bar{x}, \bar{t})$ can be picked modulo 2π. At other points, it is fixed by continuity requirements.

Let us begin with static localised solutions. Our general considerations tell us that for a single scalar field in one space-dimension, static solutions must connect only neighbouring minima of $U(\bar{\phi})$. That is, they must carry $Q = \pm 1$. Explicit solutions are easily obtained using (2.23):

$$\bar{x} - \bar{x}_0 = \pm \int_{\bar{\phi}(\bar{x}_0)}^{\bar{\phi}(\bar{x})} \frac{d\phi}{\sqrt{2U(\phi)}} = \pm \int_{\bar{\phi}(\bar{x}_0)}^{\bar{\phi}(\bar{x})} \frac{d\phi}{2 \sin(\phi/2)}. \tag{2.71}$$

This is easily integrated to give

$$\bar{\phi}(x) = 4 \tan^{-1}[\exp(\bar{x} - \bar{x}_0)] \equiv \phi_{sol}(\bar{x} - \bar{x}_0) \qquad (2.72a)$$

or

$$\bar{\phi}(x) = -4 \tan^{-1}[\exp(\bar{x} - \bar{x}_0)] \equiv \bar{\phi}_{antisol}(\bar{x} - \bar{x}_0) = -\bar{\phi}_{sol}. \qquad (2.72b)$$

The solution with the plus sign (2.72a) goes from $\bar{\phi} = 0$ to $\bar{\phi} = 2\pi$ (fig. 5(a)), or equivalently from 2π to 4π, 4π to 6π etc. It corresponds to $Q = 1$, and is often called the soliton of the system. The other solution (2.72b) has $Q = -1$ and is called the antisoliton. Each has energy $M_s = 8m^3/\lambda$ as calculated by inserting (2.72) into (2.66). Moving soliton solutions can, as before, be obtained on Lorentz-transforming (2.72a), i.e. on replacing

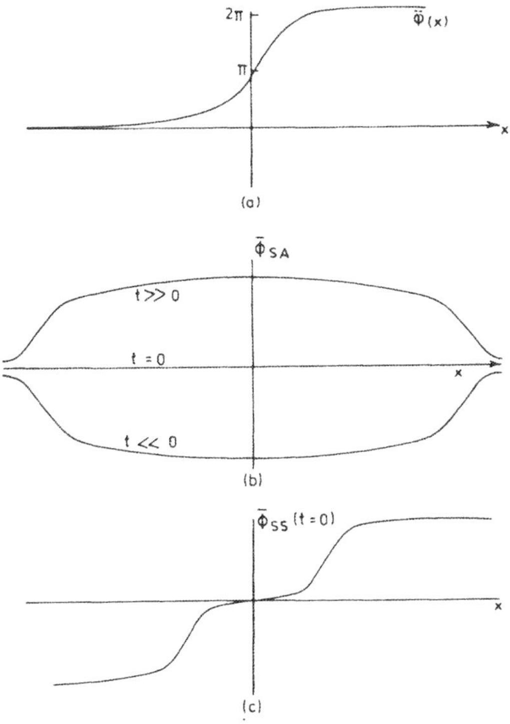

Fig. 5. (a) A sketch of the sine–Gordon soliton (eq. (2.72a)). (b) Three profiles of the soliton–antisoliton scattering solution $\bar{\phi}_{SA}(x, t)$ at t large negative, $t = 0$, and t large positive. At $t = 0$, $\bar{\phi}_{SA}$ vanishes. (c) A sketch of the soliton–soliton solution at $t = 0$.

$(\bar{x}-\bar{x}_0)$ by $[(\bar{x}-\bar{x}_0-u\bar{t})/\sqrt{1-u^2}]$. The solution (2.72a) is roughly similar in shape to the 'kink' discussed earlier, although the function, in detail, is different.

However, unlike the 'kink', we assert that the solution (2.72a) is a genuine soliton as per the stringent requirements (2.9)–(2.10) given earlier. Similarly the $Q = -1$ solution (2.72b) is also a genuine soliton in the generic sense of the word. It is called an antisoliton here partly to distinguish it from the $Q = 1$ solution (2.72a) and partly because it is related to the latter by the symmetry $\bar{\phi} \leftrightarrow -\bar{\phi}$.

In fact this system permits a third type of soliton called the doublet or breather, which we will describe later. Altogether then, the field equation (2.65) yields not just one but three different types of solitons.

The assertion that all these are indeed solitons has to be established by looking at exact time-dependent solutions of the field equations, comprising arbitrary numbers of these objects under collision. Fortunately such solutions have in principle been obtained for this equation. Let us begin with $\bar{\phi}_{SA}$, the soliton–antisoliton scattering solution (Seeger et al. 1953, Perring and Skyrme 1962; see also Rubinstein 1970). We will not derive this solution or the few others that follow. That they are indeed exact solutions may easily be verified by substituting them into (2.65). Thus, the function

$$\bar{\phi}_{SA}(\bar{x}, \bar{t}) = 4\tan^{-1}\left(\frac{\sinh(u\bar{t}/\sqrt{1-u^2})}{u\cosh(\bar{x}/\sqrt{1-u^2})}\right) \tag{2.73}$$

may be verified as satisfying (2.65). Its asymptotic behaviour in time can be extracted quite easily to yield

$$\bar{\phi}_{SA}(\bar{x}, \bar{t}) \xrightarrow[\bar{t}\to-\infty]{} 4\tan^{-1}\left[\exp\left(\frac{\bar{x}+u(\bar{t}+\Delta/2)}{\sqrt{1-u^2}}\right)\right]$$

$$-4\tan^{-1}\left[\exp\left(\frac{\bar{x}-u(\bar{t}+\Delta/2)}{\sqrt{1-u^2}}\right)\right]$$

$$= \bar{\phi}_{\text{sol}}\left(\frac{\bar{x}+u(\bar{t}+\Delta/2)}{\sqrt{1-u^2}}\right) + \bar{\phi}_{\text{antisol}}\left(\frac{\bar{x}-u(\bar{t}+\Delta/2)}{\sqrt{1-u^2}}\right) \tag{2.74}$$

where

$$\Delta \equiv ((1-u^2)/u)\ln u. \tag{2.75}$$

The solution therefore corresponds to a soliton and an antisoliton far apart and approaching one another with relative velocity $2u$, in the distant past.

Similarly one can check that

$$\bar{\phi}_{SA}(x, t) \underset{\bar{t} \to +\infty}{\longrightarrow} \bar{\phi}_{\text{sol}}\left(\frac{\bar{x} + u(\bar{t} - \Delta/2)}{\sqrt{1 - u^2}}\right) + \bar{\phi}_{\text{antisol}}\left(\frac{\bar{x} - u(\bar{t} - \Delta/2)}{\sqrt{1 - u^2}}\right) \quad (2.76)$$

where we again interpret $\bar{\phi}_{\text{sol}}$ and $\bar{\phi}_{\text{antisol}}$ modulo 2π. In the distant future, the solution $\bar{\phi}_{\text{SA}}$ corresponds to the same soliton–antisoliton pair, with the same shapes and velocities! The only change from the initial configuration (2.74) is the time delay Δ which remains as the sole residual effect of the collision between the soliton and the antisoliton. As they approach one another, they tend to annihilate each other until at $\bar{t} = 0$, the field vanishes everywhere (fig. 5(b)). But it re-emerges for positive $\bar{t}$, and asymptotically grows and separates into the same pair as if the collision had never taken place, except for the time delay. Since we are using units where $c = 1$, the velocity $u < 1$, and hence the time 'delay' Δ is negative. This indicates, qualitatively speaking, that a soliton and an antisoliton attract each other, and speed up in each other's vicinity. Note that (2.74)–(2.76) satisfy the general requirement (2.9)–(2.10) on solitons for one particular case, namely when one soliton and one antisoliton collide.

There is a similar two-soliton exact solution:

$$\bar{\phi}_{\text{SS}}(\bar{x}, \bar{t}) = 4 \tan^{-1}\left(\frac{u \sinh(\bar{x}/\sqrt{1 - u^2})}{\cosh(u\bar{t}/\sqrt{1 - u^2})}\right) \quad (2.77)$$

which is depicted in fig. 5(c). At any instant $\bar{t}$, it goes from -2π to $+2\pi$ as x goes from $-\infty$ to $+\infty$, and consequently it belongs to the $Q = 2$ sector. Note that our earlier general discussion prohibited only *static* solutions from belonging to $Q \neq 1$ sectors. Time-dependent solutions can exist in all sectors. The solution (2.77) also has a simple asymptotic behaviour as $\bar{t} \to \pm\infty$. The same simple algebra shows that analogous to $\bar{\phi}_{\text{SA}}$, $\bar{\phi}_{\text{SS}}$ corresponds to two $Q = 1$ solitons approaching one another at $\bar{t} = -\infty$ and after collision, again to two $Q = 1$ solitons going away from each other with the same speeds, but with some time delay. Strictly speaking, in (2.77), the two solitons bounce back with velocities equal and opposite to their initial velocities. This is a case of backward scattering. But if we continue to think of fields modulo 2π, and not distinguish between the first soliton which goes from $\bar{\phi} = -2\pi$ to $\bar{\phi} = 0$ from the second one which goes from $\bar{\phi} = 0$ to $\bar{\phi} = 2\pi$, then there is no difference asymptotically between backward and forward scattering.

Finally, by the $\bar{\phi} \leftrightarrow -\bar{\phi}$ symmetry, $\bar{\phi}_{\text{AA}} \equiv -\bar{\phi}_{\text{SS}}$ is the anti-soliton–antisoliton solution. These exact solutions (2.74) and (2.77) indicate

indicate that what we have termed the soliton and the antisoliton of this system may both be genuine solitons. But (2.74) and (2.77) correspond to cases where only *two* of these objects collide. The full definition (2.9)–(2.10) demands that a similar behaviour must exist for an arbitrary number of them. We will return to this point after describing the so-called doublet or breather solutions, $\bar{\phi}_v(\bar{x}, \bar{t})$. It may be verified that

$$\bar{\phi}_v(\bar{x}, \bar{t}) = 4\tan^{-1}\left[\frac{\sin(v\bar{t}/\sqrt{1+v^2})}{v\cosh(\bar{x}/\sqrt{1+v^2})}\right] \tag{2.78}$$

is also a solution of the field equations. Here, v is a parameter rather like u in (2.73). Indeed, if we convert the real parameter u in $\bar{\phi}_{\text{SA}}$ into an imaginary one by setting $u = \mathrm{i}v$, we get (2.78) which still gives a real function $\bar{\phi}_v$. Considering that u represented the asymptotic velocities of the soliton and the antisoliton in $\bar{\phi}_{\text{SA}}$, one can interpret the doublet (2.78) as a 'bound' solution of a soliton–antisoliton pair. The doublet is clearly a periodic solution with period

$$\bar{\tau} = (2\pi\sqrt{1+v^2})/v.$$

The soliton and antisoliton oscillate with respect to one another with this period. Profiles of $\phi_v(\bar{x}, \bar{t})$ for $0 > \bar{t} > \bar{\tau}/4$, $\bar{t} = 0$, and $0 < \bar{t} < -\bar{\tau}/4$ look very similar to fig. 5(b) drawn for $\bar{\phi}_{\text{SA}}$. But, whereas in the latter the S–A pair separate arbitrarily far apart as $\bar{t} \to \pm\infty$, in the former they separate only up to some finite distance, never become fully free of distortion in their shapes due to each other's presence, and oscillate about each other. That such a bound solution exists is qualitatively consistent with the fact that in the scattering solution $\bar{\phi}_{\text{SA}}$ the pair experience a negative time delay (2.75), indicating on the average an attractive force due to one another. By contrast, a soliton bounces back on collision with another soliton as evident from the asymptotic behaviour of $\bar{\phi}_{\text{SS}}$, indicating a repulsive force. Correspondingly, a bound soliton–soliton pair does not exist. An attempt to set $u = \mathrm{i}v$ in $\bar{\phi}_{\text{SS}}$ yields a complex rather than real solution, and is not admissible for our real scalar field system. This notion of attractive and repulsive forces borrowed from particle-mechanics is at best an approximate concept here, valid only for large separations, but it is interesting that it fits in with the presence of S–A bound solutions $\bar{\phi}_v$, and the absence of S–S bound solutions. We will return to the notion of such forces in chapter 3. A word on the status of the doublet solution as a solitary wave: note that its energy density, computed using the integrand in (2.66) will not have a space–time dependence of the $(x - ut)$ form required in the working definition (2.8) for solitary waves. In (2.78), the solution has been given in its overall rest frame; that is, the solution is centred at $\bar{x} = 0$ for all $\bar{t}$. A travelling doublet can be obtained by Lorentz-transforming (2.78). But even in its rest frame (2.78), the doublet has non-trivial time dependence. At any given $\bar{x}$, it oscillates in time. It is in this regard that it differs from the soliton (2.72), the kink, and some other solutions in this chapter, which are static in the rest frame and satisfy the requirement (2.8) in any moving frame of reference. Nevertheless, the doublet does share the basic intuitive demands on solitary waves: it has localised, finite energy; it is also essentially non-dispersive. In the rest frame, the field remains confined within the envelope

$$\pm 4 \tan^{-1}\{(1/v)\,\mathrm{sech}(\bar{x}/\sqrt{1+v^2})\}.$$

We will therefore include it in our set of solitary waves, but this example reveals the inadequacy of a simple working definition like (2.8) in covering all interesting cases.

Coming back to our assertion that (2.72a), (2.72b) and (2.78) are not just solitary waves, but also solitons, we require that time-dependent solutions must exist, where arbitrary numbers of solitons, antisolitons and doublets scatter and re-emerge unscathed as $\bar{t} \to +\infty$. The solutions $\bar{\phi}_{SA}$ and $\bar{\phi}_{SS}$ are just two pieces of the required information. That such general solutions exist has been demonstrated by the work of Ablowitz et al. (1973) and that of Faddeev and colloborators (Faddeev and Takhtajan 1974, Zakharov et al. 1974, Arefieva and Korepin 1974, Faddeev et al. Korepin 1975). As a result of this work, in a sense all solutions of the time-dependent sine–Gordon equation are known. This is a very impressive achievement, attained through the use of the so-called inverse scattering method (Gardner et al. 1967). The many-soliton solutions, unlike $\bar{\phi}_{SA}$ and $\bar{\phi}_{SS}$, are not simple expressions when written as functions $\bar{\phi}(\bar{x}, \bar{t})$. We will not even write them down here, let alone present their derivation using the inverse scattering method. This method as well as the other formal and elegant features of this system, such as the use of Backlund transformations, the presence of an infinite number of conserved quantities etc., are substantive matters whose full discussion would take us too far afield; see the disclaimer at the beginning of this chapter. It is tempting however, to offer some brief and qualitative comments, without proof, on some of these techniques.

The inverse scattering method essentially brings to bear the use of action–angle variables–a familiar technique in classical mechanics described in any advanced text on that subject (as for instance, Goldstein 1950). The sine–Gordon system offers a concrete realisation of this formal technique, as generalised to infinite degrees of freedom. In the place of the field variables $\bar{\phi}$ one changes variable to a set $\{\theta(p), -\infty < p < \infty; q_a, a = 1, 2, \ldots; q_b \text{ and } \beta_b, b = 1, 2, \ldots\}$ with range $0 \leqslant \theta \leqslant 2\pi, -\infty < q_a, q_b < \infty, 0 \leqslant \beta_b \leqslant 32\pi\, m^2/\lambda$. These are the 'angle' variables. The corresponding conjugate 'action' variables, which take the place of the original canonical momentum variables $\pi(\bar{x}, \bar{t}) \equiv (m^4/\lambda)\, \partial\bar{\phi}(\bar{x}, \bar{t})/\partial\bar{t}$, form the set

$$\{\rho(p) \text{ lying in } [0, \infty];\ P_a, P_b, \text{ both in } [-\infty, \infty]; \text{ and } \alpha_b \text{ in } [0, \pi/2]\}.$$

In terms of such variables, it can be shown that the Hamiltonian of the system (essentially the energy integral (2.66)) has the form

$$H = \int_{-\infty}^{\infty} (p^2 + m^2)^{1/2} \rho(p)\, \mathrm{d}p + \sum_a (P_a^2 + M_s^2)^{1/2} + \sum_b [P_b^2 + M_\mathrm{D}^2(\alpha_b)]^{1/2} \tag{2.79}$$

where M_s is just the soliton mass, i.e. its static energy $8m^3/\lambda$; $M_\mathrm{D}(\alpha_b)$ is the energy of the doublet solution in its overall rest frame. The variable α_b refers to the internal motion of the doublet solution. We recall that, even in the rest frame, the doublet has an oscillatory behaviour with a period $\bar{\tau}$. Doublets form a one-parameter set characterised by $\bar{\tau}$. This degree of freedom is represented by the conjugate variables α_b and β_b. The difficult task is to find these action–angle variables as functionals of the original variables $\bar{\phi}(\bar{x}, \bar{t})$ and $\pi(\bar{x}, \bar{t})$, transform to them and show that the Hamiltonian is indeed of the form (2.79). Even in the mechanics of a single particle, action–angle variables can be found explicitly only for a few cases. This formidable problem for the sine–Gordon field system is tackled by mapping it on to an inverse linear problem. This linear problem is analogous to, though not exactly the same as, finding the potential of a Schrödinger wave equation from the scattering and bound-state data of its solutions rather than the other way around. Hence the name inverse scattering method. What we omit here is the identification and solution of this inverse problem. Once the form (2.79) has been achieved, the remaining steps are trivial. Notice that the Hamiltonian depends, as required, only on the action variables and not on the angles. It is furthermore a separable sum over Hamiltonians of free relativistic particles. When the action–angle variables are written in terms of the original field variable $\bar{\phi}$, the three terms in (2.79) have a simple interpretation. The first term represents the contribution of plane-wave solutions. Recall that in the weak-field limit (small ϕ), this system becomes just the Klein–Gordon system (eq. (2.62)) and hence permits plane waves of the form $\phi_0 \exp[i(px - \sqrt{(p^2+m^2)}t)]$ with an arbitrarily small amplitude ϕ_0, but spread out all over space. Each such wave carries an energy proportional to $\sqrt{p^2+m^2}$, resulting in the first term of (2.79). This is the non-solitary-wave, non-soliton component of the system. The second term represents the energy of the solitons (and antisolitons, which have the same mass M_s). Correspondingly, the index a in (2.79) runs over individual solitons and antisolitons, whose momenta are the P_a. Finally the index b in the last term runs over doublets, whose momenta are P_b and whose masses $M_D(\alpha_b)$ are their rest-frame energies depending on the period of their internal motion. The proof that the so-called soliton, antisoliton and doublet solutions are all true solitons lies in the separability of the Hamiltonian into this set of free-particle forms, each associated uniquely with either a plane wave, a soliton (or antisoliton), or a doublet.

Another important feature of this system is the presence of Backlund transformations. As applied to the sine–Gordon system (Lamb 1971, Barnard 1973) these transformations provide a way of generating N-soliton solutions, starting from solutions with fewer solitons. Furthermore, only first-order differential equations need to be solved in the process, in the place of the parent second-order field equations. Note that the sine–Gordon equation (2.65), when written in terms of 'light cone' coordinates $\sigma \equiv \frac{1}{2}(\bar{x}+\bar{t})$ and $\rho \equiv \frac{1}{2}(\bar{x}-\bar{t})$ becomes

$$\partial^2\bar{\phi}/\partial\sigma\partial\rho - \sin\bar{\phi} = 0. \tag{2.80}$$

Let $\phi_0(\sigma,\rho)$ be a given solution of this equation. Consider $\phi_1(\sigma,\rho)$ generated by the Backlund transformation equations

$$\tfrac{1}{2}(\partial/\partial\sigma)(\phi_1-\phi_0) = a\sin[\tfrac{1}{2}(\phi_1+\phi_0)] \tag{2.81a}$$

and

$$\tfrac{1}{2}(\partial/\partial\rho)(\phi_1+\phi_0) = (1/a)\sin[\tfrac{1}{2}(\phi_1-\phi_0)] \tag{2.81b}$$

where a is a real parameter. On operating $\partial/\partial\rho$ on (2.81a) and using (2.81b), we have

$$\begin{aligned}\tfrac{1}{2}(\partial^2/\partial\sigma\partial\rho)(\phi_1-\phi_0) &= \cos[\tfrac{1}{2}(\phi_1+\phi_0)]\sin[\tfrac{1}{2}(\phi_1-\phi_0)]\\ &= \tfrac{1}{2}\sin\phi_1 - \tfrac{1}{2}\sin\phi_0.\end{aligned} \tag{2.82}$$

Given that $\phi_0(\sigma, \rho)$ is a solution of (2.80), so is $\phi_1(\sigma, \rho)$! The explicit function ϕ_1 is obtained by integrating (2.81) using ϕ_0 as input. For instance take $\phi_0(\sigma, \rho) = 0$. This is the trivial (no-soliton) solution of the sine–Gordon equation. Insertion of this ϕ_0 into (2.81) yields

$$\tfrac{1}{2}\,\partial\phi_1/\partial\sigma = \tfrac{1}{2}a^2\,\partial\phi_1/\partial\rho = a\sin(\tfrac{1}{2}\phi_1).$$

These are easily integrated to yield

$$\phi_1(\sigma, \rho) = 4\tan^{-1}[\exp(a\sigma + \rho/a)].$$

In terms of the original coordinates $\{\bar{x}, \bar{t}\}$, this gives

$$\bar{\phi}(\bar{x}, \bar{t}) = 4\tan^{-1}\left[\exp\left(\frac{\bar{x} - u\bar{t}}{\sqrt{1-u^2}}\right)\right]$$

with $u = (1 - a^2)/(1 + a^2)$. This is just the one-soliton solution (2.72a) described in a frame where it moves with velocity u. Thus, starting from the no-soliton solution, the Backlund transformation generates the one-soliton solution. In the same way, many-soliton solutions can be progressively obtained. The substantive part of this development, omitted in our discussion, really comes prior to (2.81), namely the demonstration that the sine–Gordon system does permit Backlund transformations and the identification of (2.81) as the equations that do the job. Another feature of the sine–Gordon system is that its solutions carry an infinite number of conserved quantities. We shall return to this feature in chapter 7. Readers interested in a proper discussion of the inverse scattering method, Backlund transformations, the conserved quantities etc. are referred to the review by Scott et al. (1973).

To the extent that we will use the sine–Gordon system to illustrate the semiclassical method in later chapters it is sufficient to know that (2.72) and (2.78) are exact solutions of the field equation (2.65). This the reader can easily verify by substituting these solutions into (2.65). In chapter 7, we shall illustrate the WKB method for quantising time-dependent classical solutions, by using the doublet solution (2.78) as an example. We choose the doublet for this illustration, largely because it happens to be one of the few periodic localised solutions in relativistic field theory available in exact analytic form. That the solutions (2.78) and (2.72) also happen to be solitons in the strict sense is only an added bonus. It results in the further consequence that some WKB results turn out to be exact and an exact S-matrix is available, for this system (see chapter 7). But the basic semiclassical method described in chapters 5 and 6 for obtaining quantum bound states from classical solutions only requires that these solutions be solitary waves, i.e. localised non-dissipative solutions, and not necessarily that they be solitons in the strict sense. This is perhaps one reason why the term solitons has often been enlarged in the literature to include solitary waves as well, whether or not they satisfy the stringent requirements (2.9) and (2.10).

CHAPTER 3

MONOPOLES AND SUCH

3.1. On solutions in higher dimensions

In the preceding chapter, we dealt basically with scalar fields in $(1+1)$ dimensions. We would now like to study more complicated systems, and in higher dimensions. The methods available for solving non-linear equations are so inadequate that even for scalar fields in $(1+1)$ dimensions, all time-dependent solutions of a given system cannot be obtained in general. The sine–Gordon system was an exception rather than the rule and its solubility was a special feature related to its having a completely separable free-particle Hamiltonian in terms of suitably chosen coordinates. Similar success is currently beyond reach for more typical systems like the two-dimensional ϕ^4 theory. Of course if in a two-dimensional theory we restrict ourselves to solutions which are static in some coordinate frame, then in that frame essentially only one (space) dimension is involved, and the field equations become ordinary differential equations. Furthermore, for translationally invariant systems, they are 'autonomous' equations, i.e. the independent variable x does not appear explicitly in the equations – only the fields and the derivatives with respect to x do. Autonomous ordinary second-order differential equations, even coupled ones, are not so hard to solve, as evidenced by the several one-dimensional static solutions we presented in the last chapter.

However, in two or more space dimensions, even static solutions obey *partial* differential equations. Such non-linear equations are much harder to solve. Most realistic systems, as for instance in particle physics, involve scalar, vector and spinor fields coupled non-linearly in $(3+1)$ dimensions. For a typical system of this sort, no systematic method of obtaining even a single non-trivial classical solution in analytic form is available as yet. If, despite this remark, we present exact solutions of several complicated higher-dimensional systems in this chapter and the next, that is only because the systems chosen are special, and clever tricks have been invented that simplify the solution of their field equations. Non-trivial solutions no doubt exist for many other models, but deriving them analytically may be a formidable task. For such models, one may have to be satisfied with whatever general features one can ferret out about their solutions without explicitly solving the field equations. Consequently, in the following sections we shall place considerable emphasis on some general features of classical solutions, before presenting exact solutions wherever we can. An important general feature for a large class of systems is the availability of homotopy classification and topological charges. Another useful general result is a Virial theorem. Let us begin with the latter.

3.2. A Virial theorem

This very useful theorem (Derrick 1964, Hobart 1963) gives basically a negative result. In progressing from scalar fields in (1 + 1) dimensions, which we studied in the last chapter, towards more realistic systems, a natural step is the set of purely scalar fields in higher dimensions. This Virial theorem severely restricts the possibility of having non-trivial static solitary waves of such systems. It tells us that no such solutions can exist when the space dimensionality is three or more and when the Lagrangian has the standard relativistic form

$$\mathscr{L}(x, t) = \tfrac{1}{2}(\partial_\mu \boldsymbol{\phi})\cdot(\partial^\mu \boldsymbol{\phi}) - U(\boldsymbol{\phi}(x, t)). \tag{3.1}$$

Here $\boldsymbol{\phi}(x, t) = [\phi_i(x, t); i = 1, \ldots, N]$ is a set of N coupled scalar fields in D space plus one time dimensions. A scalar product between $\partial_\mu \boldsymbol{\phi}$ and $\partial^\mu \boldsymbol{\phi}$, i.e. a summation over the field index i, is implied in (3.1). The potential $U(\boldsymbol{\phi}(x, t))$ is, as before, some non-negative function, vanishing only at its absolute minima. A static solution $\boldsymbol{\phi}(x)$ obeys

$$\nabla^2 \boldsymbol{\phi} = \frac{\partial U}{\partial \boldsymbol{\phi}}(x) \tag{3.2}$$

where ∇^2 is the Laplacian in D dimensions. This equation is clearly the extremum condition $\delta W = 0$ for the static energy functional

$$\begin{aligned} W[\boldsymbol{\phi}] &\equiv \int d^D x [\tfrac{1}{2} \nabla_i \boldsymbol{\phi} \cdot \nabla_i \boldsymbol{\phi} + U(\boldsymbol{\phi}(x))] \\ &\equiv V_1[\boldsymbol{\phi}] + V_2[\boldsymbol{\phi}] \end{aligned} \tag{3.3}$$

where the functionals V_1 and V_2 stand for the two terms on the right-hand side. Note that not only W but also V_1 and V_2 are non-negative. Now, let $\phi_1(x)$ be a static solution. Consider the one-parameter family of configurations

$$\phi_\lambda(x) = \phi_1(\lambda x). \tag{3.4}$$

It is easy to check that

$$\begin{aligned} W[\phi_\lambda] &= V_1[\phi_\lambda] + V_2[\phi_\lambda] \\ &= \lambda^{2-D} V_1[\phi_1] + \lambda^{-D} V_2[\phi_1]. \end{aligned} \tag{3.5}$$

Since $\phi_1(x)$ is an extremum of $W[\phi]$, it must in particular make $W[\phi_\lambda]$ stationary with respect to variations in λ; that is,

$$(\mathrm{d}/\mathrm{d}\lambda)\, W[\phi_\lambda] = 0 \qquad \text{at } \lambda = 1. \tag{3.6}$$

Differentiating (3.5) using (3.6) gives us

$$(2-D)\, V_1[\phi_1] = D\, V_2[\phi_1]. \tag{3.7}$$

Since V_1 and V_2 are non-negative, (3.7) cannot be satisfied for $D \geqslant 3$ unless $V_1[\phi_1] = V_2[\phi_1] = 0$. This means that $\phi_1(x)$ has to be space-independent and equal to one of the zeros of $U[\phi]$. This is just a trivial solution and the theorem precludes non-trivial space-dependent solutions. Note that the result holds only for *static* solutions and for Lagrangians of the form (3.1). Time-dependent solitary waves can exist for scalar fields in $(D+1)$ dimensions even when $D \geqslant 3$ (see chapter 8). The theorem does motivate us, however, if we are seeking static solutions in the real $(3+1)$ dimensional world, to consider models with more than just scalar fields. In section 3.4 we shall study such a model involving scalar and vector fields. Before doing that, it will be helpful to study a simpler system, where some important ideas occuring in later sections are already present. This system involves only scalar fields, but is in $(2+1)$ dimensions.

3.3. The non-linear O(3) model: The isotropic ferromagnet

Equation (3.7) tells us that if $D = 2$, $V_2[\phi_1] = 0$. That is, $\phi_1(x)$, for all x, must be one of the zeros (which are also the absolute minima) of the potential $U(\phi)$. If $U(\phi)$ had only discrete minima then, by continuity, $\phi_1(x)$ must be the same minimum of $U(\phi)$ for all x, i.e. it would be independent of x and trivial. If $U(\phi)$ had a continuous set of minima, then (3.7) would permit, when $D = 2$, a possible x-dependent solution where ϕ changes continuously within this set of minima.

The simplest example of $U(\boldsymbol{\phi})$ with continuous minima is of course $U(\boldsymbol{\phi}) = 0$. However this makes the model too simple. The equation obeyed by a static solution, as derived from the Lagrangian (3.1), would be

$$\nabla^2 \boldsymbol{\phi} = 0 \tag{3.8}$$

whose only non-singular solutions are constants. But we can introduce an innocent looking but non-trivial twist to the model by imposing the constraint $\boldsymbol{\phi} \cdot \boldsymbol{\phi} = 1$. Such a model is called the non-linear O(N) model. We will now study the $N = 3$ case and find that it does yield interesting solutions (Belavin and Polyakov 1975, Skyrme 1961, Faddeev 1974).

The non-linear O(3) model consists of three real scalar fields $\boldsymbol{\phi}(\boldsymbol{x}, t) \equiv \{\phi_a(\boldsymbol{x}, t); a = 1, 2, 3\}$ with the constraint that at all $(\boldsymbol{x}, t)$

$$\sum_a \phi_a^2(\boldsymbol{x}, t) \equiv \boldsymbol{\phi} \cdot \boldsymbol{\phi} = 1. \tag{3.9}$$

The dynamics is determined, subject to the above constraint, by the Lagrangian density

$$\mathscr{L} = \tfrac{1}{2} \sum_\mu \sum_a (\partial_\mu \phi_a)(\partial^\mu \phi_a) \equiv \tfrac{1}{2} (\partial_\mu \boldsymbol{\phi}) \cdot (\partial^\mu \boldsymbol{\phi}). \tag{3.10}$$

Note that $\boldsymbol{\phi}$ can be considered as a vector in 'internal space' i.e. the three-dimensional field-space, and its components are labelled above by the index a. This is to be distinguished from vectors in coordinate space, which are labelled by Lorentz indices, such as μ in (3.10). Thus, in (3.10), a scalar product is implied between $\partial_\mu \boldsymbol{\phi}$ and itself, in internal space (as indicated by the dot) as well as in coordinate space as indicated by the repeated index μ. We emphasise this because we will be repeatedly dealing with objects that are tensors in coordinate space, as well as in some internal space and the distinction between the two spaces should be carefully borne in mind. Finally, we note that both the Lagrangian (3.10) and the constraint (3.9) are invariant under global O(3) rotations in internal space.

The field equation is obtained by applying the Euler–Lagrange variational principle to the action, with the constraint imposed through a Lagrange multiplier. That is, we extremise

$$S[\boldsymbol{\phi}] \equiv \int \mathrm{d}\boldsymbol{x} \int \mathrm{d}t \, [\tfrac{1}{2}(\partial_\mu \boldsymbol{\phi}) \cdot (\partial^\mu \boldsymbol{\phi}) + \lambda(\boldsymbol{x}, t)(\boldsymbol{\phi} \cdot \boldsymbol{\phi} - 1)]. \tag{3.11}$$

The resulting field equation is

$$\partial_\mu \partial^\mu \boldsymbol{\phi} + \lambda \boldsymbol{\phi} = (\Box + \lambda) \boldsymbol{\phi} = 0. \tag{3.12}$$

The Lagrange multiplier $\lambda(\boldsymbol{x}, t)$ is eliminated by using the constraint (3.9).

$$\lambda(\boldsymbol{x}, t) = \lambda\boldsymbol{\phi}\cdot\boldsymbol{\phi} = -\boldsymbol{\phi}\cdot\Box\boldsymbol{\phi}. \tag{3.13}$$

Let us now restrict ourselves to two space dimensions and to *static* solutions. The field equation then reduces to

$$\nabla^2\boldsymbol{\phi} - (\boldsymbol{\phi}\cdot\nabla^2\boldsymbol{\phi})\boldsymbol{\phi} = 0, \tag{3.14}$$

upon inserting (3.13).

Equation (3.14) is to be contrasted with (3.8) which would have resulted in the absence of the constraint. We shall see that, unlike (3.8), equation (3.14) does yield interesting non-singular solutions in two dimensions. Furthermore the solutions can be classified into homotopy sectors, characterised by a topological index. The basic principles behind such classification are just a generalisation of our earlier discussion in the simpler one-dimensional context in chapter 2.

The energy of a static solution, as obtained from the Lagrangian (3.10), is clearly

$$E = \tfrac{1}{2}\int(\partial_\sigma\boldsymbol{\phi})\cdot(\partial_\sigma\boldsymbol{\phi})\,\mathrm{d}^2x; \qquad \sigma = 1, 2. \tag{3.15}$$

Consider first the zero-energy solutions (the 'classical vacua'). Clearly they must satisfy for all $\boldsymbol{x}$ the conditon $\partial_\sigma\boldsymbol{\phi} = 0$. That is, $\boldsymbol{\phi}(\boldsymbol{x}) = \boldsymbol{\phi}^{(0)}$, which is any unit vector in internal space. While $\boldsymbol{\phi}^{(0)}$ must be $\boldsymbol{x}$-independent in an $E = 0$ solution, it could point in any direction in internal space, as long as it is a unit vector (because of the constraint (3.9)). Thus we have a degenerate continuous family of $E = 0$ solutions, corresponding to the different directions in which $\boldsymbol{\phi}^{(0)}$ could point. As in the kink system (eq. (2.24)), this is once again a case of 'spontaneous symmetry breaking' at the classical level. Whereas in the kink problem the symmetry in question was discrete (under $\phi \leftrightarrow -\phi$), and correspondingly there were two discrete $E = 0$ solutions related to one another by that symmetry, here we have continuous O(3) symmetry and correspondingly a continuous family of degenerate classical minima, once again related to one another by O(3) rotations in internal space.

Next, we proceed to soliton solutions, i.e. those with non-zero but finite E. From (3.15), it is clear that they must satisfy, using polar coordinates (r, θ) in $\boldsymbol{x}$-space,

$$r\|\mathrm{grad}\,\boldsymbol{\phi}\| \to 0 \qquad \text{as } r \to \infty \tag{3.16}$$

or

$$\lim_{r\to\infty} \boldsymbol{\phi}(\boldsymbol{x}) = \boldsymbol{\phi}^{(0)} \tag{3.17}$$

where $\boldsymbol{\phi}^{(0)}$ is again some unit vector in internal space. Note that as we tend to infinity in coordinate space in different directions, $\boldsymbol{\phi}(\boldsymbol{x})$ must approach the *same* limit $\boldsymbol{\phi}^{(0)}$. Otherwise $\boldsymbol{\phi}(\boldsymbol{x})$ will depend on the coordinate angle θ even at $r = \infty$, and the angular component of the gradient, $(1/r)(\partial\boldsymbol{\phi}/\partial\theta)$, will not satisfy (3.16)

Since $\boldsymbol{\phi}(\boldsymbol{x})$ approaches the same value $\boldsymbol{\phi}^{(0)}$ at all points at infinity, the physical coordinate plane R_2 is essentially compacted into a spherical surface, which we will call $S_2^{(\text{phy})}$. That is, the plane R_2 may be folded into a spherical surface, with the circle at infinity reduced to the north pole of the sphere. (In more precise terms, this can be done by a stereographic mapping; see below.) Meanwhile, the 'internal space', i.e. the space of fields ϕ_a, subject to $\sum_{a=1}^{3} \phi_a^2 = 1$, is also a spherical surface, of unit radius. Let us call this $S_2^{(\text{int})}$. Then, any finite-energy static configuration $\boldsymbol{\phi}(\boldsymbol{x})$ is just a mapping of $S_2^{(\text{phy})}$ *into* $S_2^{(\text{int})}$.

Now, we will state (without giving the proof) a result well known in topology. All non-singular mappings of one spherical surface S_2 into another S_2 can be classified into homotopy sectors. Mappings within one sector can be continuously deformed into one another, whereas mappings from two different sectors cannot be continuously deformed into one another. Furthermore, there is a denumerable infinity of such homotopy sectors or classes, which can be characterised by the set of integers, positive, negative and zero. More precisely, these homotopy classes themselves form a group which is isomorphic to the group of integers. Formally all this is written in the compact form

$$\pi_2(S_2) = Z \tag{3.18}$$

where $\pi_n(S_m)$ stands for the homotopy group associated with mappings $S_n \to S_m$, and Z is the group of integers (see the classic text by Steenrod (1951) for a collection of such results and their proof).

We will on several occasions be using results such as (3.18). Readers unfamiliar with them may derive some comfort by considering the simpler case of mapping circles into circles. Consider a circle S_1 (characterised by an angle θ defined modulo 2π) mapped into another circle S_1 (characterised by Λ). A mapping is given by a continuous function $\Lambda(\theta)$, modulo 2π. Consider for instance two such mappings or functions

$$\Lambda_0(\theta) = 0 \qquad \text{for all } \theta \tag{3.19}$$

and

$$\Lambda_0'(\theta) = \begin{cases} t\theta & \text{for } 0 \leqslant \theta < \pi \\ t(2\pi - \theta) & \text{for } \pi \leqslant \theta < 2\pi \end{cases} \tag{3.20}$$

where t is some real parameter in the range $[0, 1]$. It is clear that by varying the parameter t continuously down to zero the second mapping can be continuously deformed into the first. This is also intuitively evident from fig. 6(b) and 6(c). These two mappings therefore belong to the same homotopy class. By contrast, consider the mapping

$$\Lambda_1(\theta) = \theta \qquad \text{for all } \theta. \tag{3.21}$$

This is also a continuous mapping modulo 2π, since as θ completes a full circle so does Λ_1. However, it cannot be continuously deformed into (3.19) or (3.20). This should be intuitively evident from fig. 6(d), which cannot be distorted into 6(c) or 6(b) without snipping the Λ-circle somewhere. The reason obviously is that in (3.21) the second circle is wound once around the first circle, whereas in (3.19) and (3.20) it is effectively wound zero times. Thus (3.21) belongs to a different homotopy class from (3.19) or (3.20). Indeed the integer distinguishing the two classes is just the 'winding number',

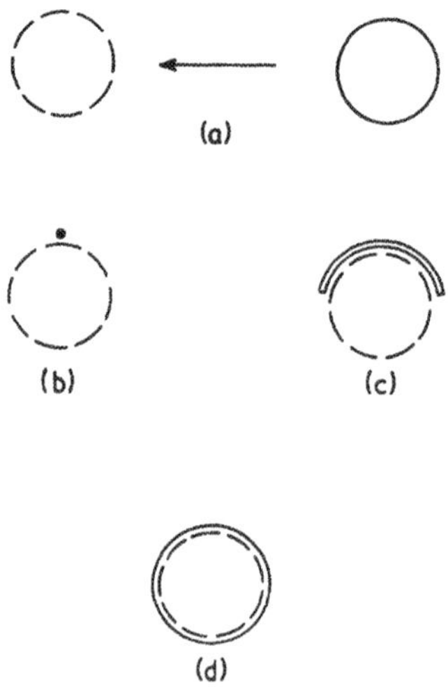

Fig. 6. Mappings of one circle, indicated by a full line in (a), into another, indicated by a dashed line. In (b) the trivial mapping (eq. (3.19)) is given, where the entire first circle maps into one point in the second circle. In (c), a non-trivial mapping (eq. (3.20)) is given. Both (b) and (c) carry zero winding number ($Q = 0$), and can obviously be continuously deformed into one another by shrinking the undashed loop. By contrast the mapping in (d), eq. (3.21), cannot be deformed into (b) or (c) without cutting the undashed circle.

$$\tilde{Q} = \frac{1}{2\pi} \int_0^{2\pi} \frac{d\Lambda}{d\theta} d\theta \tag{3.22}$$

which is clearly zero for (3.19)–(3.20), but equal to unity for (3.21). It is also clear that by winding the second circle an arbitrary number of times one can generate a denumerable infinity of homotopy classes. Thus

$$\Lambda_n(\theta) = n\theta \tag{3.23}$$

is the prototype mapping belonging to the $\tilde{Q} = n$ class. Negative values of $\tilde{Q}$ are obtained by doing the winding in the opposite sense, for instance by replacing θ by $-\theta$ in (3.23). These qualitative remarks do not constitute a proof, but hopefully they render plausible the result

$$\pi_1(S_1) = Z. \tag{3.24}$$

Equation (3.18) is just a generalisation of this result for the mappings $S_2 \to S_2$. The integer characterising the homotopy classes of $S_2 \to S_2$ is again the generalised winding number, i.e. the number of times one of the spheres has been wrapped around the other.

Returning to our O(3) model, in summary, finite-energy static configurations $\boldsymbol{\phi}(\boldsymbol{x})$ in two space dimensions can be classified into homotopy sectors, characterised by some integer which we will label Q. Note that the nature of the homotopy classification here differs from those in the kink or sine–Gordon problems in chapter 2. In those examples, the different sectors differed in the boundary values of the field at spatial infinity. In the present O(3) model, the only role of the boundary condition (3.17) is to compactify coordinate space into a spherical surface $S_2^{(\text{phy})}$. True, the boundary condition (3.17) does not uniquely specify the boundary value $\boldsymbol{\phi}^{(0)}$. While $\boldsymbol{\phi}^{(0)}$ must be the same at all points at spatial infinity, it could point in any 'internal' direction. However, field configurations with different directions of $\boldsymbol{\phi}^{(0)}$ can be obtained from one another *continuously* through O(3) rotations in internal space. Therefore the different choices of $\boldsymbol{\phi}^{(0)}$ do not, in themselves, lead to different homotopy sectors for this O(3) model. The sectors arise, instead, from the behaviour of the fields throughout space, including the interior. It will be helpful to write Q as an integral over the field function $\boldsymbol{\phi}(\boldsymbol{x})$, analogous to the integral in (3.22). Such an expression is given by

$$Q = \frac{1}{8\pi} \int \varepsilon_{\mu\nu} \boldsymbol{\phi} \cdot (\partial_\mu \boldsymbol{\phi} \times \partial_\nu \boldsymbol{\phi}) \, \mathrm{d}^2 x \tag{3.25}$$

where $\mu = 1, 2$ and $\nu = 1, 2$ refer to cartesian coordinate-space indices while the dot and cross product refer to vectors in internal space. That this expression does provide the winding number (also sometimes called the topological index or number) may be seen as follows:

The sphere $S_2^{(\text{int})}$ in internal space can be described by two variables $\{\xi_1, \xi_2\}$ such as polar angles, instead of the three cartesian variables ϕ_a subject to $\sum_a \phi_a^2 = 1$. There is a well-known expression which relates the surface area element as written in terms of cartesian and spherical variables:

$$dS_a^{(\text{int})} = d^2\xi \left(\frac{1}{2} \varepsilon_{rs} \varepsilon_{abc} \frac{\partial \phi_b}{\partial \xi_r} \frac{\partial \phi_c}{\partial \xi_s} \right). \tag{3.26}$$

Now,

$$\begin{aligned} Q &= \frac{1}{8\pi} \int \varepsilon_{\mu\nu} \varepsilon_{abc} \, \phi_a \frac{\partial \phi_b}{\partial x^\mu} \frac{\partial \phi_c}{\partial x^\nu} d^2x \\ &= \frac{1}{8\pi} \int \varepsilon_{\mu\nu} \varepsilon_{abc} \, \phi_a \frac{\partial \phi_b}{\partial \xi_r} \frac{\partial \xi_r}{\partial x^\mu} \frac{\partial \phi_c}{\partial \xi_s} \frac{\partial \xi_s}{\partial x^\nu} d^2x \\ &= \frac{1}{8\pi} \int \varepsilon_{rs} \varepsilon_{abc} \, \phi_a \frac{\partial \phi_b}{\partial \xi_r} \frac{\partial \phi_c}{\partial \xi_s} d^2\xi \end{aligned} \tag{3.27}$$

since the Jacobian of the change of variables from $\{x_1, x_2\}$ to $\{\xi_1, \xi_2\}$ is given by

$$\varepsilon_{rs} d^2\xi = \varepsilon_{\mu\nu} \frac{\partial \xi_r}{\partial x_\mu} \frac{\partial \xi_s}{\partial x_\nu} d^2x. \tag{3.28}$$

Inserting (3.26) into (3.27), we have

$$Q = \frac{1}{4\pi} \int dS_a^{(\text{int})} \cdot \phi_a = \frac{1}{4\pi} \int dS^{(\text{int})} \tag{3.29}$$

since ϕ_a is just a unit vector normal to the surface. Recalling that $S_2^{(\text{int})}$ is the surface of a unit sphere, with area 4π, we clearly see that Q gives the number of times the internal sphere is traversed as we span the coordinate space R_2 as compacted into $S_2^{(\text{phy})}$.

This homotopy classification is valid for any static field configuration for which the energy functional (3.15) is finite. It does not require that the fields be solutions of the field equation (3.14). Of course, finite-energy solutions are subsets of finite-energy configurations, and the same classification holds for them. In order to actually find some solutions in any given Q-sector, we employ an ingenious trick (Belavin and Polyakov 1975).

We begin with the identity

$$\int d^2x[(\partial_\mu\boldsymbol{\phi}\pm\varepsilon_{\mu\nu}\boldsymbol{\phi}\times\partial_\nu\boldsymbol{\phi})\cdot(\partial_\mu\boldsymbol{\phi}\pm\varepsilon_{\mu\sigma}\boldsymbol{\phi}\times\partial_\sigma\boldsymbol{\phi})]\geqslant 0. \tag{3.30}$$

The identity holds since the integrand is just the scalar product of the vector in brackets with itself. Upon expanding, this becomes

$$\int d^2x[(\partial_\mu\boldsymbol{\phi})\cdot(\partial_\mu\boldsymbol{\phi})+\varepsilon_{\mu\nu}(\boldsymbol{\phi}\times\partial_\nu\boldsymbol{\phi})\cdot\varepsilon_{\mu\sigma}(\boldsymbol{\phi}\times\partial_\sigma\boldsymbol{\phi})]\geqslant$$
$$\pm 2\int d^2x[\varepsilon_{\mu\nu}\boldsymbol{\phi}\cdot(\partial_\mu\boldsymbol{\phi}\times\partial_\nu\boldsymbol{\phi})].$$

The two terms on the left-hand side are actually equal to each other since

$$\varepsilon_{\mu\nu}\varepsilon_{\mu\sigma}(\boldsymbol{\phi}\times\partial_\nu\boldsymbol{\phi})\cdot(\boldsymbol{\phi}\times\partial_\sigma\boldsymbol{\phi})$$
$$=\delta_{\nu\sigma}[(\boldsymbol{\phi}\cdot\boldsymbol{\phi})(\partial_\nu\boldsymbol{\phi}\cdot\partial_\sigma\boldsymbol{\phi})+(\boldsymbol{\phi}\cdot\partial_\nu\boldsymbol{\phi})(\boldsymbol{\phi}\cdot\partial_\sigma\boldsymbol{\phi})]$$
$$=(\partial_\nu\boldsymbol{\phi})\cdot(\partial_\nu\boldsymbol{\phi})$$

where we have used the constraint $\boldsymbol{\phi}\cdot\boldsymbol{\phi}=1$ and its derivative $\boldsymbol{\phi}\cdot(\partial_\nu\boldsymbol{\phi})=0$. Thus,

$$2\int d^2x(\partial_\mu\boldsymbol{\phi})\cdot(\partial_\mu\boldsymbol{\phi})\geqslant\pm 2\int d^2x\varepsilon_{\mu\nu}\boldsymbol{\phi}\cdot(\partial_\mu\boldsymbol{\phi}\times\partial_\nu\boldsymbol{\phi})$$

or

$$E\geqslant 4\pi|Q|. \tag{3.31}$$

This inequality sets a lower bound for the energy of any static configuration in a given Q-sector. Now, the static field equation (3.14) is obtained by extremising the static energy functional (3.15), subject to the constraint (3.9). Since a configuration from one sector cannot, under continuous variation, move into another sector, the extremisation can be done in each sector separately. In any given Q-sector, the energy is minimised when the equality (3.31) is satisfied. This in turn implies that the equality (3.30) be satisfied, which will happen if and only if

$$\partial_\mu\boldsymbol{\phi}=\pm\varepsilon_{\mu\nu}\boldsymbol{\phi}\times(\partial_\nu\boldsymbol{\phi}). \tag{3.32}$$

Any field configuration that satisfies (3.32) as well as the constraint (3.9), will minimise E in some Q-sector, and will therefore automatically satisfy the extremum condition given by the field equation (3.14). This can be explicitly verified. For any configuration satisfying (3.32),

$$\nabla^2\boldsymbol{\phi}=\partial_\mu\partial_\mu\boldsymbol{\phi}=\pm\partial_\mu(\varepsilon_{\mu\nu}\boldsymbol{\phi}\times\partial_\nu\boldsymbol{\phi})$$
$$=\pm\varepsilon_{\mu\nu}(\partial_\mu\boldsymbol{\phi})\times(\partial_\nu\boldsymbol{\phi})$$
$$=\varepsilon_{\mu\nu}(\varepsilon_{\mu\sigma}\boldsymbol{\phi}\times\partial_\sigma\boldsymbol{\phi})\times\partial_\nu\boldsymbol{\phi}$$

$$= \delta_{\nu\sigma}[\partial_\sigma \boldsymbol{\phi}(\boldsymbol{\phi}\cdot\partial_\nu \boldsymbol{\phi}) - \boldsymbol{\phi}(\partial_\nu \boldsymbol{\phi}\cdot\partial_\sigma \boldsymbol{\phi})]$$
$$= \boldsymbol{\phi}(\boldsymbol{\phi}\cdot\nabla^2 \boldsymbol{\phi}).$$

which is just the field equation. In the last step, we have used

$$\boldsymbol{\phi}\cdot\partial_\nu \boldsymbol{\phi} = 0 = \partial_\nu \boldsymbol{\phi}\cdot\partial_\nu \boldsymbol{\phi} + \boldsymbol{\phi}\cdot\nabla^2 \boldsymbol{\phi}$$

which result from differentiating the constraint (3.9). Even though any field that satisfies (3.32) will also satisfy (3.14), the converse need not be true. One could in principle have solutions of (3.14) which do not satisfy (3.32). These would not represent absolute minima of E in the corresponding Q-sector, but some higher valued extrema of E, such as local minima. We will remain content with finding solutions of (3.32). Note that this is an easier equation to solve than the parent field equation (3.14). It is a first-order differential equation while the latter is a second-order equation.

In fact (3.32) can be further simplified by a change of variables. We recall that the allowed values of $\boldsymbol{\phi}$, subject to $\boldsymbol{\phi}\cdot\boldsymbol{\phi} = 1$, form the surface of the unit sphere, $S_2^{(\text{int})}$. Let us stereographically project the surface of this sphere onto a plane, and represent points on the former by giving the cartesian coordinates ω_1 and ω_2 of the corresponding points on the plane. The variables ω_1 and ω_2 are related to the variables ϕ_a by

$$\omega_1 = 2\phi_1/(1-\phi_3) \qquad \text{and} \qquad \omega_2 = 2\phi_2/(1-\phi_3) \tag{3.33}$$

where the plane on which the projection is made is parallel to the $\{\phi_1, \phi_2\}$ plane and contains the south pole. It will also be useful to construct the complex quantity

$$\omega \equiv \omega_1 + \mathrm{i}\omega_2 = 2(\phi_1 + \mathrm{i}\phi_2)/(1-\phi_3) = 2\phi/(1-\phi_3),$$

where $\phi \equiv \phi_1 + \mathrm{i}\phi_2$. Then

$$\partial_1 \omega \equiv \frac{\partial \omega}{\partial x_1} = 2[(1-\phi_3)\partial_1 \phi + \phi\partial_1 \phi_3]/(1-\phi_3)^2$$
$$= [2/(1-\phi_3)^2](\partial_1 \phi + \phi\overleftrightarrow{\partial_1}\phi_3). \tag{3.34}$$

Now, equation (3.32) tells us that

$$\partial_1 \phi = \mp \mathrm{i}\phi\overleftrightarrow{\partial_2}\phi_3 \qquad \text{and} \qquad \partial_2 \phi = \pm \mathrm{i}\phi\overleftrightarrow{\partial_1}\phi_3. \tag{3.35}$$

Substituting this into (3.34), we have

$$\partial_1 \omega = \mp \mathrm{i}\partial_2 \omega. \tag{3.36}$$

In terms of ω_1 and ω_2, this gives

$$\frac{\partial \omega_1}{\partial x_1} = \pm \frac{\partial \omega_2}{\partial x_2} \quad \text{and} \quad \frac{\partial \omega_1}{\partial x_2} = \mp \frac{\partial \omega_2}{\partial x_1}. \tag{3.37}$$

(We remind the reader: x_1 and x_2 are cartesian coordinates of our original two-dimensional physical space; ω_1 and ω_2 describe the plane in 'internal space' on which $S_2^{(\text{int})}$ has been projected stereographically.)

Equation (3.37) is all too familiar as the Cauchy–Reimann condition for ω being an analytic function of z^* (for the upper signs) or z (for the lower signs), where $z \equiv x_1 + \mathrm{i}x_2$. Thus, any analytic function $\omega(z)$ or $\omega(z^*)$ automatically solves (3.32) and therefore also the field equation, when written in terms of the original variables ϕ_a and $\boldsymbol{x}$. Furthermore, while ω must be analytic in either z or z^*, it need not be an entire function. While cuts are prohibited by the single-valuedness of $\phi_a(\boldsymbol{x})$, isolated poles in $\omega(z)$ are permitted. That ω diverges at the poles need not cause concern. $\omega \to \infty$ merely corresponds to the 'north pole' in $S_2^{(\text{int})}$, i.e. to $\phi_3 = 1$.

It will be useful to write down the expressions for E and Q in terms of ω for the case when ω is analytic in, say, z. They are given by (the derivation is straightforward)

$$E = \int \mathrm{d}^2x \frac{|\mathrm{d}\omega/\mathrm{d}z|^2}{(1 + |\omega|^2/4)^2} \quad \text{and} \quad |Q| = \frac{E}{4\pi}. \tag{3.38}$$

A prototype solution for arbitrary positive Q is given by

$$\omega(z) = [(z - z_0)/\lambda]^n \tag{3.39}$$

where n is any positive integer, λ is any real number and z_0 is any complex number. Since (3.39) is an analytic function, our analysis assures us that it will yield an exact static solution of the field equation when rewritten in terms of ϕ_a and $\boldsymbol{x}$. Upon Lorentz transformation into a moving frame, it will yield exact time-dependent solutions which move undistorted in shape. Note that the parent theory is Lorentz-invariant.

In (3.39), ω represents a point in field space, while z stands for a point in coordinate space. Clearly (3.39) allows n roots for z, for a given ω. Therefore it must correspond to the $Q = n$ sector. This may be verified by substituting (3.39) into (3.38). We have

$$Q = \frac{1}{4\pi} \int \mathrm{d}^2x \frac{n^2 |z - z_0|^{2n-2} \lambda^{2n}}{(\lambda^{2n} + \frac{1}{4}|z - z_0|^{2n})^2}. \tag{3.40}$$

Using

$$z - z_0 = \rho e^{i\theta} \qquad \text{and} \qquad d^2x = \rho\, d\rho\, d\theta \tag{3.41}$$

the integration is trivial and yields $Q = n$. Hence $E = 4\pi Q = 4\pi n$ is finite. Clearly, then, these are explicit solitary-wave solutions, for any positive integer n.

The constants λ and z_0 (which stands for a pair of coordinates $(x_1)_0$ and $(x_2)_0$) refer to the size and location of the soliton solution. The fact that the solution exists for arbitrary λ and z_0 and the fact that neither Q nor E depend on these constants is a reflection of scale and translational invariance. Notice that $E[\phi]$ in (3.15) is obviously invariant under $\boldsymbol{x} \to \boldsymbol{x} - \boldsymbol{a}$ and $\boldsymbol{x} \to \lambda \boldsymbol{x}$.

The O(3) model is also interesting in (1 + 1) dimensions. It has been shown (Pohlmeyer 1976, Luscher and Pohlmeyer 1978) that, like the sine–Gordon system, the O(3) model in (1 + 1) dimensions is also characterised by an infinite number of conserved quantities and by Backlund transformations for generating solutions. [In the quantised version of the theory, it has been shown that it is asymptotically free and that the conserved quantities exist free of anomalies (Luscher 1978, Polyakov 1977a).] Finally, an exact factorised S-matrix has been constructed using the existence of these infinite conserved quantities (Zamolodchikov and Zamolodchikov 1979). We shall discuss this last aspect in a later chapter. The static solutions we obtained in (2 + 1) dimensions are also relevant to the model in (1 + 1) dimensions. They serve as instantons of the latter (see the next chapter). Lastly, this O(3) model and its solutions are also relevant in describing the statistical mechanics of an isotropic ferromagnet (see section 3.6).

3.4. The 'tHooft–Polyakov monopole

We are finally ready to discuss static soliton solutions for the realistic case of (3 + 1) dimensions. The virial theorem of section 3.2 tells us that spin-zero fields alone cannot yield such a solution; higher-spin fields must be involved. We will not consider spin-$\frac{1}{2}$ fields at this stage. They are Fermi fields, whose 'classical limit' requires special interpretation and will be discussed in chapter 9. Therefore we go on to spin-1 fields, in particular to gauge fields, which have attained considerable importance in particle physics.

The simplest example of a gauge theory in (3 + 1) dimensions in the free electromagnetic system whose gauge group is U(1). This involves only linear (Maxwell's) equations, and is trivial to solve. As is well known, unlike the (1 + 1) dimensional linear wave equation (2.1), free Maxwell's equations in (3 + 1) dimensions do not yield solitary waves. Any localised packet will necessarily dissipate itself. When the electromagnetic field is coupled to charged scalar fields, soliton solutions do emerge, but in (2 + 1) dimensions. We will return to this case later. If we shift our interest to non-abelian gauge groups, then the simplest candidate is SU (2). This group calls for a triplet of gauge fields known as the Yang–Mills fields (Yang and Mills 1954, Shaw 1955). This triplet is self-coupled to form a non-linear system which is far from trivial, but it has been shown that a set of pure Yang–Mills fields in (3 + 1) dimensions also fails to yield any solitary waves (Coleman 1977b, Deser 1976, Pagels 1977). It does yield interesting singular solutions (Yang and Wu 1968) but we are concerned only with non-singular finite-energy solutions.

Therefore we enlarge the system further by coupling the Yang–Mills fields to a triplet of scalar fields as was done by Georgi and Glashow (1972). It has been shown through the pioneering work of 'tHooft (1974b) and Polyakov (1974) that this model does yield a non-singular localised static solution with some remarkable properties. We devote this section to a discussion of this model and its static soliton solution. Non-abelian gauge theories, of which the model under question is an example, have been under study for over twenty years. Although our presentation will be reasonably self-contained it would be helpful if the reader were familiar with the basics of such theories. (There are several good review articles and books, e.g. Abers and Lee (1973), Bernstein (1974), Taylor (1976) and Faddeev and Slavnov (1980).)

The model consists of scalar fields $\phi^a(\boldsymbol{x}, t)$ and vector fields $A^a_\mu(\boldsymbol{x}, t)$ in (3 + 1) dimensions. The index $a = 1, 2, 3$, is an internal space index, which will transform according to local (space–time dependent) SU(2) transformations given below. For any given a, ϕ^a is a scalar and A^a_μ ($\mu = 0, 1, 2, 3$) is a vector under Lorentz transformations. The Lagrangian density is

$$\mathscr{L}(\boldsymbol{x}, t) = -\tfrac{1}{4}G^a_{\mu\nu}G^{a\mu\nu} + \tfrac{1}{2}(\mathrm{D}_\mu\phi^a)(\mathrm{D}^\mu\phi^a) - \tfrac{1}{4}\lambda(\phi^a\phi^a - F^2)^2. \qquad (3.42)$$

Here, the 'field tensor' $G^a_{\mu\nu}$ is defined by

$$G^a_{\mu\nu} \equiv \partial_\mu A^a_\nu - \partial_\nu A^a_\mu + g\varepsilon^{abc} A^b_\mu A^c_\nu, \qquad (3.43)$$

and the 'covariant derivative' $\mathrm{D}_\mu\phi^a$ by

$$\mathrm{D}_\mu\phi^a \equiv \partial_\mu\phi^a + g\varepsilon^{abc} A^b_\mu\phi^c. \qquad (3.44)$$

(The covariant derivatives of other triplets of fields such as A_μ^a or $G_{\mu\nu}^a$ are defined analogously.) The real constants g, $\lambda > 0$, and F are parameters of the model. The A_μ^a are called gauge fields and the ϕ^a, in such a context, are called Higgs fields since they lead to the so-called Higgs mechanism (see section 10.2).

To readers unfamiliar with non-abelian gauge theories, this Lagrangian may look complicated and contrived. It has however been designed to be 'gauge invariant' i.e. invariant under the set of independent SU(2) transformations at each space–time point. These transformations are defined by

$$\phi^a(\boldsymbol{x}, t) \to [U(\boldsymbol{x}, t)]_{ab}\,\phi^b(\boldsymbol{x}, t) \tag{3.45}$$

and

$$(L^a A_\mu^a)_{bc} \to U_{bd}[L^a A_\mu^a + (\mathrm{i}/g)I\partial_\mu]_{de}(U^{-1})_{ec} \tag{3.46}$$

where

$$[U(\boldsymbol{x}, t)]_{bc} = [\exp\{-\mathrm{i}\,L^a\theta^a(\boldsymbol{x}, t)\}]_{bc} \tag{3.47}$$

is any member of the group SU(2) at each space–time point $(\boldsymbol{x}, t)$, written here in its 3×3 representation. $(L^a)_{bc} = \mathrm{i}\varepsilon_{abc}$ are the three generators of SU(2) in 3×3 matrix representation, I is the identity matrix and the group parameters θ^a vary in space–time. One can verify by direct substitution that the Lagrangian (3.42) is invariant under these gauge transformations. These equations are generalisations to the group SU(2) of corresponding statements in electrodynamics for which the gauge group is U(1).

In this section we merely discuss some static classical solutions of this system – an impressive problem in its own right considering that 15 coupled non-linear fields are involved in (3 + 1) dimensions.

The equations of motion that flow from the Lagrangian (3.42) are

$$\mathrm{D}_\mu G^{a\mu\nu} = g\varepsilon^{abc}(\mathrm{D}^\nu\phi^b)\phi^c \tag{3.48}$$

and

$$\mathrm{D}_\mu \mathrm{D}^\mu \phi^a = -\lambda(\phi^b\phi^b)\phi^a + \lambda F^2\phi^a. \tag{3.49}$$

We shall restrict ourselves to solutions that are (i) static and (ii) satisfy $A_0^a(\boldsymbol{x}) = 0$ for all $\boldsymbol{x}$, a. For these, the field equations reduce to

$$\mathrm{D}_i G^{aij} = g\varepsilon^{abc}(\mathrm{D}^j\phi^b)\phi^c \tag{3.50}$$

and

$$\mathrm{D}_i \mathrm{D}^i \phi^a = -\lambda(\phi^b\phi^b)\phi^a + \lambda F^2\phi^a \tag{3.51}$$

where $i, j = 1, 2, 3$ are purely spatial indices.

We are looking for finite-energy solutions of (3.50)–(3.51), which are more complicated than the field equations we have encountered so far, but our approach to the problem will be the same as before; that is, we will first find the classical vacuum (zero-energy) solutions. This information will then be used to identify the set of allowed boundary conditions that any finite-energy configuration must satisfy. Next, we will make a homotopy classification of these boundary conditions. Lastly, amongst configurations of a given homotopy sector, we will look for a finite-energy solution. First we look for zeroes of the energy. The expression for the conserved energy of the system can be obtained from the Lagrangian as usual and reduces, for static solutions with $A_0^a = 0$, to

$$E = \int d^3x[\tfrac{1}{4}G_{ij}^a G^{aij} + \tfrac{1}{2}D_i\phi^a D_i\phi^a + \tfrac{1}{4}\lambda(\phi^a\phi^a - F^2)^2]. \tag{3.52}$$

This energy reaches a minimum and vanishes when

$$A_i^a(\boldsymbol{x}) = 0 \tag{3.53}$$

$$\phi^a(\boldsymbol{x})\phi^a(\boldsymbol{x}) = F^2 \tag{3.54}$$

and

$$D_i\phi^a = 0, \tag{3.55}$$

which amounts to $\partial_i\phi^a = 0$ in view of (3.53). (In fact the first condition (3.53) is sufficient, but not necessary. There are several other solutions for A_i^a, related to $A_i^a = 0$ by gauge transformations, which also lead to $E = 0$. We will return to this question in the next chapter. For the moment let us take $A_i^a = 0$, and concentrate on the conditions (3.54)–(3.55) on ϕ^a.)

The conditions on the ϕ^a are similar to those in the O(3) model in the last section. There is a degenerate continuous family of $E = 0$ solutions. In each of these, $\boldsymbol{\phi} \equiv \{\phi^a\}$ must have fixed magnitude F, but can point in any $\boldsymbol{x}$-independent direction in internal space. Recall that the local SU(2) gauge symmetry of this system contains in it global ($\boldsymbol{x}$-independent) SU(2) symmetry which in turn amounts to internal rotational symmetry for our real scalar fields $\boldsymbol{\phi}$. The family of $E = 0$ solutions permitted by (3.53)–(3.55) are related to each other by this symmetry.

Let us move on to configurations with non-zero but finite energy E. The similarity with the non-linear O(3) model still persists to some extent, but now there are important differences. The condition for finite E is, as before, that the fields approach some $E = 0$ configuration at spatial infinity sufficiently fast. We can see from (3.52) that this condition for the field $\boldsymbol{\phi}$ is, as $r \equiv |\boldsymbol{x}| \to \infty$,

$$r^{3/2} \mathrm{D}_i \boldsymbol{\phi} \to 0 \tag{3.56}$$

and

$$\boldsymbol{\phi} \cdot \boldsymbol{\phi} \to F^2. \tag{3.57}$$

As in the O(3) model, the magnitude of $\boldsymbol{\phi}$ must approach the 'vacuum' value F. But unlike the O(3) model, the internal space direction of $\boldsymbol{\phi}$ need not be the same when we go to spatial infinity in different directions. This is because, unlike eq. (3.16) in the O(3) model, the corresponding condition (3.56) here requires the vanishing of the *covariant* derivative $\mathrm{D}_i \boldsymbol{\phi}$ and not the *ordinary* derivative $\partial_i \boldsymbol{\phi}$. Consider the expression (3.44) for the covariant derivative, and express it in terms of spherical polar coordinates (r, θ, ϕ) and corresponding components. The θ-component of $(\mathrm{D}\phi^a)$ is given by

$$(\mathrm{D}\phi^a)_\theta = \frac{1}{r}\frac{\partial \phi^a}{\partial \theta} + g\varepsilon^{abc} A^b_\theta \phi^c. \tag{3.58}$$

As long as this combination falls off fast enough to satisfy (3.56), $\partial\phi^a/\partial\theta$ itself need not vanish as $r \to \infty$. A non-zero $\partial\phi^a/\partial\theta$ is permitted as $r \to \infty$, as long as A^b_θ, the θ-component of the gauge field, is matched with $\partial\phi^a/\partial\theta$ in such a way that the combination (3.58) goes to zero as $r \to \infty$. This in turn implies that A^b_θ falls off to zero only as fast as $1/r$. A similar statement obviously holds for the azimuthal components of (grad ϕ^a) and A^b_μ. The fact that some components of A^a_μ fall off only as slowly as $1/r$ as $r \to \infty$ is consistent with the finiteness of E. Note that the integrand in E involves $G_{ij} \cdot G^{ij}$, which will then decrease as $1/r^4$ and will be integrable.

Therefore unlike the O(3) model, in this case finiteness of E permits fields where $\phi^a(\boldsymbol{x})$ points in different internal directions at different points on the boundary of space, as long as $\sum_a \phi^a \phi^a \to F^2$. Thus, the allowed values of ϕ^a at the boundary lie on a spherical surface of radius F in internal space. Let us call this surface $S_2^{(\mathrm{int})}$. Space is three-dimensional in this model, so that its boundary is another spherical surface $S_2^{(\mathrm{phy})}$. Hence, the set of boundary conditions on $\boldsymbol{\phi}$ permitted by finiteness of E are the set of all non-singular mappings of $S_2^{(\mathrm{phy})} \to S_2^{(\mathrm{int})}$. We have already observed that such mappings fall into a denumerable infinity of homotopy classes which form the group $\pi_2(S_2) = Z$. Field configurations from one sector cannot be continuously deformed into another sector. Each sector can be characterised by an integer Q, which is the topological number of this model, and describes the number of times that $S_2^{(\mathrm{int})}$ is covered when $S_2^{(\mathrm{phy})}$ is traversed once. In a $Q = 0$ configuration, the field $\boldsymbol{\phi}$ will either tend to the same value, as $x \to \infty$ in any direction (fig. 7(a)), or will tend to some angle-dependent value that can be deformed so as to be angle-independent. The trivial vacuum solution $\phi^a(x) = \delta_{3a}F$ belongs to this $Q = 0$ sector. These are analogues of functions like (3.19)–(3.20) in the $S_1 \to S_1$ case. A prototype $Q = 1$ configuration would have $\boldsymbol{\phi}(\boldsymbol{x} \to \infty)$ pointing radially outward, i.e. with its internal directions parallel to the coordinate vector (fig. 7(b)). This would be the analogue of the mapping (3.21). We will give an explicit example of an $Q = 1$ solution later on.

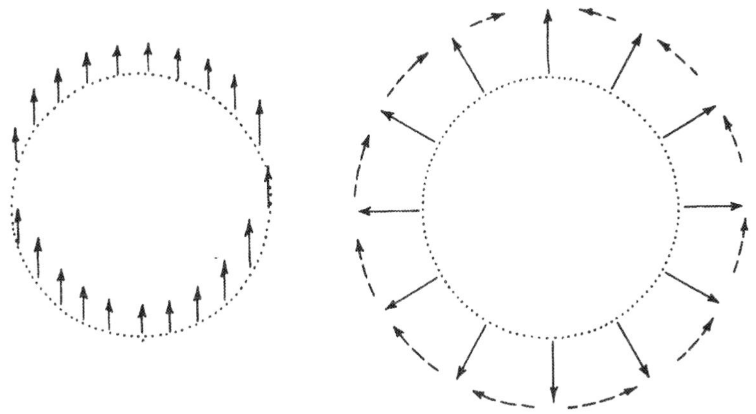

(a) (b)

Fig. 7. Cross sections of prototype $n = 0$ and $n = 1$ mappings of $S_2^{(\text{phy})} \to S_2^{(\text{int})}$ are given in (a) and (b). The dotted surface stands for $S_2^{(\text{phy})}$ and the full arrows give the directions of the corresponding points on $S_2^{(\text{int})}$, i.e. the directions of the isovector ϕ^a. Notice that if ϕ^a were rotated in (b), along the dashed arrows, in an attempt to point them all upwards as in (a), the rotation angle would be discontinuous at the south pole. This is a pictorial way of understanding why the two mappings belong in different homotopy sectors.

Note that the homotopy classification of finite-energy configurations of this model arose entirely from the boundary conditions on the fields. In this respect, this classification differs from that in the O(3) model (although both involve the same homotopy group $\pi_2(S_2)$), and is closer in spirit to the ϕ^4 and sine–Gordon models. In the latter examples, space was one-dimensional, and the different sectors corresponded to the different values that the field could take as x approached $+\infty$ or $-\infty$. In the kink solution for instance, $\phi(x)$ approached different limits as $x \to \pm\infty$. A similar thing happens in the $Q \neq 0$ sectors of our gauge model. As we approach $r \to \infty$, but in different directions, the field $\boldsymbol{\phi}$ approaches different limits.

We have been restricting ourselves to static configurations with $A_0^a(\boldsymbol{x}) = 0$. The solution we obtain later on will also come into this category. However, given a static solution with $A_0^a(\boldsymbol{x}) = 0$, one can obtain others that are time-dependent and carry non-zero A_0^a, by employing the gauge transformations (3.45)–(3.46). Since the Lagrangian is invariant and the field equations covariant under these gauge transformations, the transform of a solution will also be a solution. It will carry the same energy and topological charge, since the energy is gauge invariant and so, as we will see, is Q.

As we did with earlier models, we can again write the topological index as an explicit functional of the field. In fact, one can define a topological current k_μ (Arafune et al. 1975) by

$$k_\mu = (1/8\pi)\varepsilon_{\mu\nu\rho\sigma}\varepsilon_{abc}\partial^\nu\hat{\phi}^a\partial^\rho\hat{\phi}^b\partial^\sigma\hat{\phi}^c \tag{3.59}$$

where

$$\hat{\phi}_a = \phi^a/|\phi|; \qquad |\phi| = \left(\sum_a \phi^a\phi^a\right)^{1/2}.$$

From the antisymmetry of $\varepsilon_{\mu\nu\rho\sigma}$, it follows that

$$\partial^\mu k_\mu = 0. \tag{3.60}$$

It is clear that the conservation of this current follows just from the construction of k_μ in (3.59) and not from the dynamics, i.e. the explicit form of the Lagrangian or equations of motion. The same was true of the current (2.59). This is again a reflection of the fact stressed in the last chapter that topological charges and currents are different from the familiar Noether currents and charges. The conserved charge corresponding to (3.59) is

$$\begin{aligned} Q &= \int \mathrm{d}^3x\, k_0 \\ &= \frac{1}{8\pi}\int (\varepsilon_{ijk}\,\varepsilon^{abc}\,\partial_i\hat{\phi}^a\,\partial_j\hat{\phi}^b\,\partial_k\hat{\phi}^c)\,\mathrm{d}^3x \\ &= \frac{1}{8\pi}\int \varepsilon_{ijk}\,\varepsilon^{abc}\,\partial_i(\hat{\phi}^a\,\partial_j\hat{\phi}^b\,\partial_k\hat{\phi}^c)\,\mathrm{d}^3x \\ &= \frac{1}{8\pi}\int_{S_2^{(\mathrm{phy})}} \mathrm{d}^2\sigma_i(\varepsilon_{ijk}\varepsilon^{abc}\hat{\phi}^a\partial_j\hat{\phi}^b\partial_k\hat{\phi}^c), \end{aligned} \tag{3.61}$$

where $i, j, k = 1, 2, 3$ are space-indices and $S_2^{(\mathrm{phy})}$ is our sphere at infinity in coordinate space. To see that (3.61) is in fact the topological mapping index for $S_2^{(\text{phy})} \to S_2^{(\text{int})}$, let us introduce parameters (α_1, α_2) to describe the surface $S_2^{(\text{phy})}$. They could for instance be polar and azimuthal angles. We use the identity (3.26) rewritten for our coordinate space:

$$d^2\sigma_i = \frac{1}{2} d^2\alpha\, \varepsilon_{imn} \varepsilon_{pq} \frac{\partial x^m}{\partial \alpha_p} \frac{\partial x^n}{\partial \alpha_q}; \qquad p, q = 1, 2.$$

Also,

$$\partial_i \hat{\phi}^a = \frac{\partial \hat{\phi}^a}{\partial \alpha_p} \frac{\partial \alpha_p}{\partial x^i}. \tag{3.62}$$

Hence,

$$Q = \frac{1}{8\pi} \int d^2\alpha \left(\frac{1}{2} \varepsilon_{ijk} \varepsilon_{imn} \varepsilon^{abc} \varepsilon_{pq} \frac{\partial x^m}{\partial \alpha_p} \frac{\partial x^n}{\partial \alpha_q} \hat{\phi}^a \frac{\partial \hat{\phi}^b}{\partial \alpha_r} \frac{\partial \alpha_r}{\partial x^j} \frac{\partial \hat{\phi}^c}{\partial \alpha_s} \frac{\partial \alpha_s}{\partial x^k} \right)$$

$$= \frac{1}{8\pi} \int d^2\alpha\, \varepsilon^{abc} \varepsilon_{pq} \hat{\phi}^a \frac{\partial \hat{\phi}^b}{\partial \alpha_p} \frac{\partial \hat{\phi}^c}{\partial \alpha_q}. \tag{3.63}$$

Equation (3.63) has exactly the same form as (3.25). We need only to repeat the subsequent steps in section (3.3) to see that Q gives the desired topological winding number.

Our next step is to explain why the name 'monopoles' is used for solutions of this model. It will be seen that one can associate a magnetic monopole charge with these solutions. Further, the monopole charge is just proportional to Q, the topological number characterising the solution.

The subject of magnetic monopoles has a long history, both in terms of theoretical studies (Dirac 1931, 1948, Schwinger 1966) and experimental searches. We will not delve into this literature here other than to note the following well known statement. In conventional electrodynamics, Maxwell's equations for the vector potential A_μ read

$$\partial_\mu F^{\mu\nu} = 4\pi j^\nu \tag{3.64}$$

where

$$F^{\mu\nu} = \partial^\mu A^\nu - \partial^\nu A^\mu, \tag{3.65}$$

$$\tfrac{1}{2} \varepsilon_{\mu\nu\rho\sigma} \partial^\nu F^{\rho\sigma} = 0, \tag{3.66}$$

and where j^ν is the electric current. Introduction of magnetic charges and currents requires that a magnetic current term be added by hand to the right-hand side of eq. (3.66). In our non-Abelian SU(2) gauge theory, such a magnetic current term can be shown to be already present without having to alter the Lagrangian (3.42) or the field equations (3.48)–(3.49).

To start with, we had made no physical connection between electromagnetism and the gauge model (3.42), other than to note that as a theory (3.42) is a generalisation to the SU(2) group of what happens in electromagnetism, which is a U(1) gauge theory. However U(1) is a subgroup of SU(2), and it is possible to imbed an electromagnetic system as part of the richer system (3.42). At first glance, one might try to associate one of the three gauge fields, say A^3_μ, with the electromagnetic field. But this would not be a gauge-invariant association since the different A^a_μ mix into one another under the gauge transformations (3.46). 'tHooft therefore presented a gauge-invariant definition for the electromagnetic field tensor $F_{\mu\nu}$ in terms of the parent fields:

$$F_{\mu\nu} \equiv \hat{\phi}^a G^a_{\mu\nu} - (1/g)\varepsilon^{abc}\hat{\phi}^a D_\mu \hat{\phi}^b D_\nu \hat{\phi}^c. \tag{3.67}$$

It can be verified that (i) this expression is gauge invariant, and (ii) it does reduce in regions where $\hat{\phi}^a = (0, 0, 1)$ to

$$\mathrm{F}_{\mu\nu} = \partial_\mu A^3_\nu - \partial_\nu A^3_\mu. \tag{3.68}$$

The second statement says that if in a particular gauge ϕ^a always points in the same internal direction, then the vector field along that direction may be considered as the electromagnetic field. In general, (3.67) gives the field tensor. Unlike the usual electromagnetic system (3.64)–(3.66), the tensor in (3.67) has a dual with non-zero divergence. A little bit of algebra shows that (3.67) gives

$$\begin{aligned} \tfrac{1}{2}\varepsilon_{\mu\nu\rho\sigma}\partial^\nu F^{\rho\sigma} &= (1/2g)\varepsilon_{\mu\nu\rho\sigma}\varepsilon_{abc}\partial^\nu\hat{\phi}^a\partial^\rho\hat{\phi}^b\partial^\sigma\hat{\phi}^c \\ &= (4\pi/g)k_\mu, \end{aligned} \tag{3.69}$$

where k_μ is nothing but the topological current (3.59). By comparing with the electric current in (3.64), we see here that the magnetic current is $(1/g)k_\mu$. The magnetic field defined as usual by $B_i \equiv \frac{1}{2}\varepsilon_{ijk}F^{jk}$ satisfies

$$\nabla \cdot \boldsymbol{B} = 4\pi(k_0/g). \tag{3.70}$$

Hence, the total magnetic monopole charge is equal to

$$m = \int \frac{k_0}{g}\,\mathrm{d}^3x = Q/g \tag{3.71}$$

where Q is the topological charge.

Notice that (3.71) is similar to the Schwinger quantisation condition on the allowed values of the magnetic charge (Schwinger 1966). It is, however, not precisely the same as the Schwinger condition, nor should it be. The latter is derived from quantum theory whereas our considerations have so far been strictly classical. The Schwinger condition is

$$m = n\hbar/q; \qquad n = 0, 1, 2, \ldots \tag{3.72}$$

where $\hbar \equiv h/2\pi$, h is Planck's constant and q is the electric charge of the electron. By contrast, in (3.71) $\hbar$ never appears and g is just a coupling constant in the Lagrangian (3.42)–(3.44). It is only if some quantum theoretic condition forced the electric charge of the particles related to the field ϕ^a to be $q = g\hbar$ that our condition (3.71) would become the Schwinger condition.

Since $F^{\mu\nu}$ in (3.67) is gauge-invariant by construction, so are k_μ and the magnetic charge. The reader may be concerned that while (3.69) gives a non-zero divergence to the dual of $F^{\mu\nu}$, the form (3.68) into which $F^{\mu\nu}$ can sometimes be cast is a pure curl and has vanishing divergence for its dual. However, (3.68) follows from (3.67) only in regions where ϕ^a points in the same internal direction. Our discussion on homotopy classes tells us that starting from a $Q \neq 0$ solution, ϕ^a cannot everywhere be rotated through non-singular gauge transformations to point in the same direction, for it would then end up in a $Q = 0$ configuration. Therefore $Q \neq 0$ configurations cannot be written everywhere in the form (3.68) and they must have a non-zero dual-divergence in some regions, consistent with the existence of a non-zero magnetic charge.

These general considerations are explicitly illustrated in the $Q = 1$ example developed by 'tHooft and Polyakov. Consider the following ansatz for a static solution:

$$\phi^a(\boldsymbol{x}) = \delta_{ia}(x^i/r)F(r) \tag{3.73}$$

$$A_i^a(\boldsymbol{x}) = \varepsilon_{aij}(x^j/r)\,W(r); \qquad A_0^a(x) = 0 \tag{3.74}$$

where $r \equiv |\boldsymbol{x}|$ and $i, j, a = 1, 2, 3$. Let $F(r)$ and $W(r)$ be subject to the boundary conditions:

$$F(r) \to F \qquad \text{and} \qquad W(r) \to 1/gr \tag{3.75}$$

as $r \to \infty$. The functions $F(r)$ and $W(r)$ have to be such that the field equations (3.50)–(3.51) are satisfied. F is the same constant that appears in the Lagrangian. We see that with (3.75) the ansatz satisfies all earlier requirements including the boundary conditions (3.56)–(3.57). It is evident that (3.73)–(3.74) gives a $Q = 1$ configuration. In (3.73), the internal direction of ϕ^a is parallel to the coordinate direction x^a as in fig. 7(b). This is clearly a one-to-one mapping of $S_2^{(\text{phy})}$ into $S_2^{(\text{int})}$. It should correspond to $Q = 1$. This can be checked explicitly by inserting (3.73) into the defining integrals (3.61) for Q. Furthermore one can calculate the magnetic field $B_i = \frac{1}{2}\varepsilon_{ijk}F^{jk}$ as $r \to \infty$, corresponding to this configuration. Since $D_i\phi^a \to 0$ as $r \to \infty$, the definition of $F_{\mu\nu}$ in (3.67) yields quite easily, when (3.73)–(3.75) are inserted, the result

$$\boldsymbol{B}(\boldsymbol{x}) \xrightarrow[r \to \infty]{} \boldsymbol{x}/g\,r^3. \tag{3.76}$$

This clearly corresponds to a magnetic pole of strength $1/g$ in conformity with $Q = 1$. Since $A_0^a = 0$ and all fields are time-independent, $F_{0i} = 0$ and the electric field vanishes. This solution therefore carries only magnetic and no electric charge.

Notice that we have yet to solve explicitly the field equations (3.50)–(3.51). This is because these equations, like most coupled non-linear equations in three space dimensions, are not easy to solve. The bulk of our discussion has been a topological analysis based solely on boundary conditions. It is remarkable that so much information could be extracted about some features of the solutions without actually solving the equations. In particular, for the $Q = 1$ ansatz (3.73)–(3.74), we have so far exploited only its special tensorial form and the boundary conditions (3.75). To obtain the functions $F(r)$ and $W(r)$, one must substitute the ansatz into the field equations. The ansatz is 'spherically symmetric', i.e. except for simple tensorial factors explicitly shown it is dependent only on the radial variable r. Thanks to this simplification, the partial differential equations (3.50)–(3.51) reduce to ordinary differential equations, for this particular ansatz. On substituting (3.73)–(3.74) into the field equations, a little algebra yields

$$r^2 \frac{\mathrm{d}^2 K(r)}{\mathrm{d}r^2} = K(r)(K^2(r) - 1) + H^2(r)K(r) \tag{3.77}$$

and

$$r^2 \frac{\mathrm{d}^2 H(r)}{\mathrm{d}r^2} = 2H(r)K^2(r) + \lambda H(r)\left(\frac{H^2(r)}{g^2} - r^2F^2\right) \tag{3.78}$$

where

$$K(r) \equiv 1 - gr\,W(r) \qquad \text{and} \qquad H(r) \equiv gr\,F(r). \tag{3.79}$$

This is a set of coupled 'non-autonomous' differential equations, in technical parlance. Although much simpler than the parent field equations, these are still not easy to solve. However, in the limit $\lambda \to 0$, with g and F fixed, it has been shown (Prasad and Sommerfield 1975, Bogomol'nyi 1976) that the equations (3.77)–(3.78) are solved by the particularly simple functions

$$K(r) = \frac{rgF}{\sinh(gFr)} \quad \text{and} \quad H(r) = \frac{rgF}{\tanh(gFr)} - 1 \tag{3.80}$$

Note that although the last term in the Lagrangian (3.42) vanishes when $\lambda \to 0$, its memory lingers in the solution through the boundary conditions (3.57) and (3.75). The functions (3.80) can be inserted into the ansatz (3.79), (3.73) and (3.74), to obtain an exact static solution in the $\lambda \to 0$ limit, belonging to the $Q = 1$ sector and carrying a magnetic charge $1/g$.

In the limit $\lambda \to 0$, our model enjoys other nice features besides yielding the simple analytic solution (3.80) for the unit-monopole. In particular Bogomol'nyi (1976) has derived an inequality relating the energy of a static configuration to its topological index, very similar to the result (3.31) for the O(3) model. When $\lambda = 0$, the energy given in (3.52) reduces, for a static solution with $A_0^a = 0$, to

$$\begin{aligned} E &= \int d^3x \left(\tfrac{1}{4} G_{ij}^a G_{ij}^a + \tfrac{1}{2} D_k\phi^a D_k\phi^a\right) \\ &= \int d^3x \sum_{i,j,a} \tfrac{1}{4}(G_{ij}^a - \varepsilon_{ijk} D_k\phi^a)^2 + \int d^3x\, \tfrac{1}{2}\varepsilon_{ijk} G_{ij}^a D_k\phi^a. \end{aligned} \tag{3.81}$$

The second term may be written as

$$\begin{aligned} \int d^3x\, \tfrac{1}{2}\varepsilon_{ijk} G_{ij}^a D_k\phi^a &= \int d^3x\, \partial_k\left(\tfrac{1}{2}\varepsilon_{ijk} G_{ij}^a \phi^a\right) \\ &= \oint_{S_2} d\sigma_k \left(\tfrac{1}{2}\varepsilon_{kij} G_{ij}^a \phi^a\right) \end{aligned} \tag{3.82}$$

if we use the identity $D_\mu \tilde{G}^{a\mu\nu} = 0$ where $\tilde{G}^{a\mu\nu}$ is the dual field $\frac{1}{2}\varepsilon^{\mu\nu\rho\sigma} G_{\rho\sigma}^a$. This identity follows from (3.43)–(3.44) (see the derivation of the Euclidean analogue (4.20) in chapter 4).

Now, consider the gauge-invariant electromagnetic tensor $F_{\mu\nu}$ in (3.67). As $r \to \infty$, $D_\mu\phi^a \to 0$ and $\hat{\phi}^a \to \phi^a/F$ for any finite-energy configuration. Hence the magnetic field obtained from (3.67) becomes, asymptotically,

$$B_k = \tfrac{1}{2}\varepsilon_{kij} F_{ij} \to (1/2F)\varepsilon_{kij} G_{ij}^a \phi^a.$$

Using this, we can write (3.82) as

$$F \oint_{S_2} d\sigma_k B_k = 4\pi m F = 4\pi (Q/g) F$$

where m is the monopole charge, related to the homotopy index Q by (3.71). Thus the energy in (3.81) can be written as

$$E = \frac{4\pi QF}{g} + \int d^3x \sum_{i,j,a} \tfrac{1}{4}(G^a_{ij} - \varepsilon_{ijk} D_k \phi^a)^2 \geq \frac{4\pi QF}{g}. \tag{3.83}$$

In any given Q-sector, the energy is clearly minimised if and only if the fields satisfy the 'Bogomol'nyi condition':

$$G^a_{ij} = \varepsilon_{ijk} D_k \phi^a. \tag{3.84}$$

These results are clearly similar to eq. (3.31)–(3.32) obtained for the O(3) model. If a field configuration satisfies the Bogomol'nyi condition (3.84), then it minimises the static energy in that Q-sector and is therefore a static classical solution in that sector. In particular, for $Q = 1$, it may be verified that the Prasad–Sommerfield solution (3.80) when inserted into the ansatz (3.73)–(3.74) does satisfy (3.84). Accordingly, the equality in (3.83) tells us that the monopole will have a mass $4\pi F/g$. Notice that like the kink solution in chapter 2, the monopole is heavy when the coupling constant g is small.

For $Q > 1$ or $\lambda \neq 0$, no analytical solutions are available so far. The numerical work and suggestive arguments given by 'tHooft and Polyakov in the original papers certainly indicate that a non-singular $Q = 1$ solution exists even when $\lambda \neq 0$, with some finite energy. A more rigorous study of the existence of such solutions has been carried out by Tyupkin et al. (1976).

Apart from all its other interesting properties, the monopole solution is, at the very least, a solitary-wave solution of the model (3.42), which is a fairly complicated coupled field system in (3 + 1) dimensions. There is no reason, however, to believe that it is a soliton in the strict sense defined in chapter 2.

3.5. More on monopoles and dyons

In the preceding section we presented the homotopy considerations that led to multi-monopole sectors of the model (3.42). We also discussed the single-monopole solution in some detail. Subsequent to these developments a great deal more work has been done in the general area of monopole-like solutions, both for the model (3.42) and for its generalisations. Space does not permit us to discuss all that work in the same detail as we discussed the basic monopole solution. We shall be content with making a collective mention of some of these developments, along with a few qualitative remarks and references. (For more discussion and references see the reviews by Marciano and Pagels (1978), and Actor (1979).)

Soon after the single-monopole solution of the model (3.42) was proposed, Julia and Zee (1975) pointed out that the same model also yields dyons, i.e., objects carrying both electric and magnetic charge. Note that the monopole solution described by the ansatz (3.73)–(3.74) carries no net electrical charge. Since the fields are time-independent, and further $A_0^a = 0$, the electric field

$$E_i = -F_{0i} = 0$$

where $F_{\mu\nu}$ is the gauge-invariant electromagnetic tensor (3.68). Julia and Zee proposed solutions which are still static, but have non-zero A_0^a. They generalised the ansatz to include

$$A_0^a = x^a J(r)/gr^2$$

with the boundary condition $J(r) \to 0$ as $r \to 0$. The fields A_i^a and ϕ^a continue to have the form (3.73)–(3.74). In the place of (3.77)–(3.78), the equations obeyed by $J(r)$, $H(r)$ and $K(r)$ are

$$r^2 \, \mathrm{d}^2K/\mathrm{d}r^2 = K(K^2 - J^2 + H^2 - 1)$$

$$r^2 \, \mathrm{d}^2H/\mathrm{d}r^2 = 2HK^2 + \lambda H[(H^2/g^2) - r^2F^2]$$

$$r^2 \, \mathrm{d}^2J/\mathrm{d}r^2 = 2JK^2.$$

In terms of these functions, the electric charge is

$$q = \int \mathrm{d}^3x \, (\mathrm{div}\, \mathrm{E})$$

$$= -\frac{8\pi}{g} \int_0^\infty \mathrm{d}r \, \frac{JK^2}{r}. \qquad (3.85)$$

In the $\lambda \to 0$ limit, exact solutions can again be found (Prasad and Sommerfield 1975, Bogomol'nyi 1976). These are

$$K(r) = \frac{rgF}{\sinh(rgF)}; \qquad H(r) = \cosh\gamma \left(\frac{rgF}{\tanh(rgF)} - 1 \right)$$

and

$$J(r) = \sinh\gamma \left(\frac{rgF}{\tanh(rgF)} - 1 \right) \qquad (3.86)$$

where γ is an arbitrary real constant. Upon inserting this solution into (3.85) we get

$$q = (4\pi/g)\sinh\gamma. \tag{3.87}$$

The asymptotic magnetic field continues to be the same as (3.76) so that the solution still carries a magnetic charge $m = 1/g$. It is clear that all the above equations reduce to those of the 'tHooft–Polyakov monopole when $\gamma = 0$. Schwinger (1966) and Zwanziger (1968) have argued that if different species of particles exist carrying electric and magnetic charges (q_i and m_i respectively), they must obey

$$q_i m_j - q_j m_i = N\hbar. \tag{3.88}$$

If we apply these conditions to our monopole ($q = 0, m = 1/g$) and dyon ($q = (4\pi/g)$ sinh γ, $m = 1/g$) solutions, we would get

$$(4\pi/g^2)\sinh\gamma = N\hbar, \tag{3.89}$$

which is not necessarily satisfied because γ could be any real number in the solution (3.86). This is, again, obviously because our considerations have been classical, whereas the Schwinger–Zwanziger conditions are obtained using quantum principles which would apply only when our classical field theory is quantised (see Jackiw 1976, Tomboulis and Woo 1976).

Our monopole and dyon solutions have been static. As we argued earlier, the basic idea of topological-monopole sectors is not limited to static solutions alone. In a recent work, time-dependent periodic solutions with magnetic and electric charges has been discussed by Christ and Jackiw (1980), for which the topological number can be related directly to the behaviour of the gauge fields A^a_μ instead of the scalar fields ϕ^a. Rossi (1979) relates the monopole solution, after a time-dependent gauge transformation, to a sequence of Yang–Mills instantons.

The fact that in solutions like (3.73)–(3.74) the internal index a is coupled to the space indices i and j, has led to an interesting line of investigation. This deals with such possibilities as fermionic systems of half-integral spin emerging out of boson fields, and the metamorphosis of isospin into spin. We should begin with a conservative remark that the mere mixing of internal indices with spatial indices in a specific solution need not, in itself, cause concern or imply a deeper significance. The index a after all serves to distinguish between the three scalar fields ϕ^a. At any point x, all the three fields ϕ^a could in general be present, and as x varies there is no reason why the relative proportion of these three fields should not vary. In that case, the iso-vector ϕ^a would vary in direction as x changes. Cases where a remains the same at all x, while perhaps more familiar until these topologically non-trivial solutions came along, are more the exception than the rule as far as allowed field configurations go.

This general remark notwithstanding, for the monopole sectors of such gauge theories this mixing of spin and internal indices does seem to herald deeper consequences. As shown by Jackiw and Rebbi (1976a) and Hasenfratz and 'tHooft (1976), when such a monopole system is combined with a charged scalar iso-doublet system, the resulting object can have half-integral spin, even though the parent fields are all bosons! The consistency of this interesting result with the usual spin-statistics relationship has been discussed by Goldhaber (1976). We refer the reader to these papers for the derivation and discussion of these results. *Apropos* our earlier comment, it should be mentioned that these consequences require much more than the mere fact that a solution such as (3.73)–(3.74) has coupled internal and spatial indices. Many other ingredients in the model – the fact that the internal group SU(2) is spontaneously broken, that this group has the same structure as the rotation group, and that the resulting solution is a magnetic monopole – all contribute to these results. Indeed, these results are a generalisation to SU(2) gauge theory monopoles of the long familiar feature that point monopoles, when combined with point charges, can have such anomalous angular momenta (see e.g. Saha 1936).

The ansatz (3.73)–(3.74) which led to the single monopole ($Q = 1$) solution was spherically symmetric. It has been shown by Guth and Weinberg (1976) (see also Frampton 1976) that higher solutions cannot be obtained from such an ansatz. One must search for them among functions with higher partial waves (Michel et al. 1977, Cremmer et al. 1976, O'Raifeartaigh 1977). Patrascioiu (1975) has argued that, in order to have non-trivial homotopy sectors of the kind we encountered, one must necessarily have some long-range fields. We recall that in our monopole solution, effectively one of the three gauge fields A^a_μ decreased as $1/r$ when $r \to \infty$ in any given direction. The larger question of when gauge theories yield non-trivial homotopy sectors has been discussed by Coleman in his lecture notes (Coleman 1977a) along with how solutions from different sectors can be 'patched together'. The extension to cases where the scalar fields form higher representations of the SO(3) group (instead of the triplet vector representation in (3.42)) has been studied by Shankar (1976).

Extension of these monopole-like solutions to higher SU(n) gauge theories has been studied by a number of workers including Marciano and Pagels (1975), Chakrabarti (1975), Corrigan et al. (1976), Horvath and Palla (1976), Sinha (1976), Jacobs (1976) and Bais and Weldon (1978). More recently Weinberg (1980) has discussed in detail monopoles of arbitrary, compact simple Lie groups.

Finally, in a very interesting study of generalised monopole solutions of higher gauge groups, Olive and colloborators (Olive 1976, Goddard et al. 1977, Montonen and Olive 1977, Goddard and Olive 1978) point out the existence of a magnetic charge group, which is dual to the original gauge group. The two groups seem to be symmetric with respect to each other. The dual of the dual magnetic group is the original guage group, and it is proposed that Noether and topological charges exchange roles in the two groups.

We hope that this modest collection of references, along with the detailed introduction in the preceding section will give the interested reader a reasonable entry into the growing literature in this area.

3.6. Homotopy considerations in condensed-matter systems

In section 3.4 we discussed in some detail the solitary-wave solutions of a SU(2) gauge theory in four dimensions. A similar analysis can be done for many other theories. While the availability of exact solutions in analytic form depends on the particular model, the homotopy classification of finite-energy configurations can be carried out for many models. Such classification is also pertinent to some condensed-matter systems like liquid crystals, ferromagnets, liquid helium and superconductors with flux lines in them. Indeed, non-trivial boundary conditions were entertained for some condensed matter systems long before they were taken up by particle physicists and generalised to relativistic systems and non-abelian gauge theories.

We shall not discuss each of these models in detail, since the same principles apply for them as were used in the preceding sections. We will just mention some of them and comment on the homotopy classification of their solutions.

For this purpose let us gather together the relevant results on homotopy groups from the book by Steenrod (1951). These are

$$\pi_n(S_n) = Z \tag{3.90a}$$

$$\pi_n(S_m) = 0 \qquad \text{for } n < m \tag{3.90b}$$

and

$$\pi_n(S_1) = 0 \qquad \text{for } n > 1 \tag{3.90c}$$

where $\pi_n(S_m)$ refers to the homotopy group for the mapping of S_n into S_m, and Z refers to the group of integers. That is, mappings of $S_n \to S_n$ come in a discrete infinity of homotopy classes, each characterised by an integer. This is true for all positive integers n. In (3.90b) and (3.90c), the zero on the right-hand side stands for the fact that the group is trivial, i.e. that in those cases all mappings can be deformed into one another.

A classic example in condensed-matter physics where such homotopy considerations apply is the system of flux-tubes in type-II superconductors. As a prelude to considering this system, we recall that we chose, for a gauge-model in (3 + 1) dimensions, the non-abelian theory (3.42). What would have happened if we had chosen instead a simpler abelian gauge-model? The abelian analogue of (3.42) is described by the Lagrangian, in (D + 1) dimensions,

$$\mathscr{L}(x, t) = -\tfrac{1}{4}F^{\mu\nu}F_{\mu\nu} + \tfrac{1}{2}(\mathrm{D}_\mu\phi)^*(\mathrm{D}^\mu\phi) - \tfrac{1}{4}\lambda(|\phi|^2 - F^2)^2 \tag{3.91}$$

where $\phi(x, t)$ is a *complex* scalar field, $F_{\mu\nu}$ is the familiar electromagnetic tensor

$$F_{\mu\nu} = \partial_\mu A_\nu - \partial_\nu A_\mu$$

and

$$\mathrm{D}_\mu\phi \equiv (\partial_\mu - \mathrm{i}eA_\mu)\phi. \tag{3.92}$$

Following the same analysis as before, one can see that zero-energy configurations must have $|\phi(x, t)| = F$ and hence finite-energy configurations must satisfy

$$|\phi| \to F \qquad \text{as } \boldsymbol{x} \to \infty. \tag{3.93}$$

along with

$$\mathrm{D}_\mu\phi = (\partial_\mu\phi - \mathrm{i}e\,A_\mu\phi) \to 0 \qquad \text{as } \boldsymbol{x} \to \infty. \tag{3.94}$$

The phase of the complex field is not fixed by these boundary conditions at infinity. Since the phase angles α of the field $\phi \equiv |\phi|\mathrm{e}^{\mathrm{i}\alpha}$ form points on a circle S_1, the boundary conditions map the surface at spatial infinity into this S_1. So far we have not specified the space-dimensionality D. For the physically most interesting case of $D = 3$, the surface at infinity is S_2. However, (3.90c) tells us that $\pi_2(S_1) = 0$, so that all allowed boundary conditions fall in the same homotopy class. Therefore there are no non-trivial homotopy sectors of solutions for the abelian model (3.91) when $D = 3$. (This is one reason why in section 3.4 we chose to discuss in detail the non-abelian SU(2) system in three spatial dimensions.) For the abelian model, the same situation holds for all D $\geqslant 3$ since $\pi_{D-1}(S_1) = 0$ for $D \geqslant 3$. It is only when $D = 2$ (a two-space one-time system) that the spatial boundary is also S_1, leading to a non-trivial homotopy group $\pi_1(S_1)$. This is the familiar group of winding numbers. For the reasons discussed earlier, this winding number is a conserved quantity and will remain constant even for time-dependent solutions of the system (3.91) in $(2+1)$ dimensions.

This abelian model in $(2+1)$ dimensions was discussed in detail by Nielsen and Olesen (1973) whose work can be considered a precursor to the 'tHooft–Polyakov monopole. Nielsen and Olesen were motivated to study this system since it simulates the dual-string picture for elementary particles. The Lagrangian (3.91) is also the relativistic dynamical analogue of what happens in the statistical mechanics of a type-II superconductor placed in a magnetic field (see for instance Schrieffer 1964). There, the complex scalar field ϕ stands for the order parameter and A_μ is again the electromagnetic field. The free energy of the system is a functional $G[\phi, A_\mu]$, given in the Landau–Ginzburg model by

$$G[\phi, A_\mu] = \int \mathrm{d}\boldsymbol{x}\left(\tfrac{1}{2}|(\nabla - \mathrm{ie}\,\boldsymbol{A})\phi|^2 + \tfrac{1}{4}[|\phi|^2 - F^2(T)]^2 + \frac{\mathrm{B}\cdot\mathrm{B}}{8\pi}\right) \quad (3.95)$$

when T, the temperature, is below the critical temperature T_c; $\boldsymbol{B} \equiv \nabla \times \boldsymbol{A}$, is the magnetic field. Notice the analogy to the Hamiltonian implied by (3.91) for static (or equilibrium) configurations with $A_0 = 0$. Furthermore, if we set up the above system to be invariant along the z-axis (as happens, approximately, for a slab of superconducting material placed between the flat pole pieces of a magnet with $\boldsymbol{B}$ along the z-axis), then the problem is essentially two-space-dimensional in the xy plane. The requirement of finite energy is replaced here by that of a finite free energy G. For such configurations in the xy plane, the boundary condition implied by (3.95) is identical to (3.93)–(3.94), and the same homotopy group $\pi_1(S_1)$ applies. Furthermore, the existence of an integral winding number n is related to the well-known flux quantisation condition. Note that the winding number describes the number of times the phase α is wound around the circle at spatial infinity. Let θ parametrise the latter: then

$$n = \frac{1}{2\pi}\int_0^{2\pi} \frac{\mathrm{d}\alpha}{\mathrm{d}\theta}\mathrm{d}\theta. \quad (3.96)$$

But the conditions (3.93)–(3.94) tells us that as $|\boldsymbol{x}| = r \to \infty$,

$$\phi(r, \theta) \to F\mathrm{e}^{\mathrm{i}\alpha(\theta)}; \qquad A_\mu \to (-\mathrm{i}/e)(\delta_\mu\phi/\phi). \quad (3.97)$$

Hence A_θ, the tangential component at $r \to \infty$, becomes

$$A_\theta = \frac{-\mathrm{i}}{e}\left(\frac{1}{F\,\mathrm{e}^{\mathrm{i}\alpha}}\right)\frac{1}{r}\frac{\mathrm{d}}{\mathrm{d}\theta}(F\,\mathrm{e}^{\mathrm{i}\alpha(\theta)}) = +\frac{1}{er}\frac{\mathrm{d}\alpha}{\mathrm{d}\theta}. \tag{3.98}$$

Hence

$$n = +\frac{e}{2\pi}\int_0^{2\pi} A_\theta r\,\mathrm{d}\theta = +\frac{e}{2\pi}\oint \boldsymbol{A}.\mathrm{d}\boldsymbol{l} = +\frac{e}{2\pi}\times(\text{flux}) \tag{3.99}$$

where (flux) is the total magnetic flux across the xy plane. Thus we see that the requirement that the winding number n be an integer is tantamount to 'quantising' the total flux across the xy plane in the superconductor to be $+2\pi/e$ times that integer.

Flux quantisation in superconductors is a well-known result. But it may appear strange that we have obtained it here purely from classical considerations. The explanation is similar to what we offered in the context of the monopole quantisation condition (3.72). As Coleman (1977a) has pointed out, (3.99) is not exactly the usual flux-quantisation condition in superconductivity. The latter gives (Schrieffer 1964)

$$\text{flux} = 2\pi n\,\hbar/2q \tag{3.100}$$

where $2q$ is the charge of the Cooper pair of two electrons. (The velocity of light is set equal to unity as we have always done.) By contrast, in (3.99), $\hbar$ never appears and e is just a coupling constant in the Lagrangian (3.91). Of course, the Lagrangian (3.91) does permit a conserved charge

$$q = \mathrm{i}e\int\left(\phi^*\frac{\partial}{\partial t}\phi - \phi\frac{\partial}{\partial t}\phi^*\right)\mathrm{d}\boldsymbol{x} \tag{3.101}$$

but this need not be an integral multiple of e at the classical level. It is only in the quantum version that the coupling constant e and the charge of the Cooper pair $2q$ are related through $2q = e\hbar$ and (3.99) turns into the full flux quantisation condition (3.100).

Aside from these general topological statements, one would like to find explicit solutions in the different sectors. Analogous to the monopole ansatz (3.73)–(3.74), we can try an ansatz which, for this problem, would have to be cylindrically rather than spherically symmetric

$$A_r = A_0 = 0, \qquad A_\theta = A(r),$$

$$\phi = F(r)\mathrm{e}^{\mathrm{i}n\theta}, \tag{3.102}$$

where we use polar coordinates (r, θ) in the (x, y) plane. The boundary conditions (3.97) imply

$$F(r) \to F \qquad \text{and} \qquad A(r) \to n/er. \tag{3.103}$$

Clearly, this ansatz will lead to a winding number n. The field equations resulting from (3.95) reduce, using the ansatz (3.102), to

$$-\frac{1}{r}\frac{\mathrm{d}}{\mathrm{d}r}\left(r\frac{\mathrm{d}}{\mathrm{d}r}\mathrm{F}(r)\right)+\left[\left(\frac{\mathrm{n}}{\mathrm{r}}-e\,A(r)\right)^2+\lambda[F^2(r)-F^2]\right]F(r)=0 \tag{3.104a}$$

$$-\frac{\mathrm{d}}{\mathrm{d}r}\left(\frac{1}{r}\frac{\mathrm{d}}{dr}[r\,A(r)]\right)+\left(e^2\,A(r)-\frac{ne}{r}\right)F^2(r)=0. \tag{3.104b}$$

No analytic solution is known to these coupled non-linear equations for arbitrary λ. However, it may be seen that solutions, if they do exist, would approach the boundary values (3.103) exponentially fast. As $r \to \infty$, one can substitute $F(r) = F$ in (3.104b). This makes that equation linear and soluble, and yields the asymptotic behaviour:

$$A(r) \xrightarrow[r\to\infty]{} (n/er) + (\text{const.}) \times \mathrm{e}^{-eFr}/\sqrt{r}. \tag{3.105a}$$

Inserting this into (3.104a) gives

$$F(r) \xrightarrow[r\to\infty]{} F + (\text{const.}) \times \exp\,(-\sqrt{\lambda}Fr). \tag{3.105b}$$

Numerical work supports the existence of non-singular solutions with these asymptotic properties, where at least the $n = 1$ solution (the single vortex-line, or 'flux-tube') seems to be stable. A more general and rigorous examination of the existence of such solutions has been carried out by Taubes (1980).

Note that these solutions have finite 'size' in the two-dimensional plane. That is, both A_μ and ϕ approach their boundary 'vacuum' values exponentially fast as $r \to \infty$, as evident from (3.105). The magnetic field obtained by inserting (3.105a) into (3.102) and (3.92) also falls to zero exponentially fast in r. All these exponentials have a characteristic range proportional to F^{-1}. This finiteness of size is related to the fact that unlike the O(3) model in section 3.3, or the pure Yang–Mills and CP_n models to be studied in the next chapter, the system (3.91) is *not* scale-invariant. The existence of a non-trivial solution of finite size in the $n = 1$ sector of this model will prove to be a very useful result even if we do not know its exact analytic form. We will use this solution, in its role as an instanton, to illustrate the quantum effect of instantons in chapter 10.

Such considerations also apply to vortices in helium II, where the order parameter is again a complex scalar field. That problem is also essentially two-dimensional, in a plane perpendicular to the vortex line, so that the non-trivial homotopy group $\pi_1(S_1)$ classifies configurations with such vortex lines in their interior. These are the so called 'line-singularities'. At the same time, since $\pi_2(S_1) = 0$, there will be no non-trivial three-dimensional sectors. Thus 'point' singularities do not exist in He II. By contrast, for isotropic ferromagnetic systems, where the order parameter is the magnetisation vector $\boldsymbol{M}$, the boundary condition permits different directions of the unit vector $\hat{\boldsymbol{m}} = \boldsymbol{M}/|\boldsymbol{M}|$. These form the surface S_2 and, since $\pi_2(S_2) = Z$, point singularities are possible. These are the analogues of the monopole solutions of the SU(2) gauge theories. But the ferromagnet will not permit 'line-singularities', since $\pi_1(S_2) = 0$ from (3.90b). Finally, another interesting system is the nematic liquid crystal, which is characterised by a 'director' vector $\boldsymbol{d}$ with the proviso that $\boldsymbol{d}$ and $-\boldsymbol{d}$ are indistinguishable. The surface of degeneracy is now the sphere S_2, but with diametrically opposite points identified. Such a domain may be called $P_2 \equiv S_2/Z_2$, and it can be shown that $\pi_1(P_2) = Z_2$. Thus, line singularities can exist in such liquid crystals, but they fall into only two homotopy classes and not a discrete infinity of classes. For a detailed discussion of the homotopy properties of these systems see for instance Volovik and Mineev (1977), Toulouse and Kleman (1976), and Mermin (1979).

There is one important difference between the examples mentioned in the above paragraph and earlier examples which contained gauge fields. In the gauge theory examples, the energy or free-energy functional involved the *covariant* derivatives of the scalar fields. By suitably adjusting the gauge fields at infinity, the covariant derivatives could be made to vanish at infinity even for angle-dependent boundary conditions (see for instance (3.94), (3.97)). Thus finite-energy configurations were possible with non-trivial boundary conditions. By contrast, in liquid crystals, ferromagnets etc., no gauge fields are present. We have not listed here the expressions for the free energies of these systems, but they are readily accessible from standard literature in those subjects. It can be seen that the energy functional involves the ordinary derivatives of the fields. These will not vanish at infinity sufficiently fast for non-trivial boundary conditions, and the energy of such solutions will diverge in the infinite volume limit. Of course, real condensed-matter systems involve finite samples, and the energies of these configurations remains finite. They are therefore physically accessible.

Another possibility is that homotopy classification arises, not from the boundary conditions, but from the behaviour of the fields in the interior. This is what happened in the O(3) model in section 3.3. There again the presence of a (grad $\phi)^2$ term in the energy precluded angle-dependent boundary conditions in the infinite volume limit, but homotopy classification was nevertheless possible, based on the behaviour of $\phi(x)$ at finite x in two space dimensions. It should be pointed out that the O(3) model is also a useful model in solid-state physics. It represents the continuum limit of a two-dimensional isotropic ferromagnet. The field $\phi(x)$ would stand for the spins of that system, and (3.15) would give the Hamiltonian. Belavin and Polyakov (1975), who obtained the solutions of the O(3) model, point out that these solutions represent metastable 'pseudo-particles' since they extremise the Hamiltonian. Since they have finite energy, a finite density of them (of order $\exp(-E/kT)$) will be excited at any temperature, however small. This is indeed the sense in which finite-energy static soliton solutions become relevant in the statistical mechanics of condensed matter systems. For the O(3) model in two dimensions, these metastable states become particularly important. Even though there is a small density of them at low temperature, each 'pseudo-particle' can have arbitrarily large size, thanks to the scale invariance of the Hamiltonian (3.15). Consequently, they can occupy all of space. Further, each $Q = 1$ 'pseudoparticle' has the spin ϕ pointing in all different directions as x varies. Consequently, Belavin and Polyakov argue, long-range order is destroyed at any temperature, however small, for this system.

That solitons play an important role in the thermodynamics of continuum systems by acting as elementary excitations was pointed out in quantitative terms by Krumhansl and Schrieffer (1975). This has been further elaborated upon by several workers. For a review, see Bishop et al. (1980). We will be dealing with essentially the same ideas in chapter 10, where the impact of instantons in Euclidean quantum theory will be treated as if they constituted a gas.

3.7. Soliton–soliton interactions

We conclude our study of classical solitons with a brief discussion of their interaction. The resemblance of a single soliton solution to an extended particle has already been emphasised. We now look for some similar resemblance between multi-soliton solutions and a set of particles. In particular let us see if the dynamics of a set of two or more solitons can be described, at least approximately, in terms of an interaction potential which depends on their relative separation, analogous to what we do for particles.

Although several models yield one-soliton solutions, exact multi-soliton solutions are unavailable for most of them. (Counterexamples like the sine–Gordon model are rare.) It is therefore both necessary and useful to gain whatever qualitative understanding we can about systems of two or more solitons, in terms of intuitive arguments and approximations. Since our field equations are all non-linear, a superposition of single-soliton functions will in general not be a solution. But if the solitons are far enough apart their overlap will be small. The distortion in each soliton caused by non-linear effects due to the presence of the others should also be small. That is, one may expect solutions to exist, whether or not we can find them in analytical form, corresponding to widely separated solitons which retain their individual identity except for some small distortion. In general such solutions may not be static. One would expect that each soliton would exert some force on the others, in addition to distorting (polarising) them, and cause them to accelerate. To hold the solitons stationary in the face of their mutual interaction, additional external forces may have to be added.

These expectations, which picture a collection of widely separated solitons as being similar to a set of widely separated extended particles, can be realised in concrete form for specific models. As a simple example, consider the kinks of the ϕ^4 model in eq. (2.25). Boundary conditions in this model do permit kink–antikink configurations, but no non-trivial exact solution of this type is known. Consider the time-independent field equation of this model:

$$\phi'' - \lambda\phi^3 + m^2\phi = 0. \qquad (3.106)$$

As discussed in section 2.3, the only finite-energy solutions of this equation are the kink, the antikink and the trivial solutions $\phi(x) = \pm m/\sqrt{\lambda}$. It has no solutions that can be considered as a widely-separated static kink–antikink pair. This is consistent with our expectation that the kink and antikink will exert some force on one another and will not remain static. But if they could be 'held' stationary by applying some external force, a static kink–antikink solution may exist. Such 'holding' can be done through the following device (Perring and Skyrme 1962).

Instead of (3.106), consider the modified equation

$$\phi'' - \lambda\phi^3 + m^2\phi = \alpha(R)\{\delta(x - R/2) + \delta(x + R/2)\}. \qquad (3.107)$$

The right-hand side may be viewed as two external point-forces applied at $x = \pm R/2$. Unlike (3.106), eq. (3.107) *does* yield a finite-energy solution that looks like a kink–antikink pair separated by a distance R. This solution will, of course, have a slope discontinuity at the points $x = \pm R/2$. Such a solution emerges only if $\alpha(R)$, the strength of the applied external forces, is chosen appropriately as a function of R. The kink and the antikink in this solution are seen to be distorted are compared to their individual shape, but this distortion reduces to zero as $R \to \infty$. All these features conform to our expectations and to the analogy with extended particles.

We will not present the details of this kink–antikink solution or its derivation (see Rajaraman 1977). It is a straightforward matter to integrate (3.107) and obtain this solution in terms of elliptic integrals. When this static solution is inserted into the energy integral (2.19), its energy comes out as (Rajaraman 1977)

$$E(R) = \frac{4\sqrt{2}}{3}\frac{m^3}{\lambda} - 8\sqrt{2}\frac{m^3}{\lambda}\exp(-\sqrt{2}mR) + \text{lower order terms in } R. \tag{3.108}$$

The first term is just the rest mass of a free kink plus a free antikink. The remaining terms may be attributed to the interaction potential energy $V(R)$ of the kink–antikink pair, as a function of their separation R. Thus

$$V(R) \underset{\text{large } R}{\sim} -(8\sqrt{2}m^3/\lambda)\exp(-\sqrt{2}mR). \tag{3.109}$$

Notice that this interaction is (i) attractive, and (ii) strong when the coupling λ/m^2 is weak. A similar attractive potential can also be obtained by the same procedure for the soliton–antisoliton pair of the sine–Gordon model. We display the potential (3.109) only for large R, since at small R, the kink and antikink loose their identity and $V(R)$ is not a useful object.

[The potential (3.109) is clearly reminiscent of a one-meson exchange potential extracted from quantum field theory. In fact, in chapter 5 where we quantise this theory, it will be seen to carry a meson of mass $\sqrt{2}\,m$, and to yield a kink–kink–meson vertex of order $1/\sqrt{\lambda}$. If we remember that two such vertices are involved when a meson is exchanged between kinks, and put in sufficient factors of m to meet dimensional requirements, the one-meson-exchange Born amplitude must clearly yield a potential

$$V(R) = c\,(m^3/\lambda)\exp(-\sqrt{2}\,mR),$$

where c is a dimensionless number. Our potential $V(R)$ in (3.109) has the same form, although it was obtained from a purely classical field system. This is an example of a feature that will be further discussed in chapter 5, namely that properties of non-perturbative solitons in quantum theory receive their leading contributions from their classical analogues.]

Manton (1977, 1979) has approached the problem of inter-soliton forces from a different point of view. Instead of obtaining them from a static potential, he multiplies the mass of a soliton by its acceleration in the presence of another soliton. This acceleration is evaluated by constructing a solution of the form $\phi(x - \frac{1}{2}at^2)$, where $\phi(x)$ is the static soliton function. Taking the acceleration a to be small (a reasonable assumption when the solitons are far apart), he finds its value by the requirement that the field equations be satisfied to lowest order in a. He also shows that, for the cases where a static potential such as (3.109) can be found, the force obtained from that potential agrees with the mass times acceleration of the soliton. This strengthens the analogy to Newtonian particles, and lends more credibility to such a potential.

In the first of his papers Manton applied his method directly to the force between two 'tHooft–Polyakov monopoles – a much more complicated problem than dealing with kinks. In particular he found the following interesting result. We would intuitively expect two monopoles to have a long-range force m^2/R^2, where m is the magnetic charge of each. Instead, he found that in the Prasad–Sommerfield limit ($\lambda \to 0$ in (3.42)), the long-range force vanishes between monopoles of equal charge, whereas between monopoles of opposite charge the force is double the expected magnetostatic value. The reason for this is that in the $\lambda \to 0$ limit of (3.42), the scalar field is also massless. Exchange of this scalar boson will lead to an attractive $O(1/R^2)$ force, which just cancels the magnetostatic repulsion between like-monopoles, and augments the magnetostatic attraction between unlike monopoles. These issues have been further clarified, and generalised to monopole solutions of arbitrary compact groups with non-adjoint Higgs scalar fields by O'Raifeartaigh et al. (1979). For a study of monopole–dyon interactions, see Jersak et al. (1977). The forces between vortices of the abelian Higgs model discussed earlier have been obtained by Schaposnik (1978). Raj Lakshmi (1979) has investigated the force between charged solitons of complex scalar fields.

CHAPTER 4

CLASSICAL INSTANTON SOLUTIONS

4.1. What are instantons?

The term instantons has come to refer to localised finite-action solutions of the classical Euclidean field equations of a theory. The Euclidean version of a theory, or a set of field equations, involves replacing the Minkowskian metric $g^{\mu\nu}$ ($g^{00} = 1$; $g^{ij} = -\delta_{ij}$ for $i, j \neq 0$; all other $g^{\mu\nu} = 0$), by the Euclidean metric $\delta^{\mu\nu}$. The space–time coordinate vector $(x^\mu)_{\text{Min}}$ ($\mu = 0, 1, 2, 3$ in four dimensions) is replaced by $(x^\mu)_{\text{Euc}}$, ($\mu = 1, 2, 3, 4$). The components of $(x^\mu)_{\text{Euc}}$ are still real, but the Euclidean theory is left invariant by 0(4) transformations rather than Lorentz transformations. Thus

$$(x^\mu x^\mu)_{\text{Euc}} = \sum_{\mu=1}^{4} (x^\mu)^2_{\text{Euc}}$$

is an invariant. Correspondingly, any other Lorentz tensor is replaced by an 0(4) tensor. For instance, a real vector field $(A^\mu)_{\text{Min}}$ would be replaced by a real $(A^\mu)_{\text{Euc}}$, where 0(4) transformations leave $(A^\mu A^\mu)_{\text{Euc}}$ invariant. 'Internal-space' indices, such as the index a in eq. (3.10), are to be unaffected by this change. Obviously, because of the metric $\delta^{\mu\nu}$, there is no difference between upper and lower space–time indices in a Euclidean system.

Note that this is not just a change of notation. The Euclidean system is not the same as the Minkowskian system rewritten in terms of different coordinates. For instance while $(x^\mu)_{\text{Min}}$ has four real components with $(x_0^2 - \boldsymbol{x}^2)$ invariant, $(x^\mu)_{\text{Euc}}$ also has four real components, but with $(\sum_1^4 x_\mu^2)$ invariant. The two systems are different. However, it is possible to think of the Euclidean system as an analytical continuation of the Minkowskian system. If we take real time $t = x_0$ of the Minkowskian system and continue it to imaginary values, then $(x_4)_{\text{Euc}} \equiv it$ can serve as a real fourth component of the Euclidean coordinate vector with the remaining coordinates x_1, x_2, x_3 unchanged. Then $(x^\mu x^\mu)_{\text{Euc}}$ will be related to the Minkowskian inner product *after* such a continuation by $(x^\mu x^\mu)_{\text{Euc}} = -(x^\mu x_\mu)_{\text{Min}}$. A similar continuation has to be done on the time components of all four vectors. To get the Euclidean action, we analytically continue the Minkowskian action, and further multiply by $(-\text{i})$. That is, $S_{\text{Euc}} = -\text{i}(S_{\text{Min}})_{\text{continued}}$. The extra factor of $(-\text{i})$, which will not affect the classical equations, is introduced for later convenience in the quantum theoretic context. Also, with the factor $(-\text{i})$, S_{Euc} will be typically non-negative and will resemble the energy functional of static configurations (see below).

As an illustration consider the Minkowskian Klein–Gordon system, described by the action

$$S_{\text{Min}} = \int \mathrm{d}t \int \mathrm{d}\boldsymbol{x} \left[\frac{1}{2}\left(\frac{\partial \phi}{\partial t}\right)^2 - \frac{1}{2}(\nabla\phi)^2 - m^2\phi^2 \right], \tag{4.1}$$

which yields the field equation

$$\left(\frac{\partial^2}{\partial t^2} - \nabla^2\right)\phi + m^2\phi = 0. \tag{4.2}$$

To obtain the Euclidean system, we continue from real t to $t = -\text{i}x_4$, and from the S_{Min} so obtained get S_{Euc} by using $S_{\text{Euc}} = -\text{i}S_{\text{Min}}$. The resulting Euclidean action is

$$S_{\text{Euc}} = \int \mathrm{d}x_4 \int \mathrm{d}\boldsymbol{x} \left[\frac{1}{2}\left(\frac{\partial \phi}{\partial x_4}\right)^2 + \frac{1}{2}(\nabla\phi)^2 + m^2\phi^2 \right] \tag{4.3}$$

which leads to the Euclidean field equation

$$(\partial^2/\partial x_4^2 + \nabla^2)\phi - m^2\phi = 0. \tag{4.4}$$

Unlike eq. (4.2), which has a hyperbolic differential operator, eq. (4.4) has an elliptic operator, which permits a different set of real non-singular solutions.

Instantons are localised solutions of such Euclidean field equations with finite Euclidean action. Why, one might ask, do we wish to indulge in this apparent perversion of going Euclidean, when systems of physical interest are imbedded in Minkowskian space–time? The answer will be discussed in chapter 10. It will be seen that certain features of Minkowskian quantum field theories can be conveniently explored by starting from classical solutions of the Euclidean version of its field equations. Notable among these features is the phenomenon of quantum tunnelling between degenerate classical ground states. Euclidean instanton solutions are just as important to Minkowskian quantum field theories as are classical

Minkowskian soliton solutions. Also, as with quantum solitons, the role of instantons in quantum field theory will be developed in a semiclassical expansion. The starting point is again a set of classical solutions – but to Euclidean field equations.

In this chapter, we shall discuss some examples of such instanton solutions at the purely classical level. The larger question of their role in quantum theory will be dealt with in chapters 10 and 11. At the classical level, instantons are not very different from static solutions of Minkowskian equations. This is obviously because static solutions involve only the spatial coordinates, i.e. the Euclidean subspace of Minkowskian space–time. Indeed, some of the static solutions we have already obtained can be directly used as instantons of lower dimensional systems (see section 4.4). The only difference is that the requirement of finiteness of energy of solitons will be replaced by the requirement of finiteness of Euclidean action in the case of instantons. Note that the Euclidean action, such as in the example (4.3), has the same structure as the energy of a static field configuration, in one-higher dimension. Aside from this, when we calculate the quantum effects of instantons in chapters 10 and 11 it will be seen that the contribution of a given classical instanton is proportional to $\exp(-S_{\mathrm{Euc}})$. If S_{Euc} were infinite this contribution would vanish. Hence, only finite-action Euclidean solutions matter, at least in the semiclassical expansion which we will use. Apart from this replacement of the role of static energy by the role of Euclidean action, our discussion of classical instantons will closely parallel our earlier discussions of static soliton solutions.

In the next two sections, we shall discuss instantons of the Yang–Mills system. It was with this example that the work of Belavin et al. (1975) pioneered the study of instantons. Moreover, non-abelian gauge theories, of which the Yang–Mills system is the prototype example, are of crucial importance in the current theories of elementary particles. In sections 4.4 and 4.5 we shall discuss instantons of other systems. Throughout this chapter, unless otherwise mentioned, we will be dealing only with Euclidean systems. Hence the subscript 'Euc' will be dropped and taken as understood.

4.2. Euclidean Yang–Mills configurations

We have already encountered the Yang–Mills fields A_μ^a in the last chapter in connection with the monopole solution. There, the system also included 'matter fields' ϕ^a. We will omit these now, and just consider the SU(2) gauge fields A_μ^a, self-coupled to one another. To make the notation simpler, it will be helpful to represent the three vector-fields A^a_μ, $a = 1, 2, 3$, by a matrix-valued vector-field A_μ defined by

$$A_\mu(x) \equiv \sum_a g \frac{\sigma^a}{2\mathrm{i}} A^a_\mu(x) \tag{4.5}$$

where x stands for (x_1, x_2, x_3, x_4). Here g is the coupling constant given in (3.43) and σ^a are the familiar Pauli spin matrices,

$$\sigma^1 = \begin{pmatrix} 0 & 1 \\ 1 & 0 \end{pmatrix}, \quad \sigma^2 = \begin{pmatrix} 0 & -\mathrm{i} \\ \mathrm{i} & 0 \end{pmatrix} \quad \text{and} \quad \sigma^3 = \begin{pmatrix} 1 & 0 \\ 0 & -1 \end{pmatrix}.$$

More pertinently $\sigma^a/2$ form the three generators of the two-dimensional representation of the group SU(2), which is the gauge group of our system. This 2×2 anti-Hermitian matrix $A_\mu(x)$ represents for each μ and at each point (x), the same information as the three Yang–Mills fields $A^a_\mu(x)$. Correspondingly, we define a matrix-valued field tensor:

$$G_{\mu\nu} \equiv \sum_a g \frac{\sigma^a}{2\mathrm{i}} G^a_{\mu\nu}. \tag{4.6}$$

In terms of these matrices, the relations (3.43) and (3.45)–(3.46) become simpler. It is easy to check, using the Lie algebra $[\sigma^a/2, \sigma^b/2] = \mathrm{i}\varepsilon^{abc}\sigma^c/2$ for the group SU(2), that

$$G_{\mu\nu} = \partial_\mu A_\nu - \partial_\nu A_\mu + [A_\mu, A_\nu]. \tag{4.7}$$

The gauge transformations (3.45)–(3.46) are not formally altered by going to the Euclidean metric. In terms of the matrix field A_μ they become

$$A_\mu \to U A_\mu U^{-1} + U \partial_\mu U^{-1} \tag{4.8a}$$

and

$$G_{\mu\nu} \to U G_{\mu\nu} U^{-1}. \tag{4.8b}$$

Note that whereas (3.45)–(3.46) was written in the 3×3 representation of the group elements $U(\boldsymbol{x}, t)$ and the generators, the same transformation holds in any representation. Here, to be consistent with the definition (4.5), we use the 2×2 representation. We shall also often use the identity:

$$\partial_\mu(U U^{-1}) = 0 = (\partial_\mu U) U^{-1} + U \partial_\mu (U^{-1}). \tag{4.9}$$

Another advantage of this matrix notation is that it reveals in an obvious way the parallel to electromagnetism where the gauge group is the abelian U(1). In that case the group elements are $U = \mathrm{e}^{i\alpha(x)}$, the fields $A_\mu(x)$ are just numbers instead of matrices, and (4.7) and (4.8) reduce to familiar equations in electromagnetism.

Returning to the Euclidean Yang–Mills system, we have four matrix-valued fields A_μ, ($\mu = 1, 2, 3, 4$) at each Euclidean space point x_μ ($\mu = 1, 2, 3, 4$). The Euclidean action is obtained from the Minkowskian Lagrangian (3.42) by the prescription outlined above, after discarding the scalar field ϕ^a, and using the matrix notation for A_μ. This action is

$$S = -\frac{1}{2g^2}\int d^4x \, \mathrm{Tr}\,[G_{\mu\nu}\, G_{\mu\nu}]. \tag{4.10}$$

Notice that the coupling constant g appears explicitly only in the overall factor $-1/2g^2$ multiplying the action. The remaining g dependence has been absorbed into the scale of the field A_μ in the definition (4.5). In the classical context, therefore, the value of the coupling is not very significant. In the quantum context, the size of the action compared with $\hbar$ could be important. Hence we shall retain the factor $-1/2g^2$ in (4.10). The resulting Euclidean Yang–Mills equation is

$$D_\mu G_{\mu\nu} \equiv \partial_\mu G_{\mu\nu} + [A_\mu, G_{\mu\nu}] = 0. \tag{4.11}$$

In this matrix notation, it is obvious that this action and the field equation are invariant under the gauge transformations (4.8).

The Yang–Mills instantons are finite-action solutions of (4.11). To find them, we proceed just as we did in the case of the monopole and other static solutions of earlier examples. First we identify the boundary conditions to be satisfied by any finite-action field configurations (of which finite-action solutions of (4.11) are a subset). Based on these boundary conditions, we shall make a homotopy classification. This procedure will occupy all of this section. Actual solutions will be obtained only in the next section.

As a first step towards identifying finite-action configurations, we consider zero-action configurations. From (4.10) we see that $S = 0$ if and only if $G_{\mu\nu} = 0$. This allows an infinite set of possibilities for the parent field A_μ. Note that $G_{\mu\nu} = 0$ is a gauge-invariant condition (see (4.8b)). Thus, it will be satisfied not only by $A_\mu = 0$, but by any gauge-transformed field obtained from $A_\mu = 0$. These fields, called 'pure gauges', are given by (see (4.8a))

$$A_\mu(x) = U(x)\partial_\mu(U^{-1}(x)) \tag{4.12}$$

where $U(x)$, at each x, is any element of the group SU(2) in its 2×2 representation. The reader can also directly verify, using (4.7), that (4.12) yields $G_{\mu\nu} = 0$. The converse, that $G_{\mu\nu} = 0$ everywhere implies (4.12), can also be proved.

Turning to finite-action configurations, it is clear from (4.10) that $G_{\mu\nu}$ must vanish on the boundary of Euclidean four-space, i.e. on the three-dimensional spherical surface $S_3^{(\text{phy})}$ at $r = \infty$ where $r \equiv |x| = (x_1^2 + x_2^2 + x_3^2 + x_4^2)^{1/2}$ is the radius in four dimensions. In fact $G_{\mu\nu}$ must fall to zero faster than $1/r^2$ as $r \to \infty$. The vanishing of $G_{\mu\nu}$ on $S_3^{(\text{phy})}$ in turn implies, as per (4.12), the following boundary conditions on A_μ:

$$\lim_{r \to \infty} A_\mu(x) = \lim_{r \to \infty} U \, \partial_\mu U^{-1} \tag{4.13}$$

where U is some group-element-valued matrix function. Based on (4.13) we would like to associate with every finite-action configuration $A_\mu(x)$ a group function U, defined on the surface at infinity, $S_3^{(\text{phy})}$. This surface may be parametrised by three variables α_1, α_2 and α_3, which could for instance be the polar angles in four dimensions. However, if the function U depended only on α_1, α_2 and α_3, one could not define its radial derivative, whereas a general $A_\mu(x)$ may have a radial component as well. One way of overcoming this difficulty is to gauge transform A_μ such that its radial component vanishes everywhere. That is, suppose we are given an $A_\mu(x)$ with non-zero radial component $A_r(x)$. Then perform a gauge transformation according to (4.8a), using as a gauge function the object

$$\tilde{U}(\mathrm{x}) = \mathrm{P}\Big(\exp \int_0^r \mathrm{d}r' \, A_r(x')\Big). \tag{4.14}$$

The notation in (4.14) has to be explained. This function is an example of 'path ordered' exponentials – a very useful construction in gauge theory (see Wu and Yang 1975). Given a point x, consider a path, which in the case of (4.14) runs from the origin to x along the radius vector. The component of A_μ parallel to this path is A_r. Thus, (4.14) involves a line-integral of A_μ along this radial path. Now since $A_\mu(x')$ at each x' is a matrix, these matrices at different x' will not commute in general and their ordering has to be defined. In (4.14) this ordering is specified by defining $\tilde{U}$ to be a product of the exponentials $\exp(A_r \, \mathrm{d}r')$ over each infinitesimal element of the path, arranged sequentially as we go from the starting point ($x = 0$) to the end-point x. The 'path ordering' prefix P stands for this definition. In general, a path ordered exponential has the form $\mathrm{P}(\exp \int \mathrm{d}x_\mu \, A_\mu)$, of which (4.14) is a special case. Notice that since $A_\mu(x)$ is an element of the Lie algebra (see (4.5)), $\tilde{U}(x)$ is an element of the gauge group for each x. Thus, it may be used to perform gauge transformations.

Given an $A_\mu(x)$, we gauge transform it using the gauge function $\tilde{U}(x)$. The radial component of the transformed field $A'_\mu(x)$, obtained from (4.8a), clearly vanishes. We have

$$A'_r(x) = \tilde{U} A_r (\tilde{U})^{-1} - (\partial_r \tilde{U})(\tilde{U})^{-1}$$
$$= \tilde{U}[A_r - A_r](\tilde{U})^{-1} = 0.$$

Further, since the action is gauge invariant, if A_μ has finite action, so does A'_μ. It must also tend to a pure gauge at infinity. Combining this with the fact that $A'_r \equiv 0$, we can write the boundary condition (4.13) as

$$(A'_\mu(x))_{S_3^{(\mathrm{phy})}} = U(\alpha_1, \alpha_2, \alpha_3)\, \partial_\mu U^{-1}(\alpha_1, \alpha_2, \alpha_3) \tag{4.15}$$

where $U(\alpha_1, \alpha_2, \alpha_3)$ needs to be defined only on $S_3^{(\mathrm{phy})}$. We will shortly perform a homotopy classification of the functions $U(\alpha_1, \alpha_2, \alpha_3)$. It will be seen (eq. (4.18) below) that this homotopy index Q can be written as a *gauge invariant* integral over the fields. In that integral, therefore, we can directly use the original field $A_\mu(x)$ instead of its gauge-transformed partner $A'_\mu(x)$. In this way, a homotopy index Q may be associated with any finite-action configuration $A_\mu(x)$.

[It must be remembered that this $U(\alpha_1, \alpha_2, \alpha_3)$, being specified only on $S_3^{(\mathrm{phy})}$, cannot be used as a gauge function for performing non-singular gauge transformations. For the latter, we need a $U(x)$ defined non-singularly throughout four-space. In fact, in general our $U(\alpha_1, \alpha_2, \alpha_3)$ will not even be the boundary value of a non-singular $U(x)$. This is an important point, stressed by Coleman (1979) and we shall return to it later.]

The boundary condition (4.15) permits an infinite number of possibilities, related to different choices of $U(\alpha_1, \alpha_2, \alpha_3)$. This function can be classified by homotopy considerations. Since the matrices U form the two-dimensional representation of SU(2), the function $U(\alpha_1, \alpha_2, \alpha_3)$ represents a mapping of $S_3^{(\mathrm{phy})}$ into the group space of SU(2). To classify such mappings, let us look at the topology of this group space. By definition of SU(2), the matrices U are the set of all 2×2 unitary unimodular matrices. Such matrices can be written uniquely in the form:

$$U = \sum_{\mu=1}^{4} a_\mu s_\mu \tag{4.16}$$

where $s_4 = I$, the unit 2×2 matrix,

$$s_{1,2,3} = \mathrm{i}\sigma_{1,2,3},$$

and a_μ are any four real numbers satisfying

$$\sum_\mu a_\mu a_\mu = 1. \tag{4.17}$$

The group is thus parametrised by these four real variables a_μ, subject to the constraint (4.17). The group space is therefore the three-dimensional surface of a unit sphere in four dimensions. We will call this surface $S_3^{(\text{int})}$. The function $U(\alpha_1\ \alpha_2\ \alpha_3)$ is therefore a mapping of $S_3^{(\text{phy})}$ into $S_3^{(\text{int})}$.

Notice again the similarity to our discussion of the monopole system in section 3.4. There, the boundary conditions involved a mapping of S_2 into S_2. Here, we have the corresponding situation for three-dimensional spherical surfaces. From eq. (3.90) we see that the corresponding homotopy group $\pi_3(S_3)$ is also isomorphic to the group of integers Z.

Therefore the mappings of $S_3^{(\text{phy})} \to S_3^{(\text{int})}$ can be divided into a discrete infinity of homotopy classes, each characterised by an integer Q. This integer is often called the Pontryagin index. Mappings from one class cannot be continuously deformed into a mapping from another class.

Given that $U(\alpha_1, \alpha_2, \alpha_3)$ can be classified into homotopy classes, our finite-action configurations $A_\mu(x)$ can also be classified into sectors, characterised by the same index Q. Any given sector corresponds to a given homotopy class of the $U(\alpha_1, \alpha_2, \alpha_3)$ occuring in the boundary condition (4.15) for A_μ. A field $A_\mu(x)$ in four-space belonging to a given sector Q cannot be continuously deformed so as to fall into another sector without violating finiteness of action. For, if this were possible, the $U(\alpha_1, \alpha_2, \alpha_3)$ related to the boundary value of $A_\mu(x)$ would have also deformed from one homotopy class to another, which cannot happen by definition. We can write an explicit integral for the Pontryagin index Q in terms of the fields, just as we did for the index Q of the monopole system. This expression is

$$\begin{aligned} Q &\equiv \textstyle\int Q(x)\,d^4x \\ &= -\frac{1}{16\pi^2}\int d^4x\, \mathrm{Tr}[\tilde{G}_{\mu\nu} G_{\mu\nu}] \end{aligned} \tag{4.18}$$

where $\tilde{G}_{\mu\nu}$, the dual, is defined by

$$\tilde{G}_{\mu\nu} = \tfrac{1}{2}\varepsilon_{\mu\nu\rho\sigma} G_{\rho\sigma}. \tag{4.19}$$

Let us now prove that the integral in (4.18) does indeed give the homotopy index Q. This demonstration is a little lengthy but the central idea connecting the arguments is that Q is essentially the 'winding number' of $S_3^{(\text{phy})}$ into $S_3^{(\text{int})}$, i.e. it is the number of times the group space of SU(2) is wrapped around $S_3^{(\text{phy})}$, the surface at infinity. We wish to show that the integral on the right-hand side of (4.18) gives this winding number. The first step is to write this integral as a surface integral over $S_3^{(\text{phy})}$. We begin with a very useful identity

$$\begin{aligned} D_\mu \tilde{G}_{\mu\nu} &\equiv \partial_\mu \tilde{G}_{\mu\nu} + [A_\mu, \tilde{G}_{\mu\nu}] \\ &= \varepsilon_{\mu\nu\alpha\beta}\{\partial_\mu(\partial_\alpha A_\beta + A_\alpha A_\beta) + [A_\mu, (\partial_\alpha A_\beta + A_\alpha A_\beta)]\} = 0 \end{aligned} \tag{4.20}$$

where the last equality follows simply from expanding out the terms and using the antisymmetry of $\varepsilon_{\mu\nu\alpha\beta}$. This identity requires only that $G_{\mu\nu}$ obey the defining equation (4.7), and not necessarily that it obey the field equation (4.11). Notice also that when the gauge group is the abelian $U(1)$, and A_μ are numbers instead of matrices, the identity (4.20) reduces to the Euclidean version of the second of the covariant Maxwell's equations (3.66). We now use (4.20) to write the Pontryagin 'density' $Q(x)$ as a divergence. We have,

$$\begin{aligned} -16\pi^2 Q(x) &= \mathrm{Tr}[G_{\mu\nu}\tilde{G}_{\mu\nu}] \\ &= \mathrm{Tr}[(\partial_\mu A_\nu - \partial_\nu A_\mu)\ \tilde{G}_{\mu\nu} + (A_\mu A_\nu - A_\nu A_\mu)\tilde{G}_{\mu\nu}] \\ &= \mathrm{Tr}\{(\partial_\mu A_\nu - \partial_\nu A_\mu)\tilde{G}_{\mu\nu} + A_\mu[A_\nu, \tilde{G}_{\mu\nu}]\}, \\ &= \mathrm{Tr}[(\partial_\mu A_\nu - \partial_\nu A_\mu)\tilde{G}_{\mu\nu} - A_\mu \partial_\nu \tilde{G}_{\mu\nu}] \\ &= \mathrm{Tr}[\partial_\mu A_\nu \tilde{G}_{\mu\nu} - \partial_\nu(A_\mu \tilde{G}_{\mu\nu})] \end{aligned} \tag{4.21}$$

where the cyclic property of the trace as well as (4.20) have been used. Next we expand out $\tilde{G}_{\mu\nu}$ to write

$$\begin{aligned} -16\pi^2 Q(x) &= \mathrm{Tr}\{\varepsilon_{\mu\nu\alpha\beta}[(\partial_\mu A_\nu)(\partial_\alpha A_\beta + A_\alpha A_\beta) \\ &\qquad - \partial_\nu(A_\mu \partial_\alpha A_\beta + A_\mu A_\alpha A_\beta)]\} \\ &= \mathrm{Tr}\{\varepsilon_{\mu\nu\alpha\beta}[2\partial_\mu(A_\nu \partial_\alpha A_\beta + \tfrac{2}{3} A_\nu A_\alpha A_\beta)]\} \end{aligned} \tag{4.22}$$

where we have used

$$\mathrm{Tr}[\varepsilon_{\mu\nu\alpha\beta}(\partial_\mu A_\nu) A_\alpha A_\beta] = \tfrac{1}{3}\mathrm{Tr}[\varepsilon_{\mu\nu\alpha\beta}\partial_\mu(A_\nu A_\alpha A_\beta)]$$

which again follows from the cyclicity of trace and the antisymmetry of $\varepsilon_{\mu\nu\alpha\beta}$. Thus

$$Q(x) = \partial_\mu j_\mu \tag{4.23}$$

where

$$j_\mu \equiv -(1/8\pi^2)\varepsilon_{\mu\nu\alpha\beta}\mathrm{Tr}[A_\nu(\partial_\alpha A_\beta + \tfrac{2}{3} A_\alpha A_\beta)]. \tag{4.24}$$

Hence

$$Q = \int Q(x) \mathrm{d}^4 x = \oint_{S_3^{(\mathrm{phy})}} \mathrm{d}\sigma_\mu j_\mu . \tag{4.25}$$

Further, on the surface at infinity $S_3^{(\mathrm{phy})}$, our finite-action configurations have $G_{\alpha\beta} = 0$ and hence $\varepsilon_{\mu\nu\alpha\beta}\partial_\alpha A_\beta = -\varepsilon_{\mu\nu\alpha\beta}A_\alpha A_\beta$. Thus,

$$Q = +\frac{1}{24\pi^2} \oint_{S_3^{(\mathrm{phy})}} \mathrm{d}\sigma_\mu \varepsilon_{\mu\nu\alpha\beta} \mathrm{Tr}[A_\nu A_\alpha A_\beta]. \tag{4.26}$$

Finally, inserting the asymptotic behaviour (4.15) of the fields A_μ, we have

$$Q = -\frac{1}{24\pi^2} \oint \mathrm{d}\sigma_\mu \varepsilon_{\mu\nu\alpha\beta} \mathrm{Tr}[(\partial_\nu U)U^{-1}(\partial_\alpha U)U^{-1}(\partial_\beta U)U^{-1}]. \tag{4.27}$$

Thus, we have written the volume integral in (4.18) as a surface integral, with the integrand directly in terms of the group-element-valued function U on $S_3^{(\mathrm{phy})}$, corresponding to any given finite-action configuration. The next ingredient in the proof is to identify an expression for measuring the group volume, i.e. a group measure, so that we can look for how many times the group has been wrapped around $S_3^{(\mathrm{phy})}$ by a given gauge function U. We have already noted that the parameter space of $SU(2)$ is also a hypersphere, $S_3^{(\mathrm{int})}$. It can be parametrised either by the four cartesian variables a_μ in (4.16) subject to the constraint (4.17), or equivalently by three independent variables ξ_1, ξ_2 and ξ_3, which could for instance be polar angles in four-space. A given group element U corresponds to some value (ξ_1, ξ_2, ξ_3) of these parameters. In the vicinity of U, the group measure may be written as

$$\mathrm{d}\mu(U) = \rho(\xi_1, \xi_2, \xi_3)\mathrm{d}\xi_1 \mathrm{d}\xi_2 \mathrm{d}\xi_3 \tag{4.28}$$

where the density $\rho(\xi_1, \xi_2, \xi_3)$ must be such that the measure $\mathrm{d}\mu(U)$ is invariant under group translations. To illustrate this, we multiply U by any fixed group element $\tilde{U}$ to get

$$U' = \tilde{U} U. \tag{4.29}$$

Let U' correspond to parameters (ξ'_1, ξ'_2, ξ'_3). Similarly let the set of elements in the infinitesimal volume $\mathrm{d}\xi_1\, \mathrm{d}\xi_2\, \mathrm{d}\xi_3$ near (ξ_1, ξ_2, ξ_3) when multiplied by the same $\tilde{U}$, fall in some volume $\mathrm{d}\xi'_1 \mathrm{d}\xi'_2 \mathrm{d}\xi'_3$. Of course

$d\xi_1 d\xi_2 d\xi_3$ need not equal $d\xi'_1 d\xi'_2 d\xi'_3$. But the density ρ should be such that

$$\begin{aligned} d\mu(U) &\equiv \rho(\xi_1, \xi_2, \xi_3) d\xi_1\, d\xi_2\, d\xi_3 \\ &= d\mu(U') \\ &= \rho(\xi'_1, \xi'_2, \xi'_3) d\xi'_1\, d\xi'_2\, d\xi'_3. \end{aligned} \tag{4.30}$$

This requirement is just the generalisation to continuous groups, of the notion from finite groups that the *number* of group elements does not change when they are all multiplied by some given element. (We have multiplied on the left by $\tilde{U}$ here. In fact the results that follow hold in the case of the group $SU(2)$, for multiplication on the right as well.) The following choice of $\rho(\xi_1, \xi_2, \xi_3)$ satisfies this requirement:

$$\rho(\xi_1, \xi_2, \xi_3) = \varepsilon_{ijk} \operatorname{Tr}\left(U^{-1} \frac{\partial U}{\partial \xi_i} U^{-1} \frac{\partial U}{\partial \xi_j} U^{-1} \frac{\partial U}{\partial \xi_k} \right), \tag{4.31}$$

where the group elements are written in their matrix representation. On multiplication on the left by $\tilde{U}$,

$$\begin{aligned} &U(\xi_1, \xi_2, \xi_3) \to U'(\xi'_1, \xi'_2, \xi'_3) \equiv \tilde{U} U(\xi_1, \xi_2, \xi_3) \\ &U = \tilde{U}^{-1} U' \qquad \text{and} \qquad U^{-1} = (U')^{-1} \tilde{U}. \end{aligned}$$

Thus

$$\begin{aligned} \rho(\xi_1, \xi_2, \xi_3) = \varepsilon_{ijk} \operatorname{Tr}\Bigg(&(U')^{-1} \tilde{U} \tilde{U}^{-1} \frac{\partial U'}{\partial \xi'_p} \frac{\partial \xi'_p}{\partial \xi_i} (U')^{-1} \\ &\tilde{U} \tilde{U}^{-1} \frac{\partial U'}{\partial \xi'_q} \frac{\partial \xi'_q}{\partial \xi_j} (U')^{-1} \tilde{U} \tilde{U}^{-1} \frac{\partial U'}{\partial \xi'_r} \frac{\partial \xi'_r}{\partial \xi_k} \Bigg) \\ = \varepsilon_{ijk} \operatorname{Tr}\Bigg(&(U')^{-1} \frac{\partial U'}{\partial \xi'_p} (U')^{-1} \frac{\partial U'}{\partial \xi'_q} (U')^{-1} \frac{\partial U'}{\partial \xi'_r} \Bigg) \\ &\times \frac{\partial \xi'_p}{\partial \xi_i} \frac{\partial \xi'_q}{\partial \xi_j} \frac{\partial \xi'_r}{\partial \xi_k}. \end{aligned} \tag{4.32}$$

But

$$\varepsilon_{ijk} \frac{\partial \xi'_p}{\partial \xi_i} \frac{\partial \xi'_q}{\partial \xi_j} \frac{\partial \xi'_r}{\partial \xi_k} = \varepsilon_{pqr} \operatorname{Det} \left| \frac{\partial \xi'}{\partial \xi} \right| \tag{4.33}$$

where Det $|\partial \xi' / \partial \xi|$ stands for the Jacobian of the transformation from

(ξ_1, ξ_2, ξ_3) to (ξ'_1, ξ'_2, ξ_3). Thus, (4.32) reduces to

$$\rho(\xi_1, \xi_2, \xi_3) = \rho(\xi'_1, \xi'_2, \xi'_3)\,\mathrm{Det}\,|\partial\xi'/\partial\xi|.$$

Consequently

$$\rho(\xi_1, \xi_2, \xi_3)\mathrm{d}\xi_1\,\mathrm{d}\xi_2\,\mathrm{d}\xi_3 = \rho(\xi'_1, \xi'_2, \xi'_3)\mathrm{d}\xi'_1\,\mathrm{d}\xi'_2\,\mathrm{d}\xi'_3. \tag{4.34}$$

Therefore we see that the choice (4.31) satisfies the requirement (4.30). When (4.31) is inserted, (4.30) gives the group measure.

Our last step is to show that the surface integral (4.27) for Q reduces just to an integral over the group measure. We already notice some resemblance between (4.31) and (4.27). Note however that it is the derivatives $\partial U/\partial\xi_i$ which occur in (4.31) whereas in (4.27), the derivatives $\partial U/\partial x_\mu$ occur. To forestall confusion, let us recapitulate that the three parameters ξ_i describe the group space. The 2×2 matrices U vary as the ξ_i vary, and the corresponding derivatives occur in the measure (4.31). By contrast, (4.27) deals with a given field configuration, whose boundary conditions (4.15) introduce a gauge function U, defined on $S_3^{(\mathrm{phy})}$. Here, the group matrices, and therefore also the corresponding group parameters ξ_i, are themselves functions of the spatial coordinates x_μ of the points on $S_3^{(\mathrm{phy})}$. Therefore, in (4.27), the matrices U may be considered as functions of ξ_i, which in turn are functions of x_μ. Keeping this in mind, we can write (4.27) as

$$Q = -\frac{1}{24\pi^2} \oint_{S_3^{(\mathrm{phy})}} \mathrm{d}\sigma_\mu \left[\varepsilon_{\mu\nu\alpha\beta} \mathrm{Tr}\left(U^{-1}\frac{\partial U}{\partial \xi_i} U^{-1}\frac{\partial U}{\partial \xi_j} U^{-1}\frac{\partial U}{\partial \xi_k} \right) \frac{\partial \xi_i}{\partial x_\nu}\frac{\partial \xi_j}{\partial x_\alpha}\frac{\partial \xi_k}{\partial x_\beta} \right]. \tag{4.35}$$

Since (4.35) is in terms of cartesian coordinates x_μ and cartesian tensors, it will be helpful to distort the hypersphere $S_3^{(\mathrm{phy})}$ at infinity into the surface of an infinite hypercube. The eight sides of this cube are the surfaces $x_\mu = \pm\infty$, for $\mu = 1, 2, 3$ and 4. Note that Gauss's theorem used in (4.25) is unaffected by this smooth distortion, as are the homotopy arguments. Take any one of the surfaces of this hypercube, say, the one at $x_4 = -\infty$. The contribution to the surface integral (4.35) can be written as

$$-\frac{1}{24\pi^2} \int \mathrm{d}x_1\,\mathrm{d}x_2\,\mathrm{d}x_3\,\varepsilon_{mnl}\,\mathrm{Tr}\left(U^{-1}\frac{\partial U}{\partial \xi_i} U^{-1}\frac{\partial U}{\partial \xi_j} U^{-1}\frac{\partial U}{\partial \xi_k} \right) \frac{\partial \xi_i}{\partial x_m}\frac{\partial \xi_j}{\partial x_n}\frac{\partial \xi_k}{\partial x_l}, \tag{4.36}$$

where m, n, l range from 1 to 3. But

$$\varepsilon_{mnl}\frac{\partial\xi_i}{\partial x_m}\frac{\partial\xi_j}{\partial x_n}\frac{\partial\xi_k}{\partial x_l}\mathrm{d}x_1\,\mathrm{d}x_2\,\mathrm{d}x_3=\varepsilon_{ijk}\,\mathrm{d}\xi_1\,\mathrm{d}\xi_2\,\mathrm{d}\xi_3. \tag{4.37}$$

Hence the contribution (4.36) from the surface $x_4 = -\infty$ is

$$\frac{-1}{24\pi^2}\int \mathrm{d}\xi_1\,\mathrm{d}\xi_2\,\mathrm{d}\xi_3\,\varepsilon_{ijk}\,\mathrm{Tr}\left(U^{-1}\frac{\partial U}{\partial\xi_i}U^{-1}\frac{\partial U}{\partial\xi_j}U^{-1}\frac{\partial U}{\partial\xi_k}\right), \tag{4.38}$$

which is nothing but an integral over the group measure, except possibly for an overall constant which we will deal with later. A similar contribution comes from all eight surfaces of the hypercube. Hence their sum (4.35) can also be written as

$$\begin{aligned} Q &\propto \int \mathrm{d}\xi_1\,\mathrm{d}\xi_2\,\mathrm{d}\xi_3\,\rho(\xi_1,\xi_2,\xi_3)\\ &\propto \int \mathrm{d}\mu(U). \end{aligned} \tag{4.39}$$

Therefore, Q, originally written in (4.18) as a functional of the field variables, reduces to just an integral over group space of the group measure. That is, Q is proportional, roughly speaking, to the volume of group space spanned by the gauge function U as it varies on $S_3^{(\mathrm{phy})}$. The input from homotopy theory, stated earlier, is that when we integrate over $S_3^{(\mathrm{phy})}$ once, the group space $S_3^{(\mathrm{int})}$ may be spanned an integral number of times. Thus, Q will be proportional to this integer. Finally, the constant $(-16\pi^2)^{-1}$ in (4.18) has been so arranged that Q will in fact equal this integer. This can be checked by the prototype example

$$U_1(x) = (x_4 + \mathrm{i}x_j\sigma_j)/|x| = \textstyle\sum_\mu \hat{x}_\mu s_\mu. \tag{4.40}$$

On comparing it with the general representation (4.16), we see that this gauge function corresponds to $a_\mu = \hat{x}_\mu$. That is, every point on $S_3^{(\mathrm{phy})}$ is mapped on to the 'corresponding' point, at the same polar angles, on $S_3^{(\mathrm{int})}$. Thus, the homotopy index Q must be equal to unity. On inserting (4.40) into (4.27), a little algebra yields

$$Q = \oint_{S_3^{(\mathrm{phy})}} j_\mu \mathrm{d}\sigma_\mu = -\frac{1}{24\pi^2}\int \mathrm{d}\sigma_\mu\left(\frac{-12x_\mu}{|x|^4}\right). \tag{4.41}$$

Using $\mathrm{d}\sigma_\mu = \mathrm{d}\Omega\, x_\mu |x|^2$, where $\mathrm{d}\Omega$ is the solid angle element in Euclidean four-space, gives

$$Q = \frac{1}{2\pi^2}\int d\Omega\, x_\mu |x|^2 \frac{x_\mu}{|x|^4}$$
$$= \frac{1}{2\pi^2}\int d\Omega = 1, \tag{4.42}$$

since the total solid angle $\int d\Omega$ in four-space is just $2\pi^2$. (Notice for later use that this surface integral equals unity not only as $|x| \to \infty$, but for any radius $|x|$.)

This ensures that, with the factor $-1/16\pi^2$ in (4.18), Q measures precisely the winding number, and completes our sketch of the proof that (4.18) gives the homotopy index.

Notice that $\mathrm{Tr}[G_{\mu\nu}\tilde{G}_{\mu\nu}]$ is gauge invariant (see 4.8b), and therefore so is Q. But the parent boundary condition (4.13) is not. Under a gauge transformation $U'(x)$, this boundary condition changes, as per (4.8a), from $A_\mu \to U\partial_\mu(U^{-1})$ to

$$A'_\mu \to U'(U\partial_\mu U^{-1})(U')^{-1} + U'\partial_\mu(U')^{-1}$$
$$= (U'U)\partial_\mu(U'U)^{-1}. \tag{4.43}$$

One might wonder whether this variation is consistent with the gauge invariance of Q. In particular, can we not choose $U' = U^{-1}$, so that $A_\mu = 0$ on $S_3^{(\mathrm{phy})}$? If this were always possible, Q could be reduced to zero (see (4.26)) by a mere gauge transformation, regardless of its original value and in violation of its gauge invariance. That we cannot choose $U' = U^{-1}$ except when U falls in the $Q = 0$ sector has been argued nicely by Coleman (1979) in his lecture notes. The central point, which we mentioned in the vicinity of eq. (4.15), is that the function U occurring in the boundary condition need be defined non-singularly only on the spherical surface $S_3^{(\mathrm{phy})}$. It is some function $U(\alpha_1, \alpha_2, \alpha_3)$ of the parameters of that surface. It need not be the boundary value of some non-singular gauge function defined in all of four-space. By contrast the function $U'(x)$ in (4.43) transforms the field throughout space, and must be non-singular at all x. Now, let us write $U'(x)$ as $U'(|x|, \alpha_1, \alpha_2, \alpha_3)$ in terms of polar variables. As $|x|$ is varied, U' must be a continuous function on a nested family of hyperspheres. That is, at any fixed $|x|$, U' is a function of α_1 α_2 and α_3. As $|x|$ is varied, these functions must deform continuously. At $|x| = 0$, $U'(0, \alpha_1, \alpha_2, \alpha_3)$ must be independent of the α_i in order to be non-singular. That is, it must be a constant matrix. Any constant SU(2) matrix can be continuously obtained from the identity matrix. Putting all this together, $U'(|x|, \alpha_1, \alpha_2, \alpha_3)$ must be continuously deformable to the identity matrix, for all $|x|$. Letting $|x| \to \infty$, the boundary value of U' on $S_3^{(\text{phy})}$ must also be continuously deformable to the identity element, i.e. it must lie in the $Q = 0$ sector. In that case, U' cannot equal U^{-1} on $S_3^{(\text{phy})}$ unless U also belonged to $Q = 0$. Conversely, when U belongs to $Q \neq 0$, it cannot be continued all the way down to $|x| = 0$ without encountering a singularity. For example the function U_1 in (4.40), although well behaved on $S_3^{(\text{phy})}$, is singular at $|x| = 0$.

More precisely, homotopy theory tells us that if U belongs to the index Q_1 and U' to Q_2 then the function $U'U$ belongs to $Q_1 + Q_2$. We have seen that if $U'(x)$ is to be a legitimate gauge transformation, its boundary value on $S_3^{(\text{phy})}$ must correspond to $Q_2 = 0$. Thus, $Q_1 + Q_2 = Q_1$ and the homotopy index will be unaffected by gauge transformations, consistent with the gauge-invariant expression (4.18).

As with the monopole system, much more can be said about the homotopy properties of the Euclidean Yang–Mills system, but we will stop at this stage. Note that our classification pertains to all finite-action configurations, whether or not they are solutions of the field equation (4.11). Nowhere has this equation been used in the above discussion. Our next task is to look, among finite-action configurations in any given homotopy sector Q, for some configurations that actually solve the field equation. The exact $Q = +1$ (or -1) solution we obtain will be called the instanton (or anti-instanton). We will also get some exact solutions with $Q = N$ (or $-N$), with $|N| > 1$. These are the multi-instanton (or anti-instanton) solutions.

4.3. The Yang–Mills instantons

The first step in obtaining the exact instanton solutions to the Euclidean Yang–Mills equation (4.11) is an observation due to Belavin et al. (1975) on the significance of self-dual and self-antidual configurations. We used a similar argument, again due to Belavin and Polyakov (1975), in the simpler context of the O(3) model. We begin with the trivial identity

$$-\int d^4x \, \mathrm{Tr}[(G_{\mu\nu} \pm \tilde{G}_{\mu\nu})^2] \geqslant 0. \tag{4.44}$$

(Note that $G_{\mu\nu}$ in (4.6) is anti-hermitian for real fields $G^a_{\mu\nu}$.) Using $\mathrm{Tr}[G_{\mu\nu} G_{\mu\nu}] = \mathrm{Tr}[\tilde{G}_{\mu\nu} \tilde{G}_{\mu\nu}]$, this gives

$$-\int d^4x \, \mathrm{Tr}[G_{\mu\nu} G_{\mu\nu}] \geqslant \mp \int d^4x \, \mathrm{Tr}[\tilde{G}_{\mu\nu} G_{\mu\nu}]$$

or

$$S \geqslant (8\pi^2/g^2)|Q|. \tag{4.45}$$

where S is the Euclidean action (4.10) and Q the Pontryagin homotopy index (4.18). Equation (4.45) again holds for any finite-action configuration. Now, the Yang–Mills equation (4.11), like any field equation, is obtained by extremising the action S through the variational principle

$$\delta S[A_\mu] = 0. \tag{4.46}$$

The field equation (4.11) is just an explicit form of (4.46). Such a variation can clearly be done in each homotopy sector separately since small variations will keep the fields within the same sector. Fields which extremise the action in any given homotopy sector are solutions of (4.11) falling in that sector. From eq. (4.44)–(4.45) we see that the absolute minimum value of $S(S = (8\pi^2/g^2)|Q|)$ is attained in any given sector Q when

$$\tilde{G}_{\mu\nu} = \tfrac{1}{2}\,\varepsilon_{\mu\nu\rho\sigma}\,G_{\rho\sigma} = \pm G_{\mu\nu}. \tag{4.47}$$

Thus, self-dual and self-antidual configurations extremise S, and hence solve (4.11). Of course, the absolute minima of S need not be its only extrema. Therefore our derivation does not prove that all solutions of (4.11) are self-dual or self-antidual, but only the converse. We will limit ourselves to self-dual and self-antidual solutions, in which case we need to solve only (4.47) rather than the (possibly) more general (4.11). That fields satisfying (4.47) also satisfy (4.11) can be seen more directly by recalling the identity proved earlier (4.20),

$$D_\mu\,\tilde{G}_{\mu\nu} = 0. \tag{4.48}$$

This is satisfied by any $G_{\mu\nu}$ of the form (4.7), but when $\tilde{G}_{\mu\nu} = \pm G_{\mu\nu}$, (4.48) reduces to the field equation. The alternate proof using (4.45) brings out the added fact that such solutions in fact give the absolute minimum of the action S in any given Q-sector. Notice the similarity between the derivation of (4.47) and of the corresponding condition (3.32) in the O(3) model.

Before proceeding to solve (4.47), it is worth emphasising the role of the Euclidean metric in this context. Note that the dual of the dual tensor is

$$\begin{aligned}\tilde{\tilde{G}}_{\mu\nu} &\equiv \tfrac{1}{2}\,\varepsilon_{\mu\nu\rho\sigma}\,\tilde{G}_{\rho\sigma} \\ &= \tfrac{1}{4}\,\varepsilon_{\mu\nu\rho\sigma}\,\varepsilon_{\rho\sigma\alpha\beta}\,G_{\alpha\beta}.\end{aligned} \tag{4.49}$$

In a Euclidean metric, we have

$$\varepsilon_{\mu\nu\rho\sigma}\,\varepsilon_{\rho\sigma\alpha\beta} = 2[\delta_{\mu\alpha}\delta_{\nu\beta} - \delta_{\mu\beta}\delta_{\nu\alpha}]$$

so that

$$\tilde{\tilde{G}}_{\mu\nu} = G_{\mu\nu}. \tag{4.50}$$

Formally (4.49)–(4.50) may be written symbolically as

$$\tilde{\tilde{G}} = \varepsilon^2 G = G \tag{4.51}$$

so that the eigenvalues of the dualising operator ε are ± 1. Hence self-dual and self-antidual configurations can exist. By contrast, if the metric were Minkowskian, (4.49) would be replaced by

$$\tilde{\tilde{G}}_{\mu\nu} = \tfrac{1}{4}\,\varepsilon_{\mu\nu}{}^{\rho\sigma}\,\varepsilon_{\rho\sigma}{}^{\alpha\beta}\,G_{\alpha\beta} = -G_{\alpha\beta}. \tag{4.52}$$

The eigenvalues of the dualising operator will now be $\pm$ i, and no self-dual or self-antidual configurations can exist.

Let us now look for some solutions of the self-duality (or self-antiduality) condition (4.47). Belavin et al. offered a simple derivation of the one-instanton solution of (4.47). We will however follow a method evolved subsequently in stages by several authors ('tHooft 1976a, Wilczeck 1977, Corrigan and Fairlie 1977, Jackiw et al. 1977, 1978). This method yields in one stroke the one- and many-instanton solutions. Multi-instanton solutions were also obtained in a different way by Witten (1977). We will use the following ansatz for the gauge field:

$$A_\mu(x) = \mathrm{i}\,\bar{\Sigma}_{\mu\nu}\,\partial_\nu(\ln\phi(x)) \tag{4.53}$$

where $\phi(x)$ is a scalar function to be obtained, and $\bar{\Sigma}_{\mu\nu}$ are the components of a matrix built from Pauli matrices as follows:

$$\bar{\Sigma}_{\mu\nu} = \frac{1}{2}\begin{bmatrix} 0 & +\sigma_3 & -\sigma_2 & -\sigma_1 \\ -\sigma_3 & 0 & +\sigma_1 & -\sigma_2 \\ +\sigma_2 & -\sigma_1 & 0 & -\sigma_3 \\ +\sigma_1 & +\sigma_2 & +\sigma_3 & 0 \end{bmatrix}. \tag{4.54}$$

This can be written compactly in the form ('tHooft 1976a):

$$\bar{\Sigma}_{\mu\nu} = \bar{\eta}^{i\mu\nu}\,\sigma^i/2; \qquad i = 1, 2, 3 \tag{4.55}$$

where

$$\bar{\eta}^{i\mu\nu} = -\bar{\eta}^{i\nu\mu} = \begin{cases} \varepsilon^{i\mu\nu} & \text{for } \mu, \nu = 1, 2, 3 \\ -\delta^{i\mu} & \text{for } \nu = 4. \end{cases} \tag{4.56}$$

Notice that $\bar{\Sigma}_{\mu\nu}$ is both antisymmetric and self-antidual in its indices. The following easily verifiable properties of $\bar{\Sigma}_{\mu\nu}$ will be of relevance to us:

$$[\bar{\Sigma}_{\mu\sigma}, \bar{\Sigma}_{\nu\rho}] = \mathrm{i}[\delta_{\mu\nu}\bar{\Sigma}_{\sigma\rho} + \delta_{\rho\sigma}\bar{\Sigma}_{\mu\nu} - \delta_{\mu\rho}\bar{\Sigma}_{\sigma\nu} - \delta_{\nu\sigma}\bar{\Sigma}_{\mu\rho}] \tag{4.57a}$$

$$\varepsilon_{\mu\nu\alpha\beta}\,\bar{\Sigma}_{\beta\sigma} = [\delta_{\mu\sigma}\,\bar{\Sigma}_{\nu\alpha} + \delta_{\nu\sigma}\,\bar{\Sigma}_{\alpha\mu} + \delta_{\alpha\sigma}\,\bar{\Sigma}_{\mu\nu}] \tag{4.57b}$$

and

$$\tfrac{1}{2}\,\varepsilon_{\mu\nu\alpha\beta}\,\bar{\Sigma}_{\alpha\beta} = -\bar{\Sigma}_{\mu\nu}. \tag{4.57c}$$

Using these identities on the ansatz (4.53) for A_μ, we get for the field tensor:

$$\begin{aligned} G_{\mu\nu} &= \mathrm{i}\,\bar{\Sigma}_{\nu\sigma}(\partial_\mu\partial_\sigma \ln\phi - \partial_\mu(\ln\phi)\,\partial_\sigma(\ln\phi)) \\ &\quad -\mathrm{i}\,\bar{\Sigma}_{\mu\sigma}(\partial_\nu\partial_\sigma \ln\phi - \partial_\nu(\ln\phi)\,\partial_\sigma(\ln\phi)) \\ &\quad -\mathrm{i}\,\bar{\Sigma}_{\mu\nu}\,(\partial_\sigma \ln\phi)^2. \end{aligned} \tag{4.58}$$

Taking its dual, we get

$$\begin{aligned} \tilde{G}_{\mu\nu} &= \mathrm{i}\,\varepsilon_{\mu\nu\alpha\beta}\,[\bar{\Sigma}_{\beta\sigma}(\partial_\alpha\partial_\sigma \ln\phi - \partial_\alpha\,(\ln\phi)\partial_\sigma\,(\ln\phi)) \\ &\quad -\tfrac{1}{2}\bar{\Sigma}_{\alpha\beta}\,(\partial_\sigma \ln\phi)^2] \\ &= \mathrm{i}\,\bar{\Sigma}_{\nu\alpha}(\,\partial_\alpha\partial_\mu \ln\phi - \partial_\alpha\,(\ln\phi)\,\partial_\mu(\ln\phi)) - (\mu \leftrightarrow \nu) \\ &\quad +\mathrm{i}\,\bar{\Sigma}_{\mu\nu}\partial_\sigma\partial_\sigma\,(\ln\phi). \end{aligned} \tag{4.59}$$

Now, we demand self-duality, $\tilde{G}_{\mu\nu} = G_{\mu\nu}$. This yields

$$\begin{aligned} &\partial_\mu(\partial_\sigma \ln\phi) - (\partial_\mu \ln\phi)\,(\partial_\sigma \ln\phi) \\ &\quad = \partial_\sigma(\partial_\mu \ln\phi) - (\partial_\sigma \ln\phi)\,(\partial_\mu \ln\phi) \end{aligned} \tag{4.60a}$$

and

$$\partial_\sigma\partial_\sigma\,(\ln\phi) + (\partial_\sigma \ln\phi)^2 = 0. \tag{4.60b}$$

Equation (4.60a) is automatically satisfied, while (4.60b) can be written compactly as

$$\Box\phi/\phi = 0 \tag{4.61}$$

where $\Box\phi$ in Euclidean metric is $\partial_\sigma\partial_\sigma\phi$. The remaining task is to solve (4.61) for $\phi(x)$ and insert it in (4.53). When ϕ is non-singular, (4.61) reduces to $\Box\phi = 0$, which permits only the trivial solution $\phi =$ constant, leading to $A_\mu = 0$; but when singular $\phi(x)$ are considered we get interesting and ultimately non-singular solutions for the gauge field. For example, consider $\phi(x) = 1/|\,x\,|^2$. At $x \neq 0$, this satisfies

$$\frac{1}{\phi}\Box\phi = \frac{1}{\phi}\,\partial_\sigma\!\left(-\frac{2x_\sigma}{|\,x\,|^4}\right) = 0. \tag{4.62}$$

At $x = 0$, $\Box(1/|x|^2) = -4\pi^2\,\delta^4(x)$, consistent with Gauss's theorem:

$$\int d^4x\,\Box(1/|x|^2) = \oint d\sigma_\mu\,\partial_\mu(1/|x|^2)$$

$$= \int d\Omega \lim_{x\to\infty}\left[|x|^3\frac{\partial}{\partial|x|}\left(\frac{1}{|x|^2}\right)\right] = -4\pi^2. \tag{4.63}$$

Thus, even at $x = 0$,

$$(1/\phi)\Box\phi = -4\pi^2\,|x|^2\,\delta^4(x) = 0. \tag{4.64}$$

Therefore $\phi(x) = 1/|x|^2$ solves (4.61) and so does the more general form

$$\phi(x) = 1 + \sum_{i=1}^{N}\frac{\lambda_i^2}{|x_\mu - a_{i\mu}|^2}, \tag{4.65}$$

where $a_{i\mu}$ and λ_i are any real constants.† These solutions when inserted in (4.53) will yield, after some gauge transformation, the N-instanton solutions with homotopy index $Q = N$. Let us begin with $N = 0$, which corresponds to $\phi(x) = 1$ and $A_\mu = 0$. This is the trivial solution and belongs to the $Q = 0$ sector (see either (4.18) or (4.26)).

4.3.1. *The instanton*

Next consider $N = 1$ in (4.65). Using $y_\mu \equiv (x_\mu - a_{1\mu})$ and $y^2 \equiv y_\mu y_\mu$, we have

$$\phi(x) = 1 + \lambda_1^2/y^2, \tag{4.66}$$

leading to

$$A_\mu(x) = -2i\lambda_1^2\,\Sigma_{\mu\nu}\,\frac{y_\nu}{y^2(y^2+\lambda_1^2)}. \tag{4.67}$$

We notice that this solution is singular at $y = 0$. However, the singularity can be removed by a correspondingly singular gauge transformation. Since the self-duality condition and the equation of motion are gauge covariant,

---

† Jackiw et al. (1977) write $\phi(x)$ in the form

$$\sum_{i=1}^{N+1}\frac{\lambda_i^2}{|x-a_i|^2},$$

which is more general and reduces to (4.65) when $a_{N+1}\to\infty$, $\lambda_{N+1}\to\infty$ with $\lambda_{N+1}^2/a_{N+1}^2 = 1$. This form has the advantage that it is conformally covariant. We will however use (4.65), attributed to 'tHooft (private communication).

the resulting function will continue to satisfy them. The required gauge function is precisely $U_1(y)$ where U_1 is given in (4.40). This is because

$$[U_1(y)]^{-1}\frac{\partial}{\partial y_\mu}[U_1(y)] = -2\mathrm{i}\,\bar{\Sigma}_{\mu\nu}(y_\nu/y^2) \tag{4.68}$$

as can be verified by substituting (4.40). Thus (4.67) can be written as

$$A_\mu(x) = [U_1(y)]^{-1}\,\partial_\mu[U_1(y)]\,\frac{\lambda_1^2}{y^2+\lambda_1^2}. \tag{4.69}$$

Upon gauge transforming with $U_1(y)$, we get

$$\begin{aligned} A_\mu(x) &\to A'_\mu(x) \\ &= [U_1(y)][A_\mu+\partial_\mu][U_1(y)]^{-1} \\ &= (\partial_\mu U_1)U_1^{-1}\left(\frac{\lambda_1^2}{y^2+\lambda_1^2}-1\right), \qquad \text{using (4.9)}, \\ &= -(\partial_\mu U_1)U_1^{-1}\frac{y^2}{y^2+\lambda_1^2} = U_1(\partial_\mu U_1^{-1})\frac{y^2}{y^2+\lambda_1^2}. \end{aligned} \tag{4.70}$$

To write the space dependence more explicitly, one can use a tensor $\Sigma_{\mu\nu}$, which is the self-dual analogue of $\bar{\Sigma}_{\mu\nu}$. It is defined by

$$\Sigma_{\mu\nu} = \eta^{i\mu\nu}\sigma_i/2 \tag{4.71}$$

$$\text{with } \eta^{i\mu\nu} = -\eta^{i\nu\mu} = \begin{cases} \varepsilon^{i\mu\nu} & \text{for } \mu,\nu = 1,2,3 \\ +\delta^{i\mu} & \text{for } \nu = 4. \end{cases} \tag{4.72}$$

It satisfies, in analogy to (4.68), the relation

$$[U_1(y)]\frac{\partial}{\partial y_\mu}[U_1(y)]^{-1} = -2\mathrm{i}\,\Sigma_{\mu\nu}\frac{y_\nu}{y^2}. \tag{4.73}$$

In terms of this $\Sigma_{\mu\nu}$, the gauge transformed solution (4.70) becomes

$$A'_\mu(x) = -2\mathrm{i}\,\Sigma_{\mu\nu}\frac{y_\nu}{y^2+\lambda_1^2} = -2\mathrm{i}\,\Sigma_{\mu\nu}\frac{(x-a_1)_\nu}{|(x-a_1)|^2+\lambda_1^2}. \tag{4.74}$$

This is the one-instanton solution. Although we adopted the device of first obtaining a singular solution (4.67) and then gauge transforming it with a singular gauge function, the final solution (4.74) has no singularities at any x for any given $\lambda_1 \neq 0$. It has the following features:

(i) When substituted into (4.7) it leads to a field tensor

$$G'_{\mu\nu} = 4\mathrm{i}\,\Sigma_{\mu\nu}\frac{\lambda_1^2}{[|x-a_1|^2+\lambda_1^2]^2}. \tag{4.75}$$

Since $\Sigma_{\mu\nu}$ is self-dual, so is $G'_{\mu\nu}$. This verifies that (4.74) is a solution of the Euclidean Yang–Mills equation.

(ii) As $x \to \infty$, the solution behaves as a 'pure gauge',

$$A'_\mu(x) \to U_1(x)\partial_\mu[U_1(x)]^{-1}, \tag{4.76}$$

consistent with the boundary condition required by finiteness of action. In fact, the action of this solution is

$$\begin{aligned} S &= -\frac{1}{2g^2}\int \mathrm{d}^4x\,\mathrm{Tr}[G'_{\mu\nu}G'_{\mu\nu}] = \frac{48\lambda_1^4}{g^2}\int\frac{\mathrm{d}^4y}{(y^2+\lambda_1^2)^4} \\ &= +8\pi^2/g^2 \end{aligned} \tag{4.77}$$

(iii) Since $G'_{\mu\nu} = \tilde{G}'_{\mu\nu}$ for this solution, $S = (8\pi^2/g^2)Q$ and hence $Q = +1$. This again is consistent with our earlier result (4.42) that $U_1(x)$ belongs to the $Q = +1$ sector.

(iv) An analogous but self-antidual solution can be derived in exactly the same way, by replacing $\bar{\Sigma}_{\mu\nu}$ by $\Sigma_{\mu\nu}$ in the ansatz (4.53). This leads to the replacement of $\Sigma_{\mu\nu}$ by $\bar{\Sigma}_{\mu\nu}$ in the final result and gives the solution

$$\begin{aligned} A'_\mu(x) &= -2\mathrm{i}\,\bar{\Sigma}_{\mu\nu}\frac{(x-a_1)_\nu}{|(x-a_1)|^2+\lambda_1^2} \\ &= U_1^{-1}\partial_\mu(U_1)\frac{|(x-a_1)|^2}{|(x-a_1)|^2+\lambda_1^2}. \end{aligned} \tag{4.78}$$

This solution is self-antidual, has $Q = -1$ and may be called the anti-instanton.

(v) It is a matter of convention as to which of the two solutions (4.74) and (4.78) we call the instanton and which the anti-instanton. Less trivially, it is also a matter of convention as to whether the self-dual solution (4.74) is attributed a $Q = +1$ or the self-antidual solution (4.78). By our definition (4.18), the self-dual solution has $Q = +1$. But we could just as easily have defined Q in (4.18) with an extra minus sign. Recall that Q is essentially a winding number, and the arbitrariness in the sign of Q is related to which of the two senses of winding is deemed positive.

(vi) The four-vector $a_{1\mu}$ in the solution represents the 'location' of the instanton. The action density in (4.77) is localised around $y_\mu = (x-a_1)_\mu$

$= 0$. The point a_1 can be chosen arbitrarily because of the translational invariance of the Yang–Mills equation. Similarly the constant λ_1, which represents the 'size' of the instanton in terms of action density, can be arbitrary as long as it is non-zero. This freedom is related to the scale invariance of the Yang–Mills system, under the transformation $x_\mu \to \lambda x_\mu$ and $A_\mu \to \lambda A_\mu$ for any $\lambda \neq 0$. More explicitly, if $A_\mu(x)$ is a solution of the Yang–Mills equation, so is $\lambda A_\mu(\lambda x)$. It may be seen that solutions with different λ_1 in (4.74) are related to one another by such scaling. We will return to such degrees of freedom later. Recall that solutions of the O(3) model had similar degrees of freedom.

(vii) The one-instanton solution is 'essentially' spherically symmetric about the point a_1. Apart from the tensor $\Sigma_{\mu\nu}(x - a_1)_\nu$ it depends only on $(x - a_1)^2$. The gauge function U_1 corresponding to its behaviour at $x \to \infty$ is also spherically symmetric. The direction in group space $S_3^{(\text{int})}$ of U_1 is the same as that of radius vector x_μ in coordinate space. In this sense, the instanton resembles the $n = 1$ monopole solution discussed earlier, where in three space dimensions, the isovector ϕ^a had the same direction as the coordinate vector $\boldsymbol{x}$, as $\boldsymbol{x} \to \infty$.

(viii) The untransformed singular solution (4.67)–(4.69), even though singular at $x = a_1$, behaves as a 'pure gauge' near that point, with a gauge function U_1^{-1}.

$$A_\mu(x) \to U_1^{-1}(x - a_1)\ \partial_\mu U_1(x - a_1) \qquad \text{as } x \to a_1.$$

That is why, by the appropriate inverse gauge transformation U_1, we were able to get rid of the singularity. The resulting $A'_\mu(x)$ in (4.74), in fact vanishes at $x = a_1$.

(ix) The name 'instantons' was coined by 'tHooft (1976a) because, unlike solitons of Minkowskian systems which are not localised in time, these solutions are localised in x_4 (the imaginary time coordinate) as well.

4.3.2. *N-instanton solutions*

These are obtained by inserting the function (4.65) for arbitrary N into the ansatz (4.53). The corresponding gauge field is

$$\begin{aligned} A_\mu(x) &= i\,\bar{\Sigma}_{\mu\nu}\,\partial_\nu\left[\ln\left(1 + \sum_{i=1}^{N} \frac{\lambda_i^2}{|y_i|^2}\right)\right] \\ &= -2i\bar{\Sigma}_{\mu\nu}\left(\sum_i \frac{\lambda_i^2\, y_{i\nu}}{|y_i|^4}\right)\Big/\left(1 + \sum_j \frac{\lambda_j^2}{|y_j|^2}\right), \end{aligned} \tag{4.79}$$

where $(y_i)_\mu \equiv (x-a_i)_\mu$; $i, j = 1, 2, \ldots, N$.

Using the identity (4.68) for each y_i, this can be cast in the form

$$A_\mu(x) = \sum_{i=1}^{N} U_1^{-1}(y_i)\partial_\mu U_1(y_i) f_i(x) \tag{4.80}$$

where

$$f_i(x) = \left(\frac{\lambda_i^2}{(x-a_i)^2}\right) \Bigg/ \left(1 + \sum_j \frac{\lambda_j^2}{(x-a_j)^2}\right) \tag{4.81}$$

and $U_1(y_i)$ is the same prototype $Q = 1$ gauge function (4.40) with $y_i = (x-a_i)$ as the argument. Note that

$$\partial_\mu \equiv \frac{\partial}{\partial x_\mu} = \frac{\partial}{\partial(x-a_i)_\mu}.$$

The field (4.79) or (4.80) is formally a solution of the Yang–Mills equation since it has been derived from the self-duality condition. But, like (4.67), it is singular. It has N singularities, at the points $x = a_i$; $i = 1, \ldots, N$. Since

$$f_j(x) \to \delta_{ij} \qquad \text{as } x \to a_i, \tag{4.82}$$

$$A_\mu(x) \to U_1^{-1}(x-a_i)\,\partial_\mu[U_1(x-a_i)] \qquad \text{as } x \to a_i. \tag{4.83}$$

Thus, near any one of its singularities the N-instanton solution behaves like a pure gauge, just as did the one-instanton solution (4.69) near its centre. In fact, the gauge function U_1^{-1} is the same as in the one-instanton case. Therefore, the N-instanton solution can be thought of as a collection of N individual single instantons, whose shapes at any arbitrary x are 'polarised' due to the presence of the remaining instantons. Near the core of the ith instanton ($x \simeq a_i$), the effect of the other instantons is negligible, and the N-instanton solution is dominated by that single instanton.

The singularities in (4.79) have to be removed by a judicious gauge transformation (Giambiaggi and Rothe 1977, Sciuto 1979), as we did for the one-instanton case. Equation (4.83) offers the clue that the appropriate gauge function must reduce to $U_1(x-a_i)$ when x approaches any of the a_i. One such function is

$$U_N(x) = (Z_4 + i\boldsymbol{Z}\cdot\boldsymbol{\sigma})/\sqrt{Z^2} \tag{4.84}$$

with

$$Z_\mu(x) = \sum_{i=1}^{N} \beta_i \left(\frac{y_{i\mu}}{y_i^2}\right); \qquad y_i = x - a_i \tag{4.85}$$

where β_i are some real constants. As $y_i \to 0$ for some i,

$$Z_\mu \to \beta_i y_{i\mu}/y_i^2; \qquad Z^2 \to \beta_i^2/y_i^2$$

and

$$U_N(x) \to \beta_i \left(\frac{y_{i4} + i\boldsymbol{\sigma}\cdot\mathbf{y}_i}{y_i^2} \right) \frac{y_i}{\beta_i} = U_1(y_i) \tag{4.86}$$

Gauge transformation of $A_\mu(x)$ by $U_N(x)$ will remove all the singularities at $x = a_i$. We have

$$A'_\mu(x) = U_N(x)(A_\mu + \partial_\mu)U_N^{-1}(x) \tag{4.87}$$

and

$$\begin{aligned} A'_\mu(x) \xrightarrow[x \to a_i]{} & U_1(y_i)[U_1^{-1}(y_i)\partial_\mu U_1(y_i)]U_1^{-1}(y_i) \\ & + U_1(y_i)\partial_\mu U_1^{-1}(y_i) + \text{possible finite terms} \\ & = \partial_\mu[U_1(y_i)U_1^{-1}(y_i)] + \text{finite terms,} \end{aligned} \tag{4.88}$$

upon using (4.83) and (4.86). The first term vanishes, and the singularity at $x = a_i$ is removed . This holds for each a_i, $i = 1 \ldots N$. However, there is still the danger that new singularities, not present in the original function (4.79), may have crept in because of the gauge function $U_N(x)$. It is evident that $U_N(x)$ has, in addition to the singularities at $y_{i\mu} = 0$, another one at $Z_\mu = 0$. Of course we still have the arbitrary real constants β_i in the definition (4.85) of Z_μ, to play with. It has been shown by a simple but clever exercise (Giambiaggi and Rothe 1977) that if the points a_i all lie on a straight line in Euclidean four-space, then β_i can be chosen such that Z_μ never vanishes. In that case, the gauge transformed field $A'_\mu(x)$ in (4.87) will be non-singular.

To show that this solution does correspond to a Pontryagin index $Q = N$ (apart from the earlier qualitative observation that it is a collection of N single instantons) we proceed as follows. Once $A'_\mu(x)$ is non-singular, we can use without ambiguity the earlier definition

$$Q = -\frac{1}{16\pi^2} \int \mathrm{Tr}\,[G'_{\mu\nu}\tilde{G}'_{\mu\nu}]\,\mathrm{d}^4x \tag{4.89}$$

where $G'_{\mu\nu}$ is the field tensor corresponding to A'_μ. Since the integrand is non-singular, we can exclude from the integration region N small spheres with centres at $x = a_i$ and radii ε each, where $\varepsilon \to 0$. For any finite small ε, the integration is over V_ε which stands for all space minus these N spheres (fig. 8). The point of this exclusion is that in V_ε even the original solution $A_\mu(x)$ in (4.79) is non-singular. Further, in this region,

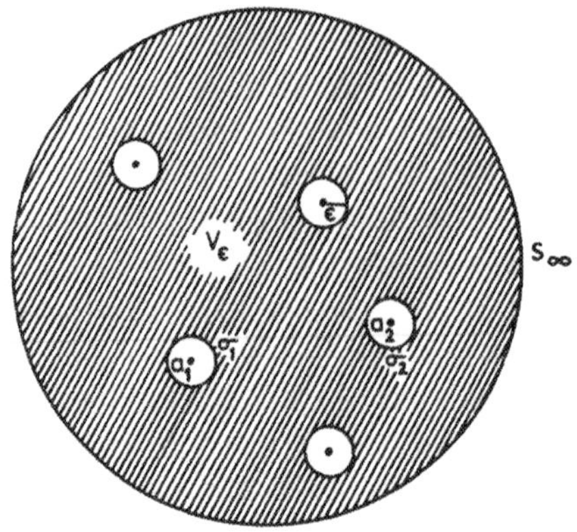

Fig. 8. The volume V_ε, which is all of four-space excluding small spheres of radius ε centred at the points a_i. The total surface of V_ε is the spherical surface at infinity S_∞, plus the n spherical surfaces σ_i around the a_i.

$$\mathrm{Tr}\,[G_{\mu\nu}\tilde{G}_{\mu\nu}] = \mathrm{Tr}\,[G'_{\mu\nu}\tilde{G}'_{\mu\nu}] \tag{4.90}$$

where $G_{\mu\nu}$ corresponds to A_μ, since A_μ and A'_μ are related by gauge transformations. Hence

$$Q = \lim_{\varepsilon\to 0} -\frac{1}{16\pi^2}\int_{V_\varepsilon} \mathrm{Tr}\,[G_{\mu\nu}\tilde{G}_{\mu\nu}]\mathrm{d}^4x. \tag{4.91}$$

Equation (4.91) has the advantage that we need not explicitly evaluate $A'_\mu(x)$ in (4.87) to get Q. The tensor $G_{\mu\nu}$ depends only on the simpler expression (4.79) for A_μ. Further, the singularities of $A_\mu(x)$ are excluded from V_ε, so that we can unambiguously use Gauss's theorem with the help of (4.23). Note that the surfaces which bound V_ε are the spherical surface S_∞ at $|x| = \infty$ and the surfaces σ_i of the N tiny spheres. Thus,

$$Q = \lim_{\varepsilon\to 0}\Big(\oint_{S_\infty} j_\mu \mathrm{d}\sigma_\mu - \oint_{\sigma_1} j_\mu \mathrm{d}\sigma_\mu - \oint_{\sigma_2} j_\mu \mathrm{d}\sigma_\mu \ldots - \oint_{\sigma_N} j_\mu \mathrm{d}\sigma_\mu\Big), \tag{4.92}$$

where j_μ is the current defined in (4.24) and the minus signs for the inner-surface contributions is in the convention that $\mathrm{d}\sigma_\mu$ points radially outwards in each sphere. Now, as $|x| \to \infty$, $A_\mu(x)$ behaves as $1/|x|^3$ and hence j_μ vanishes rapidly enough for $\int_{S_\infty} j_\mu \mathrm{d}\sigma_\mu$ to be zero. On the tiny spheres, where x approaches one of the a_i, A_μ becomes a pure gauge (4.83) with a gauge function $U_1^{-1}(x - a_i)$. When A_μ is a pure gauge, $\oint j_\mu \mathrm{d}\sigma_\mu$ has the form (4.27).

Recall that this surface integral equalled unity when the gauge function was U_1. Note that this result (4.42) holds for a spherical surface of any radius. When the gauge function is U_1^{-1}, as happens here, the corresponding value of $\oint j_\mu \, d\sigma_\mu$ on each tiny sphere is obviously -1. Putting all this together, (4.92) gives

$$Q = \lim_{\varepsilon \to 0} [0-(-1)-(-1) \ldots -(-1)].$$

The limit $\varepsilon \to 0$ can be taken now trivially, yielding $Q = N$.

Since these N-instanton solutions are self-dual, their action $S = 8\pi^2 Q/g^2 = 8\pi^2 N/g^2$, which is N times the action of a single instanton. This is a remarkable property for solutions of a non-linear field equation. Consider the more familiar context of static solitary waves in a Minkowskian metric. A static solution with N solitary waves separated by finite distances, assuming it exists, would be expected generally to have a different energy from N such isolated solitary waves. The difference in energy would be attributed, in physical terms, to the interaction between these solitary waves. For our Euclidean system, the energy concept does not exist, but the action functional S, which is somewhat analogous, seems to have the remarkable property

$$S(N \text{ instantons}) = N\, S(\text{one instanton}).$$

Superficially, this happens because all these solutions are self-dual and as a result $S \propto Q$, but perhaps a deeper understanding of this feature is possible.

This essentially completes our discussion of the Yang–Mills instanton solutions. Anti-self-dual, N-anti-instanton solutions are obtained in the same way, by replacing $\bar{\Sigma}_{\mu\nu}$ by $\Sigma_{\mu\nu}$ in the ansatz (4.53). Gauge theories have been the object of vigorous investigations in recent years, and a great deal more is known about them and about their instanton solutions, even at the classical level. Our presentation has been at best introductory. An important property of instantons not presented here is their behaviour under conformal transformations. This has been discussed in detail by Jackiw and Rebbi (1976b) and Jackiw et al. (1977). We have only touched the surface of this property in pointing out the dilatation invariance of the Yang–Mills system. In fact it possesses symmetry under the full four-dimensional conformal group, which has 15 generators corresponding to (i) scale or dilatation transformations, (ii) the four translations, (iii) the six O(4) rotations, and (iv) the four special conformal transformations

$$x_\mu \to \left(\frac{x_\mu - c_\mu x^2}{1 - 2c_\mu x_\mu + x^2 c^2} \right).$$

All this is in addition to gauge transformations whose effects partially overlap with the action of the conformal group (Jackiw and Rebbi 1976b).

A related important question deals with the number of degrees of freedom of the N-instanton solutions. That is, given a particular N-instanton solution, how many more physically distinct solutions can be obtained from it by exploiting the full symmetries of the system? Or, equivalently, how many free parameters should a general N-instanton solution have? To start with, consider the one-instanton solution. We have already obtained five free parameters, λ and a_μ, corresponding to dilatation and translation symmetry. The effect of the remaining generators of the conformal group (O(4) rotations and special conformal transformations) can, as was shown by Jackiw and Rebbi, be undone by gauge transformations and translations. As far as gauge freedom goes, conventional wisdom based on physical considerations says that local gauge transformations are not genuine symmetries unlike, say, spatial rotations. Their action produces, not a physically distinct system, but the same system in a different (gauge) convention. Correspondingly, one 'fixes' the gauge by some gauge condition and then derives physical consequences. An exception to this is the set of global (i.e. x-independent) gauge transformations. It has been argued (see Coleman 1979) that these are genuine symmetries since they lead to conservation laws etc. Recall for instance familiar cases of isotopic spin, or the global charge (U(1)) symmetry in electrodynamics. For our SU(2) theory, global symmetry then gives three more degrees of freedom, making a total of eight free parameters for the single instanton. Of these, the solution (4.74) reveals only five, because of the limitations of the ansatz (4.53) which in turn is related to the specific gauge function $U_1(x)$ (see (4.68)). The remaining three degrees of global gauge freedom can be generated by replacing $U_1(x)$ by its globally rotated cousins in group space.

If a single instanton has eight parameters, we would expect $8N$ parameters in an N-instanton solution. We have already seen that the N-instanton solution looks like a one-instanton solution near each of the a_i. Equivalently, when the sizes λ_i of the individual instantons are small compared with their separations $\sqrt{|a_i - a_j|^2}$, the N-instanton solution will look like a collection of N (almost) non-overlapping single instantons. This would lead us to expect $8N$ parameters. Since it is a discrete integer, we would expect this number not to change when the instantons are brought closer together with increasing overlap, by smoothly varying the a_i and the λ_i. This count of $8N$ degrees of freedom could be reduced to $8N$-3 if we decide on physical grounds to ignore global transformations as symmetries. Note that, even then, the relative orientations in internal space of the N

instantons would be meaningful, so that the number of parameters would be $8N$-3 and not $5N$. Once again, the specific solution (4.79) reveals only $5N$ of these parameters because of the limitations of the ansatz (4.53). Our arguments are far too qualitative to merit being called a derivation. However, the same number of degrees of freedom for self-dual solutions have been obtained rigorously and elegantly using powerful mathematical techniques (Atiyah and Ward 1977). Readers interested in a deeper appreciation of classical instantons must study the papers cited above, as well as Atiyah et al. (1978) and Christ et al. (1978).

Finally, the solutions we presented were all self-dual or self-antidual. As emphasised earlier, the derivation of (4.47) shows only that such functions also solve the Euclidean Yang–Mills field equation, but not that all solutions of the latter are self-dual or self-antidual. One could imagine such examples as an 'instanton–anti-instanton bound solution', which, if it exists, would have $Q = 0$, but $S > 0$. As far as we know, no such solutions have been found so far, nor on the other hand is there any proof that self-dual and self-antidual fields are the only finite-action solutions of the Euclidean Yang–Mills equations.

It is evident that Yang–Mills instantons are fascinating, even at the classical level, as interesting exact solutions of highly non-trivial, non-linear field equations. But they play an even more important role in quantum gauge theories. That part of the story, which essentially begins with the work of 'tHooft (1976a) will be discussed in the last two chapters.

4.4. Instantons of other models

As we defined it in section 4.1, the term instanton is generic. It may be used for the finite-action Euclidean solutions of any model. Such solutions play the same basic role in the corresponding quantum theory – they lead to vacuum tunnelling and related phenomena. In the preceding section, we discussed instantons of the Yang–Mills system in four dimensions, characterised by a homotopy index. Similar solutions exist for many other models, some more complicated than the Yang–Mills system, and others much simpler.

Perhaps the simplest example deals with a unit-mass particle in a double-well potential $V(q)$. In real time, t, the Lagrangian is

$$L = \tfrac{1}{2}(\mathrm{d}q/\mathrm{d}t)^2 - \tfrac{1}{4}(q^2-1)^2. \tag{4.93}$$

The corresponding 'Euclidean' action, and Euclidean equation of motion are obtained by continuing to $t = -\mathrm{i}\tau$, and setting $S_{\mathrm{Euc}} = -\mathrm{i}S_{\mathrm{Min}}$.

$$S_{\text{Euc}} = \int_{-\infty}^{\infty} \mathrm{d}\tau \, [\tfrac{1}{2}(\mathrm{d}q/\mathrm{d}\tau)^2 + \tfrac{1}{4}(q^2-1)^2], \tag{4.94}$$

leading to

$$\mathrm{d}^2q/\mathrm{d}\tau^2 = \mathrm{d}V/\mathrm{d}q = q^3 - q. \tag{4.95}$$

Notice that this last equation is identical to that obeyed by static solutions of the ϕ^4 Higgs theory in $(1+1)$ dimensions considered in chapter 2 (see eq. (2.27), set $\lambda = m = 1$ and replace ϕ by q and x by τ). The action (4.94) is also just the energy of a static configuration in that ϕ^4 theory. Thus, non-trivial finite-action solutions of (4.95) are the same functions as the finite-energy static solutions of the ϕ^4 theory, namely, the kink and the antikink:

$$q(\tau) = \pm \tanh[(\tau - \tau_0)/\sqrt{2}]. \tag{4.96}$$

These are the instanton and anti-instanton solutions of the double-well problem. They carry, as discussed in chapter 2, a topological index of ± 1 respectively. This index is the analogue of the Pontryagin index Q in (4.18). Based on ideas to be developed in chapter 10, we will see that this solution is related to the quantum amplitude for the particle to tunnel from the potential minimum $q = -1$ to the other minimum $q = +1$.

Similarly, consider the quantum pendulum, i.e. a particle in a potential $V(q) = 1 - \cos q$. The classical Euclidean equation for this system is

$$\mathrm{d}^2q/\mathrm{d}\tau^2 = \sin q. \tag{4.97}$$

This is identical to the static field equation of the two-dimensional sine–Gordon system (i.e. the time-independent case of eq. (2.65)). Similarly, the Euclidean action of the pendulum

$$S = \int_{-\infty}^{\infty} \mathrm{d}\tau [\tfrac{1}{2}(\mathrm{d}q/\mathrm{d}\tau)^2 + 1 - \cos q] \tag{4.98}$$

is the same as the energy of a static sine–Gordon configuration. Consequently the instanton (anti-instanton) solution of (4.97) will be the same function as the static soliton (antisoliton) of the sine–Gordon system. We shall use the pendulum as a simple pedagogical tool in chapter 10 to illustrate the significance of instantons in quantum theory.

Notice that both these particle-mechanics examples may be thought of as one-dimensional (zero-space, one-time) scalar field theories. Their instantons are just the static solitons of the corresponding field theory in two $(1+1)$ dimensions. This pattern may be seen in many other instances.

Consider for example the O(3) model in $(2+1)$ dimensions discussed in chapter 3. We obtained its static soliton solutions with finite energy. Consider the same model, but in $(1+1)$ dimensions. Its Euclidean field equation would obviously be same as the static field equation (3.14) of the $(2+1)$ dimensional system, with a Euclidean action given by the static energy (3.15) of the latter system. Consequently the static solitons we obtained for the Minkowskian $(2+1)$-dimensional O(3) model also serve as instantons of the $(1+1)$ dimensional O(3) model. What would happen if we generalised from the O(3) model to the O(N) model with arbitrary $N > 3$? That is, directly generalise the O(3) model to N real scalar fields $\boldsymbol{\phi} = \{\phi_a;\ a = 1, \ldots, N\}$ with $\boldsymbol{\phi}\cdot\boldsymbol{\phi} = \sum_{a=1}^{N} \phi_a^2 = 1$. Consider static solitons in two space dimensions. (The arguments are equally applicable to instantons of the corresponding $(1+1)$-dimensional theory.) The logic proceeds exactly as in section 3.3, and the boundary condition (3.17) again compactifies Euclidean two-space into the surface of a sphere $S_2^{(\text{phy})}$. However, the allowed values of the field, subject to $\sum_1^N \phi_a^2 = 1$, now fall on the surface of a hypersphere $S_{N-1}^{(\text{int})}$ imbedded in N dimensions. Consequently the homotopy group of localised solutions would now be $\Pi_2(S_{N-1})$. From the results (3.90) we see that this group is trivial except when $N = 3$. Consequently non-trivial instanton sectors for the two-dimensional O(N) model will exist only when $N = 3$. The situation does not improve if we go to higher space dimensions, because the dynamics, as evident from the virial theorem in section 3.2 will rule out non-trivial solutions. Consequently, the O(3) model's results cannot be generalised by going to an O(N) model with $N > 3$ (see Din and Zakrewski: 1980). We will return to this point later.

Turning to gauge fields, let us consider instantons of electromagnetic systems, which are simpler and more familiar than the Yang–Mills fields already studied. The free electromagnetic field with no sources is of course too simple. It is linear and exactly soluble, unlike the Yang–Mills field which, even in the absence of other source-fields, is self-coupled. Consider therefore the electromagnetic field coupled to some other charged field, as for instance the abelian Higgs model (3.91) in $(D+1)$ dimensions. It is evident from (3.91) that the Euclidean action of that system will be finite only if (3.93) and (3.94) are satisfied at the boundary of the $(D+1)$-dimensional Euclidean domain. The range of boundary values is therefore again characterised by $S_1^{(\text{int})}$, the circle of phases of ϕ. On the other hand, the boundary of Euclidean $(D+1)$ space is $S_D^{(\text{phy})}$. The homotopy classification of instantons is therefore based on the group $\pi_D(S_1)$. From (3.90), we see that this is non-trivial only for $D = 1$. Thus, in the realistic case of $(3+1)$ dimensions, where the Yang–Mills instantons are so interesting, the abelian system (3.91) has no non-trivial instanton sectors! This is one reason why we chose the Yang–Mills system rather than electrodynamics to discuss in detail instantons in four dimensions. It is only in $(1+1)$ dimensions that the electrodynamic instantons exist with integral homotopy indices. Notice again that the Euclidean field equations in $(1+1)$ dimensions arising from (3.91) would be the same as the static field equations of the same system in $(2+1)$ dimensions, if we work in the $A_0 = 0$ gauge of the latter theory. One identifies the $(x\text{-}y)$ coordinates and (A_x, A_y) fields of the latter with the (x, it) coordinates and (A_x, iA_0) fields, respectively, of the former. Therefore, in effect, we have already studied instantons of the $(1+1)$-dimensional abelian Higgs model in the guise of the flux-vortex line solutions in a superconductor. From the discussion in section 3.6, we know almost as much about these instantons, as we do about the Yang–Mills instantons, short of knowing the explicit analytic solutions. Recall from section (3.6) that numerical calculations and the work of Bogomol'nyi (1976) and Taubes (1980) show that a non-singular instanton does exist for this abelian model. We know its asymptotic behaviour (eq. (3.105)) and an explicit expression for the homotopy index (eq. (3.99)). We also know that it has finite size proportional to F^{-1}, unlike the Yang–Mills or O(3) instantons which can be arbitrarily large because of their arbitrary scale factor (see eq. (4.74)). This last feature of the abelian Higgs model instantons will prove to be an asset in calculating their quantum effects (see chapter 10).

The next more complicated gauge theory after electromagnetism is the SU(2) Yang-Mills theory, whose instantons we have already discussed in detail. For gauge theories involving higher non-abelian groups like SU(3), SU(4) etc., one need not discuss each case individually as far as homotopy classification of their instantons in four dimensions is concerned. As Coleman (1979) has pointed out, we can exploit a certain powerful theorem in homotopy theory (Bott 1956). The theorem states that any continuous mapping of S_3 into any general simple Lie group G can be continuously deformed into a mapping of S_3 into an SU(2) subgroup of G. Since we have already classified the latter mappings using $\pi_3(S_3) = Z$, we have effectively classified the former, in four Euclidean dimensions. Once an arbitrary mapping of $S_3 \to G$ has been deformed into a mapping $S_3 \to \mathrm{SU}(2)$, all our earlier work for classifying the latter, including the formula for Pontryagin index etc. can be directly used. This result, valid for an arbitrary simple Lie group, is in particular valid for all SU(n). Also, if a group is not simple, but a direct product of simple groups, we only need to perform a separate analysis for each factor simple group i.e. assign a set of independent, essentially SU(2) Pontryagin indices for each such factor group in classifying mappings of S_3 into that product group. Further since any general compact Lie group is locally a direct product of an abelian group and a string of simple groups, we can see that our analysis of the SU(2) instantons in four dimensions actually suffices for a very much larger family of groups.

It should be pointed out that instantons of SU(N) gauge groups, for arbitrary N, are of more than academic interest even though existing theories of elementary particles call for SU(N) gauge groups with only some fairly small and finite N. (The currently accepted theory of weak and electromagnetic interactions uses an $SU(2) \times U(1)$ gauge group, while strong interaction theory uses the SU(3) 'colour' group; some theories that attempt to unify all these interactions use SU(4), SU(5) or SU(7) groups.) The reason why an SU(N) gauge theory with an arbitrarily large N may nevertheless be important is because of the possibility of the so called $1/N$ expansion. It has been persuasively argued ('tHooft 1974a, Witten 1979a, 1979b) that a good approximation scheme for the quantum SU(N) gauge theories may be an expansion in powers of $1/N$. In that case, the leading term in such an expansion would correspond to an SU (N) theory with $N \to \infty$.

Finally, the concept of solitons and instantons has also been extended to general relativistic theories. See for instance Salam and Strathdee (1976), Wilczek (1976), Hawking (1977), Marciano et al. (1977), Gibbons and Perry (1978), Gibbons and Hawking (1979) and Eguchi and Hanson (1978).

4.5. The CP_N model

We mentioned in chapter 3 that the O(3) model in (1 + 1) dimensions has several interesting properties, many of them similar to the Yang–Mills theory in (3 + 1) dimensions. Both systems yield instantons characterised by integer-valued topological indices. The procedures for obtaining these solutions in sections 3.3 and 4.3 were also similar; in both cases we obtained inequalities that gave a lower bound to the action S by a value proportional to the homotopy index Q. The minimality of S led to first-order differential equations (the 'self-duality' conditions) which we then solved. Both models were scale invariant and yielded instantons of arbitrary size. (At the quantum level, the similarities persist. Although it will be outside the scope of this book to discuss such matters in any detail, it should be mentioned that both the O(3) and Yang–Mills theories are renormalisable and asymptotically free.) At the same time, the O(3) model is comparatively simple. It consists of only three scalar fields in two dimensions with a simple Lagrangian. Consequently, apart from its relevance to two-dimensional ferromagnets, the O(3) model is a useful toy model from which further insights may be gained about the more complicated non-abelian gauge theories. However, the similarity fails in one respect. We mentioned in the last section that the SU(2) Yang–Mills system could be imbedded in higher SU(N) gauge theories, and that instanton solutions will continue to exist for arbitrarily large N. This allows the possibility of comparing quantum instanton effects (along lines to be developed in chapter 10) with the large-N expansion (Witten 1979a, b). Unfortunately, as we pointed out, the corresponding generalisation from O(3) to O(N) models does not yield instantons when $N > 3$.

However, there exists another family of scalar field theories, again in $(1+1)$ dimensions, which shares all the above mentioned features of the O(3) model, while at the same time yielding instanton solutions even when N, the number of fields, becomes arbitrarily large. These are the so called CP_N models (Eichenherr 1978, Golo and Perelomov 1978, Cremmer and Scherk 1978). We will see that for $N = 1$ the CP_1 model is essentially the same as the O(3) model. For higher N, it is an appropriate generalisation, more so than the O(N) model, in the sense that it continues to have instanton solutions. Considerable work has been done on this model despite its recent inception, because of its attractive features. Therefore we will devote this section to stating the model and obtaining its instanton solutions (D'Adda et al. 1978, Witten 1979a). Our discussion will closely parallel those in earlier sections.

The CP_N model consists of $N+1$ *complex* scalar fields $n_a(x)$, $a = 1, \ldots, N+1$, in two dimensions. We will collectively denote the fields by $\boldsymbol{n}(x) = \{n_a(x)\}$. Since we are interested in instantons, we will work in Euclidean two-dimensional space with coordinates $x = \{x_1, x_2\}$. The fields are subject to the constraint

$$(\boldsymbol{n}(x))^* \cdot \boldsymbol{n}(x) = \sum_{a=1}^{N+1} n_a^*(x) n_a(x) = 1 \tag{4.99}$$

with a Lagrangian density

$$\mathscr{L}(x) = (\partial_\mu \boldsymbol{n})^* \cdot (\partial_\mu \boldsymbol{n}) + (\boldsymbol{n}^* \cdot \partial_\mu \boldsymbol{n})(\boldsymbol{n}^* \cdot \partial_\mu \boldsymbol{n}) \tag{4.100}$$

with $\mu = 1, 2$. The constraint (4.99), when differentiated with respect to x_μ, yields

$$\boldsymbol{n}^* \cdot \partial_\mu \boldsymbol{n} + \partial_\mu \boldsymbol{n}^* \cdot \boldsymbol{n} = 2\mathrm{Re}\,(\boldsymbol{n}^* \cdot \partial_\mu \boldsymbol{n}) = 0 \tag{4.101}$$

Hence $\boldsymbol{n}^* \cdot \partial_\mu \boldsymbol{n}$ is pure imaginary, a fact that will be repeatedly used.

Now, consider the set of space-dependent transformations

$$n_a(x) \to n_a(x) \exp(\mathrm{i}\Lambda(x)). \tag{4.102}$$

The phase $\Lambda(x)$, while space-dependent, is independent of the index a. All $N+1$ fields are multiplied by the same phase factor. Under this transformation,

$$\partial_\mu \boldsymbol{n} \to (\partial_\mu \boldsymbol{n} + \mathrm{i}\partial_\mu \Lambda \boldsymbol{n}) \mathrm{e}^{\mathrm{i}\Lambda}$$

$$\boldsymbol{n}^* \cdot \partial_\mu \boldsymbol{n} \to \boldsymbol{n}^* \cdot \partial_\mu \boldsymbol{n} + \mathrm{i}\partial_\mu \Lambda$$

and

$$\mathscr{L}(x) \to \mathscr{L}(x). \tag{4.103}$$

From these equations it is clear that the system enjoys local gauge invariance under the U(1) transformations (4.102). The situation is somewhat similar to abelian electrodynamics, and as in that system we identify all fields which differ from one another only by the transformations (4.102), as 'gauge equivalent'. The Lagrangian is gauge invariant. The set of $N+1$ complex numbers, subject to this equivalence identification and also to the constraint (4.99), form an N-dimensional complex projective space. Hence the acronym CP_N.

This gauge invariance is brought out more clearly by introducing an auxiliary vector field $A_\mu(x)$ and rewriting the Lagrangian as

$$\mathscr{L}(x) = \partial_\mu \boldsymbol{n}^* \cdot \partial_\mu \boldsymbol{n} + A_\mu^2 - 2A_\mu\,(\mathrm{i}\boldsymbol{n}^* \cdot \partial_\mu \boldsymbol{n}). \tag{4.104}$$

Upon extremising the action $S = \int \mathrm{d}^2 x \mathscr{L}(x)$ with respect to $A_\mu(x)$, the latter obeys

$$0 = \partial S/\partial A_\mu = 2(A_\mu - \mathrm{i}\boldsymbol{n}^* \cdot \partial_\mu \boldsymbol{n}).$$

Hence

$$A_\mu = \mathrm{i}(\boldsymbol{n}^* . \partial_\mu \boldsymbol{n}). \tag{4.105}$$

This is clearly a constraint equation for $A_\mu(x)$. The field A_μ does not represent independent degrees of freedom, but is entirely determined in terms of $\boldsymbol{n}(x)$ through (4.105). When (4.105) is inserted into the Lagrangian (4.104) (a legitimate dynamical step in treating such simple constraints (see Dirac 1950)), we obtain the Lagrangian (4.100). The two Lagrangians (4.100) and (4.104) are therefore equivalent. Recall that $\boldsymbol{n}^* \cdot \partial_\mu \boldsymbol{n}$ was pure imaginary, so that $A_\mu(x)$ is real. [The systems (4.100) and (4.104) are in fact equivalent even after quantisation. This can be shown by using functional integrals and integrating out the quadratic dependence on A_μ in the action. We will not go into this point further since our discussion in this section will remain at the classical level, but such functional techniques will be discussed in later chapters.]

Using (4.105) and (4.101), the Lagrangian (4.104) can be compactly written as

$$\mathscr{L}(x) = (\mathrm{D}_\mu \boldsymbol{n})^* \cdot (\mathrm{D}_\mu \boldsymbol{n}) \tag{4.106}$$

where

$$\mathrm{D}_\mu \boldsymbol{n} \equiv (\partial_\mu + \mathrm{i}A_\mu)\boldsymbol{n}. \tag{4.107}$$

Under the gauge transformation (4.102),

$$A_\mu(x) \to A_\mu(x) - \partial_\mu \Lambda(x)$$

and

$$\mathrm{D}_\mu \boldsymbol{n} \to (\mathrm{D}_\mu \boldsymbol{n})\mathrm{e}^{\mathrm{i}\Lambda}. \tag{4.108}$$

The gauge invariance of the Lagrangian written in the form (4.106) is now transparent, as is the extent of similarity to scalar electrodynamics. Notice, however, that (4.106) is not quite the Lagrangian of complex scalar fields interacting with an electromagnetic field. The kinetic terms corresponding to A_μ are absent in (4.106). This is of course because $A_\mu(x)$ here is not an independent field, but completely determined by (4.105) in terms of $\boldsymbol{n}(x)$.

The Euclidean field equation is obtained by extremising the action

$$S = \int \mathrm{d}^2x (\mathrm{D}_\mu \boldsymbol{n})^* \cdot (\mathrm{D}_\mu \boldsymbol{n}) \tag{4.109}$$

with respect to $\boldsymbol{n}(x)$ subject to the constraint (4.99). The constraint is introduced in the variational process as usual with a Lagrange multiplier. That is, we extremise $S + \int \mathrm{d}^2x \lambda(x)(\boldsymbol{n}^* \cdot \boldsymbol{n} - 1)$. The resulting field equation is

$$\mathrm{D}_\mu \mathrm{D}_\mu \boldsymbol{n} + \lambda \boldsymbol{n} = 0. \tag{4.110}$$

The Lagrange multiplier $\lambda(x)$ is eliminated using

$$\lambda = \lambda \boldsymbol{n}^* \cdot \boldsymbol{n} = -\boldsymbol{n}^* \cdot \mathrm{D}_\mu \mathrm{D}_\mu \boldsymbol{n}$$

to yield

$$\mathrm{D}_\mu \mathrm{D}_\mu \boldsymbol{n} - (\boldsymbol{n}^* \cdot \mathrm{D}_\mu \mathrm{D}_\mu \boldsymbol{n})\boldsymbol{n} = 0. \tag{4.111}$$

The instantons we are seeking are finite-action solutions of this equation. We obtain them indirectly, as we did in earlier examples, by first classifying finite-action configurations.

Finiteness of the action (4.109) requires that as $r \equiv |\boldsymbol{x}| \to \infty$,

$$\mathrm{D}_\mu \boldsymbol{n} \equiv \partial_\mu \boldsymbol{n} + \mathrm{i} A_\mu \boldsymbol{n} = 0. \tag{4.112}$$

Taking each component $n_a \equiv |n_a| \mathrm{e}^{\mathrm{i}\phi_a}$ separately, this yields

$$-iA_\mu = \frac{\partial_\mu n_a}{n_a} = \frac{\partial_\mu |n_a|}{|n_a|} + i\,\partial_\mu \phi_a. \tag{4.113}$$

Notice that $-\mathrm{i}A_\mu$ in (4.113) is pure imaginary, and independent of a. This means that when $r \to \infty$, (i) $\partial_\mu |n_a| = 0$, and (ii) $\partial_\mu \phi_a$ is independent of a. Consequently, the boundary condition (4.112) implies that as $r \to \infty$,

$$\boldsymbol{n}(x) \to \boldsymbol{n}^{(0)} \mathrm{e}^{\mathrm{i}\phi(\theta)} \tag{4.114}$$

where $\boldsymbol{n}^{(0)}$ is any fixed complex vector with $(\boldsymbol{n}^{(0)})^* \cdot \boldsymbol{n}^{(0)} = 1$ and ϕ is a common phase angle. This ϕ can depend on θ, the angle in coordinate space that parametrises the boundary of space, which in this problem is a circle $S_1^{(\mathrm{phy})}$. Meanwhile the allowed values of the phase ϕ also form a circle, $S_1^{(\mathrm{int})}$. Consequently, the set of boundary conditions permitted by (4.114) are mappings of $S_1^{(\mathrm{phy})} \to S_1^{(\mathrm{int})}$. We already know that such mappings fall into homotopy classes characterised by a winding number Q. We also know an expression for Q in the form

$$Q = \frac{1}{2\pi} \int \mathrm{d}\theta \, \frac{\mathrm{d}\phi}{\mathrm{d}\theta}. \tag{4.115}$$

The freedom in the choice of the constant normalised complex vector $\boldsymbol{n}^{(0)}$ does not introduce further homotopy classification. Notice that our Lagrangian also enjoys global (x-independent) $\mathrm{SU}(N+1)$ symmetry, in addition to that fact that a U(1) subgroup of this $\mathrm{SU}(N+1)$ is a local gauge symmetry. Through such global $\mathrm{SU}(N+1)$ rotations, one can continuously change from one value of $\boldsymbol{n}^{(0)}$ to any other.

The expression (4.115) can be rewritten in terms of A_μ. The θ-component of A_μ is given by

$$A_\theta = \frac{\mathrm{i}}{r} \boldsymbol{n}^* \cdot \frac{\partial}{\partial\theta} \boldsymbol{n} = -\frac{1}{r}\frac{\mathrm{d}\phi}{\mathrm{d}\theta}, \qquad \text{on } S_1^{(\mathrm{phy})}.$$

Hence

$$Q = -\frac{1}{2\pi} \int_{S_1^{(\mathrm{phy})}} \mathrm{d}\theta \, r A_\theta = -\frac{1}{2\pi} \int_{S_1^{(\mathrm{phy})}} \mathrm{d}\boldsymbol{l} \cdot \boldsymbol{A}$$

$$= -\frac{1}{2\pi}\int d^2x\, \varepsilon_{\mu\nu}\,\partial_\mu A_\nu. \tag{4.116}$$

One can also rewrite this as

$$Q = -\frac{i}{2\pi}\int d^2x\, \varepsilon_{\mu\nu}(D_\mu \boldsymbol{n})^* \cdot (D_\nu \boldsymbol{n}). \tag{4.117}$$

The equality of (4.117) to (4.116) is a matter of trivial algebra, using (4.105).

Next we use an appropriate inequality as in preceding examples. We have,

$$\int d^2x\,(D_\mu \boldsymbol{n} \pm i\varepsilon_{\mu\nu} D_\nu \boldsymbol{n})^* \cdot (D_\mu \boldsymbol{n} \pm i\varepsilon_{\mu\nu} D_\nu \boldsymbol{n}) \geqslant 0. \tag{4.118}$$

On expanding this,

$$2\int d^2x\,\{(D_\mu \boldsymbol{n})^* \cdot (D_\mu \boldsymbol{n}) \pm i\varepsilon_{\mu\nu}(D_\mu \boldsymbol{n})^* \cdot (D_\nu \boldsymbol{n})\} \geqslant 0$$

or

$$S \geqslant 2\pi |Q|. \tag{4.119}$$

Equations (4.112)–(4.119) hold for any finite-action configuration. However, when the equality in (4.118) holds, i.e. when S reaches its minimum value in any Q-sector, the field will extremise the action in that sector and satisfy the field equation. The condition for equality in (4.118) is clearly

$$D_\mu \boldsymbol{n} = \pm i\varepsilon_{\mu\nu} D_\nu \boldsymbol{n}. \tag{4.120}$$

This condition is obviously analogous to the self-duality condition (4.47), or to the relation (3.32) in the O (3) model. All the qualifying remarks made there also hold here. For instance (4.120) is a first-order differential equation unlike the parent field equation (4.111) and easier to solve. Solutions of (4.120), by minimising S, will automatically solve (4.111), but the converse need not be true.

To solve (4.120), we change to a set of gauge-invariant variables in the place of $\boldsymbol{n}$. Remember that because of the constraint $\boldsymbol{n}^* \cdot \boldsymbol{n} = 1$, all the fields $n_a(x)$ cannot vanish at any point. Consider a region R_1 in coordinate space, where say, $n_1 \neq 0$. In that region, define

$$\boldsymbol{w}(x) \equiv \boldsymbol{n}(x)/n_1(x). \tag{4.121}$$

Notice that the fields $\boldsymbol{w}(x) \equiv \{w_a(x),\ a = 1, \ldots, N+1\}$ are gauge invariant. The transformation (4.102) introduces the same factor $e^{i\Lambda(x)}$ in the numerator and denominator of (4.121). Also $w_1(x) = 1$ for all $x \in R_1$ so that there are only N independent gauge invariant complex fields w_a. This is consistent with the dimensionality of CP_N. As we move away from region R_1, n_1 may vanish and the definition (4.121) will not hold. However, because of $\boldsymbol{n}^* \cdot \boldsymbol{n} = 1$, there must be a contiguous region R_2, where some other component, say n_2, is non-vanishing. There, define

$$\boldsymbol{w}' \equiv \boldsymbol{n}/n_2. \tag{4.122}$$

One can satisfy oneself that n_2 may be so chosen that R_1 and R_2 must overlap in a sub-region where neither n_1 nor n_2 vanish. In that overlap region, $\boldsymbol{w}'$ and $\boldsymbol{w}$ are related by

$$\boldsymbol{w}' = \boldsymbol{w}\, n_1/n_2. \tag{4.123}$$

In this fashion, we can define fields $\boldsymbol{w}$ (or $\boldsymbol{w}'$, $\boldsymbol{w}''$ etc.) patchwise, by dividing x-space into overlapping regions, with the different definitions analytically continued from one another using relations such as (4.123) in overlapping portions. This procedure of defining fields patchwise is akin to what is done for gauge fields in the presence of monopoles (see Wu and Yang (1975)).

Our equation (4.120) is local, and we will solve it in any one region, say R_1, where $\boldsymbol{w}$ is defined as per (4.121). The same arguments will hold in any other region and it is understood that solutions can be continued from one region to another using relations such as (4.123). Let us rewrite eq. (4.120) in terms of $\boldsymbol{w}$. We have,

$$\boldsymbol{n} = n_1 \boldsymbol{w}.$$

Hence (4.120) becomes

$$\mathrm{D}_\mu (w_a n_1) = \pm \mathrm{i}\varepsilon_{\mu\nu} \mathrm{D}_\nu (w_a n_1); \qquad a = 1, \ldots, N+1$$

or

$$w_a \mathrm{D}_\mu n_1 + n_1 \partial_\mu w_a = \pm \mathrm{i}\varepsilon_{\mu\nu} (w_a \mathrm{D}_\nu n_1 + n_1 \partial_\nu w_a). \tag{4.124}$$

Use (4.120) for n_1 to cancel the first terms of the two sides of (4.124). In the remaining second terms, we divide by $n_1 \neq 0$ to get

$$\partial_\mu w_a = \pm \mathrm{i}\varepsilon_{\mu\nu} \partial_\nu w_a; \qquad \mu, \nu = 1, 2. \tag{4.125}$$

Note that in (4.125) it is the ordinary derivatives $\partial_\mu w_a$ that occur and not the covariant derivatives. Further, just as in the O (3) model, eq. (4.125) can be recognised as the Cauchy–Reimann condition. With the minus sign in (4.125), it states that each w_a is an analytic function of $z \equiv x_1 + \mathrm{i}x_2$. The plus sign requires each w_a to be analytic in z^*.

Thus, we have found our instantons. Any set of analytic functions $w_a(z)$; $a = 2, \ldots, N+1$, with $w_1 = 1$, will solve (4.120). So will any analytic functions $w_a(z^*)$. By the arguments given above, these functions when simply rewritten in terms of $\boldsymbol{n}$ and x_μ, will give explicit exact finite-action solutions of the Euclidean field equation (4.111). These then are the instantons of this model.

As an explicit example, we will present the one-instanton solution. We choose

$$\boldsymbol{w}(z) = \boldsymbol{u} + [(z - z_0)/\lambda]\,\boldsymbol{v} \tag{4.126}$$

where $\boldsymbol{u}$, $\boldsymbol{v}$ are any pair of orthonormal complex vectors satisfying $u_1 = 1$, $v_1 = 0$,

$$\boldsymbol{u}^* \cdot \boldsymbol{u} = \boldsymbol{v}^* \cdot \boldsymbol{v} = 1 \qquad \text{and} \qquad \boldsymbol{u}^* \cdot \boldsymbol{v} = 0.$$

Clearly (4.126) is a set of analytic functions and hence a solution. The constant λ is real and represents the 'size' of the instanton while the complex number $z_0 \equiv (x_1)_0 + \mathrm{i}\,(x_2)_0$ represents its location in the $z = x_1 + \mathrm{i}x_2$ plane. The freedom to choose λ and z_0 arbitrarily reflects the scale and translational invariance of the action (4.109). To write this solution more explicitly, we invert (4.121). Modulo a phase factor which can in any case be gauged away, (4.121) gives

$$\boldsymbol{n} = n_1\,\boldsymbol{w} = \boldsymbol{w}/|\boldsymbol{w}|$$

where

$$|\boldsymbol{w}| = (\boldsymbol{w}^* \cdot \boldsymbol{w})^{1/2} = |n_1|^{-1}.$$

Inserting (4.126), the solution becomes

$$\boldsymbol{n}(z) = \frac{\lambda \boldsymbol{u} + (z - z_0)\,\boldsymbol{v}}{(\lambda^2 + |z - z_0|^2)^{1/2}}. \tag{4.127}$$

It is straightforward to verify explicitly that it solves the field equation (4.111). Also, at spatial infinity as $z \to \infty$,

$$\boldsymbol{n}(z) \to (z/|z|)\,\boldsymbol{v} = \mathrm{e}^{\mathrm{i}\theta}\boldsymbol{v}.$$

This satisfies the boundary condition (4.114) with a phase angle $\phi(\theta) = \theta$. Clearly, it belongs to the $Q = 1$ sector and represents a single instanton. The anti-instanton is obtained by replacing z by z^*. [Note that if we had omitted the constant vector $\boldsymbol{u}$ in (4.126), the corresponding $\boldsymbol{n} = (z - z_0)\,\boldsymbol{v}/|z - z_0|$, would have been ill defined at $z = z_0$.] Multi-instantons come from more complicated choices of $w(z)$.

It will be useful to write down the expression for A_μ and Q directly in terms of $\boldsymbol{w}(z)$. We have

$$A_\mu = \mathrm{i}(\boldsymbol{n}^* \cdot \partial_\mu \boldsymbol{n})$$
$$= (\mathrm{i}/2|\boldsymbol{w}|^2)(\boldsymbol{w}^* \cdot \partial_\mu \boldsymbol{w} - \boldsymbol{w} \cdot \partial_\mu \boldsymbol{w}^*).$$

Upon using the Cauchy–Reimann relations (4.125), this reduces to

$$A_\mu = \pm \varepsilon_{\mu\nu} \frac{\boldsymbol{w}^* \cdot \partial_\nu \boldsymbol{w} + \boldsymbol{w} \cdot \partial_\nu \boldsymbol{w}^*}{2|\boldsymbol{w}|^2} = \pm \tfrac{1}{2} \varepsilon_{\mu\nu} \partial_\nu \ln |\boldsymbol{w}|^2. \tag{4.128}$$

Then,

$$Q = -\frac{1}{2\pi} \int \mathrm{d}^2 x \, \varepsilon_{\mu\nu} \partial_\mu A_\nu = \pm \frac{1}{4\pi} \int \mathrm{d}^2 x \, \Box \ln |\boldsymbol{w}|^2. \tag{4.129}$$

The action, as per the equality in (4.119) is $S = 2\pi|Q|$. It is a matter of simple algebra to insert the example (4.126) and explicitly verify that (4.129) yields $Q = 1$, and $S = 2\pi$.

We conclude our discussion by relating the CP_N models to the O(3) model. When $N = 1$, the CP_1 model has two complex fields n_1 and n_2 subject to $|n_1|^2 + |n_2|^2 = 1$. We construct three fields ϕ^a, $a = 1, 2, 3$ by

$$\phi^a \equiv n_p^* (\sigma^a)_{pq} n_q \tag{4.130}$$

where σ^a are the three Pauli matrices. Explicitly,

$$\phi^1 = 2 \operatorname{Re}(n_1^* n_2), \qquad \phi^2 = 2 \operatorname{Im}(n_1^* n_2).$$

and

$$\phi^3 = |n_1|^2 - |n_2|^2.$$

Clearly the fields ϕ^a are real and satisfy

$$\sum_{a=1}^{3} \phi^a \phi^a = (|n_1|^2 + |n_2|^2)^2 = 1.$$

A little more algebra reduces the Lagrangian density (4.100) to $\sum_a (\partial_\mu \phi^a)(\partial_\mu \phi^a)$. Therefore the CP_1 model is essentially the O(3) model when written in terms of the ϕ^a. For higher N, the CP_N models are more appropriate generalisations of the O(3) model than are O(N) models, by virtue of continuing to yield instantons for arbitrary N, in two dimensions.

In summary, we see that the CP_N model, even though it consists only of scalar fields, enjoys local gauge invariance. It carries instantons which exhibit many similarities to instantons of the four-dimensional Yang–Mills system. For a more general discussion of non-linear models behaving as gauge theories, see Balachandran et al. (1979) and Brezin et al. (1980).

CHAPTER 5

QUANTISATION OF STATIC SOLUTIONS

5.1. Introduction

The preceding chapters offered illustrative examples of localised classical solutions of non-linear relativistic field equations. We are now ready to discuss the relevance of such classical solutions to the corresponding quantum field theories. It will be seen that under certain favourable circumstances, one can relate classical Minkowskian solutions to bound and scattering states of the quantum theory. In particular, solitary waves and solitons (henceforth jointly termed solitons for convenience) can be associated with quantum extended-particle states. The next few chapters will be devoted to exploring this association, which is of more than conceptual interest. It will yield us new and concrete information about the quantum theory. That is to say, certain properties of these quantum states, as for instance their energy or form factors, can be expanded in a semiclassical series. The leading terms in this series will be seen to be related to the corresponding classical solutions. In this way, knowledge of the classical soliton solutions will yield some information about quantum particle states, in a systematic semiclassical expansion. Moreover, this information will be non-perturbative in the non-linear couplings since, in most cases of interest, the corresponding classical solutions are themselves non-perturbative. Note for instance that most classical solutions described in the preceding chapters are non-perturbative. They become singular when the non-linear terms in the field equations tend to zero.

The Euclidean solutions (instantons) will also yield non-perturbative information on the corresponding quantum theory – but on a different aspect of the theory, dealing with the vacuum state and the tunnelling phenomenon. These matters will be discussed, as promised, in the last two chapters.

We begin in this chapter with the simplest application of classical solutions in quantum field theory, namely the quantisation of *static* soliton solutions to obtain extended, non-perturbative, quantum particle states. The formalism for executing such quantisation has been developed in the original papers through a variety of techniques (Dashen et al. 1974b, Cahill 1974, Goldstone and Jackiw 1975, Polyakov 1975a, b, Christ and Lee 1975). These different techniques, which include functional methods, canonical operator methods etc., all lead to the same basic physical phenomenon. A choice amongst these techniques is partly a matter of taste, rather like a choice between the use of the Schrödinger differential equation, the Feynman path-integral, or the Heisenberg commutator methods to tackle the harmonic oscillator problem. We will be adopting a suitable mixture of these techniques to help exhibit different aspects of the problem.

We have already noted that a single-soliton solution resembles, even at the classical level, an extended particle. (By an 'extended' particle, we mean a particle of finite, non-zero size. Most particles in nature, even the so-called elementary particles, seem to be extended, although they are sometimes approximated by point particles in theoretical calculations). Like extended particles, our soliton solutions are localised, finite-energy objects. In their rest frames they are static, and in moving frames they move without dispersion. Their shapes in motion are Lorentz contracted, (see for instance eq. (2.32)) and their energy E in motion is related to their static energy M through the familiar relativistic formula $E = M/\sqrt{1-v^2}$ (see for instance eq. (2.33)). These features of the classical solitons strongly tempt us to expect that they should have something to do with particles in the corresponding quantum theory as well.

This expectation turns out to be true, but the connection between the classical solution and the quantum particle is not completely trivial. A hasty identification of the classical solution with, for instance, something like the wavefunction of the quantum particle would be quite incorrect, for a host of reasons. The correct interpretation, although simple in its own way, is somewhat more subtle. Indeed, having pointed out the resemblance of classical solitons to some aspects of particles, we should balance the picture by cautioning that the definition of a particle in quantum field theory is, superficially, quite far removed from the classical solutions of its field equations. To appreciate this, let us recall the basic differences between classical and quantum field theory. In the former, the fields are c-number functions of space–time. The fields themselves specify the states of the classical system, and their dynamics is governed by the non-linear partial differential field equations. Our solitons are merely solutions of these equations. Their resemblence to physical particles notwithstanding, the concept of a 'particle' does not exist in classical field theory.

By contrast, in quantum field theory where the notion of a particle *is* viable, the fields themselves have a different status. They are not c-numbers, but operator functions of space–time, acting on vectors in a Hilbert space. The fields do obey Heisenberg field equations, but the solutions of these equations would again be operators. They do not specify states of the quantum system. The latter are specified by vectors in that Hilbert space. In particular, 'particles' are special state-vectors obeying requirements we would expect from particle states in the relativistic quantum context. Loosely speaking, a particle in quantum field theory would correspond to a discrete hyperboloid in the spectrum, i.e. a family of simultaneous eigenvectors of the Hamiltonian and momentum operators, with eigenvalues E and $\boldsymbol{P}$ obeying $E^2 - \boldsymbol{P}^2 = M^2$ with some fixed discrete M. One would further require that appropriate form-factors in these states be localised, reflecting the feature that a particle should be a localised lump of matter. The dynamics of these particle states, like the dynamics of any other state-vectors, will be governed by the field theoretical Schrödinger equation, which is superficially quite different from the field equations. Viewed in this perspective, the concept of a particle in quantum field theory seems quite divorced from classical solutions of its field equations! In fact, in treatments of quantum field theory in standard textbooks there is little mention of the relation between the particles of the theory on the one hand and solutions of its field equations on the other, except for the trivial case of free field theory.

Of course, the two entities are related. This book owes its existence to their relationship! The cautionary paragraph above is meant only to discourage us from taking the relationship as self-evident merely because of some resemblance between particles and classical solitons. We need to do a little more work to establish the connection between them.

Fortunately, the basic principles needed to establish the correct connection are, in their own right, quite simple. All we need to do is to borrow from ordinary quantum mechanics such familiar concepts as the Correspondence Principle and the semiclassical expansion, which relate quantum levels to classical orbits in a systematic approximation. We will then generalise that procedure to field theory. Perhaps the most prominent example of such a relationship in non-relativistic quantum mechanics is the WKB method and the resultant Bohr–Sommerfeld quantisation condition. This condition associates selected classical periodic orbits with quantum energy levels. In later chapters, we will generalise this method to field theory and exploit periodic time-dependent solutions of classical field equations. In this chapter we are concerned with the simpler problem of quantising *static* solutions. The analogous relationship between static solutions of Newton's equation of motion and non-relativistic quantum energy levels is even more elementary than the WKB method. We begin the next section by recalling relevant aspects of this well known relationship in ordinary quantum mechanics. A direct generalisation of these simple results to field theory will form the central idea behind quantising static solitons.

5.2. The central idea

Consider a non-relativistic, unit-mass particle in one dimension under the influence of a potential $V(x)$. Classically, the particle is described by giving x as a function of time, and this is obtained by solving Newton's equation

$$\mathrm{d}^2x/\mathrm{d}t^2 = -\mathrm{d}V/\mathrm{d}x. \tag{5.1}$$

In quantum theory, the particle is described, not by giving the value of x, but of the wavefunction $\psi(x)$. In particular, energy eigenstates $\psi_n(x)$ obey

$$H\psi_n = (\tfrac{1}{2}\hat{p}^2 + V(x))\psi_n(x) = E_n\psi_n(x) \tag{5.2}$$

where $\hat{p} \equiv -\mathrm{i}\hbar\,\partial/\partial x$ is the momentum operator. As a pedagogical starting point for the quantisation of static solitons, let us recall features of the approximate relationship between static solutions of (5.1) and the quantum energy eigenstates determined by (5.2).

(i) Static (time-independent) solutions of the classical equation (5.1) are clearly the extrema of $V(x)$. Consider, for purposes of illustration, the potential $V(x)$ in fig. 9. It has three extrema, at $x = a$, $x = b$ and $x = c$. The particle, placed at any of these three points, will stay there. Thus, the

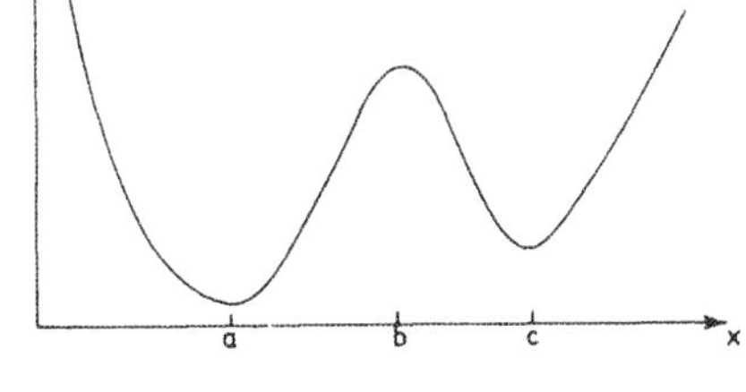

Fig. 9. An illustrative potential $V(x)$ for a one-dimensional unit-mass particle.

classical static solutions are

$$x(t) = a \tag{5.3a}$$

$$x(t) = b \tag{5.3b}$$

and

$$x(t) = c \tag{5.3c}$$

Of these, a and c are minima ($d^2V/dx^2 > 0$) and represent stable static solutions, while the solution at b is unstable. Let us begin with the first solution (5.3a). It may be called the 'classical ground state' in that it represents the lowest-energy solution permitted by classical mechanics for this particle. Its total energy is

$$E_0^{\text{cl}} = V(a). \tag{5.4}$$

(ii) In quantum theory, such a state is not allowed. The uncertainty principle will not permit the particle to have both zero momentum and a fixed position. Consequently, even in the ground state (i.e. the state of lowest eigenvalue in (5.2)), the particle will fluctuate around $x = a$, leading to a ground state energy

$$E_0 = E_0^{\text{cl}} + \Delta_0 = V(a) + \Delta_0 \tag{5.5}$$

where Δ_0 represents the quantum correction due to zero-point motion.

(iii) Furthermore, if the potential is approximately harmonic near $x = a$, we can attempt a 'weak-coupling expansion'. That is, let us make a Taylor expansion of $V(x)$ near $x = a$. Let

$$V(x) = V(a) + \frac{1}{2}\omega^2(x-a)^2 + \frac{1}{3!}\lambda_3(x-a)^3 + \frac{1}{4!}\lambda_4(x-a)^4 + \dots \tag{5.6}$$

Then, for those wavefunctions that satisfy

$$\lambda_r \langle (x-a)^r \rangle \ll \omega^2 \langle (x-a)^2 \rangle, \qquad r = 3, 4, \dots \tag{5.7}$$

where $\langle \dots \rangle$ stands for the expectation value, the effects of the anharmonic terms of $V(x)$ will be small. For λ_r sufficiently small, the low-lying energy eigenstates whose spread is localised in the vicinity of $x = a$ will satisfy (5.7), and for these, the potential will be dominantly that of a harmonic oscillator, whose properties are only too well known. The energies of these low-lying states can therefore be written as

$$E_n = V(a) + (n + \tfrac{1}{2})\hbar\omega + O(\lambda_r). \tag{5.8}$$

In particular, the quantum ground state energy becomes, in this weak-coupling approximation,

$$E_0 = V(a) + \tfrac{1}{2}\hbar\omega + O(\lambda_r). \tag{5.9}$$

(iv) Equation (5.9) represents the simplest example of a relation between quantum states and classical solutions. The left-hand side is the energy of the quantum ground state. The terms on the right-hand side are all related to the corresponding static classical solution (5.3a). The first term is the energy of this classical solution. In the second term, which represents the leading quantum correction, ω is the classical stability frequency of this solution. The $O(\lambda_r)$ corrections, which can be obtained by standard perturbation theory, will once again involve only the constants λ_r and ω, which give derivatives of the potential *at* the classical solution $x = a$.

(v) In fact, not only the ground state, but the energies of a tower of low-lying states are similarly related to this classical solution, through (5.8).

(vi) The energy of a quantum state represents only one of its features. The full information about the state is contained in its wavefunction. But the classical solution also gives, in the same approximation, other features of the ground state wavefunction. The basic reason behind this is that even though this wavefunction $\psi_0(x)$ has a spread in x, it will still be typically localised around the classical solution $x = a$. For instance its position expectation value is given by

$$\langle x \rangle \equiv \int |\psi_0(x)|^2 x \mathrm{d}x = a + \ldots \tag{5.10}$$

where the dots represent corrections due to the anharmonic constants λ_r. Recall that the quantity a in (5.10) may just be a number, but it also represents the static classical solution in (5.3a), whereas the left-hand side is a property of the quantum ground state.

(vii) Let us now repeat the same procedure for the more interesting case of the solution $x(t) = c$ in (5.3c). Classically this solution has energy

$$\tilde{E}^{\mathrm{cl}} = V(c), \tag{5.11}$$

which is higher than $V(a)$. This solution is the analogue of the classical static solitons in field theory. They too are static solutions, but with higher energies than the classical vacua in the corresponding field theories. Even though $x = c$ is only a local minimum of $V(x)$, one can again attempt a harmonic oscillator, or 'weak-coupling' approximation near it. Let

$$V(x) = V(c) + \tfrac{1}{2}(\omega')^2 (x-c)^2 + \sum_{r=3} \frac{\lambda'_r}{r!}(x-c)^r. \tag{5.12}$$

(viii) Once again, if the λ'_r are sufficiently small, then near $x = c$ the anharmonic affects will be small. One can try to construct a family of approximate harmonic oscillator states centred at $x = c$, with energies

$$\tilde{E}_{n'} = V(c) + \hbar\omega'(n' + \tfrac{1}{2}) + O(\lambda'_r). \tag{5.13a}$$

The lowest of these states, if this approximation were valid, would have an energy

$$\tilde{E}_0 = V(c) + \tfrac{1}{2}\hbar\omega' + O(\lambda'_r) \tag{5.13b}$$

and an expectation value

$$\langle x \rangle = c + O(\lambda'_r). \tag{5.14}$$

Thus, we would again have an approximation to a set of energy eigenstates, whose energy is related to $V(c)$, the classical energy of a static solution, and whose $\langle x \rangle$ is related to the solution c itself. However, see point (ix) below.

(ix) Such perturbation theory starting from a harmonic oscillator approximation, when applied to the local minimum $x = c$, treats the local potential well near $x = c$ as if the other, deeper well near $x = a$ did not exist. In actuality, we know that wave packets built in the potential well around $x = c$ will tunnel into the well around $x = a$ and vice versa. Consequently, the two subsets of energy levels will mix. But, if the λ_r and λ'_r are all small, the two minima $x = a$ and $x = c$ will be widely separated, with a large potential barrier in between them. Therefore, the tunnelling will be slow, and the resultant change in energy eigenvalues due to tunnelling will be small. To put it differently, tunnelling, if any, will be a non-perturbative phenomenon in the parameters λ_r and λ'_r. To any finite order in the weak-coupling expansion, the set of levels (5.13) around $x = c$ can be considered separately from the set (5.8) around $x = a$.

In our field theoretical applications we will encounter examples of both kinds – ones where such tunnelling takes place and ones where it does not. Examples of the tunnelling phenomenon, associated with instantons, will be discussed in chapter 10. By contrast, states built around most of our soliton solutions will be seen not to decay through tunnelling. This is related to the non-zero topological index carried by these solutions, which effectively places an infinite energy barrier between them and the vacuum solutions. This feature will be explained in section 5.6.

(x) It is obvious that there will not be a set of levels built around the static solution $x(t) = b$, because it is unstable. The corresponding frequency $[(d^2V/dx^2)_b]^{1/2}$ would be imaginary.

(xi) It is evident that when the harmonic constant ω vanishes this whole procedure runs into trouble. No matter how small the λ_r may be, the 'weak-coupling' condition (5.7) cannot be satisfied. In particular the special case where $V(x)$ is independent of x is noteworthy. Even though it is a trivial problem, it offers valuable lessons regarding the semiclassical method outlined thus far. Here, all the Taylor coefficients ω and λ_r vanish and the weak-coupling condition (5.7) is meaningless. Correspondingly, the physics behind the association of a single classical static solution with a tower of quantum levels breaks down. Although any point $x = x_0$ is now a classical static solution, the quantum energy eigenfunctions are not localised near x_0, because the potential now has no curvature near x_0 to localise the wavefunction. Indeed, the correct energy eigenfunctions e^{ikx} span all points on the x-axis, i.e. they are to be associated not with a single static solution, but a whole continuous family of static solutions. This is related to the existence of translational symmetry, implied by the x-independence of V.

From this example, it is clear that when the potential is independent of a coordinate, the corresponding harmonic frequency ω will vanish, and the semiclassical expansion outlined above has to be modified. We will discuss the required modification at a later stage. The same problem will arise in field theory while quantising solitons, for the same reason – the presence of continuous symmetries. The treatment of the resultant zero-frequency modes will be quite complicated, and we will devote chapter 8 to just this issue. But the basic physics behind the existence of these zero-frequency modes and their effect on the semiclassical method will be the same as in the trivial example given here. For later use, note that the exact energy levels of our simple problem, when V is independent of x, are

$$E_n = V + \tfrac{1}{2}(p_n)^2 \tag{5.15}$$

where p_n are momenta. Thus, the quantum correction term involves the momentum rather than the frequency ω, which in this case vanishes. Note that momentum is the conserved quantum number resulting from the translational symmetry which gave rise to $\omega = 0$ in the first place. The same pattern will be seen in field theory, for every zero-frequency mode caused by a continuous symmetry.

(xii) Returning to a general x-dependent $V(x)$ we see that, aside from possible problems when $\omega = 0$, one can associate a tower of energy eigenstates with each static stable classical solution. The quantisation of static solitons is based on a generalisation of the same ideas to field theory. The specific points highlighted above, trivial as they seem, contain the nucleus of important features of soliton quantisation. It should be emphasised that the relationships (5.8)–(5.10) and (5.13)–(5.14) are valid *only in the weak coupling expansion*, where the anharmonic terms are treated as small. Note that anharmonic terms in $V(x)$ also lead to non-linear terms in the equation of motion (5.1). Correspondingly, quantisation of static solitons in field theory will be valid only when the non-linear couplings are small. Nevertheless, the results may be non-perturbative. This is because the classical solution itself may be non-perturbative. Its properties form the leading terms, as for instance in the expansions (5.13)–(5.14). It is only the quantum corrections which are treated perturbatively.

As an intermediate step towards generalising these ideas to field theory, consider a system with a large (but finite) number of degrees of freedom. Take a bunch of unit-mass non-relativistic particles in any number of dimensions, collectively described by cartesian coordinates $\boldsymbol{x} \equiv (x_1, \ldots, x_N)$, with a potential $V(x_1, \ldots, x_N)$. Let $\boldsymbol{x} = \boldsymbol{a}$ be a (local or absolute) minimum of V in the N-dimensional space. Then $\boldsymbol{x}(t) = \boldsymbol{a}$ will be a stable static classical solution. Expand $V(x_1, \ldots, x_N)$ around $x_i = a_i$.

$$V(\boldsymbol{x}) = V(\boldsymbol{a}) + \tfrac{1}{2}(x_i - a_i)(x_j - a_j)\left[\frac{\partial^2 V}{\partial x_i \partial x_j}\right]_{\boldsymbol{x}=\boldsymbol{a}} + \frac{1}{3!}(x_i - a_i)(x_j - a_j)(x_k - a_k)\left[\frac{\partial^3 V}{\partial x_i \partial x_j \partial x_k}\right]_{\boldsymbol{x}=\boldsymbol{a}} + \ldots \tag{5.16}$$

If the third- and higher-order derivatives of V at $\boldsymbol{x} = \boldsymbol{a}$ are small (weak-coupling limit), then one can construct a tower of low-lying harmonic-oscillator states around the point $\boldsymbol{x} = \boldsymbol{a}$. For this purpose, diagonalise the matrix of second derivatives

$$\left[\frac{\partial^2 V}{\partial x_i \partial x_j}\right]_{\boldsymbol{x}=\boldsymbol{a}},$$

by changing variables to some normal modes ξ_i. Then, in terms of ξ_i, the problem reduces to a set of oscillators of frequencies ω_i, where ω_i^2 are the eigenvalues of this matrix. (If $\boldsymbol{x} = \boldsymbol{a}$ is a true minimum, all the ω_i^2 will be positive.) These modes are coupled through the (weak) cubic and higher terms in (5.16). The lowest-energy state constructed around $\boldsymbol{x} = \boldsymbol{a}$ will have energy

$$E_0 = V(\boldsymbol{a}) + \sum_{i=1}^{N} \tfrac{1}{2}\hbar\omega_i + \text{corrections}. \tag{5.17}$$

Higher excitations (but not so high as to violate the weak-coupling approximation) will be given by

$$E_{\{n_i\}} = V(a) + \hbar \sum_{i=1}^{N} (n_i + \tfrac{1}{2})\,\omega_i + \text{ corrections.} \tag{5.18}$$

Corrections to these formulae are obtained by perturbation theory. These results are a simple generalisation of the one-degree-of-freedom problem, with the set of eigenvalues $(\omega_i)^2$ of

$$\left[\frac{\partial^2 V}{\partial x_i\,\partial x_j}\right]_{x=a}$$

replacing the single number ω^2. All our earlier remarks apply, trivially generalised. In particular, if one of the ω_i is zero (a neutral stability mode) then the above method is not directly applicable. That mode would have to be handled differently.

Moving on to field theory, let us begin with a scalar field $\phi(\boldsymbol{x}, t)$ governed by a Lagrangian

$$L = \int \mathrm{d}\boldsymbol{x}\left[\frac{1}{2}\left(\frac{\partial\phi}{\partial t}\right)^2 - \frac{1}{2}(\nabla\phi)^2 - U(\phi)\right]$$

where $U(\phi)$ is, as in chapter 2, any function of ϕ bounded from below. Instead of a finite number of degrees of freedom $x_1, \ldots, x_N$, we now have a continuous infinity of degrees of freedom $\phi(\boldsymbol{x})$, i.e. the value of ϕ at each space point $\boldsymbol{x}$. But for this difference, the classical dynamics of this system is quite similar to particle mechanics. For instance, the Lagrangian has the standard form

$$L = T[\phi] - V[\phi] \tag{5.19}$$

with kinetic energy

$$T[\phi] \equiv \frac{1}{2}\int \mathrm{d}\boldsymbol{x}\left(\frac{\partial\phi}{\partial t}\right)^2 \tag{5.20a}$$

and potential energy

$$V[\phi] \equiv \int \mathrm{d}\boldsymbol{x}\,[\tfrac{1}{2}(\nabla\phi)^2 + U(\phi)]. \tag{5.20b}$$

The Euler–Lagrange equation of motion resulting from this is

$$\partial^2\phi(\boldsymbol{x},t)/\partial t^2 = -\delta V[\phi]/\delta\phi(\boldsymbol{x},t) \tag{5.21}$$

where the right-hand side is a functional derivative. Recall that the coordinates of this system are the fields $\phi(\boldsymbol{x})$ and consequently the potential energy is a function of the functions $\phi(\boldsymbol{x})$, i.e. a functional. Notice the similarity of the field equation written in the form (5.21), to Newton's equation (5.1). The static solutions of (5.21) are once again the extreme of the potential in field-space, satisfying

$$\delta V[\phi]/\delta\phi(\boldsymbol{x}) = 0. \tag{5.22}$$

In particular, stable static solutions are minima of V, just as in particle mechanics.

Let $\phi(\boldsymbol{x}) = \phi_0(\boldsymbol{x})$ be one such minimum. We can make a functional Taylor expansion of V about ϕ_0:

$$V[\phi] = V[\phi_0] + \int \mathrm{d}\boldsymbol{x}\, \frac{1}{2}\left\{\eta(\boldsymbol{x})\left[-\nabla^2 + \left(\frac{\mathrm{d}^2 U}{\mathrm{d}\phi^2}\right)_{\phi_0(\boldsymbol{x})}\right]\eta(\boldsymbol{x}) + \ldots\right\} \tag{5.23}$$

where $\eta(\boldsymbol{x}) \equiv \phi(\boldsymbol{x}) - \phi_0(\boldsymbol{x})$, integration by parts has been used, and the dots represent cubic and higher terms in η. This functional Taylor series is obtained by taking functional derivatives of $V[\phi]$. The operator $(-\nabla^2 + \mathrm{d}^2U/\mathrm{d}\phi^2)$ evaluated at $\phi(\boldsymbol{x}) = \phi_0(\boldsymbol{x})$ is the generalisation of the matrix of second derivatives of V occurring in (5.16). Its eigenvalues and eigenfunctions are given by the differential equation

$$\left[-\nabla^2 + \left(\frac{\mathrm{d}^2 U}{\mathrm{d}\phi^2}\right)_{\phi_0(\boldsymbol{x})}\right]\eta_i(\boldsymbol{x}) = \omega_i^2\,\eta_i(\boldsymbol{x}) \tag{5.24}$$

where $\eta_i(\boldsymbol{x})$ are the orthonormal 'normal modes' of fluctuations around $\phi_0(\boldsymbol{x})$. Next we write (Creutz 1975)

$$\begin{aligned}\eta(\boldsymbol{x},t) &\equiv \phi(\boldsymbol{x},t) - \phi_0(\boldsymbol{x})\\ &\equiv \sum_i c_i(t)\,\eta_i(\boldsymbol{x}).\end{aligned} \tag{5.25}$$

Then, the Lagrangian (5.18) becomes, on using (5.23) and (5.24),

$$L = \tfrac{1}{2}\sum_i [\dot{c}_i(t)]^2 - \left(V[\phi_0] + \tfrac{1}{2}\sum_i [c_i(t)]^2\,\omega_i^2\right) + \text{corrections} \tag{5.26}$$

where $\dot{c}_i \equiv \mathrm{d}c_i/\mathrm{d}t$, and the corrections are the contributions of the third and higher derivatives in the Taylor series (5.23). The 'weak-coupling approximation' refers, in this context, to treating these correction terms in perturbation. This would require that the magnitudes of the fluctuation $\eta(\boldsymbol{x})$, as well as the third and higher derivatives of $V[\phi]$ at ϕ_0, be small. To lowest order in this approximation, the corrections in (5.26) are neglected altogether. Then the Lagrangian (5.26) clearly reduces to that of a set of harmonic oscillations, one for each normal mode, apart from a constant term $V[\phi_0]$.

Correspondingly, in quantum theory one can construct a set of approximate harmonic-oscillator states, spread in field space around $\phi_0(x)$, by quantising the normal-mode coefficients c_i. The energies of these states would be

$$E_{\{n_i\}} = V[\phi_0] + \hbar \sum_i (n_i + \tfrac{1}{2})\,\omega_i + \text{corrections} \tag{5.27}$$

where n_i is the excitation number of the ith normal mode. This is the analogue of eq. (5.13). It relates, approximately, the energies of certain quantum levels, to the classical solution $\phi_0(x)$. The first term is just the classical energy of the static solution $\phi_0(x)$. The second term involves ω_i, which are the stability frequencies of $\phi_0(x)$. Note that (5.24) is just the classical linear stability equation. The correction terms in (5.27) are to be obtained by standard perturbation techniques in powers of the anharmonic correction terms in (5.26). Once again, we emphasise that if any of the ω_i are equal to zero, the simple ideas given here need to be modified.

This is the basic idea behind the promised association of a set of quantum levels with any given stable static classical solution $\phi_0(\boldsymbol{x})$. These results are valid only in the weak-coupling approximation stated above. We will work throughout this chapter under that approximation. The solution ϕ_0 could either be an absolute minimum or a local minimum of $V[\phi_0]$. If ϕ_0 is an absolute (but possibly degenerate) minimum, then it is the 'classical vacuum' of the system. Because of the $(\nabla\phi)^2$ term in $V[\phi]$, such an absolute minimum would have to be space-independent. The quantum level constructed by the above procedure around such an absolute minimum would be the vacuum state and the familiar quanta of that field theory. To illustrate this, consider that typical example, the usual ϕ^4 field theory, given by

$$L = \int d\boldsymbol{x}\,[\tfrac{1}{2}(\partial\phi/\partial t)^2 - \tfrac{1}{2}(\nabla\phi)^2 - \tfrac{1}{2}m^2\phi^2 - \tfrac{1}{4}\lambda\phi^4]. \tag{5.28}$$

Then

$$\begin{aligned} V[\phi] &= \int d\boldsymbol{x}\,[\tfrac{1}{2}(\nabla\phi)^2 + \tfrac{1}{2}m^2\phi^2 + \tfrac{1}{4}\lambda\phi^4] \\ &= \int d\boldsymbol{x}\,[-\tfrac{1}{2}\phi\nabla^2\phi + \tfrac{1}{2}m^2\phi^2 + \tfrac{1}{4}\lambda\phi^4] \end{aligned} \tag{5.29}$$

and a static solution obeys,

$$\begin{aligned} \delta V[\phi]/\delta\phi(\boldsymbol{x}) &= -\nabla^2\phi + m^2\phi + \lambda\phi^3 \\ &= 0. \end{aligned} \tag{5.30}$$

This permits the trivial static solution

$$\phi_0(\boldsymbol{x},t)=0, \qquad \text{with } V[\phi_0=0]=0 \tag{5.31}$$

which is the absolute minimum of $V[\phi]$. The potential $V[\phi]$ as written in (5.29) is already Taylor-expanded about $\phi=0$. The second derivative of the potential at $\phi=0$ is the operator $(-\nabla^2+m^2)$, whose eigenvalue equation

$$(-\nabla^2+m^2)\eta_i(\boldsymbol{x})=\omega_i^2\eta_i(\boldsymbol{x}) \tag{5.32}$$

is trivially solved to give $\eta_i(\boldsymbol{x})=L^{-3/2}\exp(\mathrm{i}\boldsymbol{k}_i\cdot\boldsymbol{x})$; $\boldsymbol{k}_i=2\pi\boldsymbol{N}_i/L$, with $L\to\infty$, and $\omega_i^2=\boldsymbol{k}_i^2+m^2$. The weak-coupling approximation amounts to perturbing in λ. The set of approximate harmonic oscillator levels built around $\phi=0$ have energies

$$\begin{aligned}E_{\{n_i\}}&=V[\phi_0]+\hbar\sum_{\boldsymbol{k}_i}\sqrt{\boldsymbol{k}_i^2+m^2}\,(n_i+\tfrac{1}{2})+\mathrm{O}(\lambda)\\&=\hbar\sum_{\boldsymbol{k}_i}\sqrt{\boldsymbol{k}_i^2+m^2}\,(n_i+\tfrac{1}{2})+\mathrm{O}(\lambda).\end{aligned} \tag{5.33}$$

(In relativistic quantum field theory, one often works in units where $\hbar=c=1$. Since we wish to bring out the semiclassical nature of our theory, we will explicitly retain $\hbar$ for a few steps, but c will be set equal to unity. In that case, the parameter m has dimensions (mass/$\hbar$).) The lowest of these levels with $n_i=0$ is the ground state of this quantum field theory, i.e. the vacuum state. Its energy is

$$E_{\mathrm{vac}}\equiv E_{\{0\}}=\tfrac{1}{2}\hbar\sum_{\boldsymbol{k}_i}\sqrt{\boldsymbol{k}_i^2+m^2}+\mathrm{O}(\lambda). \tag{5.34}$$

The next excited state, where the $\boldsymbol{k}_i=0$ mode is excited once, has energy, $E_1\equiv E_{\{n_0=1,\,n_{i\neq 0}=0\}}$.

$$\begin{aligned}E_1&=\tfrac{1}{2}\hbar\sum_{\boldsymbol{k}_i}\sqrt{\boldsymbol{k}_i^2+m^2}+\hbar m+\mathrm{O}(\lambda)\\&=E_{\mathrm{vac}}+\hbar m+\mathrm{O}(\lambda).\end{aligned} \tag{5.35}$$

Notice that $E_{\{n_i\}}$ may formally diverge because of the infinitely many modes $\boldsymbol{k}_i$. The treatment of such divergence through renormalisation is well known. We will return to this point in later sections. For the moment, note that to the order λ^0 explicitly shown, $(E_{\{n_i\}}-E_{\mathrm{vac}})$ is finite. Thus E_1-E_{vac} represents the energy of the single-quantum particle at rest with mass $\sim\hbar m$. Other excited states will yield the quantum particle travelling with momentum $\hbar\boldsymbol{k}$ and so on. These results are only too familiar as products of standard perturbative field theory. We have recalled them here only to emphasise that standard perturbation theory can be considered as a special case of our semiclassical method – where the classical static solution used is the trivial absolute minimum of the potential $V[\phi]$. We are concerned in this chapter with newer results obtained when the same procedure is repeated for a local minimum of $V[\phi]$, i.e. a non-trivial x-dependent classical solution. Unfortunately the model (5.28) permits no such non-trivial finite-energy static solution. This is clear from the considerations of chapters 2 and 3. In the next section we therefore consider a slightly different example, which does permit a non-trivial static solution, to illustrate the new content of these methods.

First we shall make some remarks about proof. In each case, we have expanded the potential about a minimum, and diagonalised the quadratic term. This gives a sum of harmonic oscillator terms, plus anharmonic terms which are, by assumption, to be treated perturbatively. After that point, we have merely stated the leading-order results with no further 'derivation'. This is only because they are straightforward generalisations of standard harmonic oscillator results, derived in any elementary text on quantum mechanics. Higher-order corrections to the above equations are a different matter, particularly for the field theoretical applications to soliton quantisation, where they will require a careful adaptation of perturbation theory. But much of the basic physics of soliton quantisation can be explored already at the level of the leading quantum corrections, which involve only the harmonic oscillator approximation. We will therefore continue with our heuristic approach for the rest of this chapter, to keep the physics transparent. Later, we will show how these results can be formally derived using functional or operator methods.

5.3. Quantisation of the kink solution

Consider the example of a scalar field $\phi(x,t)$ in $(1+1)$ dimensions with a Lagrangian that is slightly different from (5.28), namely

$$L = \int \mathrm{d}x \left[\frac{1}{2}\left(\frac{\partial \phi}{\partial t}\right)^2 - \tfrac{1}{2}\left(\frac{\partial \phi}{\partial x}\right)^2 + \frac{1}{2} m^2 \phi^2 - \frac{\lambda}{4}\phi^4 - \frac{m^4}{4\lambda} \right]. \tag{5.36}$$

We have already studied this system at the classical level in chapter 2, and know that it yields a static kink solution. Now, let us apply the semiclassical method of the preceding section to it. The potential is

$$V[\phi] = \int \mathrm{d}x \left[\frac{1}{2}\left(\frac{\partial \phi}{\partial x}\right)^2 + \frac{\lambda}{4}\left(\phi^2 - \frac{m^2}{\lambda}\right)^2 \right]. \tag{5.37}$$

The equation obeyed by a static classical solution is

$$\frac{\partial V[\phi]}{\partial \phi(x)} = -\frac{\partial^2}{\partial x^2}\phi - m^2\phi + \lambda\phi^3 = 0. \tag{5.38}$$

This equation yields, as we saw in chapter 2, the trivial solutions $\phi(x,t) = \pm m/\sqrt{\lambda}$ as well as the non-trivial kink solutions $\phi_K(x, t) = \pm(m/\sqrt{\lambda})\tanh[m(x-a)/\sqrt{2}]$. Our semiclassical method can be applied to each of these static solutions.

5.3.1. *The vacuum and its excitations*

Let us begin with the trivial solution $\phi_1(x,t) = m/\sqrt{\lambda}$. The potential can be expanded around ϕ_1 as

$$\begin{aligned}V[\phi] &= V[\phi_1] + \int \mathrm{d}x\left(-\tfrac{1}{2}\tilde{\phi}\frac{\partial^2\tilde{\phi}}{\partial x^2} - \tfrac{1}{2}m^2\tilde{\phi}^2 \right.\\ &\quad \left. + \tfrac{3}{2}\lambda\phi_1^2\tilde{\phi}^2\right) + \lambda\int \mathrm{d}x\,(\phi_1\tilde{\phi}^3 + \tfrac{1}{4}\tilde{\phi}^4)\\ &= \int \mathrm{d}x\frac{\tilde{\phi}}{2}\left(-\frac{\partial^2}{\partial x^2} + 2m^2\right)\tilde{\phi} + m\sqrt{\lambda}\int\tilde{\phi}^3\,\mathrm{d}x + \frac{\lambda}{4}\int\tilde{\phi}^4\,\mathrm{d}x\end{aligned} \tag{5.39}$$

where

$$\tilde{\phi}(x) \equiv \phi(x) - \phi_1(x) = \phi(x) - m/\sqrt{\lambda}$$

and

$$V[\phi_1] = 0.$$

If the constant λ is sufficiently small, we can hope to treat the cubic and quartic terms in (5.39) by perturbation. In the lowest-order quadratic term, the second derivative of $V[\phi]$ at ϕ_1 is the operator $(-\partial^2/\partial x^2 + 2m^2)$ whose eigenvalues are clearly $(k_n^2 + 2m^2)$, with eigenfunctions $e^{ik_n x}$. The allowed values of k_n are obtained, in box-normalisation by

$$k_n L = 2n\pi \tag{5.40}$$

where L, the length of the box, will ultimately tend to infinity with the replacement

$$\sum_{k_n} \rightarrow \frac{L}{2\pi}\int \mathrm{d}k.$$

Using all this, we can construct a tower of approximate harmonic oscillator states around ϕ_1, the lowest of which will have energy

$$E_{\text{vac}} = 0 + \tfrac{1}{2}\hbar \sum_n (k_n^2 + 2m^2)^{1/2} + O(\lambda) \tag{5.41}$$

where the zero represents the classical energy $V[\phi_1]$. This is the vacuum state of the system. Higher excitations, analogous to (5.33) will have energies

$$E_{\{N_n\}} = \hbar \sum_n (N_n + \tfrac{1}{2})(k_n^2 + 2m^2)^{1/2} + O(\lambda). \tag{5.42}$$

These correspond to the familiar quanta of the theory, where N_n of them have momentum $\hbar k_n$. Their rest mass is $m\hbar\sqrt{2}$ to lowest order. We will call this set of states built around ϕ_1 the 'vacuum sector'. Since this procedure essentially quantises the shifted field $\tilde{\phi} = \phi - \phi_1$ as in standard perturbation methods, we can borrow the familiar result to lowest order that

$$\langle 0|\tilde{\phi}(x,t)|0\rangle = 0 \tag{5.43}$$

where $|0\rangle$ is the vacuum state. Thus

$$\langle 0|\phi(x,t)|0\rangle = \phi_1 + O(\lambda). \tag{5.44}$$

This is the analogue of (5.10) and is another relation connecting the quantum vacuum state to the classical solution ϕ_1, apart from the energy relation (5.41).

These simple results are again familiar to us from conventional perturbation theory, as applied to this system. Before the advent of soliton quantisation, the quanta derived above were the only species of particles explicitly obtained for this system. We will call these quanta, for convenience, the 'mesons' of this model. Now we will see that this same field theory yields in addition to these mesons, another species of particles, obtained when this same procedure is applied to the kink solution.

[An identical set of vacuum and many-meson states can also be built around $\phi_2 = -m/\sqrt{\lambda}$, which differs from the earlier set only in that $\langle 0|\phi(x,t)|0\rangle = \phi_2$ for this vacuum. These two vacua individually violate the $\phi \leftrightarrow -\phi$ symmetry of the parent Lagrangian. This is one of the simplest examples of 'spontaneous symmetry breaking' – a well known phenomenon in quantum field theory, which we shall not discuss in detail here. Instead we proceed to the non-trivial kink solution.]

5.3.2. *The quantum kink and its excitations*

Let us now apply the same method to the static kink solution,

$$\phi_K(x-a) = (m/\sqrt{\lambda})\tanh\left[m(x-a)/\sqrt{2}\right] \tag{5.45}$$

with energy

$$V[\phi_K] = (2\sqrt{2}/3)(m^3/\lambda). \tag{5.46}$$

This is actually a family of solutions, for different values of a. Let us, to be specific, take the case $a = 0$. Since the kink is a static solution satisfying (5.38), it is an extremum of $V[\phi]$. When this potential is expanded, this time about $\phi_K(x)$, the linear term will again be absent, thanks to (5.38). We get,

$$V[\phi] = V[\phi_K] + \int \mathrm{d}x\,\tfrac{1}{2}\eta(x)\left(-\frac{\partial^2}{\partial x^2} - m^2 + 3\lambda\phi_K^2\right)\eta(x) + \lambda\int \mathrm{d}x\,(\phi_K\eta^3 + \tfrac{1}{4}\eta^4) \tag{5.47}$$

where $\eta(x) \equiv \phi(x) - \phi_K(x)$. The eigenvalues of the second derivative of $V[\phi]$ at ϕ_K are given by the equation

$$\begin{aligned}\left(-\frac{\partial^2}{\partial x^2} - m^2 + 3\lambda\phi_K^2\right)\eta_n(x) &= \left[-\frac{\partial^2}{\partial x^2} - m^2 + 3m^2\tanh^2\left(\frac{mx}{\sqrt{2}}\right)\right]\eta_n(x)\\ &= \omega_n^2\,\eta_n(x).\end{aligned} \tag{5.48}$$

This equation, although less trivial than (5.32), is still exactly soluble. Upon changing variables to $z = mx/\sqrt{2}$, this equation reduces to a Schrödinger-like equation,

$$\left(-\frac{1}{2}\frac{\partial^2}{\partial z^2} + (3\tanh^2 z - 1)\right)\tilde{\eta}_n(z) = \frac{\omega_n^2}{m^2}\,\tilde{\eta}_n(z). \tag{5.49}$$

The eigenfunctions and eigenvalues of this Schrödinger equation are exactly known (Morse and Feshbach 1953). It has two discrete levels followed by a continuum. The discrete levels are:

$$\omega_0^2 = 0 \quad \text{with } \tilde{\eta}_0(z) = 1/\cosh^2 z \tag{5.50a}$$

$$\omega_1^2 = \tfrac{3}{2}m^2 \quad \text{with } \tilde{\eta}_1(z) = \sinh z/\cosh^2 z \tag{5.50b}$$

This is followed by a continuum of levels which we shall label by q rather than by $n \geqslant 2$. These are:

$$\omega_q^2 = m^2(\tfrac{1}{2}q^2+2)$$

with

$$\tilde{\eta}_q(z) = e^{iqz}(3\tanh^2 z - 1 - q^2 - 3iq\tanh z). \tag{5.50c}$$

The allowed values of q in (5.50c), like the allowed value of k_n in (5.40), are fixed by periodic boundary conditions in a box of length L, with $L \to \infty$. Note that $\tilde{\eta}_q(z)$ in (5.50c) has an asymptotic behaviour

$$\tilde{\eta}_q(z) \xrightarrow[z\to\pm\infty]{} \exp\left[i(qz \pm \tfrac{1}{2}\delta(q))\right] \tag{5.51}$$

where

$$\delta(q) = -2\tan^{-1}[3q/(2-q^2)] \tag{5.52}$$

is just the phase shift of the scattering states of the associated Schrödinger problem (5.49). Also, the length of the box, in terms of the variable $z = mx/\sqrt{2}$, is $mL/\sqrt{2}$. Thus, periodic boundary conditions require that

$$q_n(mL/\sqrt{2}) + \delta(q_n) = 2n\pi \tag{5.53}$$

where n is any positive or negative integer. This fixes the allowed values q_n. In the $L \to \infty$ limit, these allowed values merge into a continuum with the replacement

$$\sum_{q_n} \to \frac{1}{2\pi}\int_{-\infty}^{\infty} dq\left(\frac{mL}{\sqrt{2}} + \frac{\partial}{\partial q}[\delta(q)]\right). \tag{5.54}$$

Using these normal modes, we can again diagonalise the potential (5.47) and the Lagrangian in the vicinity of $\phi_K(x)$, up to order λ^0. All that is required is the application of (5.25)–(5.26) to this example, using (5.50) for the normal modes and their frequencies. Since $\phi_K \sim O(1/\sqrt{\lambda})$, the cubic and quartic terms in (5.47) will be of order $\sqrt{\lambda}$ and order λ respectively. These are, by assumption, to be treated perturbatively. Consequently, in quantum theory we should once again expect to construct a set of approximate harmonic oscillator states around the point $\phi_K(x)$ in field space. We should also expect the energies of these states to be, upon using (5.27) with (5.50),

$$\tilde{E}_{\{N_n\}} = V[\phi_K] + \hbar\sum_{n=0}^{\infty}(N_n+\tfrac{1}{2})\,\omega_n + O(\lambda) \tag{5.55}$$

$$= \frac{2\sqrt{2}\,\mathrm{m}^3}{3\lambda} + (N_1 + \tfrac{1}{2})\hbar\sqrt{\tfrac{3}{2}}\,m + m\hbar\sum_{q_n}(N_{q_n} + \tfrac{1}{2})\,(\tfrac{1}{2}q_n^2 + 2)^{1/2}$$

$$+ \mathrm{O}(\lambda). \tag{5.56}$$

However, there is a catch: while this analysis is essentially valid for all the $n \geqslant 1$ modes in (5.50), it does not hold for the $n = 0$ mode, because $\omega_0 = 0$ in (5.50a). We have repeatedly cautioned against the use of the harmonic oscillator expansion when $\omega = 0$ for obvious reasons. Unlike the $n \geqslant 1$ modes which are genuine vibrational modes for small fluctuations, the $n = 0$ mode is not vibrational at all. The 'spring constant' ω_0 vanishes. Correspondingly, the quantum wavefunction along the $n = 0$ mode will not be confined near a given classical solution, but will tend to spread. This same difficulty will be reflected in the form of technical problems at the computational level. If we try to naively compute the higher-order corrections to (5.55), where the $n = 0$ mode is coupled to the others, the zero frequency will occur in energy denominators, leading to divergences.

Strictly speaking, therefore, we must first learn how to treat such zero-frequency modes before proceeding further with our analysis of kink quantisation. However, even though the physical reasons behind the occurrence of these zero-modes are easy to understand (see section 5.5) their systematic treatment is technically quite involved. Indeed, all of chapter 8 will be devoted to treating such zero-modes. Fortunately, one can explore much of the physics of the states (5.55) without getting involved in details of the zero-mode's treatment. Let us, for the moment, do so in the interests of simplicity. After all, the majority of the modes (all except the $n = 0$ mode, in our example) are vibrational and the quantum field-theoretical wavefunctions will be approximately harmonic-oscillator-like along these modes. Also, the energy expressions (5.55) are not seriously affected to order λ^0, if we start with a non-zero ω_0 and let $\omega_0 \to 0$. The first term is independent of ω_n, and in the next term the contribution of the $n = 0$ mode merely vanishes as $\omega_0 \to 0$. Difficulties will arise only in the $\mathrm{O}(\lambda)$ terms not explicitly shown in (5.55). Let us therefore restrict ourselves to $\mathrm{O}(\lambda^0)$ for the present, and proceed with our simple-minded analysis. We will return to a qualitative discussion of the $n = 0$ mode in section (5.5) and to a fuller treatment of such modes in chapter 8.

With this stipulation, let us go back to the tower of states (5.55) built around the kink solution in the small-λ approximation. We offer the following interpretation of these states, to be supported by the ensuing discussion in the rest of this chapter.

(i) The lowest-energy state in the family (5.55), which corresponds to

$\{N_n = 0\}$, will be interpreted as the state of the quantum kink particle, at rest. Note that although this state has lowest energy in the 'kink sector', i.e. in the family (5.55), it is not the absolute ground state or vacuum of this theory. The vacuum of this theory has already been identified as the lowest-energy state in the 'vacuum sector', i.e. in the family (5.42) built around ϕ_1. That state has an energy given in (5.41), which has a lower value than the energy of all the other states in both families (5.42) and (5.55), to the order λ^0 exhibited. Therefore the $\{N_n = 0\}$ state in (5.55) is not the vacuum and will be interpreted instead as the quantum kink. It will be seen to have properties expected of an extended particle.

(ii) The next higher energy level in (5.55), to order λ^0, arises when the $n = 1$ mode is excited once, i.e. when $N_1 = 1$. This has energy

$$\tilde{E}_1 \equiv \tilde{E}_{\{N_1 = 1;\, N_{q_n} = 0\}} = \tilde{E}_0 + \sqrt{\tfrac{3}{2}}\, m\hbar + O(\lambda).$$

This state may be interpreted as a discrete excited state of the kink particle. Higher excitations of this mode ($N_1 > 1$) give higher excited states of the kink. The relationship of the basic kink state to its discrete excited states is rather similar to the relationship of the ground state to the excited states of an atom. Remember that an atom, in our terminology, is also an extended particle. It has discrete energy and localised form factors.

(iii) The remaining states in the family (5.55), obtained by exciting the $n \geqslant 2$ modes (or equivalently the $N_q \neq 0$ states in (5.56)) have a different interpretation. They can be thought of as scattering states of the mesons of this theory in the presence of the kink particle. In fact, the modes $\tilde{\eta}_q$ in (5.50c) and their energies can be considered, to order λ^0, as the reduced one-particle wavefunctions and energies of the mesons when they scatter off the kink. In support of this interpretation, notice that the excitation energy of the qth mode is, from (5.50c),

$$\hbar\omega_q = \hbar\left(\tfrac{1}{2}m^2q^2 + 2m^2\right)^{1/2}$$

which is just the kinetic energy of a meson with momentum $\hbar mq/\sqrt{2}$. Recall that in this theory, the meson has mass $\sqrt{2}m\hbar$. That the asymptotic momentum of the scattered meson is written in the form $\hbar mq/\sqrt{2}$ is just an artifact of the change of variables from x to $z = mx/\sqrt{2}$. If, in the eigenfunctions $\tilde{\eta}_q(z)$ in (5.51), we go back to the variable x, they have the asymptotic form

$$\exp\left[\mathrm{i}\left((qmx/\sqrt{2}) \pm \tfrac{1}{2}\delta\right)\right],$$

consistent with an asymptotic momentum $\hbar mq/\sqrt{2}$. The phaseshift $\delta(q)$ is the scattering phaseshift for a meson interacting with the kink.

It may be noticed that in this interpretation of these states as meson–kink scattering states, only the kinetic energy of the meson appears in (5.56), and not that of the kink. This is because, to $O(\lambda^0)$, the kink may be considered as static, and producing a static potential in which the mesons scatter. The reason, as will be clear in the next section, is that the kink mass M is of order $(1/\lambda)$. In the weak-coupling limit, it is very heavy and essentially static. The leading term in the kink's kinetic energy will have the form $P^2/2M$, which is of order λ since $M \sim O(1/\lambda)$. Thus, in our analysis up to order λ^0, the static limit for the kink particle is valid. Based on these observations, we suggest that when one of the continuum modes is excited once ($N_q = 1$ for some q in (5.56)), we get a meson–kink scattering state. When several such modes are excited, once or more than once, we get a state with several mesons with the corresponding asymptotic momenta, all scattering off the kink.

This is our interpretation of the family of quantum states constructed around the classical kink, in the weak-coupling approximation. In the remaining sections, we will explore some more properties of these states, which will be seen to further support such an interpretation.

5.4. The kink mass and its renormalisation

Let us now explicitly evaluate the mass of the quantum kink particle. This has been worked out carefully by Dashen et al. (1974b) in one of their pioneering papers in this field. We have associated the kink particle with the lowest energy level in the set (5.55). This energy is given by

$$\tilde{E}_0 = 2\sqrt{2}\, m^3/3\lambda + \tfrac{1}{2}\hbar m\sqrt{\tfrac{3}{2}} + \tfrac{1}{2}\sum_{q_n} \hbar m\,(\tfrac{1}{2}q_n^2 + 2)^{1/2} + O(\lambda). \tag{5.57}$$

This expression is formally divergent. The infinite series over $\sum_{q_n}$ becomes, in the continuum limit (5.54), a quadratically divergent integral. That in itself need not worry us since the energy of the vacuum (5.41) is also quadratically divergent. What matters physically is the difference in energy between any given state and the vacuum state. This difference, obtained by subtracting (5.41) from (5.57), is

$$\tilde{E}_0 - E_{\text{vac}} = 2\sqrt{2}\, m^3/3\lambda + \tfrac{1}{2}\sqrt{\tfrac{3}{2}}\,\hbar m + \tfrac{1}{2}\hbar\sum_n \left[m\,(\tfrac{1}{2}q_n^2 + 2)^{1/2} - (k_n^2 + 2m^2)^{1/2}\right] + O(\lambda). \tag{5.58}$$

Since both terms in the bracket are divergent, we must subtract them carefully so as not to lose finite pieces. Let us start with a finite box L. Then k_n and q_n are related by (5.40) and (5.53) which give

$$2n\pi = k_n L = q_n mL/\sqrt{2} + \delta(q_n). \tag{5.59}$$

Thus, the term in brackets in (5.58) becomes

$$\{[(k_n - \delta_n/L)^2 + 2m^2]^{1/2} - (k_n^2 + 2m^2)^{1/2}\}$$
$$= -(k_n\delta_n/L)(k_n^2 + 2m^2)^{-1/2} + \mathrm{O}(1/L^2). \tag{5.60}$$

Going to the $L \to \infty$ limit and using

$$\sum_{k_n} \to \frac{L}{2\pi}\int \mathrm{d}k,$$

we have

$$\tilde{E}_0 - E_{\text{vac}} = \frac{2\sqrt{2}m^3}{3\lambda} + \tfrac{1}{2}\sqrt{\tfrac{3}{2}}\hbar m - \frac{\hbar}{4\pi}\int_{-\infty}^{\infty} \mathrm{d}k \frac{k\delta(k)}{\sqrt{k^2 + 2m^2}} + \mathrm{O}(\lambda). \tag{5.61}$$

Here $\delta(k)$ written in terms of k, using the original q-dependence of (5.52) is

$$\delta(k) = -2\tan^{-1}\left(\frac{3\sqrt{2}km}{2m^2 - 2k^2}\right) + \mathrm{O}(1/L) \tag{5.62}$$

where the $\mathrm{O}(1/L)$ terms can now be dropped. Integrating by parts, we get

$$\tilde{E}_0 - E_{\text{vac}} = \frac{2\sqrt{2}m^3}{3\lambda} + \tfrac{1}{2}\sqrt{\tfrac{3}{2}}\hbar m - \frac{\hbar}{4\pi}\left[\delta(k)\sqrt{k^2 + 2m^2}\right]_{-\infty}^{\infty}$$
$$+ \frac{\hbar}{4\pi}\int_{-\infty}^{\infty} \mathrm{d}k\sqrt{k^2 + 2m^2}\,\frac{\mathrm{d}}{\mathrm{d}k}[\delta(k)] + \mathrm{O}(\lambda).$$

Using the phaseshift in (5.62), this gives, after some elementary algebra,

$$\tilde{E}_0 - E_{\text{vac}} = \frac{2\sqrt{2}m^3}{3\lambda} + \tfrac{1}{2}\sqrt{\tfrac{3}{2}}\hbar m - \frac{3\hbar m}{\pi\sqrt{2}}$$
$$- \frac{6m\hbar}{4\pi\sqrt{2}}\int_{-\infty}^{\infty} \frac{\mathrm{d}p\,(p^2 + 2)}{\sqrt{p^2 + 4}\,(p^2 + 1)} + \mathrm{O}(\lambda) \tag{5.63}$$

where in the last integral, the variable $p \equiv k\sqrt{2}/m$ has been used. While the quadratic divergence in $\tilde{E}_0$ has been removed by subtracting out E_{vac}, we notice that (5.63) still has a logarithmic divergence in the integral. It cannot be removed by merely adding a divergent constant to the Lagrangian, since (5.63) is a difference between two energy levels. Nevertheless, this divergence at this stage of the calculation need not cause us concern. In fact, we should expect it to be there, and it can be removed by normal-ordering the Hamiltonian.

It would be too much of a digression for us to introduce normal-ordering or renormalisation theory from first principles. The occurence of 'ultraviolet' divergences in quantum field theory due to the short-distance behaviour of products of field operators is well known in standard perturbation theory (Bogoliubov and Shirkov 1959). So is their removal by adding suitable 'counter terms' to the Hamiltonian. For the purposes of this section, we will expect the reader to be familiar with these time-honoured procedures. In particular, for our example, we require renormalisation only to the extent of normal ordering. For a ϕ^4 theory in two dimensions, the only divergences that occur are known to be removable by just normal-ordering. Recall that the Hamiltonian for our theory obtained from (5.36) is

$$H = \int dx \left[\frac{1}{2}\pi^2 + \frac{1}{2}\left(\frac{\partial \phi}{\partial x}\right)^2 - \frac{1}{2}m^2\phi^2 + \frac{\lambda}{4}\phi^4 + \frac{m^4}{4\lambda} \right] \tag{5.64}$$

where $\pi(x, t) = \dot{\phi}(x, t)$ is the canonical momentum. In the quantised theory, operators such as $\phi^2(x, t)$, $\phi^4(x, t)$ etc. are formally divergent and ill-defined, and therefore so is the Hamiltonian. Consequently, energy levels calculated naively from this Hamiltonian will also be divergent. This is the reason behind the divergence in (5.63).

Clearly, this problem has nothing especially to do with the kink solution or its quantisation. The Hamiltonian (5.64) is common to all the states in this field theory, including the familiar vacuum and meson states (5.41) and (5.42). We have already noted that the naive vacuum energy (5.41) was divergent. Further if we had calculated the mass of the meson to higher order, (to lowest order it was $\sqrt{2}\,m\hbar$), we would have again encountered a logarithmic divergence. Thus, the removal of such divergences is essential even in the familiar vacuum sector of states. This is accomplished by replacing the Hamiltonian by its normal-ordered from :H:. In our semiclassical presentation, it would be difficult to work directly with the normal-ordered form. However, the latter can be written as the original non-ordered form plus some counter terms, since

$$:\phi^4: = \phi^4 - A\phi^2 - B$$
$$:\phi^2: = \phi^2 - C \tag{5.65}$$

where A, B and C are constants which diverge in perturbation theory. Therefore the normal-ordered Hamiltonian may be written as

$$:H: = H - \int_{-\infty}^{\infty} \mathrm{d}x \left(\tfrac{1}{2}\partial m^2 \phi^2 + D\right) \tag{5.66}$$

where the constants ∂m^2 and D may be evaluated in perturbation theory by standard methods. For ∂m^2, the contribution to order λ comes from the Feynman graph in fig. 10, which yields

$$\partial m^2 = \frac{3\lambda\hbar}{4\pi} \int_{-m\Lambda}^{m\Lambda} \frac{\mathrm{d}p}{\sqrt{p^2+2m^2}} = \frac{3\lambda\hbar}{4\pi} \int_{-\Lambda}^{\Lambda} \frac{\mathrm{d}p}{\sqrt{p^2+2}} \tag{5.67}$$

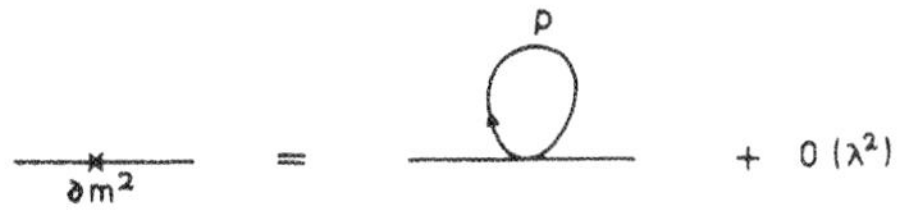

Fig. 10. The one-loop graph yielding the mass counter term ∂m^2 to order λ due to normal ordering. The numerical factors in (5.67) come from the combinatorial factors of associating each field operator in ϕ^4 (x) with each line in the graph.

where $m\Lambda$ is the momentum cut-off. Notice that consistantly with our semiclassical presentation, we have explicitly retained the correct $\hbar$-dependence of the one loop graph (see Coleman 1975a). We will not evaluate the other constant D since the kink mass involves the difference between two energy levels where the effects of D will cancel. It is well known that the counter terms in (5.66) remove, to any given order, all ultraviolet divergences in the vacuum sector of this theory, including the energies of the vacuum and one-meson states. We will now show that the same counter teams also remove the logarithmic divergence in the kink mass (5.63). This is most satisfactory because in a renormalisable theory all physical quantities should be rendered free of ultraviolet divergences by the same set of counter terms.

The replacement of H by $:H:$ as per (5.66) amounts to altering the potential (5.37) by the additional terms

$$\Delta V = -\int dx(\tfrac{1}{2}\partial m^2 \phi^2(x) + D).$$

Given functions $\phi_K(x)$ and $\phi_1(x)$, their classical energies will be altered. This in turn will affect the quantum energy levels $\tilde{E}_0$ and E_{vac}, since the classical energies constitute their leading terms. As a result, we have to add to (5.63) an amount

$$\begin{aligned}\Delta\tilde{E}_0 - \Delta E_{vac} &= \int_{-\infty}^{\infty} dx[-\tfrac{1}{2}\partial m^2(\phi_K^2 - \phi_1^2) - (D - D)] \\ &= \tfrac{1}{2}\partial m^2 \int_{-\infty}^{\infty} dx(m^2/\lambda)(1 - \tanh^2 mx/\sqrt{2}) \\ &= \partial m^2 m\sqrt{2}/\lambda. \qquad (5.68)\end{aligned}$$

This is the leading contribution of the counter terms to the kink mass. Other contributions will come because: (a) the quantum corrections involve the derivatives of the potential which has now been altered, and (b) the classical solutions themselves will change because the field equation will also get altered. These are however higher-order effects (see remark (vi) below).

Upon inserting ∂m^2 from (5.67) into (5.68) and adding this to (5.63), we get for the renormalised kink mass,

$$\begin{aligned}M &\equiv (\tilde{E}_0 + \Delta\tilde{E}_0) - (E_{vac} + \Delta E_{vac}) \\ &= \frac{2\sqrt{2}m^3}{3\lambda} + m\hbar\left(\frac{1}{2}\sqrt{\frac{3}{2}} - \frac{3}{\pi\sqrt{2}}\right) \\ &\quad - \frac{3\sqrt{2}m\hbar}{4\pi}\int_{-\Lambda}^{\Lambda} dp\left(\frac{(p^2+2)}{\sqrt{p^2+1}(p^2+4)} - \frac{1}{\sqrt{p^2+2}}\right) + O(\lambda\hbar^2) \qquad (5.69)\end{aligned}$$

where the cut-off has also been introduced into the integral in (5.63). We notice that both terms in the integrand behave as $1/p$ as $p \to \infty$, so that the logarithmic divergences cancel. The cut-off Λ can therefore be taken to infinity and the residual integral evaluated. The net result, quoted by Dashen et al. (1974b) is,

$$M = (2\sqrt{2}/3\lambda)m^3 + m\hbar(\tfrac{1}{6}\sqrt{\tfrac{3}{2}} - 3/\pi\sqrt{2}) + O(\lambda\hbar^2). \qquad (5.70)$$

This expression gives us the mass of the kink particle, i.e. the energy of the lowest state around the kink solution, after due normal-ordering. We make the following observations about this result:

(i) The first term in the mass of the quantum kink particle is the energy of the classical static kink solution. The next term represents the leading correction due to quantum fluctuations. Appropriately, the first term is of order $\hbar^0$ and the second term is of order $\hbar$.

(ii) The leading term, the energy of the classical kink, is singular as $\lambda \to 0$. Thus, our result is non-perturbative. It could not have been obtained by a perturbation expansion starting from the vacuum sector.

(iii) Nevertheless, this result is valid only in the weak-coupling approximation. The non-perturbative nature of the result is brought about solely by the classical contribution. The quantum corrections are being treated perturbatively in powers of λ.

(iv) Thus, the semiclassical expansion in powers of $\hbar$ and the weak-coupling expansion in increasing powers of λ are going hand in hand. This is to be expected. If we apply elementary dimensional analysis to the Lagrangian (5.36) (in our hybrid units where $c = 1$, but $\hbar$ is explicitly retained), the ratio $\lambda\hbar/m^2$ turns out to be dimensionless. This is in fact our dimensionless expansion parameter, and $\lambda\hbar/m^2 \ll 1$ is the weak-coupling condition. It explains why the $\hbar$-expansion and λ-expansion are the same in the quantisation of a static classical solution (see also chapter 6).

(v) In the weak-coupling limit $\lambda\hbar/m^2 \ll 1$, the mass of the kink state, thanks to the classical term, is much larger than the mass of the meson, which, to order $\hbar\lambda^{\circ}$ was $\sqrt{2}m\hbar$. The kink, in our approximation, is therefore a much heavier particle than the meson.

(vi) In computing the correction (5.68) to the kink mass, we have taken into account the change in energy of the given kink function (5.45) due to counter terms. But we did not include the fact that the kink function will itself change because of the counter terms. The latter are being added to the Lagrangian, and will alter the field equation to the extent of replacing $m^2\phi$ by $(m^2 + \delta m^2)\phi$. Thus, the kink solution will also suffer a replacement of m^2 by $(m^2 + \delta m^2)$. Let us write this change as

$$\phi_K \to \phi_K + \delta\phi_K,$$

where, since δm^2 is $O(\lambda\hbar)$, $\delta\phi_K/\phi_K$ will also be $O(\lambda\hbar)$. The change in kink energy will have an explicit and an implicit dependence on δm^2.

$$\Delta \tilde{E}_0 = \frac{\partial V(\phi_K)}{\partial m^2}\partial m^2 + \int dx \left(\frac{\delta V}{\delta\phi}\right)_{\phi_K} \delta\phi_K(x) + \text{order } (\hbar^2\lambda)$$

$$= \int \mathrm{d}x \left[\frac{1}{2} \phi_K^2(x)\,\delta m^2 + \left(\frac{\delta V}{\delta \phi}\right)_{\phi_K} \delta\phi_K(x) \right] + \mathrm{O}(\lambda\hbar^2). \tag{5.71}$$

We have included only the first term in (5.68). This is because ϕ_K is an extremum of V and hence the second term vanishes. The effect of $\delta\phi_K$ will be felt only in $\mathrm{O}(\hbar^2\lambda)$. The same arguments hold for the vacuum solution ϕ_1. Therefore the computation of the kink mass is valid as it stands in (5.70) to order $\hbar$.

This completes our discussion of the mass of the quantum kink particle. For calculations of higher-order corrections to the mass, see de Vega (1976) and Verwaest (1977).

5.5. The translation mode

Although a fuller treatment of zero-modes will have to wait until chapter 8, we can give a qualitative discussion now of the origin and physical consequences of the $\omega_0 = 0$ mode encountered in (5.50a). The presence of such a mode is not an accidental feature related specifically to the kink solution. Such modes will always arise when quantising any space-dependent static solution of a translationally invariant theory.

Recall that the Lagrangian (5.36), the potential (5.37) and the field equation (5.38) are all translationally invariant. Consider any given field configuration $\phi_0(x)$, whether or not it solves the field equation. This function $\phi_0(x)$ and its translated counterparts $\phi_0(x-a)$ for any real a, all have the same potential energy: $V[\phi_0(x-a)]$ is independent of a. The function $\phi_0(x)$ can be considered as one point in field space, i.e. in the space of functions $\phi(x)$. Similarly, the family of translated functions $\phi_0(x-a)$, for all a, form a one-parameter curve in that space. All points on it have the same potential energy $V[\phi_0(x)]$ and it is therefore an 'equipotential' curve. It is clear that the entire space of fields can be divided into such equipotential curves. (Some of these curves can be trivial, consisting of just one point. This happens if $\phi_0(x)$ is independent of x, in which case all its translated counterparts are the same as itself.)

Now consider the kink function $\phi_K(x) = (m/\sqrt{\lambda}) \tanh(mx/\sqrt{2})$. It too belongs to an equipotential curve consisting of the family of mutually translated kinks $\phi_K(x-a)$. All points on this curve have the same potential energy $V[\phi_K(x-a)] = (2\sqrt{2}m^3)/3\lambda$. Since the field equation (5.38) is translationally invariant, given that $\phi_K(x)$ is a solution to it, so are all its translated counterparts on that equipotential curve. The key point is that although $\phi_K(x)$ is an extremum of $V[\phi]$ by virtue of being a static solution, it is not a minimum of $V[\phi]$ in field space, even locally. Starting from ϕ_K, we can always move along the equipotential curve, in which case $V[\phi]$ will not change. Along this curve, a given kink solution has only 'neutral' stability. The functional second derivative (and for that matter, all derivatives) of $V[\phi]$ will vanish if we displace the field along the equipotential curve. This is the origin of the zero eigenvalue for the second derivative operator, given in (5.50a). We can also verify that under an infinitesimal translation

$$\begin{aligned}\Delta\phi_K &\equiv \phi_K(x-\delta a)-\phi_K(x) = (-\delta a)\partial\phi_K/\partial x \\ &= -\delta a(m^2/\sqrt{2\lambda})\,\mathrm{sech}^2\,(mx/\sqrt{2}).\end{aligned} \tag{5.72}$$

This is just the eigenmode associated with $\omega_0 = 0$ in (5.50a), apart from an overall normalisation constant. Thus, the zero-frequency eigenmode is just the local tangent, at $\phi_K(x)$, to the equipotential curve in field space connecting all translated solutions $\phi_K(x-a)$.

The presence of this $\omega = 0$ mode has nothing to do with 'massless' fields. In this model, the mass parameter m is non-zero. Also, this zero-mode does not correspond to a 'Goldstone boson', although its origin has some similarities to the origin of Goldstone bosons. Note that the $\omega = 0$ mode in (5.50) is a discrete member of the eigenvalue spectrum, unlike Goldstone bosons, which would yield a zero frequency as the lower limit of a continuum of frequencies. One can also understand why such a zero frequency does not occur in conventional perturbation treatments in field theory text books (in the absence of massless fields). As we have pointed out, the usual perturbation theory is equivalent to expanding about a trivial space-independent classical solution, i.e. the classical vacuum configuration. In the quantisation of the 'vacuum sector' for our model, for example, the spectrum of frequencies in (5.41) did not include a zero frequency. The lowest eigenvalue there was $2m^2$. The reason is that the classical solution involved, $\phi(x) = \phi_1 = m/\sqrt{\lambda}$ was itself space-independent. Upon translation, it stays the same, instead of yielding a different but translated function. The equipotential 'curve' containing ϕ_1 consists of the point ϕ_1 alone. There is no possibility of moving away from ϕ_1 by translations and keeping $V[\phi]$ unchanged. Hence the absence of a zero-frequency mode there. Zero-modes due to translation symmetry can arise only when quantising about classical solutions which are themselves not translationally invariant – i.e. when the solution has non-trivial space dependence.

One can easily generalise these notions to more complicated theories. If we were in three dimensions instead of one, there would in general be three zero-frequency modes corresponding to the fact that a non-trivial solution may be translated in three independent ways. Furthermore, if the theory were rotationally invariant and not the solution, then there would be more zero frequencies for the same reason: we can get other nearby solutions with the same energy by rotating the given solution. Finally, if the theory enjoyed a continuous internal symmetry, not shared by a given solution, once again there would be more zero frequencies, one for each parameter of the symmetry group. Such modes will be dealt in detail in chapter 8.

Having understood the origin of these zero-modes, we turn to the impact of such a mode on the quantum wavefunctions constructed around classical solutions. For this purpose, let us return to the example of the kink. Since that theory is in one space dimension and enjoys no continuous internal symmetry, we expect only one zero-mode, due to translations. Accordingly, in (5.50), all modes with $n \neq 0$ have non-zero frequencies. Thus, the kink is almost a local minimum of $V[\phi]$; that is, except along the translation mode. We can try to visualise $V[\phi]$ in field space. Its profile in the vicinity of ϕ_K is like a valley. At the bottom of the valley runs the equipotential curve consisting of ϕ_K and its translated cousins. In all directions orthogonal to this equipotential curve, the potential increases. Consequently, if we construct energy eigenstates in a weak-coupling expansion around ϕ_K, they would be approximately oscillator-like in all directions except along the translation mode. It is only along that mode that the potential has no confining influence and will allow the wavefunction to spread. The situation along this mode is similar to the elementary prototype example we considered in section 5.2, point (xi). There, we had a particle in one dimension under a trivial potential $V(x) = V$. Since the potential was x-independent, that problem also enjoyed translational symmetry. Any point $x = x_0$ was a static classical solution, but all other points, obtained by translating x_0, were also static classical solutions with the same energy. The entire x-axis is the equipotential curve in that problem. Correspondingly, the energy eigenfunctions $e^{ik_n x}$ spread throughout the x-axis. The energy eigenvalues of that trivial problem are

$$E_n = V + \tfrac{1}{2}(\hbar k_n)^2. \tag{5.73}$$

The first term is the classical energy of the static solution $x(t) = x_0$, and the second term may be considered as the quantum correction due to the (large) fluctuations along the x-axis. The free spread of the wavefunction along the x-axis results in adding the correct kinetic energy contribution to the energy levels.

Returning to the kink problem, since the physics of the translation mode is very similar, we would again expect that the effect of the translation mode on the energy levels would be to produce kinetic energy contributions. The kink occurs in a relativistic theory so that its energy should have the form $\sqrt{M^2 + P^2}$, where P is the momentum. But, as already pointed out, the kink has a very large mass of order $(1/\lambda)$. Thus, its energy may be expanded as

$$E(P) = \sqrt{M^2 + P^2} = M + P^2/2M + \ldots \tag{5.74}$$

The leading kinetic energy term is clearly of order λ. Therefore, we would expect the translation mode to make its impact on the energy eigenvalues only at order λ and beyond, through kinetic energy contributions. That is why we were able to ignore this mode in the last section where we displayed terms only up to $O(\lambda^0)$. By the same count, a calculation of the kink energy to order λ^0 ignores the kinetic motion of the kink particle. To this order, the quantum kink is effectively at rest. That is why we interpreted the kink energy in (5.70), to order λ^0, as the 'mass' of the quantum kink particle. These remarks are supported by the fuller discussion in chapter 8.

5.6. Stability of the kink and its form factor

Goldstone and Jackiw (1975), in their independent investigation of the quantum states built around static solitons, offered a set of postulates. Their postulates are very helpful towards consolidating the ideas we have developed so far and exploring some more consequences, particularly regarding the form factor and stability of soliton states. Let us list these postulates, as applied to the example of our ϕ^4 theory in $(1+1)$ dimensions. We will henceforth set $\hbar = 1$.

Postulate (i). The Hilbert space of states of this theory contains, in addition to the familiar 'vacuum sector' (i.e. the vacuum and multi-meson states), another set of states in the 'kink sector' or 'soliton sector'. This latter set is spanned by the following energy–momentum eigenstates:

(a) the states $|P\rangle$ of the kink particle of momentum P and energy $E = \sqrt{P^2 + M^2}$;

(b) the excited states of the kink $|P^*\rangle$, of momentum P and energy $E = \sqrt{P^2 + M^{*2}}$;

(c) the scattering states $|P, k_1, k_2, \ldots, k_m\rangle$ consisting of the kink particle and m mesons, of asymptotic momenta $P, k_1, k_2, \ldots, k_m$ (the number m can vary from 1 to infinity);

(d) the scattering states $|P^*, k_1, k_2, \ldots, k_m\rangle$ consisting of the excited kink plus m mesons, of asymptotic momenta $P, k_1, k_2, \ldots, k_m$.

Postulate (ii). This kink sector of states is orthogonal to the vacuum sector and, furthermore, not connected by any localised operator to it. In particular, the quantum kink is stable against decay into mesons.

Postulate (iii). The mass of the quantum kink behaves as (constant)/λ in the weak-coupling limit $\lambda/m^2 \to 0$.

Postulate (iv). The matrix elements of the field operator $\Phi(x, t)$ between these states depend on λ in the weak-coupling limit as

$$\langle P'|\Phi|P\rangle \sim O(1/\sqrt{\lambda}),$$

$$\langle P'|\Phi|P, k\rangle \text{ and } \langle P'|\Phi|P^*\rangle \sim O(1)$$

and in general

$$\langle P', k'_1, \ldots, k'_l|\Phi|P, k_1, \ldots, k_m\rangle_c \sim O(\lambda^{(l+m-1)/2}) \tag{5.75}$$

where the subscript c denotes the connected part of the matrix element. We have used $\Phi(x, t)$ to denote the quantum field operator compared with $\phi(x, t)$ for the classical field.

We will leave these as postulates and not attempt to derive them in any rigorous way. However, they will be seen to be consistent with the physical ideas we have already developed, and with one another. The set of states mentioned in postulate (i) are just the states we had constructed, in the weak-coupling approximation, around the kink solution. The state $|P\rangle$ is the lowest state in the sequence (5.56), and we had interpreted it as the quantum kink particle. The state $|P^*\rangle$ is the excited state where only the discrete mode $\omega_1 = \sqrt{\tfrac{3}{2}}\, m$ is excited once, i.e. $N_1 = 1$ and $N_{q_n} = 0$ in (5.56). Higher excitations of this mode ($N_1 \geq 2$) will be unstable against decay into a kink and a meson. Recall that the meson has mass $\sqrt{2}\, m\hbar$ which lies between $\hbar\omega_1$ and $2\hbar\omega_1$. The momentum dependence of the states $|P\rangle$ and $|P^*\rangle$ are not explicitly visible in (5.56) only because, as we have said, the kink has mass of order $(1/\lambda)$ and hence kinetic energy terms will appear only at order λ. Finally, the states $|P, k_1, \ldots, k_m\rangle$ and $|P^*, k_1, \ldots, k_m\rangle$ correspond to excitations of the continuum modes in (5.56) in the absence and presence of a ω_1 excitation, respectively. Note that this postulate associates no new type of states with excitations of the zero-mode ω_0. This is a reflection of the notion that excitations along this translation mode will merely restore the correct momentum dependence of any given state in the kink sector.

Postulate (ii) makes a strong statement. The kink sector of states was built around ϕ_{kink}, which is not an absolute minimum of $V[\phi]$. At best, (i.e.

except along the ω_0 mode) it is a local minimum. *A priori*, purely from energetics, we may expect that states built around ϕ_{kink} would decay into the vacuum sector of states built around the lower minimum ϕ_1. Indeed, all the kink-sector states have higher energy than any given vacuum-sector state in the weak-coupling limit, thanks to the leading $O(1/\lambda)$ term in (5.56). The kink states seem similar to the states built around $x = c$ in the baby problem discussed in section 5.2, paragraph (ix).

Yet postulate (ii) claims not only that the quantum kink is stable against decay into mesons, but that the entire kink sector is unconnected to the vacuum sector. The motivation behind this assertion lies in the notion of topological classification as generalised to quantum theory. Recall from chapter 2 that we had classified all classical finite-energy configurations of this ϕ^4 theory into four topological sectors based on the allowed boundary conditions $\phi(x = \infty, t) = \pm m/\sqrt{\lambda}$ and $\phi(-\infty, t) = \pm m/\sqrt{\lambda}$. We showed that classically, a configuration from one sector cannot, under time evolution, move into another sector. Consistent with this, we have defined a conserved topological charge Q. For the quantised theory, let us use the same definition after replacing $\phi(x, t)$ by the field operator $\Phi(x, t)$

$$Q = \int j_0 \,\mathrm{d}x = \frac{\sqrt{\lambda}}{m}[\Phi(\infty, t) - \Phi(-\infty, t)] \tag{5.76}$$

with

$$j^\mu = \varepsilon^{\mu\nu}\partial_\nu\Phi \qquad \text{and} \qquad \partial_\mu j^\mu = 0. \tag{5.77}$$

Since $\Phi(x, t)$ is a Hermitian operator, so is the charge Q. It continues to be conserved, by construction, thanks to (5.77), and may be called the topological quantum number. Now, the vacuum sector of states is built by quantising fluctuations around the function $\phi_1 = +m/\sqrt{\lambda}$. More precisely, we can define this sector as consisting of field-theoretical wavefunctionals $\Psi[\phi]$ which vanish unless $\phi(x)$ obeys $\phi(\infty) = \phi(-\infty) = m/\sqrt{\lambda}$. Consequently, the expectation value of Q for any state in the vacuum sector should be zero (see for instance (5.43) and (5.44)). Similarly the kink sector states can be defined as states whose wavefunctionals $\Psi[\phi]$ vanish unless $\phi(\pm\infty) = \pm m/\sqrt{\lambda}$. Equivalently, this sector may be defined as the sector having an expectation of Q equal to unity. Since Q is a conserved hermitian operator, the kink sector with $Q = 1$ and the vacuum sector with $Q = 0$ are not only orthogonal but cannot evolve into one another. In particular, the kink state, which is the ground state within the kink sector, is stable and cannot decay into mesons.

Furthermore, consider any localised physical observable $A(t)$. That is, let

$$A(t) = \int_{-\infty}^{\infty} a(x, t)\, \mathrm{d}x$$

where $a(x, t)$ has support only for some finite region in x at any given t. Then the equal time commutator

$$\begin{aligned}[A, Q]_t &= \lim_{L \to \infty} \frac{\sqrt{\lambda}}{m} \left\{ [A, \Phi(L)]_t - [A, \Phi(-L)]_t \right\} \\ &= 0 \end{aligned} \tag{5.78}$$

because of the causality condition on operators with space-like separation. Thus, any such operator A cannot connect sectors with different values of Q. This suggests that Q is something like a superselection quantum number, separating the kink sector from the vacuum sector (Ezawa 1978).

Postulate (iii) is simple and has been explicitly supported by our derivation of the kink mass in eq. (5.70). Turning to postulate (iv), the claim that the matrix element of Φ between kink states is of order $1/\sqrt{\lambda}$ will be explicitly demonstrated below. We may anticipate it on the grounds that the kink state is built around the kink function, which is itself of order $1/\sqrt{\lambda}$. The other portions of postulate (iv) imply that the emission or absorption of every additional meson by the kink, or the excitation of the discrete ω_1 mode, carries the penalty of an additional factor of $\sqrt{\lambda}$. This is to be expected from the form of the potential $V[\phi]$ in (5.47). The kink sector is obtained by quantising the displaced field $\eta(x) = \phi(x) - \phi_K(x)$. Emission or absorption of a meson by the kink, or excitation of the kink through the normal mode ω_1, occurs in perturbation theory through the interaction terms in (5.47). The leading interaction term $\lambda \int \mathrm{d}x\, \phi_K \eta^3$ is of order $\sqrt{\lambda}$ if we recall that ϕ_K is $\mathrm{O}(1/\sqrt{\lambda})$.

From this discussion we can see that these postulates embody the ideas developed in the preceding sections. Let us examine some further consequences that result from them, by following Goldstone and Jackiw who adapted a method developed by Kerman and Klein (1963) for the many-body problem.

We define the functions $f_1(x)$, $f_2(x; k)$ and $f_2^*(x)$ by the relations

$$\langle P | \Phi(x, 0) | P' \rangle \equiv \int_{-\infty}^{\infty} \frac{\mathrm{d}a}{2\pi} \exp[\mathrm{i}(P - P')a] f_1(x - a) \tag{5.79}$$

$$\langle P, k|\Phi(x,0)|P'\rangle \equiv \int_{-\infty}^{\infty} \frac{da}{2\pi} \exp[i(P+k-P')a]\, f_2(x-a;k) \tag{5.80a}$$

and

$$\langle P^*|\Phi(x,0)|P'\rangle \equiv \int_{-\infty}^{\infty} \frac{da}{2\pi} \exp[i(P-P')a]\, f_2^*(x-a). \tag{5.80b}$$

In these relations, the dependence on momenta solely through the momentum transfers $(P-P')$ and $(P+k-P')$ follows from translational invariance. Consider now the Heisenberg equation of motion obeyed by the field $\Phi(x,t)$:

$$\left(\frac{\partial^2}{\partial t^2} - \frac{\partial^2}{\partial x^2} - m^2\right)\Phi(x,t) = -\lambda\Phi^3(x,t). \tag{5.81}$$

This is formally similar to the classical equation, but now $\Phi(x,t)$ is an operator. Let us take matrix elements of both sides of (5.81) at $t=0$, between kink states $|P\rangle$ and $|P'\rangle$ and equate them to leading order $(1/\sqrt{\lambda})$. The first term on the left-hand side gives

$$\left(\frac{\partial^2}{\partial t^2}\langle P|e^{-iHt}\Phi(x,0)e^{+iHt}|P'\rangle\right)_{t=0}$$
$$= -(E_P - E_{P'})^2\langle P|\Phi(x,0)|P'\rangle. \tag{5.82}$$

Since

$$(E_P - E_{P'})^2 = \left(\frac{P^2 - P'^2}{2M} + \ldots\right)^2 = O(\lambda^2),$$

we shall omit this term to leading order. The remaining terms on the left-hand side of (5.81) give, on using (5.79),

$$\int_{-\infty}^{\infty} \frac{da}{2\pi}\left(-\frac{\partial^2}{\partial x^2} - m^2\right)\exp[i(P-P')a]\, f_1(x-a). \tag{5.83}$$

Let us write the right-hand side's contribution as

$$-\lambda\sum_{n,n'}\langle P|\Phi(x,0)|n\rangle\langle n|\Phi(x,0)|n'\rangle\langle n'|\Phi(x,0)|P'\rangle \tag{5.84}$$

Here $|n\rangle$ and $|n'\rangle$ refer to the complete set of states in the kink sector, listed in postulate (i). The vacuum sector of states need not be inserted since they are unconnected, as per postulate (ii). To leading order, out of the sets $|n\rangle$ and $|n'\rangle$ only the kink states contribute, as per postulate (iv). Thus, (5.84) reduces, in leading order, to

$$-\lambda \sum_{P'',P'''} \langle P|\Phi(x,0)|P''\rangle \langle P''|\Phi(x,0)|P'''\rangle \langle P'''|\Phi(x,0)|P'\rangle$$
$$= -\lambda \int \frac{\mathrm{d}a}{2\pi} \exp[\mathrm{i}(P-P')a]\,[f_1(x-a)]^3 \tag{5.85}$$

after some trivial algebra. Note that the operator $\Phi^3(x,t)$ has ultraviolet divergences but these divergences will occur only in higher order, and can be removed by normal-ordering. In the leading order we are considering, these problems do not arise.

Therefore, to leading order, (5.83) and (5.85) are equal, and for all choices of P and P'. Therefore $f_1(x-a)$ must obey

$$\left(-\frac{\mathrm{d}^2}{\mathrm{d}x^2} - m^2\right) f_1(x-a) = -\lambda [f_1(x-a)]^3. \tag{5.86}$$

This is just the classical static field equation, and its solution subject to boundary conditions appropriate to the kink sector is

$$f_1(x-a) = (m/\sqrt{\lambda})\tanh[m(x-a)/\sqrt{2}] \equiv \phi_K(x-a). \tag{5.87}$$

Hence (5.79) can be rewritten as

$$\langle P|\Phi(x,0)|P'\rangle = \int_{-\infty}^{\infty} \frac{\mathrm{d}a}{2\pi} \exp[\mathrm{i}(P-P')a]\,\phi_K(x-a)$$
$$+ \text{higher-order terms.} \tag{5.88}$$

This is an important result. Firstly, it brings out the relation between the classical solution $\phi_K(x)$ and the quantum particle state built around it. The former is related to the expectation value of the field operator in the latter. Compare this result with eq. (5.14) for the simple non-relativistic problem as well as with eq. (5.44) in the vacuum sector where again the classical solution was related, in leading order, to the quantum expectation value. Equation (5.88) is very similar, with the difference that it involves not a single solution, but an integral over the family of classical solutions $\phi_K(x-a)$. This modification is obviously due to the presence of a degenerate family of translated solutions. It also reflects our intuitive feeling that the quantum kink state should spread throughout the translation mode.

Since ϕ_K is $O(1/\sqrt{\lambda})$, eq. (5.88) also supports postulate (iv). Finally, one can invert (5.88) and write

$$\phi_K(x) = \int dQ\, e^{iQx} \langle P+Q|\Phi(0,0)|P\rangle + \text{higher-orders}. \tag{5.89}$$

This shows that the classical kink function is, to leading order in λ, the 'form factor' of the field operator Φ between quantum kink states. The more interesting form factors correspond not so much to Φ itself but to physically important operators like the energy density, topological charge density, etc. These operators involve the combinations $\partial\Phi/\partial x$ or $(\Phi^2 - m^2/\lambda)$. It is clear from (5.89) that their corresponding form factors will, to leading order, involve $(\partial\phi_K(x)/\partial x)$ and $(\phi_K^2(x) - m^2/\lambda)$. Since $\phi_K(x) \to \pm m/\sqrt{\lambda}$ as $x \to \pm\infty$, all these form factors will be localised. This is an important result since it supports the interpretation of the ground state in the kink sector as an extended particle. Such a particle should have extended but localised form factors. As an illustration, consider the energy density distribution in the kink state, i.e. the form factor

$$\mathscr{E}(x) \equiv \int d(P-P')\exp[i(P-P')x]\langle P|\mathscr{H}(0,0)|P'\rangle \tag{5.90}$$

where

$$\mathscr{H}(x,t) = \frac{1}{2}\Pi^2 + \frac{1}{2}\left(\frac{\partial\Phi}{\partial x}\right)^2 + \frac{\lambda}{4}\left(\Phi^2 - \frac{m^2}{\lambda}\right)^2 \tag{5.91}$$

is the Hamiltonian density, and $\Pi(x,t) = \partial\Phi(x,t)/\partial t$. Let us evaluate $\langle P|\mathscr{H}(0,0)|P'\rangle$ to lowest order by inserting complete sets of states between field operators in (5.91). To leading order, only the unexcited kink intermediate states will contribute as before. The non-kinetic terms in $\langle P|\mathscr{H}(0,0)|P'\rangle$ give,

$$\begin{aligned}
&\left\langle P\left|\frac{1}{2}\left(\frac{\partial\Phi}{\partial x}\right)^2 + \frac{\lambda}{4}\left(\Phi^2 - \frac{m^2}{\lambda}\right)^2\right|P'\right\rangle_{x=0,t=0} \\
&= \int\frac{da}{2\pi}\exp[i(P-P')a]\left[\frac{1}{2}\left(\frac{\partial\phi_K(-a)}{\partial x}\right)^2 + \frac{\lambda}{4}\left(\phi_K^2(-a) - \frac{m^2}{\lambda}\right)^2\right] \\
&= \int\frac{da}{2\pi}\exp[i(P-P')a]\,\mathscr{E}_{cl}(-a)
\end{aligned} \tag{5.92}$$

where we have omitted elementary steps similar to these employed in (5.84)–(5.85). The result (5.88) has been used and $\mathscr{E}_{cl}(x)$ refers to the energy density of the classical static kink solution. The kinetic term in $\langle P|\mathscr{H}(0,0)|P'\rangle$ gives

$$\begin{aligned}\langle P|\tfrac{1}{2}\Pi^2(0,0)|P'\rangle &= \tfrac{1}{2}\langle P|(\partial\Phi/\partial t)^2|P'\rangle_{x=t=0} \\ &= -\tfrac{1}{2}\sum_{P''}\langle P|\Phi|P''\rangle\langle P''|\Phi|P'\rangle(E_P - E_{P''})(E_{P''} - E_{P'}) \\ &= O(\lambda). \end{aligned} \tag{5.93}$$

This incidentally agrees with our earlier ideas that the kink's kinetic energy will make an appearance only in order λ. The contribution (5.93) can, to lowest order, be neglected compared with (5.92) which is $O(1/\lambda)$. Inserting (5.92) into (5.90) we have

$$\mathscr{E}(x) = \mathscr{E}_{\mathrm{cl}}(x). \tag{5.94}$$

Thus, to lowest order, the energy form factor in the kink state is just the classical energy density of the kink which, as we saw in chapter 2, is a localised function. The total energy E_P of the kink state is given by

$$\begin{aligned}\partial(P-P')E_P &= \langle P|H|P'\rangle \\ &= \langle P|\textstyle\int\mathscr{H}(x,0)\,\mathrm{d}x|P'\rangle \\ &= 2\pi\delta(P-P')\langle P|\mathscr{H}(0,0)|P'\rangle. \end{aligned} \tag{5.95}$$

Hence

$$E_P = 2\pi\langle P|\mathscr{H}\ (0,0)|P\rangle = \int dx\,\mathscr{E}(x) \tag{5.96}$$

where $\mathscr{E}(x)$ is the form factor in (5.90). To lowest order, $O(1/\lambda)$, (5.94) gives

$$E_P = \int \mathrm{d}x\,\mathscr{E}_{\mathrm{cl}}(x) = \frac{2\sqrt{2}m^3}{3\lambda}. \tag{5.97}$$

We have thus reproduced the result that the leading term in the quantum kink's energy is just its classical energy, consistent with (5.70).

Let us turn to the function $f_2(x-a;k)$ defined in (5.80). It may be evaluated, just as $f_1(x-a)$ was, by taking matrix elements of the operator field equation (5.81). This time, we take the matrix elements between a pure kink state $|P'\rangle$ and a kink plus one meson state $|P,k\rangle$, and evaluate it to leading order. The left-hand side of the field equation yields

$$\begin{aligned}&\langle P,k|(\partial^2/\partial t^2 - \partial^2/\partial x^2 - m^2)\Phi(x,t)|P'\rangle \\ &\quad = \langle P,k|(-(E_P+\omega_k - E_{P'})^2 - \partial^2/\partial x^2 - m^2)\Phi(x,t)|P'\rangle. \end{aligned} \tag{5.98}$$

Here ω_k enters as the energy of the meson. Now, since $E_P - E_{P'} = O(\lambda)$, the leading terms in (5.98) become, at $t = 0$,

$$\langle P, k\,|(-\omega_k^2 - \partial^2/\partial x^2 - m^2)\Phi(x,0)|\,P'\rangle$$
$$= \int \frac{da}{2\pi}\exp[i(P+k-P')a]\left(-\omega_k^2 - m^2 - \frac{\partial^2}{\partial x^2}\right) f_2(x-a;k). \quad (5.99)$$

This contribution, from postulate (iv), will be $O(\lambda^0)$. The right-hand side of the field equation, when expanded with intermediate states, will give a leading contribution

$$-3\lambda \sum_{P'',P'''} \langle P, k|\Phi(x,0)|P''\rangle\langle P''|\Phi(x,0)|P'''\rangle\langle P'''|\Phi(x,0)|P'\rangle$$

$$= -3\lambda\int\frac{da}{2\pi}\exp[i(P+k-P')a]\;\phi_K^2(x-a) f_2(x-a;k) \quad (5.100)$$

where (5.88) has been used. The factor of 3 arises because the external meson could have been emitted by any of the three field operators in the product Φ^3, all these possibilities making equal contributions. Since (5.99) equals (5.100) for arbitrary P, k and P', we have

$$[-\partial^2/\partial x^2 - m^2 + 3\lambda\phi_k^2(x-a)] f_2(x-a;k)$$
$$= \omega_k^2 f_2(x-a;k). \quad (5.101)$$

This is just the eigenvalue equation (5.48) for the second derivative $[\delta^2 V/(\delta\phi\,\delta\phi)]_{\phi_K}$. Thus, ω_k which was introduced in (5.98) in its role as the energy of the meson, becomes identified with the eigenvalue in (5.48), as we had anticipated earlier. The function $f_2(x,k)$ is proportional to the corresponding normalised eigenmode $\eta_k(x)$ occuring in (5.48). The relation (5.101) being linear cannot fix the proportionality constant. We will set, using familiar boson normalisation,

$$f_2(x,k) = \eta_k(x)/\sqrt{2\omega_k}. \quad (5.102)$$

The validity of this can be checked in equations where f_2 occurs in a non-linear way, as in the canonical commutation rule used below.

An entirely similar result holds for the function $f_2^*(x-a)$ in (5.80). We take matrix elements of the Heisenberg field equation between $|P'\rangle$ and $|P\rangle$ and proceed as before. We will get the eigenvalue equation again, but with $f_2(x-a;k)$ and ω_k replaced by $f_2^*(x-a)$ and ω_1 respectively, where $\omega_1 = M^* - M$. Correspondingly, we associate $f_2^*(x-a)$

with the discrete $n = 1$ mode in (5.50b) and set, analogously to (5.102),

$$f_2^*(x-a) = \eta_1(x)/\sqrt{2\omega_1}.$$

Note that, considered as a complete set of functions, the set η_0, η_1 and η_k satisfies

$$\eta_0^\dagger(x)\eta_0(y) + \eta_1^\dagger(x)\eta_1(y) + \sum_k \eta_k^\dagger(x)\eta_k(y)$$

$$= \delta(x-y). \tag{5.103}$$

This is because the functions η_n, *including* η_0, form a complete set of orthonormal eigenfunctions of the Hermitian operator in (5.48). Yet, in the complete set of states in the kink sector, listed in postulate (i), there is no new type of state associated with excitations of the η_0 mode. While excitation of the η_1 mode leads to $|P^*\rangle$ and the η_k modes to meson-plus-kink states, the fluctuations along η_0 are supposed to provide only the correct momentum dependence of the above states. As a test of whether this postulate is consistent with the fact that η_0 is very much needed in (5.103), consider the canonical commutation rule

$$[\Phi(x,t), (\partial/\partial t)\Phi(y,t)]_{t=0} = \mathrm{i}\delta(x-y). \tag{5.104}$$

Let us take matrix elements between kink states $|P\rangle$ and $|P'\rangle$. We get

$$\langle P|[\Phi(x,0), (\partial/\partial t)\Phi(y,0)]|P'\rangle = \mathrm{i}\,\delta(P-P')\,\delta(x-y). \tag{5.105}$$

This can be verified by evaluating the left-hand side using the complete set of intermediate states in the kink sector. To leading order, only the kink, its excited partner, and the kink-plus-one-meson states contribute. The pure-kink intermediate states give

$$\sum_Q \{\mathrm{i}(E_Q - E_{P'})\langle P|\Phi(x,0)|Q\rangle\langle Q|\Phi(y,0)|P'\rangle$$

$$-\mathrm{i}(E_P - E_Q)\langle P|\Phi(y,0)|Q\rangle\langle Q|\Phi(x,0)|P'\rangle\}$$

$$= \mathrm{i}\int\frac{\mathrm{d}a}{2\pi}\int\frac{\mathrm{d}b}{2\pi}\int\mathrm{d}Q\Bigg[\Bigg(f_1(x-a)f_1(y-b)\exp[\mathrm{i}(P-Q)a$$

$$+\mathrm{i}(Q-P')b]\frac{(Q^2-P'^2)}{2M}\Bigg) - \Bigg(f_1(y-b)f_1(x-a)\exp[\mathrm{i}(P-Q)b$$

$$+\mathrm{i}(Q-P')a]\frac{(P^2-Q^2)}{2M}\Bigg)\Bigg]. \tag{5.106}$$

In the second term we change the integration variable from Q to $\tilde{Q} \equiv P + P' - Q$. Then the exponent in the second term becomes $\mathrm{i}(P-Q)b + \mathrm{i}(Q-P')a = \mathrm{i}(P-\tilde{Q})a + \mathrm{i}(\tilde{Q}-P')b$. The two terms can therefore be combined to give,

$$\mathrm{i}\int_{-\infty}^{\infty}\frac{\mathrm{d}a}{2\pi}\int_{-\infty}^{\infty}\frac{\mathrm{d}b}{2\pi}\int_{-\infty}^{\infty}\mathrm{d}Q\exp[\mathrm{i}(P-Q)a+\mathrm{i}(Q-P')b]$$

$$\times f_1(x-a)f_1(y-b)\frac{(Q-P')(Q-P)}{M}$$

$$=\frac{\mathrm{i}}{M}\int_{-\infty}^{\infty}\frac{\mathrm{d}a}{2\pi}\left(\frac{\partial}{\partial x}f_1(x-a)\right)\left(\frac{\partial}{\partial y}f_1(y-a)\right)\exp[\mathrm{i}(P-P')a]. \tag{5.107}$$

Using (5.87) and (5.72), we see that

$$\frac{\partial}{\partial x}f_1(x-a) = \frac{\partial}{\partial x}\phi_K(x-a) = A\eta_0(x-a) \tag{5.108}$$

where the constant A is obtained by recalling that η_0 is normalised:

$$1 = \int_{-\infty}^{\infty}\mathrm{d}x\,\eta_0^2(x) = \frac{1}{A^2}\int_{-\infty}^{\infty}\mathrm{d}x\left(\frac{\mathrm{d}\phi_K(x)}{\mathrm{d}x}\right)^2$$

$$= \frac{1}{A^2}\int_{-\infty}^{\infty}\mathrm{d}x\frac{m^4}{2\lambda\cosh^4(mx/\sqrt{2})} = \frac{2\sqrt{2}\,m^3}{3\lambda A^2} = \frac{M}{A^2},\ \text{to leading order.} \tag{5.109}$$

Thus, the pure kink intermediate state contribution in (5.106) becomes

$$\mathrm{i}\int_{-\infty}^{\infty}\frac{\mathrm{d}a}{2\pi}\exp[\mathrm{i}(P-P')a]\eta_0(x-a)\,\eta_0(y-a). \tag{5.110}$$

The kink-plus-one-meson intermediate state contribution is easy to evaluate. Note that

$$\langle P,k|(\partial\Phi/\partial t)(y,0)|P'\rangle$$
$$= \mathrm{i}(E_P - E_{P'} + \omega_k)\int\frac{\mathrm{d}b}{2\pi}\exp[\mathrm{i}(P+k-P')b]f_2(y-b;k) \tag{5.111}$$

where, to lowest order, $(E_P - E_{P'})$ can be dropped in comparison with ω_k. Using this, the kink-plus-meson terms give

$$\sum_{Q,k} \langle P|\Phi(x,0)|Q,k\rangle \langle Q,k|(\partial\Phi/\partial t)(y,0)|P'\rangle$$
$$- \langle P|(\partial\Phi/\partial t)(y,0)|Q,k\rangle \langle Q,k|\Phi(x,0)|P'\rangle$$
$$= \mathrm{i}\int \frac{\mathrm{d}a}{2\pi} \sum_k 2\omega_k f_2^\dagger(x-a;k) f_2(y-a;k) \exp[\mathrm{i}(P-P')a]$$
$$= \mathrm{i}\int \frac{\mathrm{d}a}{2\pi} \exp[\mathrm{i}(P-P')a] \left(\sum_k \eta_k^\dagger(x-a)\eta_k(y-a) \right). \tag{5.112}$$

In the same way the intermediate states $|Q^*\rangle$ of the excited kink will yield a contribution

$$\mathrm{i}\int \frac{\mathrm{d}a}{2\pi} \exp[\mathrm{i}(P-P')a] [\eta_1(x-a)\eta_1(y-a)]. \tag{5.113}$$

All other intermediate states will not contribute to this order. Adding (5.110), (5.112) and (5.113), we see that, to leading order, the left-hand side of (5.105) yields

$$\mathrm{i}\int \frac{\mathrm{d}a}{2\pi} \exp[\mathrm{i}(P-P')a] [\eta_0(x-a)\eta_0(y-a) + \eta_1(x-a)\eta_1(y-a)$$
$$+ \sum_k \eta_k^\dagger(x-a)\eta_k(y-a)]$$
$$= \mathrm{i}\int \frac{\mathrm{d}a}{2\pi} \exp[\mathrm{i}(P-P')a]\, \delta(x-y)$$
$$= \mathrm{i}\,\delta(P-P')\,\delta(x-y). \tag{5.114}$$

(Recall from (5.50) that η_0 and η_1 are real.) Thus, the canonical commutation relation in (5.105) is satisfied even though the intermediate states did not include a separate set of states corresponding to the η_0 mode. The η_0 term in (5.114), needed for completeness in the sense of functions, was supplied by the kinetic energy of the kink state through eq. (5.106)–(5.110). For a fuller discussion of the representations of commutation rules in soliton sectors, see Steimann (1978).

5.7. Conclusions

The discussion in the preceding section shows that the postulates associating a soliton sector of states with the classical kink solution are mutually consistent, at least to the extent that we have explored their consequences. We had used the example of the kink solution of the model (5.36) to illustrate our ideas. But the same principles can be used to quantise static localised solutions of a large class of theories. To begin with, a stable static solution of any other scalar field theory in $(1+1)$ dimensions can be quantised exactly as we quantised the kink. The sine–Gordon system, for example, yields a static one-soliton solution in $(1+1)$ dimensions. This can be quantised to yield the one-soliton sector of states, as distinct from the vacuum sector of that theory (Dashen et al 1975a). The normal modes for this system consist of the mandatory translation mode followed by the continuum modes. There is no analogue of the discrete ω_1 mode for the sine–Gordon system, and hence no 'excited' soliton state, but aside from that the rest of our interpretation can be used directly.

If the model were in $(2+1)$ or $(3+1)$ dimensions, the algebra would become more complicated. Clearly we will have more normal modes, including more zero-modes corresponding to translations and rotations. Ultraviolet divergences will be stronger in these higher dimensions, and non-trivial renormalisation counter terms may have to be added, apart from normal-ordering. Furthermore, we know from Derrick's theorem that no static solution can exist in $(3+1)$ dimensions if scalar fields alone are involved. Higher-spin fields must be included to get a static classical solution in the first place. Such fields introduce additional complications of their own, although the principles behind the interpretation and evaluation of quantum solitons remain essentially the same. For example, the monopole solution discussed in chapter 3 involves scalar fields and vector fields with non-abelian gauge invariance. Consequently the classical field equation and the linearised eigenvalue equation for small fluctuations (the analogue of (5.24)) have to be solved after gauge fixing. These equations are technically difficult to solve. We saw in chapter 3 that the field equation (3.50)–(3.51) was solved by Prasad and Sommerfield only for the case $\lambda \to 0$ and that too by ingenious trial and error. As far as we know, no analytical solution is known for arbitrary λ. These problems make it difficult to evaluate analytically the classical and leading quantum contributions to the monopole energy, form factor etc., unlike the kink system. However, we see no reason to believe that once the gauge is fixed and the appropriate differential equations solved, the principles for quantisation would be any different from that of the kink example. While all normal modes of the monopole problem have not yet been obtained, considerable work has been done along these lines, particularly on the zero-modes of that problem (Adler 1979, Mottola 1978).

When fermion fields are present, as for instance in the SLAC bag (Bardeen et al. 1975), the situation is conceptually different. Fermion fields do not have a classical limit. Loosely speaking, this is because a large number of fermions cannot be present coherently in the same state to produce a near-classical field, thanks to the Pauli principle. The conceptual relationship between 'classical' field equations and the corresponding quantum theories is therefore different when Fermi fields are present.

We will discuss the quantisation of fermionic solitons separately in chapter 9. It will be seen that, despite this conceptual difference, many of the operational steps are similar to what we have described here, with judicious changes of minus signs.

Of course, our discussion of the soliton sector has been far from rigorous. All we did was to apply the weak-coupling approximation, which we had generalised using plausible physical arguments, from the harmonic approximation in non-relativistic quantum mechanics. Due homage was paid to some new complications introduced by continuum field theory, such as renormalisation, topological quantum numbers and zero-modes. Nevertheless our discussion has been at best 'suggestive'. In the next chapter, we will re-derive the energy levels of the soliton sector by using functional methods. That derivation, while more fancy, will not be significantly more rigorous. Truly rigorous proofs will have to come from powerful methods such as in constructive field theory. Such proofs are exceedingly difficult to construct. Even the existence of spontaneous symmetry breaking for such simple systems as our ϕ^4 theory in two dimensions was demonstrated only recently (Glimm et al. 1975; see also Dobrushin and Minlos 1973). The soliton sector has been discussed by Frohlich (1976). Such rigorous demonstrations are well beyond the scope of this book. We will stay content with the fact that, as far as we can tell, these rigorous studies support rather than contradict the ideas we have developed using more primitive techniques.

We have obtained the quantum soliton particle starting from the classical static soliton. For a related discussion in which solitons are associated with coherent states, see Taylor (1978). In a different approach, Matsumoto et al. (1979) have used solitons in quantum many-body theory in terms of a 'boson transformation' method.

CHAPTER 6

FUNCTIONAL INTEGRALS AND THE WKB METHOD

6.1. The path-integral formalism

A convenient formalism in which different aspects of the semiclassical method can be studied is the functional-integral formulation of quantum field theory. In the preceding chapter we have already presented some semiclassical results, but without invoking functional integrals. This was possible only because we had limited ourselves to quantisation of static soliton solutions of scalar fields, a procedure which reduces, in the weak-coupling approximation, to just a study of coupled harmonic oscillators. But when it comes to quantising time-dependent solitons, or to studying instanton effects, the use of functional integrals becomes crucial. Other methods, such as canonical operator techniques, can also be used for this purpose – indeed we shall use them in portions of the book – but functional integrals offer perhaps the most convenient unified formalism for studying these questions.

Functional integrals in field theory had been in use for quite some time before they were utilised in studying solitons or instantons. Their major applications include the study of symmetry breaking using the 'effective action', the generation of Green functions using the vacuum functional, the derivation of the Faddeev–Popov rules in gauge theories, and many others. It is not our intention to offer a detailed discourse on functional integrals and their various applications to quantum field theory. Rather, our philosophy will be to take the basic functional-integral formalism as given, and concentrate on the specific applications dealing with solitons and instantons. For the sake of completeness, however, a capsule introduction to functional integrals, as tailored to our needs, will be given in this section and the next. As exercises, some of the results of chapter 5 will be derived using them. In later sections and chapters, new results will be derived.

Functional integrals in field theory are generalisations of path integrals in non-relativistic one-particle quantum mechanics. Let us therefore begin our capsule introduction with the latter. Consider, once again, the simple one-particle system we had used in chapter 5, i.e. a non-relativistic particle in one dimension with a Lagrangian

$$L = \tfrac{1}{2}\dot{q}^2 - V(q). \tag{6.1}$$

The corresponding Hamiltonian in quantum theory is

$$H = \tfrac{1}{2}\hat{p}^2 + V(q) \tag{6.2}$$

where $\hat{p} = -\mathrm{i}\hbar\,\partial/\partial q$ is the momentum operator. Then the Schrödinger equation

$$\mathrm{i}\hbar\frac{\partial}{\partial t}\psi(q,t) = H\psi(q,t) \tag{6.3}$$

tells us that given a wavefunction $\psi(q,0)$ at $t = 0$, it will evolve at $t = T$ into

$$\psi(q,T) = \exp(-\mathrm{i}Ht/\hbar)\,\psi(q,0). \tag{6.4}$$

Consequently, the quantum transition amplitude for a particle to start at $q = q_a$ at $t = 0$, and end up at $q = q_b$ at $t = T$, is given by

$$K(q_b, T; q_a, 0) \equiv \langle q_b|\exp(-\mathrm{i}HT/\hbar)|q_a\rangle \tag{6.5}$$

where $|q_a\rangle$ and $|q_b\rangle$ are eigenstates of the position operator $\hat{q}$. The path-integral formulation of quantum mechanics gives an alternative but equivalent prescription for this amplitude. (See Feynman and Hibbs (1965) for a lucid and detailed discussion of the path-integral formalism.) This prescription is

$$K(q_b, T; q_a, 0) = \int \mathscr{D}[q(t)]_{q_b,q_a,T}\exp\{(\mathrm{i}/\hbar)S[q(t)]\} \tag{6.6}$$

where

$$S[q(t)] \equiv \int_0^T \mathrm{d}t\,L = \int_0^T \mathrm{d}t\,(\tfrac{1}{2}\dot{q}^2 - V(q))$$

is the action, and $\int \mathscr{D}[q(t)]_{q_b,q_a,T}$ is an integral over all paths, i.e. all functions $q(t)$ beginning at $q(0) = q_a$ and ending at $q(T) = q_b$. We shall often omit the subscripts q_b, q_a, T in $\mathscr{D}[q]_{q_b,q_a,T}$; they are taken as understood from the context. An integral over paths is a functional integral, and is a more complicated entity than ordinary integrals. However, it may be thought of as the limit of an infinite product of ordinary integrals. A path $q(t)$ is specified by giving the value of q at each t. Consequently an integral over paths may be considered, loosely speaking, as a product of integrals over $q(t)$ for each t. More precisely, replace the time interval $[0, T]$ by discrete equidistant points $t_0 = 0, t_1, t_2, \ldots, t_{N-1}, t_N = T$, with

$$t_i - t_{i-1} = \varepsilon; \qquad N\varepsilon = T$$

and

$$q(t_i) \equiv q_i, \qquad q(0) \equiv q_a, \qquad q(T) \equiv q_b.$$

The continuum limit is regained in the limit $\varepsilon \to 0$, $N \to \infty$ with $N\varepsilon = T$. Then the path integral (6.6) may be written as

$$\begin{aligned} K(q_b, T; q_a, 0) &= \int \mathscr{D}[q(t)] \exp\left(\frac{i}{\hbar}\int_0^T (\tfrac{1}{2}\dot{q}^2 - V(q))\,dt\right) \\ &= \lim_{\substack{N\to\infty \\ \varepsilon\to 0 \\ N\varepsilon = T}} \left\{ B_N(T) \int_{-\infty}^{\infty} dq_1 \int_{-\infty}^{\infty} dq_2 \ldots \int_{-\infty}^{\infty} dq_{N-1} \right. \\ &\quad \left. \times \exp\left[\frac{i}{\hbar}\sum_{i=1}^{N}\left(\frac{(q_i - q_{i-1})^2}{2\varepsilon} - V\left(\frac{q_i + q_{i-1}}{2}\right)\varepsilon\right)\right]\right\} \end{aligned} \tag{6.8}$$

where the factor $B_N(T)$, which determines the measure, will be adjusted to make the integral finite and suitably normalised as $\varepsilon \to 0$. Mathematical aspects of such path integrals, including their existence and relationship to the Wiener measure, have been discussed, for instance, in Gel'fand and Yaglom (1960). It is also well known that the path-integral formalisation of quantum mechanics is equivalent to the more familiar Schrödinger and Heisenberg formalisms. In particular, the two expressions (6.5) and (6.6) for $K(q_b, T; q_b, 0)$ give the same result. This has been explicitly demonstrated in Feynman and Hibbs (1965) (see also Abers and Lee 1973). For our purposes, we shall take as given that the path-integral formula (6.6), as well its generalisation to field theory discussed below, yield physically correct results.

The connection between the path integral in (6.6) and the energy levels of the system is of crucial interest to us. Let $|\phi_n\rangle$ be the normalised eigenfunctions of the Hamiltonian, with eigenvalues E_n respectively. Introducing the complete set of intermediate states $|\phi_n\rangle$ in (6.5) and (6.6), we have

$$\begin{aligned}\int \mathscr{D}[q(t)] \exp\{(i/\hbar) S[q(t)]\} &= \langle q_b | \exp(-iHT/\hbar) | q_a \rangle \\ &= \sum_n \langle q_b | \phi_n \rangle \langle \phi_n | q_a \rangle \exp(-iE_n T/\hbar). \end{aligned} \tag{6.9}$$

In particular, if we set $q(0) = q(T) = q_0$ and further integrate over q_0, we have

$$\begin{aligned} G(T) \equiv \mathrm{Tr}(\exp(-iHT/\hbar)) &= \sum_n \exp(-iE_n T/\hbar) \\ &= \int_{-\infty}^{\infty} dq_0 \langle q_0 | \exp(-iHT/\hbar) | q_0 \rangle \\ &= \int_{-\infty}^{\infty} dq_0 \int \mathscr{D}[q(t)]_{q_0, q_0, T} \exp\{(i/\hbar) S[q(t)]\}. \end{aligned} \tag{6.10}$$

Here, the trace has been taken in two different ways – over energy eigenstates and equivalently over position eigenstates. Notice that the integral over q_0 adds just one more integral (over the end-point value of q) to the multiple integral over values of q at all intermediate times already contained in the path integral. Our basic technique in this chapter will be to evaluate this trace-cum-path integral in some approximation, cast it in the form of the series $\sum_n \exp(-iE_n T/\hbar)$ and thereby obtain the energy levels E_n. The approximation we shall use will be the semiclassical expansion, which amounts to the well-known stationary phase approximation as applied to the path integral. To understand this approximation, consider the multiple integral

$$\int_{-\infty}^{\infty} dq_1 \int_{-\infty}^{\infty} dq_2 \ldots \int_{-\infty}^{\infty} dq_N \, g(\boldsymbol{q}) \exp[-if(\boldsymbol{q})] \tag{6.11}$$

where $\boldsymbol{q}$ stands for the set $\{q_1 \ldots q_N\}$. Let $f(\boldsymbol{q})$ have one stationary point (an extremum) at $\boldsymbol{q} = \boldsymbol{a}$ and let $g(\boldsymbol{q})$ be sufficiently slowly varying near $\boldsymbol{q} = \boldsymbol{a}$. Expand $f(\boldsymbol{q})$ about $\boldsymbol{q} = \boldsymbol{a}$. The linear term in the Taylor expansion will be absent since $\boldsymbol{a}$ is an extremum. Let

$$f(\boldsymbol{q}) = f(\boldsymbol{a}) + \tfrac{1}{2} y_i A_{ij} y_j + O(y^3) \tag{6.12}$$

where $y_i \equiv q_i - a_i$. The spirit of the stationary phase approximation (SPA) is that if the exponential $\exp[-if(\boldsymbol{q})]$ oscillates rapidly enough, then the major contribution to (6.11) will come from the vicinity of the stationary point $\boldsymbol{q} = \boldsymbol{a}$. In that case, to leading approximation one can neglect the cubic and higher terms in y_i in the Taylor series (6.12). In this approximation,

$$\begin{aligned}
&\int dq_1 \int dq_2 \ldots \int dq_N \, g(\boldsymbol{q}) \exp[-\mathrm{i} f(\boldsymbol{q})] \\
&\quad \simeq g(\boldsymbol{a}) \exp[-\mathrm{i} f(\boldsymbol{a})] \int dy_1 \ldots \int dy_N \exp(-\tfrac{1}{2}\mathrm{i}\, y_i A_{ij} y_j) \\
&\quad = g(\boldsymbol{a}) \exp[-\mathrm{i} f(\boldsymbol{a})] \left(\frac{2\pi}{\mathrm{i}}\right)^{N/2} (\mathrm{Det}\, A)^{-1/2}. \qquad (6.13)
\end{aligned}$$

Bearing in mind that the path integral is essentially a multiple integral as per eq. (6.8), the SPA can be generalised for path integrals as well. Suppose that $q_{\mathrm{cl}}(t)$ is a classical solution, which goes from $q(0) = q_a$ to $q(T) = q_b$, of the system described by the Lagrangian (6.1). It is therefore an extremum of the action $S[q(t)]$, which can be expanded in a functional Taylor series about $q_{\mathrm{cl}}(t)$. We get

$$S[q(t)] = S[q_{\mathrm{cl}}(t)] + \tfrac{1}{2} \int_0^T dt\, y(t) O(t) y(t) + O(y^3) \qquad (6.14)$$

where

$$y(t) \equiv q(t) - q_{\mathrm{cl}}(t); \qquad y(0) = y(T) = 0,$$

and

$$O(t) \equiv -\frac{\partial^2}{\partial t^2} - \left(\frac{\partial^2 V}{\partial q^2}\right)_{q_{\mathrm{cl}}(t)}$$

is the operator playing the same role as the matrix A_{ij} in (6.12). Integration by parts has been done in the time integral. Then the SPA as applied to the path integral gives

$$\begin{aligned}
&\int \mathscr{D}[q(t)] \exp(\mathrm{i}/\hbar) S[q(t)] \\
&\quad \simeq B'(T) \exp\{(\mathrm{i}/\hbar) S[q_{\mathrm{cl}}(t)]\} [\mathrm{Det}\, O(t)]^{-1/2} \qquad (6.15)
\end{aligned}$$

where all factors involving $(2\pi\mathrm{i}\hbar)$, and the measure B are absorbed in the factor $B'(T)$. The determinant of the operator $O(t)$ is most conveniently obtained from its eigenvalues as shown in the examples below. The results (6.13) and (6.15) are of course only approximate, because the cubic and higher terms in the quantum fluctuations $y(t)$ have been neglected. The validity of this relies on the hope that the exponential in the integrand oscillates rapidly enough so that the major contribution to the integral comes from the vicinity of the stationary point of the exponent. Roughly speaking, this may be expected to happen for the path integral if the action $S[q(t)]$ is large compared with $\hbar$. But a more precise criterion for the rapid convergence of the expansion of which (6.15) forms the leading term is a fairly complicated mathematical question that we shall not discuss here. Notice that in the leading term the contribution to the integral over all paths has been approximated by the contribution of the classical path $q_{cl}(t)$ and paths in its neighbourhood. This is therefore a semiclassical approximation. But quantum effects are certainly included to leading order, through the factor $B'(T)\,[\mathrm{Det}\,O(t)]^{-1/2}$. Higher-order corrections to (6.15) due to cubic and higher powers of $y(t)$ can be obtained by perturbation theory as adapted to the path-integral formalism. We shall, however, be content with the leading term (6.15). Finally, if the integrand had several stationary exponents (i.e. several classical paths) then SPA states that each stationary path will make a separate additive contribution, provided these stationary paths are widely separated. If two stationary points are very near each other, then SPA is not reliable for either of them.

As an exercise in using this path integral, let us briefly review the harmonic oscillator problem where $V(q) = \frac{1}{2}\omega^2 q^2$. Then, from (6.10),

$$G(T) = \int \mathrm{d}q_0 \int \mathscr{D}[q(t)]_{q_0,q_0,T} \exp\left(\frac{i}{\hbar}\int_0^T \mathrm{d}t\,(\tfrac{1}{2}\dot{q}^2 - \tfrac{1}{2}\omega^2 q^2)\right). \tag{6.16}$$

Let $q_{cl}(t)$ be a classical solution going from $q_{cl}(0) = q_0$ back to $q_{cl}(T) = q_0$. Elementary classical mechanics tells us explicitly what this solution is for an oscillator, and also that it has an action

$$S_{cl}[q_0, q_0, T] = -2\omega q_0^2 \frac{\sin^2(\omega T/2)}{\sin(\omega T)}. \tag{6.17}$$

Now, expand $S[q(t)]$ about $q_{cl}(t)$. Let $y(t) \equiv q(t) - q_{cl}(t)$ with $y(0) = y(T) = 0$. Then

$$S[q(t)] = S_{cl}[q_0, q_0, T] + \frac{1}{2}\int_0^T \mathrm{d}t\, y(t)\left(-\frac{\partial^2}{\partial t^2} - \omega^2\right) y(t). \tag{6.18}$$

Notice that, since the oscillator's action is quadratic in $q(t)$ to begin with, there are no cubic or higher terms in the expansion (6.18). We can therefore obtain an exact result and need not use SPA in this problem. Then, using

(6.15), which is now an exact relation,

$$G(T) = \int \mathrm{d}q_0 \exp\{(\mathrm{i}/\hbar) S_{\mathrm{cl}}[q_0, q_0, T]\} \int \mathscr{D}[y(t)] \exp\left(\frac{\mathrm{i}}{\hbar}\int_0^T \mathrm{d}t\,(\tfrac{1}{2} y\, O(t)\, y)\right)$$

$$= \int \mathrm{d}q_0 \exp\{(\mathrm{i}/\hbar) S_{\mathrm{cl}}[q_0, q_0, T]\}\,[\mathrm{Det}\, O(t)]^{-1/2} B'(T) \tag{6.19}$$

where

$$O(t) = -\partial^2/\partial t^2 - \omega^2.$$

To obtain $[\mathrm{Det}\, O(t)]$, recall that it operates on functions $y(t)$ satisfying $y(0) = y(T) = 0$. Its eigenvalue equation is clearly

$$\left(-\frac{\partial^2}{\partial t^2} - \omega^2\right) \sin\left(\frac{n\pi t}{T}\right) = \left(\frac{n^2\pi^2}{T^2} - \omega^2\right) \sin\left(\frac{n\pi t}{T}\right) \tag{6.20}$$

where $n = 1, 2, \ldots, \infty$. Hence

$$[\mathrm{Det}\, O(t)]^{-1/2} = \prod_{n=1}^{\infty} \left(\frac{n^2\pi^2}{T^2} - \omega^2\right)^{-1/2}$$

$$\equiv K(T) \prod_{1}^{\infty} \left(1 - \frac{\omega^2 T^2}{n^2\pi^2}\right)^{-1/2} \tag{6.21}$$

The infinite product in the last equation has a simple result, namely $(\sin \omega T/\omega T)^{-1/2}$. Using this, (6.19) can be written as

$$G(T) = \int \mathrm{d}q_0 \exp[(\mathrm{i}/\hbar) S_{\mathrm{cl}}(q_0, q_0, T)]\, B'(T) K(T) \left(\frac{\omega T}{\sin \omega T}\right)^{1/2}. \tag{6.22}$$

The factors $B'(T)\,K(T)$ are purely 'kinematical', i.e. they do not involve the dynamics, represented here by the oscillator frequency ω. In particular $B'(T)$ includes the original measure $\lim_{N \to \infty} B_N(T)$ introduced in the parent definition (6.8), which was left unspecified, and is meant to be adjusted to give the correct normalisation. Upon inserting (6.17) for the classical action, we have

$$G(T) = B'(T) K(T) \int \mathrm{d}q_0 \exp\left(\frac{-2\mathrm{i}\omega q_0^2}{\hbar} \frac{\sin^2(\omega T/2)}{\sin(\omega T)}\right) \left(\frac{\omega T}{\sin \omega T}\right)^{1/2}$$

$$= \frac{1}{2\mathrm{i} \sin(\omega T/2)} B'(T) K(T) (2\pi \mathrm{i} \hbar T)^{1/2}$$

$$= B'(T)K(T)\,(2\pi i\hbar T)^{1/2} \sum_{n=0}^{\infty} \exp[-i(n+\tfrac{1}{2})\omega T]$$

$$= \sum_{n} \exp(-iE_n T/\hbar). \tag{6.23}$$

It is clear that we must choose the measure $B'(T)$, as yet unspecified, such that

$$B'(T)K(T) = (2\pi i\hbar T)^{-1/2} \tag{6.24}$$

In that case, we get the familiar harmonic oscillator levels

$$E_n = (n+\tfrac{1}{2})\hbar\omega. \tag{6.25}$$

Next, suppose $V(q)$ is more general than a harmonic oscillator potential. Suppose it has a minimum at $q = a$. Then, as argued in the last chapter, $q(t) = a$ is a stable static classical solution around which one can try to construct approximately harmonic oscillator levels. These levels are obtainable from the path integral by a trivial extension of the preceding derivation. Expand $V(q)$ about $q = a$.

$$V(q) = V(a) + \tfrac{1}{2}\omega^2\tilde{q}^2 + O(\tilde{q}^3)$$

and

$$S[q(t)] = -V(a)T + \int_0^T dt\,[\tfrac{1}{2}\dot{\tilde{q}}^2 - \tfrac{1}{2}\omega^2\tilde{q}^2 + O(\tilde{q}^3)] \tag{6.26}$$

where

$$\tilde{q}(t) \equiv q(t) - a.$$

Then, for this potential,

$$G(T) = \exp[-iV(a)T/\hbar] \int d\tilde{q}_0 \int \mathscr{D}[\tilde{q}(t)] \exp\left(\frac{i}{\hbar}\int_0^T [\tfrac{1}{2}\dot{\tilde{q}}^2 - \tfrac{1}{2}\omega^2\tilde{q}^2 + O(\tilde{q}^3)]\,dt\right). \tag{6.27}$$

Unlike (6.16), cubic and higher terms in $\tilde{q}(t)$ *are* present in the action in (6.27) because the potential is not exactly that of a harmonic oscillator. We therefore resort to SPA which amounts to neglecting these cubic and higher terms. If we do that, (6.27) reduces to the same form as (6.16), except for the factor $\exp[-\mathrm{i}V(a)T/\hbar]$. Therefore, using (6.23), we get for this problem

$$\mathrm{Tr}\,[\exp(-\mathrm{i}HT/\hbar)] \simeq \frac{\exp[-\mathrm{i}V(a)T/\hbar]}{2\mathrm{i}\sin\frac{1}{2}\omega T}$$

$$= \sum_n \exp\{-(\mathrm{i}/\hbar)T[V(a)+\hbar\omega(n+\tfrac{1}{2})]\}. \qquad (6.28)$$

Hence

$$E_n \simeq V(a) + \hbar\omega\,(n+\tfrac{1}{2}). \qquad (6.29)$$

Thus we have derived in the path-integral formalism, the basic result of section (5.2), which relates quantum energy levels to static classical solutions. Note that the neglect of anharmonic terms in (6.26), implied by the SPA, is tantamount to requiring that the coefficients of the anharmonic terms be sufficiently small compared with ω^2. This is precisely the 'weak-coupling' condition imposed in section (5.2).

Before going to field theory, let us conclude this section with the notion of Euclidean path integrals. Consider

$$K_E(q_b,\tau;q_a,0) \equiv \langle q_b|\exp(-H\tau/\hbar)|q_a\rangle$$

$$= \sum_n \langle q_b|\phi_n\rangle\langle\phi_n|q_a\rangle\exp(-E_n\tau/\hbar). \qquad (6.30)$$

This object may be considered as the analytic continuation to imaginary time of the transition amplitude in (6.5). Aside from this, the operator $\exp(-H\tau/\hbar)$ is well defined in its own right for any real parameter τ, although it is not unitary. A path-integral formula may be written for this object as follows:

$$K_E(q_b,\tau;q_a,0) = \int \mathscr{D}_E[q(\tau')]\exp\{(-1/\hbar)S_E[q(\tau')]\} \qquad (6.31)$$

where

$$S_E[q(\tau')] = \int_0^\tau \mathrm{d}\tau'\left[\frac{1}{2}\left(\frac{\partial q}{\partial\tau'}\right)^2 + V(q)\right]$$

is the Euclidean action introduced in chapter 4. The paths $q(\tau')$ once again go from $q(0)=q_a$ to $q(\tau)=q_b$. Since the two sides of (6.31) are analytic continuations to imaginary time of the two sides of (6.6), we can take (6.31) as valid. The equality of (6.31) to (6.30) can also be proved *ab initio*. The

Euclidean measure $\mathscr{D}_E[q(\tau')]$ will differ from $\mathscr{D}[q(t)]$ in (6.6) by suitable factors of i. In fact, the Euclidean path integral (6.31) is, in a mathematical sense, more basic than the real-time path integral (6.6). In (6.31), the integrand is a real negative exponential and the convergence of the integral, for paths $q(\tau')$ which are either very wiggly, or very far away (large $q(\tau')$), is mathematically more evident than in the case of (6.6). Aside from this, the Euclidean path integral will be needed by us in studying instantons in quantum theory. When $\tau \to \infty$, the Euclidean path integral is related to the ground state energy in a simple way, since

$$\lim_{\tau \to \infty} K_E(q_b, \tau; q_a, 0) = \lim_{\tau \to \infty} \sum_n \langle q_b | \phi_n \rangle \langle \phi_n | q_a \rangle \exp(-E_n \tau/\hbar)$$
$$= \lim_{\tau \to \infty} \langle q_b | \phi_0 \rangle \langle \phi_0 | q_a \rangle \exp(-E_0 \tau/\hbar). \quad (6.32)$$

Hence

$$\mathrm{Tr}[\exp(-H\tau/\hbar)] = \int \mathrm{d}q_0 \int \mathscr{D}_E[q(\tau')]_{q_0, q_0, \tau} \exp\{-(1/\hbar) S_E[q(\tau')]\}$$
$$\xrightarrow[\tau \to \infty]{} \exp(-E_0 \tau/\hbar). \quad (6.33)$$

Finally, note that $\mathrm{Tr}[\exp(-H\tau/\hbar)]$ is identical in structure to the partition function $\mathrm{Tr}[\mathrm{e}^{-\beta H}]$ in statistical mechanics, where β is the inverse temperature. This analogy will be exploited in chapter 10 in discussing the so-called dilute instanton gas.

6.2. Generalisation to field theory

The notion of path integrals can be formally generalised to quantum field theory. The analogue of eq. (6.10) to field theory is, for the case of a single scalar field $\phi(\boldsymbol{x}, t)$,

$$G(T) \equiv \mathrm{Tr}[\exp(-\mathrm{i}HT/\hbar)] = \int \mathscr{D}[\phi(\boldsymbol{x}, t)] \exp\{(\mathrm{i}/\hbar) S[\phi(\boldsymbol{x}, t)]\}. \quad (6.34)$$

Here, the integral is over all field configurations $\phi(\boldsymbol{x}, t)$ subject to boundary conditions to be specified below.

$$S[\phi(\boldsymbol{x}, t)] = \int_0^T \mathrm{d}t \int \mathrm{d}\boldsymbol{x}\, \mathscr{L}(\phi(\boldsymbol{x}, t))$$

is the action. The functional integral (6.34) can also be considered, like (6.8), as the limit of a product of ordinary integrals. For this purpose we would need to replace the space–time continuum (and not time alone) by a lattice of points and go to the limit when the lattice spacing goes to zero. The existence of a unique Lorentz-invariant limit is obviously more complicated than in the case of the purely time lattice of the elementary problem (6.8), because one could consider all kinds of $(D+1)$-dimensional lattice structures in a $(D+1)$-dimensional field theory. We shall assume that the functional integral (6.34) exists uniquely and that we can perform formal manipulations on it akin to what we did in the last section. We shall also assume that the functional integral gives the 'correct answer', in the sense that it will yield $\mathrm{Tr}[\exp(-iHT/\hbar)]$, which is also obtainable in principle from canonical operator methods.

We come to the boundary conditions on the set of $\phi(\boldsymbol{x}, t)$ to be integrated. First consider boundary conditions in time. Compare (6.34) with (6.10). Since a trace is being taken on $\exp(-iHT/\hbar)$, we set $\phi(\boldsymbol{x}, 0) = \phi(\boldsymbol{x}, T) = \phi_0(\boldsymbol{x})$ and then integrate over the initial value of the field $\phi_0(\boldsymbol{x})$, just as we integrated over q_0 in (6.10). This integral over $\phi_0(\boldsymbol{x})$, not explicitly shown in (6.34) is understood to be contained in $\int \mathscr{D}[\phi(\boldsymbol{x}, t)]$ along with integrals over fields at intermediate times. Next, consider boundary conditions in space, which also need to be specified in a field theory. Here, we shall make the following requirement which seems reasonable from a physical point of view. For illustration, recall for example the familiar system in $(1+1)$ dimensions, described by the Lagrangian

$$\mathscr{L}(x, t) = \tfrac{1}{2}(\partial_\mu \phi)^2 - U(\phi) \tag{6.35}$$

where

$$U(\phi) = \tfrac{1}{2}m^2\phi^2 + \tfrac{1}{4}\lambda\,\phi^4. \tag{6.36}$$

Here, all finite-energy configurations must satisfy $\phi \to 0$ as $x \to \pm\infty$ for all t. We expect the quantum vacuum state $\Psi_0[\phi]$ to be located around $\phi = 0$, and all states $\Psi[\phi]$ in the Fock-space to exist only over functions $\phi(x)$ satisfying $\phi(\infty) = \phi(-\infty) = 0$. Correspondingly, the functions in the functional integrals should satisfy $\phi(\infty, t) = \phi(-\infty, t) = 0$ for all t. By contrast, consider a case of spontaneous symmetry breaking as for instance in the kink system, where

$$U(\phi) = \tfrac{1}{4}\lambda(\phi^2 - m^2/\lambda)^2.$$

Then, there are four sets of permissible boundary conditions, given by

$$\phi(\infty, t) = \pm m/\sqrt{\lambda} \qquad \text{with} \qquad \phi(-\infty, t) = \pm m/\sqrt{\lambda}.$$

As we have argued in the preceding chapter, correspondingly there are four sectors of quantum states, which are furthermore unconnected from one another by a superselection quantum number. In such a case, the trace in $\mathrm{Tr}[\exp(-\mathrm{i}HT/\hbar)]$ can be taken over each sector separately, and in the corresponding functional integral the field $\phi(x,t)$ should obey boundary conditions appropriate to that sector. For instance if we need the trace of $\exp(-\mathrm{i}HT/\hbar)$ over the kink sector of states, then the fields in the functional integral must obey

$$\phi(\infty, t) = -\phi(-\infty, t) = m/\sqrt{\lambda}$$

for all t. More complicated cases of symmetry breaking are treated by a straightforward generalisation of this prescription.

Finally, the functional integral (6.34) was written for the case of a single scalar field. When several scalar fields are present, one has a product of such integrations (a multiple functional integral) over all the fields. When Fermi fields or gauge fields are present, the situation is more complicated. We shall return to these cases in later chapters.

As an illustration of the use of such functional methods, let us derive the central result of the last chapter, whereby a sector of energy levels was associated with any given static soliton solution, in the weak-coupling approximation. Conisder for instance a (1 + 1)-dimensional theory of the form (6.35), with some $U(\phi)$ which permits a stable static soliton solution $\phi(x,t) = \phi_{\mathrm{cl}}(x)$. Since $\phi_{\mathrm{cl}}(x)$ is an extremum of the potential

$$V[\phi] \equiv \int \mathrm{d}x \left[\frac{1}{2}\left(\frac{\partial \phi}{\partial x}\right)^2 + U(\phi)\right],$$

and also of the action, we can expand

$$S[\phi(x,t)] = -V[\phi_{\mathrm{cl}}]T + \tfrac{1}{2}\int \mathrm{d}x \int \mathrm{d}t\, [y(x,t)\, O(x,t)\, y(x,t)] + \mathrm{O}(y^3) \tag{6.37}$$

where

$$O(x,t) \equiv -\frac{\partial^2}{\partial t^2} + \frac{\partial^2}{\partial x^2} - \left(\frac{\partial^2 U}{\partial \phi^2}\right)_{\phi = \phi_{\mathrm{cl}}(x)} \tag{6.38}$$

and

$$y(x,t) \equiv \phi(x,t) - \phi_{\mathrm{cl}}(x).$$

Recall that all the functions $\phi(x,t)$ must obey spatial boundary conditions appropriate to that sector, i.e. the same boundary conditions as $\phi_{\mathrm{cl}}(x)$. Hence $y(x,t)$ will always obey $y(\pm\infty, t) = 0$. A stationary phase approximation to the functional integral (6.34) can be made, as in the earlier section, by neglecting the cubic and higher terms in (6.37):

$$\int \mathscr{D}[\phi(x,t)] \exp\{(\mathrm{i}/\hbar) S[\phi(x,t)]\}$$

$$\simeq \exp(-\mathrm{i} V[\phi_{\mathrm{cl}}]T/\hbar) \int \mathscr{D}[y(x,t)] \exp\left(-\frac{\mathrm{i}}{\hbar}\int \mathrm{d}x \int \mathrm{d}t\, \{y \mathrm{O}\, y\}\right)$$

$$= B''(T) \exp(-\mathrm{i} V[\phi_{\mathrm{cl}}]T/\hbar)\{\mathrm{Det}[O(x,t)]\}^{1/2}. \tag{6.39}$$

Here, $B''(T)$ again is a measure factor, contained in the definition of $\int \mathscr{D}[\phi(x,t)]$, which we have as yet left unspecified. It will be adjusted to yield the correct normalisation of states. To obtain Det $[O(x,t)]$ through its eigenvalues, note that the operator $O(x,t)$ given in (6.38) is a sum of a t-dependent term and an x-dependent term. Consequently it may be diagonalised separately in t-space and x-space. Let

$$\left[-\frac{\partial^2}{\partial x^2} + \left(\frac{\partial^2 U}{\partial \phi^2}\right)_{\phi=\phi_{\mathrm{cl}}(x)}\right] y_r(x) = \omega_r^2\, y_r(x) \ldots \tag{6.40}$$

be the eigenvalue equation for the x-dependent part of the operator $O(x,t)$ under the boundary conditions $y_r(\pm\infty) = 0$. Equation (5.48) in the last chapter is an example of (6.40) for the case of the kink solution. Then

$$\{\mathrm{Det}[O(x,t)]\}^{-1/2} = \left[\prod_{r=0}^{\infty} \mathrm{Det}\left(-\frac{\partial^2}{\partial t^2} - \omega_r^2\right)\right]^{-1/2}. \tag{6.41}$$

Thus, the determinant has been reduced to a product of harmonic oscillator determinants of the type discussed in the last section. The remaining steps in that section follow, and hence we have for our field theoretical problem

$$\mathrm{Tr}[\exp(-\mathrm{i}HT/\hbar)] \simeq \exp(-\mathrm{i} V[\phi_{\mathrm{cl}}]T/\hbar) \prod_r \left(\frac{1}{2\mathrm{i}\sin(\omega_r T/2)}\right), \tag{6.42}$$

where the measure factor $B''(T)$ in (6.39) is again understood to be so chosen that (6.42) holds in SPA. Rewriting this equation as

$$\mathrm{Tr}[\exp(-\mathrm{i}HT/\hbar)] \simeq \exp(-\mathrm{i} V[\phi_{\mathrm{cl}}]T/\hbar)$$

$$\cdot \prod_r \left(\sum_{n_r} \exp[-\mathrm{i}T\omega_r(n_r + \tfrac{1}{2})]\right)$$

$$= \sum_{\{n_r\}} \exp\left[-(\mathrm{i}T/\hbar)\left(V[\phi_{\mathrm{cl}}]\right.\right.$$

$$\left.\left. + \sum_r \hbar\omega_r(n_r + \tfrac{1}{2})\right)\right] \tag{6.43}$$

where the sum $\sum_{\{n_r\}}$ is over all sets of integers $\{n_0, n_1, n_2, \ldots\}$. Hence we have for the energy levels of states built around $\phi = \phi_{cl}(x)$ in this approximation

$$E_{\{n_r\}} \simeq V[\phi_{cl}] + \sum_r \hbar\omega_r(n_r + \tfrac{1}{2}). \tag{6.44}$$

Thus we have derived, as an exercise in using the functional integral (6.34), the central result of the last chapter. The physics is of course exactly the same as in the last chapter, and all the qualifying remarks made there hold here, suitably restated in functional-integral language. For instance, the neglect, to leading order, of the cubic and higher terms in (6.37), which forms the basis of the SPA used here, is nothing but the 'weak-coupling approximation' used in the last chapter. In general, the sum $\sum_r$ over the normal modes in (6.44) will diverge and this divergence is again to be cancelled by adding counter terms to the Lagrangian. Under this addition of counter-terms

$$L = \int \mathscr{L}(x, t)\,dx \to L - V_{ct}[\phi]$$

and

$$S[\phi] \to S[\phi] - \int V_{ct}(\phi)dt. \tag{6.45}$$

In SPA, this will replace, to leading order, $V[\phi_{cl}]T$ in (6.39) by $\{V[\phi_{cl}] + V_{ct}[\phi_{cl}]\}T$, so that (6.44) will be replaced by

$$E_{\{n_r\}} \simeq V[\phi_{cl}] + V_{ct}[\phi_{cl}] + \sum_r (n_r + \tfrac{1}{2})\hbar\omega_r. \tag{6.46}$$

If the theory is renormalisable, we can expect that the counter terms will cancel the divergence in the sum over eigenfrequencies. Finally, we had pointed out in chapter 5 that if any of the ω_r vanish due to the presence of some continuous symmetry, as inevitably happens for soliton solutions, then the treatment in that chapter would have to be modified. The same problem crops up in the present derivation under those circumstances. A symptom of the problem may be noticed in (6.42) where the factor $\{2i\sin(\omega_r T/2)\}^{-1}$ would diverge as any $\omega_r \to 0$. A more basic reason lies in the fact that the SPA in its simple form cannot be justified in this situation. Given a $\phi_{cl}(x)$, it would actually belong to a continuous family of classical solutions $P(\alpha)[\phi_{cl}(x)]$, where $P(\alpha)$ is the continuous symmetry operation in question, parametrised by the variable(s) α. Thus, $\phi_{cl}(x)$ is not an isolated stationary point of $S[\phi]$, but belongs to a continuous family of stationary points, whereas in introducing the SPA we had stated that it would be a good approximation only if the stationary points were sufficiently widely separated. Thus, zero-frequency modes due to continuous symmetries will continue to plague us in the present derivation, just as they did in chapter 5. This is as it should be, since the physics behind the approximation here is the same as in chapter 5. For some discussion on how to modify these methods in the presence of zero-modes, see, as promised, chapter 8. From these remarks, we see that the functional integral (6.34) yields the energy levels associated with a static soliton, under the same conditions and with the same limitations as in the last chapter.

A Euclidean functional integral can be introduced in field theory as well, by generalising (6.33). We have,

$$\mathrm{Tr}[\exp(-H\tau/\hbar)] \equiv \int \mathscr{D}_{\mathrm{E}}[\phi(\boldsymbol{x},\tau')] \exp\{-(1/\hbar)S_{\mathrm{E}}[\phi(\boldsymbol{x},\tau')]\} \tag{6.47}$$

where $S_{\mathrm{E}}[\phi(\boldsymbol{x},\tau')]$ is the Euclidean action defined in chapter 4 and the integration over the initial and final value of the fields, $\phi_0(\boldsymbol{x}) = \phi(\boldsymbol{x},0) = \phi(\boldsymbol{x},\tau)$ is understood to be contained in the functional integral. Just as in (6.33), this trace will be dominated when $\tau \to \infty$, by the lowest-energy state in that sector. Consider for example a quantum soliton state in $(1+1)$ dimensions which is the lowest-energy state built around some static classical soliton solution $\phi_{\mathrm{cl}}(x)$. Analogously to (6.37), we can expand

$$\begin{aligned} S_{\mathrm{E}}[\phi(x,\tau')] = V[\phi_{\mathrm{cl}}]\tau + \tfrac{1}{2}\textstyle\int \mathrm{d}x \int \mathrm{d}\tau' [&y(x,\tau')O_{\mathrm{E}}(x,\tau')y(x,\tau') \\ &+ \mathrm{O}(y^3)] \end{aligned} \tag{6.48}$$

where

$$O_{\mathrm{E}}(x,\tau') \equiv -\frac{\partial^2}{\partial\tau'^2} - \frac{\partial^2}{\partial x^2} + \left(\frac{\partial^2 U}{\partial\phi^2}\right)_{\phi_{\mathrm{cl}}} \tag{6.49}$$

$$y(x,\tau') \equiv \phi(x,\tau') - \phi_{\mathrm{cl}}(x).$$

Neglect of the $\mathrm{O}(y^3)$ terms in (6.48) is the Gaussian approximation to (6.47), and is the Euclidean analogue of SPA. In this approximation, we have, analogously to (6.39),

$$\mathrm{Tr}[\exp(-H\tau/\hbar)] \simeq B_{\mathrm{E}}(\tau)\exp(-V[\phi_{\mathrm{cl}}]\tau/\hbar)\{\mathrm{Det}[O_{\mathrm{E}}(x,\tau')]\}^{-1/2}, \tag{6.50}$$

where $B_{\mathrm{E}}(\tau)$ is the Euclidean measure, similar to $B''(T)$ in (6.39). We can rewrite (6.50) as

$$\begin{aligned} \mathrm{Tr}[\exp(-H\tau/\hbar)] &\simeq B_{\mathrm{E}}(\tau)\exp\{-[V[\phi_{\mathrm{cl}}]\tau/\hbar + \tfrac{1}{2}\ln\mathrm{Det}\,O_{\mathrm{E}}]\} \\ &= B_{\mathrm{E}}(\tau)\exp\left[-\frac{\tau}{\hbar}\left(V[\phi_{\mathrm{cl}}] + \frac{\hbar}{2\tau}\mathrm{tr}(\ln O_{\mathrm{E}})\right)\right] \end{aligned}$$

$$= B_E(\tau)\exp\left\{-\frac{\tau}{\hbar}\left[V[\phi_{cl}] + \sum_r \frac{\hbar}{2\tau}\operatorname{tr}\ln\left(\frac{-\partial^2}{\partial\tau'^2} + \omega_r^2\right)\right]\right\} \quad (6.51)$$

where the eigenfrequencies of the normal modes in (6.40) have been used. Now, the object $\operatorname{tr}\ln(-\partial^2/\partial\tau'^2 + \omega_r^2)$ is particularly easy to evaluate when $\tau \to \infty$. Recall that the eigenvalues of $-(\partial^2/\partial\tau'^2)$ for our boundary conditions are $n^2\pi^2/\tau^2$; $n = 1, 2, \ldots, \infty$. Hence

$$\operatorname{tr}\ln\left(-\frac{\partial^2}{\partial\tau'^2} + \omega_r^2\right) = \sum_{n=1}^{\infty}\ln\left(\frac{n^2\pi^2}{\tau^2} + \omega_r^2\right)$$

$$\xrightarrow[\tau\to\infty]{} \tau\int_{-\infty}^{\infty}\frac{\mathrm{d}\nu}{2\pi}\ln(\nu^2 + \omega_r^2)$$

$$= \tau\int_{-\infty}^{\infty}\frac{\mathrm{d}\nu}{2\pi}\left[\ln\left(1 + \frac{\omega_r^2}{\nu^2}\right) + \ln\nu^2\right] \quad (6.52)$$

where we have separated out the divergent part of the integral. This divergent part is purely kinematical (it does not depend on the ω_r) and is meant to be nullified by a suitable choice of the measure $B_E(\tau)$. We absorb this divergent part into the measure, thereby changing $B_E(\tau)$ into $B'_E(\tau)$; then we are left in (6.52) with

$$\tau\int_{-\infty}^{\infty}\frac{\mathrm{d}\nu}{2\pi}\ln\left(1 + \frac{\omega_r^2}{\nu^2}\right)$$

$$= \tau\int_{-\infty}^{\infty}\frac{\mathrm{d}\nu}{2\pi}\frac{2\omega_r^2}{(\nu^2 + \omega_r^2)}, \quad \text{(on integration by parts)},$$

$$= \omega_r\tau. \quad (6.53)$$

Inserting this in (6.51), we have

$$\operatorname{Tr}[\exp(-H\tau/\hbar)] \xrightarrow[\tau\to\infty]{} B'_E(\tau)\exp\left[-\frac{\tau}{\hbar}\left(V[\phi_{cl}] + \sum_r \frac{\hbar}{2\tau}\omega_r\tau\right)\right] \quad (6.54)$$

Choosing the measure such that $B'_E(\tau) = 1$, we thus get for the quantum soliton's energy

$$E_0 \simeq V[\phi_{cl}] + \sum_r \tfrac{1}{2}\hbar\omega_r. \tag{6.55}$$

This is just our old result, which can be rendered finite by adding $V_{ct}[\phi_{cl}]$.

All these derivations are meant to be exercises in the use of the semiclassical expansion for functional integrals. We have not yet obtained any new results, because we have been expanding the action, in both the Minkowskian and Euclidean cases, about static solutions. When a similar expansion is made about classical solutions which depend non-trivially on t (or τ), we shall get new results. In the Minkowskian case, this will lead to quantisation of time-dependent solitons, and will be discussed in the following sections. In the Euclidean case, this will lead to instanton physics, to be discussed in chapter 10.

6.3. The Bohr–Sommerfeld condition

The quantisation of time-dependent solutions in field theory is done by the WKB method which gives a generalisation of the Bohr–Sommerfeld condition to field theory. This condition is well known in non-relativistic quantum mechanics and is usually derived from the differential Schrödinger equation. The generalisation to field theory, however, is most conveniently done in the functional-integral formalism. As a starting point, let us briefly sketch how the Bohr–Sommerfeld condition is derived in path-integral language, for the one-particle system (Gutzwiller 1970, 1971, 1967, Maslov 1970, Berry and Mount 1972, Dashen et al. 1974a).

Consider the propagator of the system (6.2),

$$\mathscr{G}(E) \equiv \mathrm{Tr}\left(\frac{1}{E-H}\right) = \frac{\mathrm{i}}{\hbar}\int_0^\infty \mathrm{d}T\, G(T) \exp(\mathrm{i}ET/\hbar). \tag{6.56}$$

The poles of $\mathscr{G}(E)$ are clearly the bound-state energy eigenvalues. Using (6.10) for $G(T)$,

$$\mathscr{G}(E) = \frac{\mathrm{i}}{\hbar}\int_0^\infty \mathrm{d}T \exp(\mathrm{i}ET/\hbar) \int \mathrm{d}q_0 \int \mathscr{D}[q(t)]_{q(0)=q(T)=q_0} \times \exp\{(\mathrm{i}/\hbar)S[q(t)]\}. \tag{6.57}$$

The quantisation condition is obtained by repeatedly applying the SPA to this integral. That is essentially the WKB method in this language. First expand $S[q(t)]$ about some non-trivial extremum, i.e. a time-dependent solution $q_{cl}^j(t)$ satisfying

$$\frac{\delta S}{\delta[q(t)]} = \ddot{q} + \frac{\partial V}{\partial q} = 0 \tag{6.58}$$

and $q_{cl}^j(0) = q_{cl}^j(T) = q_0$. There may be more than one classical solution for a given T satisfying these boundary conditions, which are distinguished by the label j. We expand $S[q(t)]$ about $q_{cl}^j(t)$ and use the SPA on the path integral as per eqs. (6.14)–(6.15). We do this separately for each solution $q_{cl}^j(t)$, $j = 1, 2, \ldots$, assuming that they are sufficiently separated for the SPA to apply additively. Then we have

$$G(T) = \sum_j \int dq_0 \exp\left(iS_j[q_0, q_0, T]/\hbar\right)\Delta_1^j \tag{6.59}$$

where $S_j[q_0, q_0, T] \equiv S[q_{cl}^j]$ is the action of the jth classical solution beginning and ending at q_o, and

$$\Delta_1^j \propto \left\{\mathrm{Det}\left[-\frac{\partial^2}{\partial t^2} - \left(\frac{\partial^2 V}{\partial q^2}\right)_{q_{cl}^j}\right]\right\}^{-1/2}. \tag{6.60}$$

Since $(\partial^2 V/\partial q^2)_{q_{cl}^j(t)}$ is time-dependent in general, Δ_1^j is harder to evaluate than for the harmonic oscillator case in section (6.1). Its evaluation is sketched separately in Appendix A. Let us proceed to apply the SPA again, this time to the q_0 integration in (6.59). The stationary exponent must satisfy

$$\begin{aligned} 0 = \frac{\partial}{\partial q_0} S_j[q_0, q_0, T] &= \left(\frac{\partial}{\partial q_a} S_j[q_a, q_b, T]\right. \\ &\quad \left. + \frac{\partial}{\partial q_b} S_j[q_a, q_b, T]\right)_{q_a = q_b = q_0} \\ &= -p_1 + p_2. \end{aligned} \tag{6.61}$$

The last equality follows from classical mechanics, where p_1 and p_2 are the momenta at the beginning and the end of the path. Equation (6.61) shows that the path must not only have the same position q_0 at its two ends, but also the same momentum, i.e. the path is periodic. This is not surprising: after all the path integral itself is a multiple integral over positions q_i at intermediate times (6.8). A general path is jagged (fig. 11a) with discontinuous velocity. But under the SPA, the multiple integration $\int \prod_i dq_i$ picks out the classical path (fig. 11b) where the intermediate sharp corners have been smoothed out. That still leaves the edge at the point q_0. But the trace integration $\int dq_0$ performs the same function and smooths the edge at the end point as well (fig. 11c) making the path a periodic orbit. (Fig. 11 has been drawn for two space dimensions just for clarity.)

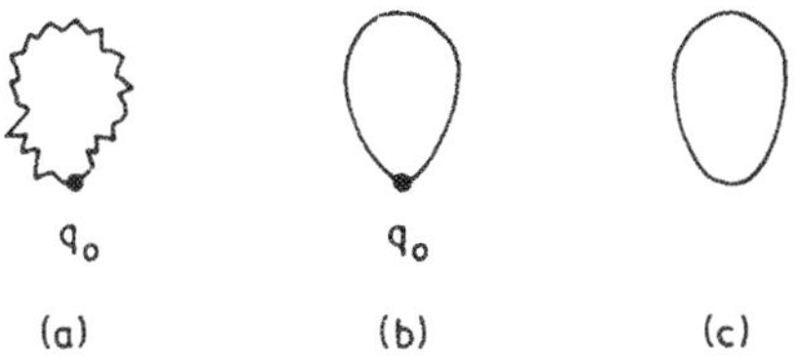

Fig. 11. (a) A general non-classical path starting and ending at q_0. (b) A classical path picked out by the stationary phase approximation (SPA) applied to the path integral. Notice that sharp edges (discontinuous velocities) have been removed by the SPA only at intermediate times because the path integral integrates only over intermediate-time values q_i. (c) A *periodic* classical path. This is picked out by SPA applied to the trace integral over q_0, which rounds out the sharp edge at q_0. This figure has been drawn for two-dimensional orbits for the sake of clarity.

Thus (6.59) reduces to a sum over all periodic orbits. Of course, the basic period (per cycle) of the orbit need not be T. It could be T/n, where n is any integer characterising the number of traverses of the same basic orbit. For a given T, $G(T)$ will receive contributions from all periodic orbits whose basic period is an integral fraction of T. We replace the sum over the label j in (6.59) by a sum over n. Thus,

$$G(T) = \sum_n \exp[(i/\hbar) n S_{cl}(T/n)] \Delta_1^{(n)}(T) \Delta_2^{(n)}(T) \tag{6.62}$$

where $\Delta_2^{(n)}$ is the factor coming from the q_0 SPA integral. It is clear that the action for n traverses is just n times the action per cycle. The factors $\Delta_1 \Delta_2$ have been computed in Appendix A and give

$$\Delta_1^{(n)}(T)\Delta_2^{(n)}(T) = \frac{T}{n}\left(-\frac{dE_{cl}}{dT}\right)^{1/2} e^{-i\pi n}\left(\frac{-i}{2\pi\hbar}\right)^{1/2} \tag{6.63}$$

where E_{cl} is the energy of the periodic orbit of period T/n. We finally get

$$\mathcal{G}(E) = \frac{i}{\hbar}\sum_n \int_0^\infty dT \exp\left(\frac{i}{\hbar}[ET + n S_{cl}(T/n)]\right)\frac{T}{n}\left(-\frac{dE_{cl}}{dT}\right)^{1/2}$$
$$\cdot\left(\frac{-i}{2\pi\hbar}\right)^{1/2}\cdot(-1)^n$$

$$= \frac{\mathrm{i}}{\hbar}\left(\frac{-\mathrm{i}}{2\pi\hbar}\right)^{1/2} \sum_n \int_0^\infty \mathrm{d}\tau \sqrt{n}\left(-\frac{\mathrm{d}E_{\mathrm{cl}}}{\mathrm{d}\tau}\right)^{1/2} \tau$$

$$\times \exp\left(\frac{\mathrm{i}n}{\hbar}[E\tau + S_{\mathrm{cl}}(\tau)]\right)(-1)^n \tag{6.64}$$

where $\tau \equiv T/n$, is the basic period per cycle. The integration over τ is performed, once again by the SPA. The stationary phase point now satisfies

$$E = -\partial S_{\mathrm{cl}}/\partial\tau = E_{\mathrm{cl}}. \tag{6.65}$$

where the last equality is again familiar in classical mechanics. Note that E is a parameter in the propagator $\mathscr{G}(E)$. The SPA then picks, for any given value of E, that periodic classical orbit whose energy is $E_{\mathrm{cl}} = E$. The period τ and the action S of this orbit are therefore also fixed by the value of E. The stationary phase approximation to (6.64) thus gives

$$\mathscr{G}(E) = \sum_n \frac{\mathrm{i}}{\hbar}\left(\frac{-\mathrm{i}}{2\pi\hbar}\right)^{1/2}\left(\frac{2\pi\hbar}{-\mathrm{i}n}\right)^{1/2}\left(\frac{\partial^2 S_{\mathrm{cl}}}{\partial\tau^2}\right)^{-1/2} \sqrt{n}\,\tau(E)$$

$$\times\left[-\left(\frac{\mathrm{d}E}{\mathrm{d}\tau}\right)\right]^{1/2}(-1)^n \times \exp\left(\frac{\mathrm{i}n}{\hbar}[E\,\tau(E) + S(\tau)]\right). \tag{6.66}$$

On noting that $\partial^2 S_{\mathrm{cl}}/\partial\tau^2 = -\mathrm{d}E/\mathrm{d}\tau$, many factors cancel and

$$\mathscr{G}(E) = \sum_{n=1}^{\infty} \mathrm{i}\frac{\tau(E)}{\hbar}(-1)^n \exp\left(\mathrm{i}n\frac{W(E)}{\hbar}\right)$$

$$= \frac{-\mathrm{i}\tau(E)}{\hbar}\,\frac{\exp[(\mathrm{i}/\hbar)W(E)]}{1+\exp[(\mathrm{i}/\hbar)W(E)]} \tag{6.67}$$

where

$$W(E) \equiv S[\tau(E)] + E\tau(E) \tag{6.68}$$

is the Legendre transform of S. $\mathscr{G}(E)$ clearly has poles at E_m which satisfy

$$W(E_m) = (2m+1)\pi\hbar \tag{6.69}$$

where m is an integer. The residue at each pole is unity since $\mathrm{d}W/\mathrm{d}E = \tau$. Thus, the energy levels are given in the WKB approximation by E_m which satisfy (6.69). This is just the usual Bohr–Sommerfeld quantisation rule

(Landau and Lifshitz 1958) because

$$W(E) = S + E\tau = \int_0^\tau dt\,(\tfrac{1}{2}\dot{q}^2 - V + E)$$

$$= \int_0^\tau dt\,\dot{q}^2 = 2\int_{q_1}^{q_2} p\,dq \tag{6.70}$$

where q_1 and q_2 are the turning points of the classical orbit.

It may appear that only the bound-state energy levels can be obtained this way as poles of $\mathcal{G}(E)$. But if the system is put in a box with periodic boundary conditions, then all energy levels become discrete, and all orbits bounded. The above method can then be used for all the levels and later the size of the box can tend to infinity.

In this derivation it has been tacitly assumed that there is only one continuous family of periodic orbits, parametrised by the basic period variable τ or equivalently by the energy $E_{cl}(\tau)$. That is, for any τ, it is assumed that there is only one orbit with that basic period τ, and that this orbit will contribute to the dT integration in (6.64) either through single traverse ($T = \tau$) or through multiple traverse ($T = n\tau$). This helped us replace $\sum_j \int dT$ by $\sum_n n \int d\tau$. For typical 'single-well' potentials in one dimension, this assumption is satisfied. If the potential V had several 'wells' near different minima, there may be several families of orbits, i.e. orbits in different regions with the same basic period τ. The same thing can happen in two or more dimensions. In such cases, if these families of orbits are sufficiently well separated, then the SPA will give an additive contribution to $\mathcal{G}(E)$ from each family. There will be an energy level when a member of any of the families satisfies the condition (6.69), to this approximation. More detailed analyses of the WKB method in the path-integral formalism are given in the papers of Gutzwiller (1967, 1970, 1971), Berry and coworkers (see Berry and Tabor 1977 and references therein), Levit (1978), and Levit and Smilansky (1977).

6.4. The WKB method in field theory

We now present the generalisation of the WKB method, first to many degrees of freedom and then to field theory considered as a system with an infinite number of degrees of freedom (Dashen et al. 1974a, 1975a). Our generalisation will be formal. The basic steps are the same as in the preceding section. A functional integral will be written for the propagator $\mathcal{G}(E) \equiv \mathrm{Tr}[1/(E-H)]$, and the SPA applied to it to pick out time-dependent periodic classical solutions. The only difference will be that the integrals over the quantum fluctuations (analogues of the factors $\Delta_1^{(n)}\,\Delta_2^{(n)}$ in eq. (6.62)) will be more complicated.

Consider for instance a two-dimensional problem – a particle in the x–y plane – with a Lagrangian

$$L = \tfrac{1}{2}\dot{x}^2 + \tfrac{1}{2}\dot{y}^2 - V(x, y).$$

Then

$$\mathscr{G}(E) = \frac{\mathrm{i}}{\hbar}\int_0^\infty \mathrm{d}T \exp(\mathrm{i}ET/\hbar)\, G(T);$$

$$G(T) = \int \mathrm{d}x_0 \int \mathrm{d}y_0 \int \mathscr{D}[x(t)] \int \mathscr{D}[y(t)] \exp\left(\frac{\mathrm{i}}{\hbar}\int_0^T \mathrm{d}t\, [\tfrac{1}{2}\dot{x}^2 + \tfrac{1}{2}\dot{y}^2 - V(x, y)]\right)$$

$$= \int \mathrm{d}\boldsymbol{r}_0 \int \mathscr{D}[\boldsymbol{r}(t)] \exp\left(\frac{\mathrm{i}}{\hbar}\int_0^T \mathrm{d}t\, [\tfrac{1}{2}\dot{\boldsymbol{r}}\cdot\dot{\boldsymbol{r}} - V(\boldsymbol{r})]\right) \qquad (6.71)$$

where the compact notation $\boldsymbol{r} = \{x, y\}$ has been used. When the SPA is applied to the path integral, classical orbits will be picked out, and the SPA as applied to the $\int \mathrm{d}\boldsymbol{r}_0$ integral will further restrict these to periodic orbits, by a trivial generalisation of (6.61). For a given period T, all periodic orbits with basic period T/n will contribute. Hence we can write, under SPA,

$$G(T) \simeq \sum_n \exp[\mathrm{i}nS_{\mathrm{cl}}(T/n)/\hbar] \int \bar{\mathscr{D}}[\boldsymbol{\rho}(t)]$$

$$\exp\left[\frac{\mathrm{i}}{\hbar}\int_0^T \frac{\mathrm{d}t}{2}\, \rho_i \left(-\frac{\partial^2}{\partial t^2} - V''\right)_{ij} \rho_j\right] \qquad (6.72)$$

where $S_{\mathrm{cl}}(T/n)$ is the action per cycle of the periodic orbit $\boldsymbol{r}_{\mathrm{cl}}^{(n)}(t)$ of basic period $\tau = T/n$, $\boldsymbol{\rho}(t) \equiv \boldsymbol{r}(t) - \boldsymbol{r}_{\mathrm{cl}}^{(n)}(t)$,

$$\left(-\frac{\partial^2}{\partial t^2} - V''\right)_{ij} \equiv \left(\delta_{ij}\frac{-\partial^2}{\partial t^2} - \frac{\partial^2 V}{\partial r_i \partial r_j}\right)_{\boldsymbol{r}_{\mathrm{cl}}^{(n)}(t)}$$

and $\mathscr{D}[\rho(t)]$ includes the trace integration over $\mathrm{d}\rho(0)$. We rewrite (6.72) as

$$G(T) \simeq \sum_n \exp[\mathrm{i}nS_{\mathrm{cl}}(T/n)/\hbar]\,\Delta^{(n)}(T) \tag{6.73}$$

where $\Delta^{(n)}(T)$ is the analogue of $\Delta_1^{(n)}\Delta_2^{(n)}$ in (6.62). It is still essentially $\{\mathrm{Det}[-\partial^2/\partial t^2 - V'']\}^{-1/2}$, but now $(-\partial_t^2 - V'')$ is a 2×2 matrix in $\{x-y\}$ space in addition to being a time-dependent differential operator. Its evaluation requires the use of some Hamilton–Jacobi theory, and the result is

$$\Delta^{(n)}(T) = \left(\frac{-\mathrm{i}}{2\pi\hbar}\right)^{1/2}\frac{T}{n}\left(-\frac{\mathrm{d}E_{\mathrm{cl}}}{\mathrm{d}T}\right)^{1/2}[2\mathrm{i}\sin(n\nu_1/2)]^{-1} \tag{6.74}$$

where ν_1 is a 'stability angle', defined below. We shall not present the derivation of this expression (see Appendix A of Dashen et al. 1975a). But it may be noticed that (6.74) is very similar to the one-dimensional result (6.63) which we have derived in our Appendix A. The result (6.74) differs from (6.63) in only two respects:

(i) The factor $\mathrm{e}^{-\mathrm{i}\pi n}$ is absent. Recall from Appendix A that this factor came from the singular points of the classical orbit. A typical periodic orbit in two dimensions (such as in fig. 11c) has no singular points and hence this factor is omitted. In special cases where the orbits do have singular points, such factors should be attached (for example, one-dimensional orbits in the two-dimensional plane).

(ii) (6.74) has an extra factor $1/[2\mathrm{i}\sin(n\nu_1/2)]$. As we will show, it represents the effect of quantum fluctuations transverse to the orbit. The stability angle ν_1 is defined as follows. Consider the zero eigenvalue equation of the operator in question $(-\partial^2/\partial t^2 - V'')$. Let

$$\left[-\delta_{ij}\frac{\partial^2}{\partial t^2} - \left(\frac{\partial^2 V}{\partial r_i \partial r_j}\right)_{r_{\mathrm{cl}}(t)}\right](\xi(t))_j = 0. \tag{6.75}$$

Here, the operator is a 2×2 matrix, and ξ is a two-component vector in x–y space. Note that this is also the 'linear stability equation' of the orbit. That is, $\boldsymbol{r}_{\mathrm{cl}}(t)$ as well as a slightly distorted classical solution $\boldsymbol{r}'_{\mathrm{cl}}(t) \equiv \boldsymbol{r}_{\mathrm{cl}}(t) + \xi(t)$ will both obey the equation of motion

$$-\ddot{\boldsymbol{r}} - \partial V/\partial \boldsymbol{r} = 0. \tag{6.76}$$

Consequently the small deformation $\xi(t)$ will obey (6.75) to leading order. Let us call the two-independent solutions of (6.75) $\xi_0(t)$ and $\xi_1(t)$. Each must satisfy

$$\xi_i(t+\tau) = \mathrm{e}^{\mathrm{i}\nu_i}\xi_i(t) \qquad \text{for all } t, \tag{6.77}$$

where ν_i are called stability angles. $\tau = T/n$ is the basic period of $r_{\text{cl}}(t)$. Equation (6.77) must hold since the operator $(-\partial_t^2 - V'')$ is itself periodic with period τ. One of these solutions is

$$\xi_0(t) = \frac{\mathrm{d}}{\mathrm{d}t} r_{\text{cl}}(t) \tag{6.78}$$

as can be verified by differentiating (6.76). Since $r_{\text{cl}}(t)$ is periodic, so is $\xi_0 = \dot{r}_{\text{cl}}$. Hence $\nu_0 = 0$. This mandatory zero-stability angle corresponding to the eigenfunction (6.78) will always exist and is related to the time-translation invariance of the equation of motion. Notice the similarity to zero-frequency modes that we encountered in earlier chapters, which were also related to some continuous symmetry. The other stability angle ν_1, corresponding to the second solution $\xi_1(t)$ will in general be non-zero, and this is what occurs in the result (6.74).

In as much as we did not present the derivation of (6.74), let us make it plausible through an example. Consider the case when the potential has the simple form

$$V(x, y) = V(x) + \tfrac{1}{2}\omega^2 y^2 \tag{6.79}$$

where $V(x)$ is any one-dimensional potential permitting a family of periodic orbits in x. Then $V(x, y)$ will permit a family of periodic orbits which lie entirely on the x-axis. That is,

$$x(t) = x_{\text{cl}}(t); \qquad y(t) = 0 \tag{6.80}$$

where $x_{\text{cl}}(t)$ has some period $\tau = T/n$. For these orbits the problem separates into an x-part and a y-part.

In particular, the matrix $(\mathrm{d}^2 V/\mathrm{d}r_i\,\mathrm{d}r_j)_{r_{\text{cl}}}$ factorises into the form

$$\begin{bmatrix} (\mathrm{d}^2 V/\mathrm{d}x^2)_{x_{\text{cl}}} & 0 \\ 0 & \omega^2 \end{bmatrix}.$$

Inserting this into (6.75), we can see that one solution of that equation is

$$\xi_1(t) = \begin{pmatrix} 0 \\ \mathrm{e}^{\mathrm{i}\omega t} \end{pmatrix},$$

satisfying $\xi_1(t+\tau) = \mathrm{e}^{\mathrm{i}\omega\tau}\xi_1(t)$. Thus, the stability angle for this case is $\nu_1 = \omega\tau$ and the fluctuation factor (6.74) reduces to

$$\left[\frac{T}{n}\left(\frac{-\mathrm{i}}{2\pi\hbar}\right)^{1/2}(-1)^n\left(\frac{-\mathrm{d}E_{\mathrm{cl}}}{\mathrm{d}\tau}\right)^{1/2}\right]\left[\frac{1}{2\mathrm{i}\sin(\omega T/2)}\right] \tag{6.81}$$

where the $(-1)^n$ has been inserted, as explained, because the orbit (6.80) must have two turning points. This is precisely the fluctuation factor we would expect, based on our earlier discussion, for the orbit (6.80) in the potential (6.79). The entire problem, including the path integral (6.71), factorises in this case and, furthermore, the orbit (6.80) lies entirely on the x-axis. Consequently we would expect the fluctuation factor to be a product of an x-factor and a y-factor. The fluctuation effect along the x-direction, for an orbit on the x-axis, has already been obtained in (6.63) and this is precisely the first factor in (6.81). Fluctuations along the y-axis must correspond to that of a harmonic oscillator (see the y-dependence of (6.79)–(6.80)). Their effect, obtained in (6.22)–(6.24), is precisely the second factor in (6.81). Thus (6.74) yields the expected answer for the special case (6.79)–(6.80).

For a general potential $V(x, y)$ which does not have the simple form (6.79), the stability angle ν_1 is the appropriate generalisation of $\omega\tau$. Just as an imaginary $\omega(\omega^2 < 0)$ leads to instability of the orbit (6.80) in the above special case, similarly an imaginary part in ν_1 will lead to instability in the general case. This is evident from (6.77). If ν_1 has an imaginary part, $\xi_1(t)$ will grow with every cycle, as t tends to either $+\infty$ or $-\infty$. The small deformation will no longer remain small. On physical grounds, we expect to obtain stable quantum bound states only from stable orbits. This will be reflected in the energy levels obtained below, which will become complex if ν_1 had an imaginary part.

Returning to the WKB method, (6.74) may be rewritten as

$$\Delta^{(n)}(T) = \left(\frac{-\mathrm{i}}{2\pi\hbar}\right)^{1/2}\tau\left(-\frac{\mathrm{d}E_{\mathrm{cl}}}{\mathrm{d}\tau}\right)^{1/2}\frac{1}{\sqrt{n}}\left(\sum_{p=0}^{\infty}\exp[-\mathrm{i}n(p+\tfrac{1}{2})\nu_1]\right). \tag{6.82}$$

Inserting this into (6.73) and (6.71) we get

$$\mathscr{G}(E) = \frac{\mathrm{i}}{\hbar}\left(\frac{-\mathrm{i}}{2\pi\hbar}\right)^{1/2}\sum_{n,p}\sqrt{n}\int_0^T \mathrm{d}\tau\left(-\frac{\mathrm{d}E_{\mathrm{cl}}}{\mathrm{d}\tau}\right)^{1/2}\tau$$

$$\exp\left(\frac{\mathrm{i}n}{\hbar}[S_{\mathrm{cl}}(\tau)+E\tau-(p+\tfrac{1}{2})\hbar\nu_1]\right). \tag{6.83}$$

The remaining steps are identical to (6.64)–(6.70) with appropriately modified results. SPA, when applied to the τ-integration, picks out, for a given E and p, the classical orbit satisfying

$$-\frac{\partial S_{\text{cl}}(\tau)}{\partial\tau}+(p+\tfrac{1}{2})\hbar\frac{\partial v_1}{\partial\tau}=E$$

or

$$E_{\text{cl}}+(p+\tfrac{1}{2})\hbar\frac{\partial v_1}{\partial\tau}=E. \tag{6.84}$$

Remember that v_1 depends on the orbit and hence $v_1=v_1(E_{\text{cl}}(\tau))$. Under this SPA, (6.83) yields

$$\mathscr{G}(E)=\sum_{n,p}\mathrm{i}\frac{\tau(E_{\text{cl}})}{\hbar}\exp\left(\frac{\mathrm{i}n}{\hbar}W_p(E)\right) \tag{6.85}$$

where

$$W_p(E)=S[E_{\text{cl}}]+E\,\tau(E_{\text{cl}})-(p+\tfrac{1}{2})\hbar\,v_1(E_{\text{cl}}) \tag{6.86}$$

and E_{cl} is a function of E through (6.84). The energy levels are poles of $\mathscr{G}(E)$ which occur when

$$W_p(E)=2m\pi\hbar \qquad \begin{matrix} m=1,2,\ldots \\ p=0,1,2,\ldots \end{matrix} \tag{6.87}$$

Notice that the energy levels given by (6.87) are characterised by two integers m and p. This fits in with the expectation that a two-dimensional bound state should have two quantum numbers. The quantum number m represents, in this semiclassical language, the number of full waves fitted along the chosen orbit, while p labels excitations transverse to the orbit. Notice that in (6.84) the quantum energy E does not equal E_{cl}, the energy of the corresponding classical orbit, unlike the one-dimensional case (6.65). The difference, $(p+\frac{1}{2})\hbar\,\partial v_1/\partial\tau$ may be attributed to the energy of fluctuations transverse to the orbit, a possibility that does not exist in one dimension. Recall that for the special case (6.79)–(6.80), $v_1=\omega\tau$ and this additional term becomes $(p+\frac{1}{2})\hbar\omega$, which is just what we would expect for the transverse harmonic oscillator excitations.

The generalisation of the results (6.84)–(6.87) to many degrees of freedom and to field theory is straightforward. For N non-relativistic coordinates, $\boldsymbol{r}(t)$ in (6.71) would stand for an N-vector. The SPA would again pick out periodic classical orbits $r_{cl}(t)$. The operator

$$[-\partial_t^2 - (\partial^2 V/\partial r_i \partial r_j)_{r_{cl}}]$$

would be an $N \times N$ matrix. The stability equation (6.75) would now have N independent eigenfunctions $\xi_i(t)$ with stability angles v_i. The angle v_0 would again vanish since $\xi_0 = \dot{r}_{cl}$, and barring further symmetries, the remaining $v_i, i = 1, \ldots, N-1$ would in general be non-zero. In the place of (6.74), the factor $[2i \sin(nv_1/2)]^{-1}$ would be replaced by the product

$$\prod_{i=1}^{N-1} [2i \sin(nv_i/2)]^{-1},$$

representing the effect of fluctuations in the $N-1$ directions transverse to the orbit. Following the remaining steps, the results (6.84)–(6.87) would be altered to

$$E_{cl} + \sum_{i=1}^{N-1} (p_i + \tfrac{1}{2})\hbar \frac{\partial v_i}{\partial \tau} = E \tag{6.88}$$

with the energy levels satisfying

$$W_{\{p_i\}}(E) = 2m\pi\hbar \qquad m = 0, 1, 2, \ldots$$
$$p_i = 0, 1, 2, \ldots, \text{ for each } i \tag{6.89}$$

where

$$W_{\{p_i\}}(E) = S[E_{cl}] + E\tau(E_{cl}] - \sum_{i=1}^{N-1} (p_i + \tfrac{1}{2})\hbar v_i. \tag{6.90}$$

We have mentioned that while one stability angle, v_0, would always vanish bacause of time–translation symmetry, the others would generally be non-zero, barring other symmetries. If the Lagrangian enjoys other continuous symmetries, they can lead to the vanishing of some other v_i, just as they led to the vanishing of some frequencies ω_i in the last chapter, and for the same reason. Given a particular periodic orbit $r_{cl}(t)$, one can continuously deform it using the symmetry operation, formally denoted by P, to obtain nearby periodic orbits $r'_{cl} = Pr_{cl}$ with the same period. The corresponding deformation mode $\xi(t) = r'_{cl}(t) - r_{cl}(t)$ would then also have the same period, and hence zero stability angle. Such symmetries cause difficulties in our derivation, just as they did in the earlier quantisation of static solutions. The corresponding factor $[2i \sin(nv_i/2)]^{-1}$ would then diverge. Of course, the leading order SPA results (6.88)–(6.90) remain non-singular as any of the

$\nu_i \to 0$. We will work only up to this leading order. But if higher-order corrections are desired, then the treatment would have to be modified in the presence of zero stability angles. For more discussion of this problem, see Appendix A of Dashen et al. (1975a), as well as chapter 8.

Subject to such zero-mode problems, the method is directly generalisable to field theory, viewed as a system with infinite degrees of freedom. Consider, for example, a scalar field $\phi(\boldsymbol{x}, t)$ with action

$$S_{\text{cl}}[\phi] = \int dt \left[\int \frac{1}{2} \left(\frac{\partial \phi}{\partial t} \right)^2 d\boldsymbol{x} - V[\phi] \right] \tag{6.91}$$

where

$$V[\phi] = \int d\boldsymbol{x} \left[\tfrac{1}{2} (\nabla \phi)^2 + U(\phi) \right]. \tag{6.92}$$

Then, once again,

$$\mathscr{G}(E) \equiv \text{Tr}\left(\frac{1}{E - H} \right) = \frac{\text{i}}{\hbar} \int_0^\infty dT \exp(\text{i}ET/\hbar) G(T) \tag{6.93}$$

with $G(T)$ given by the functional integral in (6.34). The remaining steps are formally the same. The SPA again picks out periodic classical solutions $\phi_{\text{cl}}(\boldsymbol{x}, t)$. The stability equation (the generalisation of (6.75)) is

$$\left(-\frac{\partial^2}{\partial t^2} - V'' \right) \xi(\boldsymbol{x}, t) \equiv \left[-\frac{\partial^2}{\partial t^2} + \nabla^2 - \left(\frac{\partial^2 U}{\partial \phi^2} \right)_{\phi_{\text{cl}}} \right] \xi(\boldsymbol{x}, t) = 0. \tag{6.94}$$

The solutions will obey, since $\phi_{\text{cl}}(\boldsymbol{x}, t)$ is periodic with some period τ,

$$\xi_i(\boldsymbol{x}, t + \tau) = e^{i\nu_i} \xi_i(\boldsymbol{x}, t) \qquad \text{for all } \boldsymbol{x}, t. \tag{6.95}$$

The main difference is that, with the usual periodic boundary conditions in a box, there will now be an infinite number of independent solutions $\xi_i(\boldsymbol{x}, t)$ with corresponding ν_i. As a result, the finite sums over the index i in (6.88)–(6.90) will now become infinite series. A resultant complication is that the sum $\sum_{i=1}^{\infty} \nu_i$ may in general diverge, just as the sum over normal-mode frequencies did in the last chapter. In a renormalisable theory, one expects this divergence to be cancelled by introducing the usual counter terms. Therefore, it is understood when adapting (6.88)–(6.90) to field theory, that counter terms are to be included in the action and the energy. Hence, we replace (6.88)–(6.90) by

$$E_{\rm cl}(\phi_{\rm cl}) + E_{\rm ct}(\phi_{\rm cl}) + \sum_0^\infty (p_i + \tfrac{1}{2})\hbar \frac{\partial \nu_i}{\partial \tau} = E \tag{6.96}$$

$$W_{\{p_i\}}(E) = 2m\pi\hbar \tag{6.97}$$

where

$$W_{\{p_i\}}(E) = S_{\rm cl}[\phi_{\rm cl}] + S_{\rm ct}[\phi_{\rm cl}] + E\tau(\phi_{\rm cl}) - \sum_0^\infty (p_i + \tfrac{1}{2})\hbar\nu_i. \tag{6.98}$$

Here 'ct' stands for counter-term contributions, whose effect will be to cancel divergences in the infinite sums. Equation (6.96) picks out the appropriate classical periodic solution $\phi_{\rm cl}$ for a given E and a given set of integers p_i. Then those values of E which satisfy (6.97)–(6.98) will give renormalised bound state energies. These equations are illustrated in the next chapter using the sine–Gordon model as an example.

6.5. Concluding remarks on the WKB method

Unlike the methods of chapter 5, the WKB method outlined here permits us to quantise time-dependent classical solutions, and not just static ones. But it is more difficult to execute than the weak-coupling, static-solution quantisation of chapter 5. Apart from the fact that the formulae are more complicated, we need as input information either all, or at least some, families of time-dependent period classical solutions in exact form. For these solutions, their classical stability angles ν_i must also be calculable by solving the linear stability equation. These are purely classical computations, but they are very difficult to do. The difficulties of solving non-linear field equatioins, even static ones let alone time-dependent ones, are already apparent from preceding chapters. Our attitude to this is that, difficult as it may be to obtain the classical solutions, the corresponding quantum field theory will generally be still harder to solve exactly. Rarely can one solve a quantum theory without being able to solve the corresponding classical theory. At least this WKB method allows us to get some approximate quantum results, once the relevant classical information is available. However, it must be admitted that a major bottleneck in using this method is the unavailability of exact periodic classical solutions at present for most field theories. The sine–Gordon system which we use in the next chapter to illustrate this method is a rare exception.

Once classical solutions are available, this method is more powerful than the weak-coupling method of chapter 5. True, this method also requires the weak-coupling condition in order for the approximation to be good, inasmuch as the method again neglects cubic and higher powers of the fluctuations. But the classical solutions, which go in as input in the leading stationary phase approximation, contain considerable non-perturbative information. Unlike static solutions, which probe only the regions of minima of the potential, the set of all periodic orbits probe the potential everywhere. For instance, take the prototype case, the anharmonic oscillator $[V(q) = \frac{1}{2}\omega^2 q^2 + \frac{1}{4}\lambda q^4]$. The high-lying energy levels (for which the WKB method is generally more accurate) are related to high-energy periodic orbits which sample regions where the anharmonic term λq^4 is larger than the harmonic term $\omega^2 q^2$. The knowledge of these classical orbits, which goes into our semiclassical formalism, while hard to obtain, imparts considerable non-perturbative content to the results. Indeed one can compare the WKB energy levels for the anharmonic oscillator with fairly accurate numerical evaluations of the same (Kilpatrick and Kilpatrick 1948). One finds that even in the strong-coupling limit ($\lambda \gg \omega^2$), the WKB approximation is incorrect by 1 % or less for all levels except the ground state which is off by about 22 %. The WKB method is of course exact in the $\lambda \to 0$ limit. Thus, at least for this example, the WKB results are quite accurate for almost all energy levels and in a wide range of λ. Whether similar accuracy will prevail in typical field theories or not is hard to predict, but the anharmonic oscillator, which is in some ways a generic example, offers cause for optimism. Indeed, for the sine–Gordon model, to which we will apply this method, the results seem to be exact. But that is undoubtedly a fortuitous example because of its special properties. The method needs to be applied to some more field theories before we can assess its accuracy. The main stumbling block in doing so is, as we have said, the unavailability of exact periodic classical solutions.

We make one final comment on the semiclassical method described here and in the preceding chapter. On the one hand, such semiclassical approximations can be good only if $\hbar$ is small compared with, say, the action of the classical solutions being used in the approximation. On the other hand, we have also emphasised that the method requires a weak-coupling condition, in order that the neglect of cubic and higher powers of the fluctuations, implied in the SPA, may be justified. The need for both these conditions – small $\hbar$ and small coupling constant – can be exhibited in a unified fashion as follows. Consider a scalar field theory, where the potential $U(\phi)$ has the form

$$U(\phi) = (1/g^2)\bar{U}(\bar{\phi}) \tag{6.109}$$

where g is a coupling constant which we are going to require to be weak, $\bar{\phi} \equiv g\phi$, and $\bar{U}(\bar{\phi})$ has no explicit g dependence. Notice that this form is true of most of the models we have considered. For instance, for the kink system (eq. (5.36)), we can write

$$U(\phi) = \tfrac{1}{4}\lambda(\phi^2 - m^2/\lambda)^2 = (1/g^2)[\tfrac{1}{4}m^2(\bar{\phi}^2 - 1)^2] \tag{6.110}$$

where $g^2 \equiv \lambda/m^2$ and $\bar{\phi} \equiv g\phi$. Similarly, for the sine–Gordon system (eq. (2.61))

$$U(\phi) = (m^4/\lambda)\{1 - \cos[(\sqrt{\lambda}/m)\phi]\} = (1/g^2)[m^2(1 - \cos\bar{\phi})]. \tag{6.111}$$

Of course $\bar{U}(\bar{\phi})$ may depend on other parameters of the Lagrangian, such as m^2. Given the form (6.109) for the potential, the action can be written as

$$\begin{aligned} S[\phi(x,t)] &= \int \mathrm{d}t \int \mathrm{d}x\, [\tfrac{1}{2}(\partial_\mu\phi)^2 - U(\phi)] \\ &= \frac{1}{g^2}\int \mathrm{d}t \int \mathrm{d}x\, [\tfrac{1}{2}(\partial_\mu\bar{\phi})^2 - \bar{U}(\bar{\phi})] = \frac{1}{g^2}\bar{S}[\bar{\phi}] \end{aligned} \tag{6.112}$$

where there is no explicit g dependence in $\bar{S}[\bar{\phi}]$. Now, consider any typical functional integral used to obtain semiclassical results, say, the one given in eq. (6.34):

$$\begin{aligned} G(T) &= \int \mathscr{D}[\phi(x,t)]\exp\{(\mathrm{i}/\hbar)S[\phi]\} \\ &= \int \mathscr{D}[\phi]\exp\left(\frac{\mathrm{i}}{\hbar g^2}\bar{S}[\bar{\phi}]\right) \\ &= A\int \mathscr{D}[\bar{\phi}]\exp(\mathrm{i}\bar{S}[\bar{\phi}]/\hbar g^2) \end{aligned} \tag{6.113}$$

where the constant A arises from the replacement of $\mathscr{D}[\phi]$ by $\mathscr{D}[\bar{\phi}]$. (A may contain infinite powers of $1/g$ but, being an overall constant multiplying the functional integral, it will not affect physical results. It can be absorbed by suitably redefining the functional integral's measure.) Looking at the last expression in eq. (6.113) we see that the entire g^2 dependence of the integrand comes in the combination $(\hbar g^2)^{-1}$ in the exponent. Clearly, the validity of the stationary phase approximation requires that this combination $\hbar g^2$ be small compared with the action $\bar{S}[\bar{\phi}_{\mathrm{cl}}]$ of the stationary paths, i.e. of the classical solutions. This is the correct criterion, which is of course satisfied when $\hbar$ and the coupling g^2 are both small.

Note that although we have used a single scalar field for illustration, this argument holds for any set of fields, as long as the action can be cast in the form (6.112). For instance, in the case of the Yang–Mills theory, the action can again be cast in this form in terms of the appropriately scaled matrix field A_μ (see eq. (4.10)). Our condition of small $\hbar g^2$ will again be required for semiclassical results to be good in this theory. Unfortunately, for the pure Yang–Mills theory, it is not possible to keep g (in the suitably renormalised sense) small. This renders semiclassical results for that system questionable. This problem will be discussed in section 10.5.

CHAPTER 7

SOME EXACT RESULTS

7.1. The bound-state spectrum of the sine–Gordon theory

We will now illustrate the WKB method developed in the last chapter by applying it to the two-dimensional sine–Gordon (SG) model. The choice of the SG model as illustration offers certain advantages but also one disadvantage. The advantages are several. We have already discussed this model at the classical level in chapter 2. All classical solutions of the model are known, and in particular a family of periodic solutions is available in simple analytical form. These are the doublet or breather solutions given in eq. (2.78). They can be used as input into the WKB method. The resulting algebra in executing the method turns out to be tractable and surprisingly simple. Furthermore, as we have mentioned in chapter 2, the SG system is a very special system with several interesting properties. As a result of these, the bound state spectrum given by the WKB method turns out to be exact. The system is also equivalent to the massive Thirring model, with the SG soliton state identifiable as a fermion. While using the SG system to illustrate the WKB method, we also take the opportunity to discuss these special features of the system. The disadvantage arises precisely because the SG system is special. Some aspects of the results we get will not be typical of the WKB method. As much as possible, we will try to delineate those features of the results which are special to the SG system from others which are more generic.

We will closely follow the work of Dashen et al. (1975a) who applied the WKB method to obtain the SG bound state spectrum. In fact, the semi-classical quantisation of the SG system was also done independently by Faddeev and collaborators in a series of papers. (Some references to these papers have already been cited in chapter 2; see Faddeev and Korepin (1978) for a complete set.) Faddeev and Takhtajan (1974) solved the classical SG equation by finding action-angle variables for this system. In terms of these variables, the classical Hamiltonian reduces to a sum of free-particle contributions (eq. (2.79)). Semiclassical quantisation then is straightforward. In particular, the discrete doublet spectrum is obtained readily by noticing that the angle variable associated with doublet solutions (labelled β_b) has a compact range $0 \leqslant \beta_b \leqslant 32\pi m^2/\lambda$ (see section 2.5). The Faddeev method is very elegant, but it is tailored to special systems like the SG model for which a complete set of action–angle variables is obtainable. We will, instead, derive the SG spectrum by the WKB method because that method is applicable to a larger class of systems. The procedure illustrated below by the SG example can be used for any other field theory, whether or not it shares the separability of the Hamiltonian and other remarkable features of the SG model, so long as a family of periodic classical solutions is available.

Recall the discussion of the classical SG model from chapter 2. The Lagrangian density (2.61), shown again here for convenience, involves a single scalar field $\phi(x, t)$ and is given by

$$\mathscr{L}(x, t) = \tfrac{1}{2}(\partial_\mu \phi)^2 - (m^4/\lambda)\{1 - \cos[(\sqrt{\lambda}/m)\phi]\}. \tag{7.1}$$

In chapter 2, we had rescaled variables to $\bar{x}_\mu = m x_\mu$ and $\bar{\phi} = (\sqrt{\lambda}/m)\phi$, so as to effectively eliminate the parameters m^2 and λ. But now we shall retain the original variables since m^2 and λ are of significance in the quantum context. It is taken as understood that the classical results from chapter 2 are to be rewritten in terms of $\phi(x, t)$ before being used in this section. The potential $U(\phi) = (m^4/\lambda)\{1 - \cos[(\sqrt{\lambda}/m)\phi]\}$ has discrete degenerate absolute minima at $\phi = (m/\sqrt{\lambda})(2n\pi)$, $n = -\infty \ldots, -1, 0, 1, \ldots \infty$. This will give rise to spontaneous symmetry breaking in the quantised theory. A vacuum sector of states can be constructed around any of these minima. To be specific, let us consider the vacuum sector built around $\phi = 0$. Then the Lagrangian can be expanded around $\phi = 0$ as per eq. (2.62). To order λ^0 in this expansion, the system (2.62) is a free field theory with boson mass m (as distinct from the kink system in chapters 2 and 5, where the boson mass was $\sqrt{2}m$). The energy of this vacuum state is clearly

$$E_{\text{vac}} = \sum_{k_n} \tfrac{1}{2}\sqrt{k_n^2 + m^2} + O(\lambda); \qquad L k_n = 2n\pi. \tag{7.2}$$

In (7.2) and henceforth, we will set $\hbar = 1$ consistent with the normal practice in quantum field theory. The $\hbar$ dependence of the semiclassical expansion has been shown explicitly in earlier chapters and can be inserted if necessary.

In chapter 2, we also presented classical solutions of the SG model. The single soliton (and antisoliton) solutions in eq. (2.72) are static. They can therefore be quantised, using the methods of chapter 5, to yield the quantum soliton and antisoliton particles. The model also yields the 'breather' or 'doublet' solutions (eq. (2.78)). These are a continuous family of periodic time-dependent solutions, and may therefore be quantised by the WKB method outlined in the last chapter, to yield bound states. In addition, the SG system also yields scattering solutions involving any number of solitons, antisolitons and doublets (such as in eq. (2.73) and (2.77)). These are not periodic solutions (except under the artifact of putting the whole system in a box with suitable periodic boundary conditions). They will correspond, in the quantised theory, to scattering states. To obtain the particle (bound state) spectrum of the SG theory, we need to quantise only the soliton, the antisoliton and the doublet solutions.

7.1.1. *Mass of the quantum soliton*

The SG soliton $\phi_{sol}(x)$ given in eq. (2.72) is a static solution of a $(1+1)$ dimensional scalar field theory carrying a topological charge. It is very similar to the kink solution of the ϕ^4 theory and can be quantised in the same way. The procedure is identical to what was done in chapter 5. We shall therefore omit details and just outline the steps. A sector of energy levels can be built around $\phi_{sol}(x)$, the lowest of which is the quantum soliton particle of the SG theory. Its mass will consist of a classical term followed by quantum corrections. The energy of the classical soliton obtained in section 2.5, by inserting (2.72) into (2.66), is

$$E_{cl}[\phi_{sol}] = 8m^3/\lambda. \tag{7.3}$$

The quantum corrections, to leading order, are obtained in terms of the normal-mode frequencies of the second derivative of the potential energy. The potential energy is

$$V[\phi] = \int_{-\infty}^{\infty} dx \left\{ \frac{1}{2}\left(\frac{\partial\phi}{\partial x}\right)^2 + \frac{m^4}{\lambda}\left[1 - \cos\left(\frac{\sqrt{\lambda}}{m}\phi\right)\right]\right\}. \tag{7.4}$$

The normal-mode frequencies ω_q are given by

$$\left[-\frac{\partial^2}{\partial x^2} + m^2 \cos\left(\frac{\sqrt{\lambda}}{m}\phi_{sol}\right)\right]\eta_q(x) = \omega_q^2 \eta_q(x). \tag{7.5}$$

This equation is the analogue of eq. (5.48). The quantum correction to the soliton mass M_{sol} to order λ^0 is given by (see section 5.4)

$$\tfrac{1}{2}\sum_q \omega_q + V_{\text{ct}}[\phi_{\text{sol}}]$$

where $[-V_{\text{ct}}(\phi)]$ is the normal-ordering counter term to be added to the Lagrangian. For a scalar field in two dimensions with non-derivative interactions, these are the only divergences. The potential here is not a polynomial, but since the cosine is just a sum of exponentials it is easily normal-ordered by standard methods. One essentially uses the well-known identity

$$e^{a+a^+} = e^{a^+} e^{a} e^{-\frac{1}{2}[a^+,a]} \qquad \text{if } [a, a^+] = c\text{-number}.$$

The result is (Dashen et al. 1975a, Coleman 1975b)

$$\frac{m^4}{\lambda} : \left[\cos\left(\frac{\sqrt{\lambda}}{m}\phi\right) - 1\right] : = (m^2 - \delta m^2)\frac{m^2}{\lambda}\left[\cos\left(\frac{\sqrt{\lambda}}{m}\phi\right) - 1\right]. \quad (7.6)$$

This result can also be understood by expanding $\cos[(\sqrt{\lambda}/m)\phi]$ as a power series (2.62) and drawing Feynman graphs for the Green functions. Both λ and m^2 pick up divergent corrections because of lack of normal-ordering, due to graphs such as fig. 12(a) and 12(b) respectively. But these corrections cancel in λ/m^2, as evident from the example of these graphs that contain the same loop integrals. Hence the only change in (7.6) due to normal-ordering occurs in the replacement of the extra factor m^2 by $(m^2 - \delta m^2)$. To lowest order, δm^2 is given by the graph in fig. 12(c), which yields

$$\delta m^2 = -\frac{\lambda}{4\pi}\int_0^{\Lambda} \frac{dk}{\sqrt{k^2 + m^2}}. \quad (7.7)$$

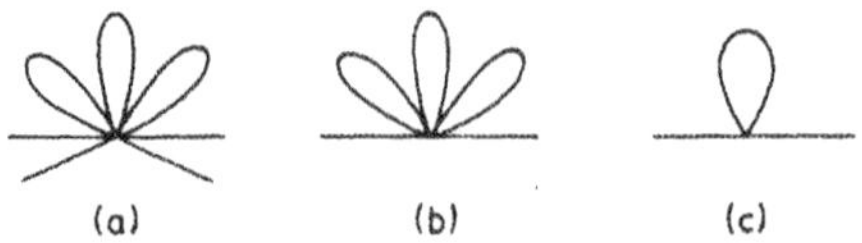

Fig. 12. (a) A typical Feynman diagram leading to a normal-ordering correction to the quartic coupling λ. (b) A corresponding correction to m^2 (c) The lowest-order contribution to δm^2.

In addition we must also subtract the vacuum energy (7.2), just as we did in chapter 5. Altogether,

$$V_{\mathrm{ct}}[\phi] = -\delta m^2 \frac{m^2}{\lambda} \int_{-\infty}^{\infty} \left[1 - \cos\left(\frac{\sqrt{\lambda}}{m}\phi\right)\right] \mathrm{d}x - E_{\mathrm{vac}}. \tag{7.8}$$

Hence the leading quantum correction to the 'soliton mass' is, using (7.2),

$$\tfrac{1}{2}\sum_q \omega_q - \delta m^2 \frac{m^2}{\lambda} \int_{-\infty}^{\infty} \left[1 - \cos\left(\frac{\sqrt{\lambda}}{m}\phi_{\mathrm{sol}}\right)\right] \mathrm{d}x - \tfrac{1}{2}\sum_{k_n} \sqrt{k_n^2 + m^2}. \tag{7.9}$$

Dashen et al. (1975a) find all solutions of (7.5), and insert the ω_q, along with δm^2 from (7.7) and ϕ_{sol} from (2.72) into (7.9). The result is that (7.9) is finite as the momentum cut-off goes to infinity and equals $(-m/\pi)$. Hence the quantum soliton's mass is

$$M_{\mathrm{sol}} = 8m^3/\lambda - m/\pi + 0(\lambda). \tag{7.10}$$

The detailed discussion and qualifying remarks in chapter 5 hold for this soliton state as well. In particular, one can also construct an antisoliton sector of states around ϕ_{antisol} in the same way. The mass of the quantum antisoliton will be, by $\phi \leftrightarrow -\phi$ symmetry, the same as M_{sol}.

7.1.2. *The doublet masses*

The doublet solutions are

$$\phi_\tau(x, t) = \frac{4m}{\sqrt{\lambda}} \tan^{-1}\left(\sqrt{\tilde{\tau}^2 - 1}\, \frac{\sin(mt/\tilde{\tau})}{\cosh(mx\sqrt{\tilde{\tau}^2 - 1}/\tilde{\tau})}\right) \tag{7.11}$$

where we have simply rewritten (2.78) in terms of $\phi(x, t)$ and the period τ. The variable $\tilde{\tau} \equiv m\tau/2\pi$ is dimensionless. Equation (7.11) gives a continuous family of periodic solutions parametrised by the period τ. These may be quantised by the WKB method developed in section 6.4. The corresponding bound state energies are given by eq. (6.96)–(6.98). All we need to do is evaluate the different terms in these equations, for the solution (7.11).

The classical action of this solution per period is

$$S_{\mathrm{cl}}[\phi_\tau] = \int_0^\tau \mathrm{d}t \int_{-\infty}^{\infty} \mathrm{d}x \left\{\tfrac{1}{2}(\partial_\mu \phi_\tau)^2 + \frac{m^4}{\lambda}\left[\cos\left(\frac{\sqrt{\lambda}}{m}\phi_\tau\right) - 1\right]\right\}. \tag{7.12}$$

Upon inserting (7.11) and integrating, this gives

$$S_{\mathrm{cl}}[\phi_\tau] = 32\pi(m^2/\lambda)[\cos^{-1}(1/\tilde{\tau}) - \sqrt{\tilde{\tau}^2 - 1}]. \tag{7.13}$$

The stability angles are obtained from solutions of the linear stability equation:

$$\left(-\frac{\partial^2}{\partial t^2} - V''[\phi_\tau]\right)\eta_i \equiv \left[-\frac{\partial^2}{\partial t^2} + \frac{\partial^2}{\partial x^2} - m^2 \cos\left(\frac{\sqrt{\lambda}}{m}\phi_\tau\right)\right]\eta_i(x, t) = 0 \tag{7.14}$$

with

$$\eta_i(x, t+\tau) = \mathrm{e}^{\mathrm{i}\nu_i}\eta_i(x, t). \tag{7.15}$$

Dashen et al. (1975a) find all solutions of (7.14) using a clever idea. Recall from section 6.4 that the linear stability equation is obeyed by any small classical perturbation to the given solution. Now, exact classical solutions to the SG model are available, consisting of an arbitrary number of doublets. Consider two-doublet solutions where one of the doublets is the given $\phi_\tau(x, t)$ and the other has a very small amplitude and arbitrary velocity. In this fashion one has available an infinite number of exact solutions, which are small perturbations on the given ϕ_τ. Using these, Dashen et al. extract all independent solutions of (7.14) and the corresponding stability angles. We will not offer more details since solving the linear equation (7.14), though not easy, is a conceptually straightforward matter. The result is

$$\nu_0 = 0, \qquad \nu_1 = 0 \tag{7.16}$$

followed by a continuum of solutions with

$$\nu_{q_n} = \tau\sqrt{m^2 + q_n^2} \tag{7.17}$$

where

$$Lq_n + \delta(q_n) = 2n\pi, \tag{7.18}$$

$$\delta(q_n) = 4\tan^{-1}(m\sqrt{\tilde{\tau}^2 - 1}/q\tilde{\tau}). \tag{7.19}$$

Here L is again the total volume, which tends to infinity. In this limit, the set of solutions characterised by q_n becomes a continuum. That two of the stability angles vanish is to be expected. One of them is related to the mandatory solution

$$\eta_0(x, t) = (\partial/\partial t)\phi_\tau(x, t)$$

discussed in the last chapter. The other is related to the space-translation symmetry of the system. Given a $\phi_\tau(x, t)$, $\phi_\tau(x+\delta a, t)$ is also an exact solution with the *same period*. Hence $\eta_1(x, t) \propto [\phi_\tau(x+\delta a, t) - \phi_\tau(x, t)]/\delta a$ will have zero stability angle ν_1. As we have repeatedly pointed out, if we work only up to the leading order of quantum corrections, such zero-modes create no problems. In the leading-order WKB result (6.96)–(6.98) for the bound state energies, we can effectively ignore zero-modes. The remaining angles ν_{q_n} given in (7.17) are non-zero.

Finally, the counter-term contribution is

$$S_{\text{ct}}[\phi_\tau] = -\int_0^\tau \mathrm{d}t\, V_{\text{ct}}[\phi_\tau] = \delta m^2 \frac{m^2}{\lambda} \int_0^\tau \mathrm{d}t$$

$$\times \int_{-\infty}^{\infty} \mathrm{d}x \left[1 - \cos\left(\frac{\sqrt{\lambda}}{m}\phi_\tau\right)\right] + \tfrac{1}{2}\tau \sum_{k_n} \sqrt{m^2 + k_n^2} \tag{7.20}$$

where $V_{\text{ct}}[\phi]$ from (7.8), (7.7) and (7.2) have been inserted. Thus we have accumulated all the ingredients needed for substituting into equations (6.96)–(6.98). Recall that these equations give energy levels characterised by a principal quantum number m and transverse quantum numbers p_i, $i = 0, 1, \ldots, \infty$. To obtain the basic doublet bound states, we can set all $p_i = 0$ for any given m. This is analogous to setting $N_n = 0$ in eq. (5.55) to obtain the basic kink state. When any of the $p_i \neq 0$, $i > 1$, we get higher excitations of the continuum modes in (7.17)–(7.18). These will correspond to scattering solutions of the doublets with the mesons of the theory, just as in the kink sector in chapter 5. When all $p_i = 0$ in (6.96)–(6.98), we are dealing with the combination

$$S_{\text{cl}}[\phi_\tau] + S_{\text{ct}}[\phi_\tau] - \tfrac{1}{2}\sum_i \nu_i.$$

Each of these terms is given explicitly above. We have,

$$\sum_i \tfrac{1}{2}\nu_i = \tfrac{1}{2}\sum_{q_n} \tau\sqrt{m^2 + q_n^2}\,. \tag{7.21}$$

As $L \to \infty$, this sum can be converted, using (7.18), into an integral which is quadratically divergent. This quadratic divergence will be cancelled by a similar divergence in the sum over k_n in S_{ct} in (7.20). A residual logarithmic divergence will be cancelled by the other term in S_{ct} which is proportional to δm^2 given in (7.7). All this can be verified, just as in section (5.4) by first evaluating $S_{ct}[\phi_c] - \frac{1}{2}\sum_{q_n} \nu_{q_n}$ with a finite L and a finite momentum cut-off Λ. The potentially divergent terms will cancel, as is to be expected in a renormalisable theory, giving a finite result as $L, \Lambda \to \infty$. This result is

$$S_{ct}[\phi_\tau] - \tfrac{1}{2}\sum_i \nu_i = 4[\sqrt{\tilde{\tau}^2 - 1} - \cos^{-1}(1/\tilde{\tau})]$$
$$= -(\lambda/8\pi m^2) S_{cl}[\phi_\tau], \tag{7.22}$$

by comparing with (7.13). Hence

$$S_{cl}[\phi_\tau] + S_{ct}[\phi_\tau] - \tfrac{1}{2}\sum_i \nu_i$$
$$= (1 - \lambda/8\pi m^2) S_{cl}[\phi_\tau]$$
$$= (32\pi/\gamma)[\cos^{-1}(1/\tilde{\tau}) - \sqrt{\tilde{\tau}^2 - 1}] \tag{7.23}$$

where

$$\gamma \equiv \frac{\lambda/m^2}{1 - \lambda/8\pi m^2}. \tag{7.24}$$

This is a remarkably simple result. The effect of the WKB quantum corrections is merely to multiply the classical action by an overall constant independent of τ. On comparing (7.23) with (7.13), we see that the coupling constant (λ/m^2) has been replaced by a 'renormalised' coupling constant γ as a result of quantum corrections. In fact the same thing had happened to the soliton mass in eq. (7.10). To leading order, we can write

$$M_{sol} = 8m^3/\lambda - m/\pi = 8m/\gamma. \tag{7.25}$$

These simplifying features are special to the SG system. They are neither typical nor essential for the working of the WKB method, but they make our analysis of the SG spectrum very simple. In presenting the results (7.13)–(7.23) we have suppressed considerable algebraic detail because no new principles are involved beyond what has been discussed in chapters 5 and 6. For more details, see Dashen et al. (1975a).

Given the result (7.23), the bound-state spectrum is easily evaluated from (6.96)–(6.98). Equation (6.96) gives, when all $p_i = 0$,

$$E = -\frac{d}{d\tau}\left(S_{cl}[\phi_\tau] + S_{ct}[\phi_\tau] - \tfrac{1}{2}\sum_i \nu_i\right)$$

$$= -\frac{\mathrm{d}}{\mathrm{d}\tau}\left(\frac{32\pi}{\gamma}[\cos^{-1}(1/\tilde{\tau}) - \sqrt{\tilde{\tau}^2 - 1}]\right)$$

$$= \frac{32\pi}{\gamma}\frac{\sqrt{\tilde{\tau}^2 - 1}}{\tau} \tag{7.26}$$

where it should be recalled that $\tilde{\tau} \equiv m\tau/2\pi$. The above equation associates with any E (the energy variable in the propagator $\mathscr{G}(E) = \mathrm{Tr}[1/(E - H)]$, a classical solution $\phi_\tau(x, t)$ whose period $\tau(E)$ is given by this equation. Remember however that E is not the classical energy of that solution. The latter is given by $E_{\mathrm{cl}} = -(\partial S_{\mathrm{cl}}/\partial\tau)$. The difference between E and E_{cl} is just the energy of zero-point fluctuations transverse to the orbit.

Equation (6.98) yields, using (7.23) and (7.26),

$$\begin{aligned} W(E) &= S_{\mathrm{cl}}[\phi_\tau] + S_{\mathrm{ct}}[\phi_\tau] - \tfrac{1}{2}\sum_i \nu_i + E\tau(E) \\ &= (32\pi/\gamma)[\cos^{-1}(1/\tilde{\tau}) - \sqrt{\tilde{\tau}^2 - 1}] + (32\pi/\gamma)\sqrt{\tilde{\tau}^2 - 1} \\ &= (32\pi/\gamma)\cos^{-1}(1/\tilde{\tau}) = (32\pi/\gamma)\sin^{-1}(E\gamma/16m). \end{aligned} \tag{7.27}$$

Finally, the quantisation condition (6.97) gives bound-state energy levels at $E = M_N$, where

$$W(M_N) = 2N\pi$$

or

$$M_N = (16m/\gamma)\sin(N\gamma/16); \qquad N = 1, 2, \ldots < 8\pi/\gamma. \tag{7.28}$$

Formally, the quantisation condition permits all integers N from 1 to ∞, but we run out of classical doublet solutions on which these bound states are based when $N > 8\pi/\gamma$. To see this, we invert (7.26) and use (7.28) to write

$$\tau(M_N) = (2\pi/m)\sec(N\gamma/16). \tag{7.29}$$

Here $\tau(M_N)$ is the period of the classical doublet corresponding to the Nth bound state. Now, the classical doublet solution corresponds to a bound soliton–antisoliton pair oscillating about one another with period τ. Doublets exist only for $0 < \tau < \infty$. As $\tau \to \infty$, the soliton–antisoliton pair tends to become unbound. We see from (7.29) that as $N \to 8\pi/\gamma$ from below, $\tau(M_N) \to \infty$. Therefore for $N > 8\pi/\gamma$ there are no corresponding classical doublets and no quantum bound states. Hence (7.28) holds only for integer values of N less than $8\pi/\gamma$.

We have thus obtained in (7.28) a finite set of bound states corresponding to the classical doublet solutions. Recall that the classical doublets carry zero topological charge ($\phi_\tau(x, t) \to 0$ as $x \to \pm\infty$). Correspondingly, the bound states in (7.28) all lie in the vacuum sector of the quantum theory. Although the classical solution $\phi_\tau(x, t)$ sketched in fig. 5 (b), looks like the wavefunction of a non-relativistic one-dimensional bound state, the function $\phi_{\tau(M_N)}(x, t)$ is *not* the wavefunction of the Nth bound state in (7.28). Remember that we are now working with a quantum field theory whose wavefunctions are functionals $\Psi[\phi]$ of fields. A similar cautionary remark was made in chapter 5. The classical solutions $\phi_{\tau(M_N)}(x, t)$ bear the same relation to the bound-state wavefunctionals $\Psi_N[\phi]$ that Bohr orbits bear to hydrogen atom wavefunctions.

This completes our illustration, using the SG system as an example, of the WKB method in field theory. The result (7.28), however, has many interesting features worth pointing out. These features will indicate strongly that the result (7.28) is exact for the mass ratios.

7.2. Interpretation of the doublet states

In weak coupling ($\lambda/m^2 \ll 1$), the mass M_N (eq. (7.28)) can be expanded as

$$\begin{aligned} M_N &= \frac{16m}{\gamma}\left(\frac{N\gamma}{16} - \frac{(N\gamma)^3}{(3!)(16)^3} + O(\gamma^5)\right) \\ &= Nm\left(1 - \frac{1}{6}\left(\frac{N\lambda}{16m^2}\right)^2 + O(\lambda^4)\right). \end{aligned} \tag{7.30}$$

The lowest of these bound states has a mass

$$M_1 = m\left(1 - \frac{1}{6}\left(\frac{\lambda}{16m^2}\right)^2 + O(\lambda^4)\right). \tag{7.31}$$

Now, the sine–Gordon Lagrangian when expanded in powers of λ/m^2 gives (see eq. (2.62)), to zeroth order in (λ/m^2), a free field theory with an 'elementary' boson of mass m. When the coupling is turned on slightly, the mass of the 'elementary' boson will be, by standard perturbation theory (after normal-ordering subtractions), equal to $m + O(\lambda^2)$. Comparing this with M_1, we identify the lowest of the doublet states with the 'elementary' boson of the theory. It seems unlikely that there would be a second discrete state in the vacuum sector, distinct from the elementary boson, whose mass also reduces to m in weak coupling. The masses M_N of the higher states can be rewritten as

$$M_N = NM_1\left[1-\tfrac{1}{6}(\lambda/16m^2)^2\,(N^2-1)+\mathrm{O}(\lambda^3)\right]. \tag{7.32}$$

Obviously, in weak coupling these states can be considered as loosely bound states of N 'elementary' bosons, with a binding energy

$$NM_1 - M_N = \tfrac{1}{6}M_1\,(\lambda/16m^2)^2\,(N^3-N)+\mathrm{O}(\lambda^3). \tag{7.33}$$

The quantum number N stands in weak coupling for the number of 'elementary' bosons in the bound state.

We have been using the word 'elementary' in quotation marks because in our derivation there is no special distinction enjoyed by the $N = 1$ state compared with the others. They all emerged as a family of bound states based on classical doublet solutions, of which the 'elementary' boson state just happens to be the lowest. The doublet solution corresponding to a given N has the form (see (7.29) and (7.11))

$$\phi_{\tau(M_N)}(x,t) = \frac{4m}{\sqrt{\lambda}}\tan^{-1}\left[\tan\left(\frac{N\gamma}{16}\right)\frac{\sin(mt\cos N\gamma/16)}{\cosh(mx\sin N\gamma/16)}\right]. \tag{7.34}$$

These are all 'breathing' field configurations with the general shape shown in fig. 5(b). There is nothing special about the $N = 1$ configuration. These results are consistent with the notion of 'nuclear democracy' championed for a long time by Chew and collaborators as part of the bootstrap approach to hadrons (Chew 1965). This approach emphasises the absence of any basic distinction between 'elementary' and 'composite' hadrons. It is particularly interesting that our method supports this idea within the framework of field theory, where one might have expected the 'elementary' boson associated with the field to have a special role.

Dashen et al. check the mass ratios M_N/M_1 given by the WKB method against calculations by other methods. By that we mean calculations carried out for the N-boson bound state by using perturbation theory and Feynman diagrams.

Firstly, to order λ, the sine–Gordon theory is an attractive $\lambda\phi^4$ theory (see (2.62)), which corresponds in the non-relativistic limit to an attractive interaction between the bosons, given by $V(x) = -(\lambda/8m^2)\delta(x)$. Since the N-boson binding energy is small $[\mathrm{O}(\lambda^2)]$ in weak coupling, one can compute it using the non-relativistic N-body Schrödinger equation with the

δ-function potential:

$$\left(-\frac{1}{2m}\sum_{i=1}^{N}\frac{\partial^2}{\partial x_i^2}-\frac{\lambda}{8m^2}\sum_{i\neq j}^{N}\delta(x_i-x_j)\right)\psi_N^{(n)}(x_1,\dots x_N)$$

$$=-\varepsilon_N^{(n)}\psi_N^{(n)}(x_1,\dots x_N). \tag{7.35}$$

(Recall that to lowest order the boson mass is m.) This equation is soluble for its ground state eigenvalue, giving

$$\varepsilon_N^{(0)}=\tfrac{1}{6}(N^3-N)(\lambda/16m^2)^2\,m \tag{7.36}$$

which agrees with the WKB binding energy in (7.33) to leading order $O(\lambda^2)$. One can carry the comparison further by computing the mass ratio $(M_2-2M_1)/M_1$ to higher order in λ by summing Feynman diagrams. One iterates the Bethe–Salpeter sequence using as kernel the elastic $2\rightarrow 2$ amplitude computed exactly to order $(\lambda/m^2)^3$ from Feynman diagrams. The diagrams summed are indicated in fig. 13. While this diagrammatic method is in principle straightforward perturbation theory, in practice it is highly non-trivial because the kernel by itself goes to 'two-

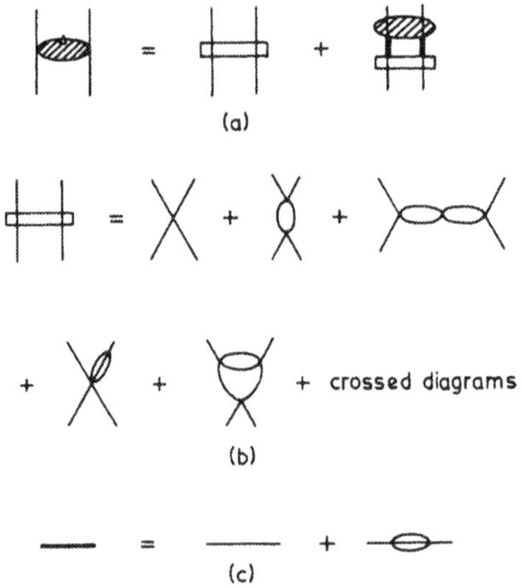

Fig. 13. The Bethe–Salpeter sequence of diagrams (a), is summed to obtain the bound state M_2 in the M_1+M_1 scattering amplitude. The kernel is given by diagram (b) to order λ^3 and the propagator is given by (c).

loop' order. The calculation is sketched in Appendix E of Dashen et al. (1975a) and the result is

$$\frac{2M_1 - M_2}{M_1} = \left(\frac{\lambda}{16m^2}\right)^2 + \frac{4}{\pi}\left(\frac{\lambda}{16m^2}\right)^3 + \left(\frac{12}{\pi^2} - \frac{1}{12}\right)\left(\frac{\lambda}{16m^2}\right)^4 + \mathrm{O}(\lambda^5). \tag{7.37}$$

The WKB answer for this quantity (eq. (7.28)) is

$$\frac{2M_1 - M_2}{M_1} = \frac{2\sin(\gamma/16) - \sin(\gamma/8)}{\sin(\gamma/16)} = 2\,[1 - \cos(\gamma/16)]. \tag{7.38}$$

When expanded this agrees exactly with the perturbation theory answer (7.37) up to and including order λ^4. This is a surprising result. By comparing our path-integral–stationary-phase derivation of the WKB approximation with the similar derivation of the perturbation 'loop' expansion by functional methods, one may expect the two methods to agree to one loop order. Agreement to two loop order is surprising, and will not happen for a general field theory. That the WKB result agrees with perturbation theory up to two loops is yet another special feature of the SG theory. This is also the first indication that these *WKB results may well be exact for this theory*. Note however that this agreement with perturbation theory works only for ratios of masses. When M_1 itself is computed in standard perturbation theory by calculating the boson self-energy to order λ^2, the result does *not* agree with the expansion of the WKB answer in (7.31). To the extent that we conjecture WKB results to be exact for the SG model, it will be only for the mass ratios, and not for the individual masses in (7.28). This limitation is to be expected. The overall mass scale of the quantum theory is set by normal-ordering which alters m^2 to $m^2 - \delta m^2$. The normal-ordering procedure has ambiguities (see Coleman 1975b) related to the mass with respect to which normal-ordering is done. This is an elementary instance of the renormalisation group. Given this ambiguity, we cannot expect the WKB method to yield the mass scale.

The SG system reveals another important property. For weak coupling, we see from the result (7.32) that M_N is only slightly smaller than NM_1. In particular $M_N > 2M_1$ for $N \geqslant 3$ and small λ/m^2. Hence the $N \geqslant 3$ levels seem to lie in the continuum of two or more $N = 1$ particles. Ordinarily we would expect these higher states to be not strictly bound, and to decay into two or more $N = 1$ particles. The topological charge does not prohibit such decays since all these states belong to the same ($Q = 0$) vacuum sector. True, a careful consideration shows that the $\phi \leftrightarrow -\phi$ symmetry will prevent, in perturbation theory, the decay for instance of $M_3 \to M_1 + M_1$. (The Feynman diagrams would have an odd number of external legs.) But even this argument does not prohibit decays such as $M_4 \to M_1 + M_1$, $M_5 \to M_2 + M_1$ etc. What does perturbation theory predict for such decay amplitudes? Consider for instance the process $M_4 \to M_1 + M_1$ in the rest frame of M_4. This decay amplitude is given in lowest-order perturbation theory by the tree graphs in fig. 14. These graphs are obtained from standard perturbation theory by expanding the cos $[(\sqrt{\lambda}/m)\phi]$ term in the Lagrangian as in (2.62). Remember that, to lowest order, the system M_4 at rest can be considered as four mesons at rest with mass m each. The energy–momentum of the two mesons in the final state are clearly $(2m, \sqrt{3}\,m)$ and $(2m, -\sqrt{3}\,m)$. Then it is trivial to see that the Feynman graphs in fig. 14(a), (b) and (c) contribute $-i\lambda^2/48m^2$, $-i\lambda^2/96m^2$ and

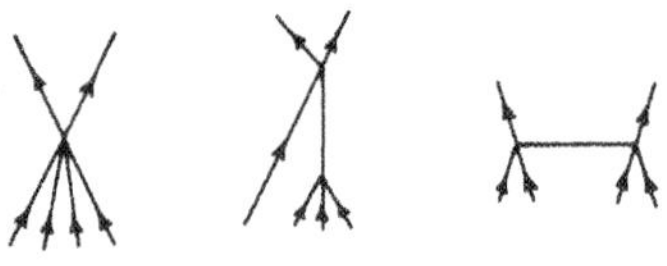

Fig. 14. Lowest-order (tree) graphs contributing to the decay $M_4 \to M_1 + M_1$. The vertices are obtained by expanding the cosine in the Lagrangian.

$i\lambda^2/32m^2$ respectively. Their sum adds up to zero! A similar null result at the tree-graph level may be verified for the other processes such as for example $M_6 \to 2M_1$, $M_8 \to 2M_1$, $M_5 \to 3M_1$ etc. If this result holds in higher orders as well (some unpublished calculations of one-loop contributions support this), we would have the remarkable result that all the bound states M_N are stable, even though topological charge and energy conservation would not prohibit the higher ones from decaying into suitable lower ones. This is once again a special feature of the SG system, presumably related to the existence of the infinite number of conserved quantities mentioned in chapter 2. Notice that, if true to all orders, this result supports our conjecture that the WKB mass ratios are exact. In (7.28), the WKB masses are real and discrete, indicating stable bound states. Instability would be revealed by imaginary contributions to M_N from higher-order corrections to the WKB result. If, however, the latter is exact, no corrections are needed and all the bound states would be stable.

Our support for the exactness of the WKB mass ratios came, so far, from comparisons with low-order perturbation results. A more convincing support, for all allowed values of the coupling λ/m^2 (or equivalently γ), comes from the S-matrix of the SG system. Recall that the classical doublet solution can be considered as a bound soliton–antisoliton pair. The same will clearly hold for the corresponding quantised states. Indeed, the masses M_N can be written in terms of the quantum soliton (or antisoliton) mass M_{sol} (eq. (7.25)) as

$$M_N = 2M_{\text{sol}} \sin(N\gamma/16) < 2M_{\text{sol}}; \qquad N = 1, 2, \ldots, < 8\pi/\gamma. \quad (7.39)$$

If (7.39) is to be exact for the mass ratios $M_N/2M_{\text{sol}}$, then the M_N should occur as the (only) poles in the physical energy sheet of the soliton–antisoliton scattering amplitude. We show in section (7.5) how this amplitude may be computed exactly, and it will be seen that poles occur precisely at the values of M_N in (7.39), for all allowed values of the coupling γ. Further evidence of the exactness of the WKB mass ratios comes from using the equivalence of the SG model to the massive Thirring model, whose discussion follows.

7.3. Equivalence of the SG and MT models; bosonisation

As early as the sixties, Skyrme (1961, 1962) studied the sine–Gordon system in a pioneering set of papers. Among other things, he suggested that the quantum SG solitons, in spite of arising from a boson field theory, may be equivalent to fermions interacting through a four-fermion interaction. But he claimed this identification only for certain values of the coupling constant and his arguments were more suggestive than rigorous.

More recently Coleman (1975b), using quite independent methods, established such an equivalence rigorously within the framework of perturbation theory. Consider the SG and massive Thirring (MT) models both in (1 + 1) dimensions, described by

$$\mathscr{L}_1 = \tfrac{1}{2}(\partial_\mu \Phi)(\partial^\mu \Phi) + m_0^2 (m^2/\lambda)\{\cos[(\sqrt{\lambda}/m)\Phi] - 1\} \quad (7.40)$$

$$\mathscr{L}_2 = \mathrm{i}\bar{\Psi}\gamma_\mu \partial^\mu \Psi - m_F \bar{\Psi}\Psi - \tfrac{1}{2} g (\bar{\Psi}\gamma^\mu \Psi)(\bar{\Psi}\gamma_\mu \Psi) \quad (7.41)$$

where Ψ is a fermi field, γ^μ the Dirac matrices in (1 + 1) dimensions, and normal-ordering counter terms have been absorbed in the parameters m_0^2 and m_F. Assuming that the two theories do exist, one can make a formal perturbation expansion of each. The *massless* Thirring model ($m_F = 0$) is well known to be exactly soluble (Thirring 1958, Glaser 1958, Johnson 1961, Sommerfield 1963). Formally, one can then expand the massive

Thirring model in powers of m_F. Such an expansion would clearly involve n-point functions of the composite operator $\bar{\Psi}\Psi$ (see Klaiber 1968), which probe the zero-fermionic-charge sector of this model. In the same fashion, the SG system can be expanded in a perturbation series in powers of m_0^2, that is, in terms of n-point functions of $\cos[(\sqrt{\lambda}/m)\Phi]$. The remaining unperturbed part of the Lagrangian $\mathscr{L}_1$ is just a free massless field and hence exactly soluble. (Note: A massless scalar field in $(1+1)$ dimensions has infrared problems which will have to be carefully regulated in the perturbation expansion.) Coleman's central result was to show explicitly that these two sets of n-point functions are equal for all n, and consequently the SG theory equivalent to the charge-zero sector of the MT model, provided one made the following identification:

$$m_0^2\,(m^2/\lambda)\cos[(\sqrt{\lambda}/m)\Phi] = -m_F\bar{\Psi}\Psi \tag{7.42a}$$

$$\lambda/4\pi m^2 = 1/(1+g/\pi) \tag{7.42b}$$

and

$$-(\sqrt{\lambda}/2\pi m)\varepsilon^{\mu\nu}\partial_\nu\Phi = \bar{\Psi}\gamma^\mu\Psi \equiv j^\mu. \tag{7.42c}$$

Coleman further showed that the Hamiltonian of the SG model is unbounded from below if $\lambda/m^2 > 8\pi$, so that the theory and the above equivalence exist only for $\lambda/m^2 < 8\pi$. Finally, based on this exact demonstration of equivalence, he resurrected Skyrme's conjecture that the quantum soliton of the SG model may be identified with the fermion of the MT model.

These are clearly remarkable results, equating a theory of Bose fields to (albeit the bosonic sector of) a theory of Fermi fields. The identification of the SG soliton, which arises in boson field theory, with a fermion does not violate the spin-statistics theorem. Remember that in $(1+1)$ dimensions, there can be no spin. Furthermore, we have argued in chapter 5 that the quantum soliton belongs to an entirely different super-selection sector from the vacuum, distinguished by a global (boundary value) quantum number. Hence the soliton cannot be created by any finite product of local field operators $\Phi(x_1)\ldots\Phi(x_N)$ acting on the vacuum. Such a soliton state therefore falls outside the scope of theorems precluding identification with a fermion. Recall that in section 3.5 we had hinted at a similar possibility in $(3+1)$ dimensions through the transmutation of isospin into spin.

We will not reproduce here Coleman's proof of this equivalence. His proof, as we have said, is exact and untainted by semiclassical methods. For the same reason it falls somewhat outside our general theme. Instead, we will take his result as given, and use it to further support and elaborate on our semiclassical considerations.

(i) The fermionic charge Q in the MT model is

$$Q = \int_{-\infty}^{\infty} j_0 \, \mathrm{d}x = \int_{-\infty}^{\infty} \bar{\Psi}\gamma_0\Psi \, \mathrm{d}x. \tag{7.43}$$

A single fermion (f) carries $Q = 1$, an antifermion ($\bar{\mathrm{f}}$) carries $Q = -1$, and bound states of the two carry $Q = 0$. Now, from the equivalence relation (7.42c), we can write Q in terms of the SG field:

$$Q = \int_{-\infty}^{\infty} -\frac{\sqrt{\lambda}}{2\pi m}\varepsilon^{01}\frac{\partial\Phi}{\partial x}\mathrm{d}x = \frac{\sqrt{\lambda}}{2\pi m}\int_{-\infty}^{\infty}\frac{\partial\Phi}{\partial x}\,\mathrm{d}x. \tag{7.44}$$

This is just the topological charge for the SG model, analogous to (5.76) of the kink problem. Its value is $+1\,(-1)$ for the soliton (antisoliton) and zero for the bound doublet states. This conforms with the identification of the SG soliton with the MT fermion. (Strictly speaking, the equivalence between the models was proved only in the $Q = 0$ sector of the MT model. In applying the relation (7.44) to a single fermion, it is understood that we consider a ($\mathrm{f}\bar{\mathrm{f}}$) pair with total $Q = 0$, widely separate them, and use (7.44) separately in the vicinity of each.)

(ii) Let us write (7.42b) in terms of the renormalised SG coupling constant γ in (7.24)

$$\gamma \equiv \frac{\lambda/m^2}{1 - \lambda/8\pi m^2} = \frac{8\pi}{1 + 2g/\pi}; \tag{7.45}$$

we see that when $0 < \gamma < 8\pi$, i.e. $0 < \lambda/m^2 < 4\pi$, the Thirring coupling g is positive. This implies, through the usual procedure of extracting the sign of the static limit force from the Lagrangian, an attraction between the fermion and the antifermion. In one space-dimension, such an attractive force, no matter how weak, will always lead to at least one boson bound state. Correspondingly we recall from (7.28) that for all γ in the range $0 < \gamma < 8\pi$, there was at least one soliton–antisoliton (doublet) bound state.

(iii) Consider the weak-coupling ($\gamma, \lambda/m^2 \ll 1$) limit in the SG model. This corresponds, by virtue of (7.45), to large g, i.e., the strong coupling region in the MT model. In this region we expect strong f–$\bar{\mathrm{f}}$ attraction and many $\mathrm{f}\bar{\mathrm{f}}$ bound states. This is in accordance with (7.28), where N, the number of soliton–antisoliton bound states is given by $[8\pi/\gamma]$ which stands for the largest integer below $8\pi/\gamma$. As γ is increased, g decreases. Correspondingly, the number of doublet states in (7.28) also decreases. Finally, when $\gamma = 8\pi$, i.e. $\lambda/m^2 = 4\pi$, eq. (7.45) gives $g = 0$. The fermions are free. Notice that this is precisely the value at which the last remaining bound state M_1 in (7.28) will cease to exist. From (7.39) we can see that as $\gamma \to 8\pi$, $M_1 \to 2M_{sol}$, i.e. the last remaining bound state becomes a free soliton–antisoliton pair. Given that the SG–MT equivalence is exact, the fact that the bound states in the WKB answer (7.28) also disappear precisely at $\gamma = 8\pi$, when the Thirring fermions become free, further supports the conjecture that the WKB result is exact.

(iv) When $\gamma > 8\pi$, g is negative. No ff bound states are expected in MT model, and correspondingly all the doublet states in (7.28) disappear. The absence of even the lowest bound state M_1 is significant. Recall that we had associated this state with the basic 'elementary meson' of the SG model. A popular notion, based on perturbation theory, used to be that for every field there would be an associated species of particles, namely, the quanta of the field excitations. With the increasing availability of non-perturbative results, such notions are no longer considered reliable in general. Clearly, $\gamma > 8\pi$ falls well in the strong-coupling domain of the SG model, where perturbation theoretical notions can be overthrown. The disappearance of the 'elementary' meson of the SG model when $\gamma > 8\pi$, is one such example. Semiclassical WKB methods have several weak points; but the availability of such important non-perturbative results represents an important virtue they have over perturbation theory. Remember that the disappearance of the M_1 state from the bound state spectrum was already given by the WKB result (7.28), without recourse to Coleman's equivalence theorem.

(v) The fact that as $\gamma \to 0$, g tends to infinity, implies qualitatively that the smaller the SG coupling constant, the stronger the soliton–antisoliton attraction. Such inverse relationship is again in contrast to the results of perturbation theory. Recall that in quantum electrodynamics treated perturbatively the forces between electrons and positrons would decrease if the fine structure constant were decreased. This inverse behaviour of intersoliton forces is not limited to the SG model and is true more generally (Faddeev 1975, Rajaraman 1977). Recall for instance that the monopole strength in section 3.4, to which magnetic intermonopole forces are proportional, also had an inverse dependence on the coupling constant g of that theory.

(vi) Finally, Coleman shows that when $\lambda/m^2 > 8\pi$, the quantum SG system is meaningless because the Hamiltonian is unbounded from below. A corresponding catastrophe is revealed by our WKB results. The renormalised coupling γ (7.24) on which subsequent results depend, diverges at $\lambda/m^2 = 8\pi$.

This completes our discussion of the SG mass spectrum, where we have used the WKB method, augmented by Coleman's equivalence theorem. Some of the findings, such as the exactness of the WKB results, the equivalence to the MT model etc., are special to the SG system. But others, including the whole procedure for obtaining the WKB spectrum, the identification of the 'elementary' boson with one of the classical periodic solutions, the strong intersoliton forces in weak-coupling theory etc. do not rely on the specific features of the SG model. Similar results could hold in other models as well.

7.3.1. *Bosonisation*

The equivalence between the SG and MT models is a dynamical result, which relies on the precise forms of the two Lagrangians. However, the identification in (7.42a) and (7.42c) in terms of which the equivalence holds can be considered as a kinematical change of variables. It could be applied to other $(1+1)$ dimensional fermion theories as well. The charge-zero sector of a fermion theory will involve only bilinear forms such as $\bar{\Psi}(x)\Psi(x)$, $\bar{\Psi}(x)\gamma^{\mu}\Psi(x)$ etc. After due ultraviolet regularisation, these can be considered as local Bose fields. In (7.42a) and (7.42c) all we have done is replace them by explicit Bose fields. We note that Φ as well as the derivative $\partial_{\nu}\Phi$ occurs in (7.42), but not the derivatives of the Fermi field Ψ. This, as far as we can see, is related to the fact that a Fermi field obeys first-order differential equations, whose boundary data require the specification of only the field itself and not the derivatives. A Bose field by contrast obeys second-order equations, requiring specification of both the field and its derivatives on the boundary. Equations (7.42a) and (7.42c) replace one set of such quantities by another equivalent set. When such a replacement is done for the MT model, the resulting boson field theory happens to be the SG model. The same step can be taken, for whatever profit it is worth, for other two-dimensional fermion models as well. This procedure is sometimes called bosonisation (see Kim and Woo 1979). Witten (1979a) discusses another model to which bosonisation has been applied.

7.3.2. *Soliton Operators*

In (7.42), a Bose field was constructed out of (bilinear forms of) a Fermi field. The reverse process is also possible. Mandelstam (1975) has constructed, starting from the SG field operator Φ, an explicit non-polynomial form that can be considered as the Fermi field of the associated MT model. Recall that the latter has two components in $(1+1)$ dimensions. Let us consider at any given time t, the two-component field defined by

$$\Psi_\alpha(x) = \begin{pmatrix} \Psi_1(x) \\ \Psi_2(x) \end{pmatrix},$$

where

$$\Psi_{1,2}(x) = C_{1,2}{:}e^{A_{1,2}(x)}{:} \tag{7.46}$$

where $C_{1,2}$ are constants and $A_{1,2}(x)$ are operators given by

$$A_{1,2}(x) = \frac{2\pi m}{i\sqrt{\lambda}}\left(\int_{-\infty}^{x} \dot{\Phi}(x')\,dx'\right) \mp \frac{i\sqrt{\lambda}}{2m}\Phi(x). \tag{7.47}$$

Here $\sqrt{\lambda}/m$ is the SG coupling constant. Define $\Psi^{\dagger}_{\alpha}(x)$ as the adjoint of $\Psi_\alpha(x)$. Starting from the canonical equal time commutation rules

$$[\Phi(x), \Phi(y)] = [\dot{\Phi}(x), \dot{\Phi}(y)] = 0$$

$$[\Phi(x), \dot{\Phi}(y)] = i\delta(x-y) \tag{7.48}$$

it is easy to check that for $x \neq y$

$$[A_1(x), A_1(y)] = i\pi[\theta(x-y) - \theta(y-x)]$$

$$= \pm i\pi \qquad \text{for } x \gtrless y. \tag{7.49}$$

where $\theta(x-y)$ is the step function. In the same fashion one can check that

$$[A_\alpha(x), A_\beta(y)] = \text{ either } i\pi \text{ or } -i\pi, \tag{7.50}$$

for all $\alpha = 1,2$; $\beta = 1,2$; and $x \neq y$. Then applying the identity

$$e^M e^N = e^N e^M e^{[M,N]} \qquad \text{if } [M,N] = c\text{-number}, \tag{7.51}$$

to the components of Ψ in (7.46), we have

$$\Psi_\alpha(x)\Psi_\beta(y) = -\Psi_\beta(y)\Psi_\alpha(x)$$

$$\Psi_\alpha(x)\Psi^{\dagger}_\beta(y) = -\Psi^{\dagger}_\beta(y)\Psi_\alpha(x) \tag{7.52}$$

when $x \neq y$, for all α, β. Thus the fields Ψ_α and $\Psi^{\dagger}_\beta$ constructed as per (7.46) anticommute when $x \neq y$. When $x = y$, the calculation is more tricky because the commutator (7.49) is not well defined and further the product of two Ψ's is singular at $x = y$. This case has to be handled by a careful limiting procedure as $x \to y$ with a suitable choice of the coefficients C_1, C_2 in (7.46). This is done by Mandelstam with the result

$$\{\Psi_\alpha(x), \Psi_\beta(y)\}_+ = 0 \qquad \text{and} \qquad \{\Psi_\alpha(x), \Psi_\beta^\dagger(y)\}_+ = z\,\delta(x-y)\delta_{\alpha\beta} \tag{7.53}$$

where $\{\ ,\ \}_+$ stands for the anticommutator and z is a renormalisation constant. Thus the field $\Psi(x)$ in (7.46) obeys local Fermi-field anticommutation rules. Furthermore, one can again easily verify, starting from the boson commutators (7.48), that

$$[\Phi(y), \Psi(x)] = (2\pi m/\sqrt{\lambda})\,\theta(x-y)\,\Psi(x) \tag{7.54}$$

for $x \neq y$. Thus, the operator $\Psi(x)$ raises the value of the field Φ by $2\pi m/\sqrt{\lambda}$ in the region to the left of x and leaves it unchanged to the right of x. Thus, when applied to the soliton state $[\Phi(\infty) - \Phi(-\infty) = 2\pi m/\sqrt{\lambda}]$ the operator $\Psi(x)$ will reduce it to a state in the vacuum sector $[\Phi(\infty) - \Phi(-\infty) = 0]$. Hence $\Psi(x)$, apart from being a Fermi field, destroys a soliton of the Bose system, consistent with the identification of the soliton with a fermion. Of course, $\Psi(x)$ alters the field Φ by a step function. The step function may be considered a 'point soliton', whereas the physical soliton state has some extended 'size', as indicated by the classical soliton solution. Mandelstam interprets his Ψ and $\Psi^\dagger$ as operators which destroy or create 'bare' point solitons, which acquire a finite extended size through interactions. The reader is referred to his paper for the demonstration that Ψ obeys the massive Thirring field equation if Φ obeyed the SG equation. For more discussion of such soliton creation operators, see Schroer and Swieca (1977), 'tHooft (1978), Mandelstam (1979) and Ezawa (1979). An alternative approach is given by Bardakci and Samuel (1978, 1979).

7.4 Factorisation of S-matrices

The vigorous study of soliton-bearing field theories has produced a biproduct of considerable theoretical interest and elegance. It has been shown that for a class of quantum field theories in (1 + 1) dimensions, the general multiparticle S-matrix can be determined exactly. Furthermore, these S-matrices are seen to enjoy several nice properties. Firstly, any multiparticle S-matrix is elastic, i.e. the number and set of momenta of particles of any given mass remains the same before and after a collision. Secondly, the general S-matrix factorises into a product of two-particle S-matrices. Thirdly, these two-particle S-matrices obey a cubic identity which, along with the time-honoured principles of unitarity, analyticity and crossing symmetry, is sufficient to yield exact analytical expressions for them in several models. These developments first began with perturbative or semiclassical investigations of the scattering of solitons of the SG system (Arefieva and Korepin 1974, Faddeev et al. 1975, Faddeev and Korepin 1975, Jackiw and Woo 1975). Subsequently the results were freed from such approximations (Zamolodchikov 1977a, Karowski et al. 1977, Zamolodchikov and Zamolodchikov 1978a) and offered as exact results starting from certain conservation laws that the theories in question enjoy. The results are not limited to the SG system alone, nor do they require that the scattering particles be solitons, so long as the particles and their theories obey a couple of assumptions given below. In this section and the next, we shall try to present a unified derivation of these results, abstracted from a recent and rapid succession of papers, and then illustrate it with the SG example. In this, we have been aided by the lucid reviews by Zamolodchikov and Zamolodchikov (1979) and Shankar (1978).

Recall that in chapter 2 we mentioned the existence of an infinite number of local conservation laws for the SG system. We did not write them down, and their explicit form is not necessary for us. All we need is that these conservation laws (i) are local and (ii) yield conserved charges which are components of Lorentz tensors of increasing rank. These features may be verified for the SG case by looking at the explicit expressions for these currents (Kulish and Nisimov 1976). Conservation laws with the same property also exist for the non-linear sigma model (i.e. the $O(N)$ model introduced in chapter 3, in $(1+1)$ dimensions) at both the classical (Pohlmeyer 1976) and quantum (Polyakov 1977a) levels. The same is true for some other two-dimensional models (see Karowski (1979)).

Our starting point will be any two-dimensional theory for which such conservation laws exist. Let Q_N be one such conserved quantity which is a component of a rank-N Lorentz tensor. Then the value of this charge, acting on the initial state of any scattering process where all particles are widely separated, will be given by

$$Q_N|p_1, p_2, \ldots, p_n\rangle_{\text{in}} = \sum_{i=1}^{n} q_N(e_i, k_i)|p_1, p_2 \ldots p_n\rangle_{\text{in}} \tag{7.55}$$

where p_i stands for the two-momentum $\{e_i, k_i\}$ of the ith particle, and $q_N(e_i, k_i)$ is some Nth-order polynomial in the energy and momentum of the ith particle. That the contributions of separated particles are additive follows from the locality of the conservation law which implies that Q_N will be the space integral of some local density. That the contribution will be a polynomial $q_N(e_i, k_i)$ follows from the Lorentz-tensor property of Q_N. Note that in (1 + 1) dimensions there is no spin, so that such tensors can be built only out of energy and momentum. In the same way, for the outgoing final state consisting of l particles

$$Q_N|p'_1, p'_2, \ldots, p'_l\rangle_{\text{out}} = \sum_{j=1}^{l} q_N(e'_j, k'_j)\,|p'_1, p'_2, \ldots, p'_l\rangle_{\text{out}}. \tag{7.56}$$

Since Q_N is conserved,

$$\sum_{j=1}^{l} q_N(e'_j, k'_j) = \sum_{i=1}^{n} q_N(e_i, k_i). \tag{7.57}$$

Since we have an infinite number of conserved charges Q_N of increasing rank N, we have an infinite number of equations of the form (7.57) with polynomials $q_N(e, k)$ of increasing rank N. They have to be satisfied by the $2(n+l)$ variables e_i, k_i, e'_j and k'_j. Clearly these equations are severely overconstrained, and will be satisfied in general only if

$$l = n \tag{7.58a}$$

$$\text{the set } \{p_1 \ldots p_n\} = \text{the set } \{p'_1 \ldots p'_n\}. \tag{7.58b}$$

For special values of the momenta it may be possible to evade (7.58) and still satisfy (7.57) for all N; but that would violate the analyticity of the scattering amplitude in the momentum variables. Hence (7.58) is the only solution. This solution tells us that the process is essentially elastic (the total number of particles cannot change). The initial set of two-momenta must be the same as the final set. This does not mean that the S-matrix is trivial. If some of the particles in the initial state have the same mass, they may exchange momenta in the final state, or be replaced by other particles of the same mass. But the number of particles of a given common mass must be separately conserved in order to preserve the same set of two-momenta.

Actually, the presence of two such tensor charges is enough to derive (7.58) (see Polyakov 1977a and Parke 1980). We have, for simplicity of presentation, assumed an infinite number only because the examples of interest which possess two also possess an infinite number. The next consequence of these conserved charges is the factorisation of the multiparticle S-matrix into a product of two-particle S-matrices. Consider for example a three-particle collision, with initial momenta $k_1 > k_2 > k_3$. By virtue of (7.58), there will only be three final particles with the same set of momenta $\{k_1, k_2, k_3\}$. The space–time diagrams of the three possible ways of collision are depicted in fig. 15(a), (b) and (c). Note that we have

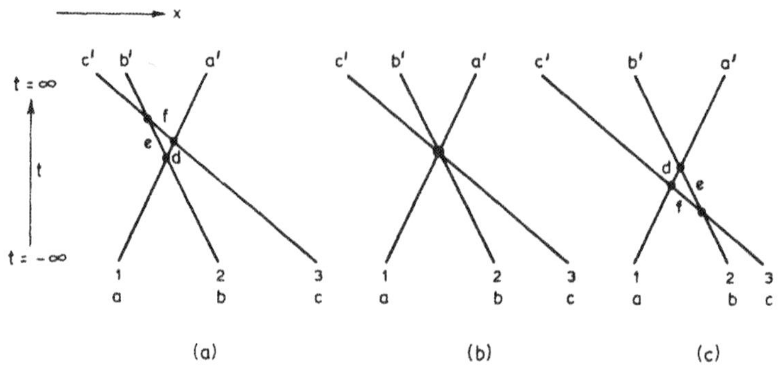

Fig. 15. World lines of the three ways in which a three-body collision can take place, when the momenta are ordered as $k_1 > k_2 > k_3$. In (a) and (c), the S-matrix is a product of three two-body S-matrices. In (b), an intrinsic three-body process is involved. The symbols a, b, c etc. refer to internal symmetry labels.

deliberately drawn thick world-lines for the particles to indicate that they have finite but non-zero size. In fig. 15(a) and (c), they collide two at a time, but in different sequences, whereas in (b) there is intrinsic three-body scattering.

Nevertheless, all three processes yield the same amplitude. The simplest way to see this is using a particle-displacement argument (Shankar and Witten 1978a). Take the initial state at some large negative time t, where the particles are widely separated. Operate on it with exp (iaQ_N), where Q_N is any of the non-trivial conserved quantities (of tensorial rank $N > 1$), and a is a real parameter. Note that in quantum theory Q_N will be an operator. Since it is the integral of a local density by assumption, Q_N will act on each of the widely separated wave packets separately. Let us write the wave packet of the ith particle as

$$|\psi_i(x)\rangle = |\int_{-\infty}^{\infty} dk \exp[ik(x - x_i)] f(k)\rangle \tag{7.59}$$

where the momentum-space wavefunction $f(k)$ is any reasonable function, say a gaussian, peaked at $k = k_i$. We are aware, of course, that these are not point particles. The variables x and k stand for the location and momentum of the centre of mass of the particle or, more precisely, to its 'collective coordinates' of position and momentum. By stationary phase approximation, one can see that $\psi_i(x)$ is peaked at $x = x_i$. Then

$$|\psi_i'(x)\rangle \equiv \exp(iaQ_N)|\psi_i(x)\rangle$$
$$= |\int dk \exp[iaq(k)] \exp[ik(x-x_i)] f(k)\rangle \qquad (7.60)$$

where the operator Q_N acting on this particle will yield the function $q_N(e, k)$, which we have written as $q(k)$, using $e^2 = k^2 + m_i^2$. By applying the stationary phase condition again to the oscillating exponent one can see that $\psi_i'(x)$ is now peaked around

$$x = x_i - a(dq/dk)_{k_i} \qquad (7.61)$$

where we recall that $f(k)$ is peaked at the mean momentum k_i. The key point now is that if Q_N has some non-trivial rank, $q(k)$ will be a non-linear function of k and hence $(dq/dk)_{k_i}$ will depend on k_i. Consequently, the operator $\exp(iaQ_N)$ displaces the ith particle by an amount $a(dq/dk)_{k_i}$ which depends on its mean momentum k_i. Therefore, when $\exp(iaQ_N)$ acts on the entire initial state, it displaces each particle by a different amount depending on its momentum. By making the parameter a arbitrarily large we can alter the relative position between any two particles by an arbitrary amount.

Now, apply this displacement to the three-body scattering in fig. 15. It is clear that by displacing line 3 relative to lines 1 and 2, we can go from fig. 15(a) to 15(b) and 15(c). This has the following important consequences:

(i) Since Q_N is conserved, it commutes with the Hamiltonian and hence the operation $\exp(iaQ_N)$ will leave the S-matrix invariant. Hence the amplitude for the three possible sequences of collision in fig. 15(a), (b) and (c) are equal!

(ii) The three body S-matrix can be written as a product of three two-body S-matrices, as per fig. 15(a) or (c). Even the case of fig. 15(b), where an intrinsic three-body S-matrix is involved, is equal to the factorised amplitude (fig. 15(a) or (c)) by virtue of the symmetry under $\exp(iaQ_N)$.

(iii) Since the processes in fig. 15(a) and (c) are equal, we can write the factorised amplitude in two different sequences which must be equal. In other words,

$$S(123) = S(23)\,S(13)\,S(12) = S(12)\,S(13)\,S(23) \qquad (7.62)$$

where $S(ij)$ are two-body S-matrices and the notation is self-evident. Therefore, not only does the three-body amplitude factorise, but the two-body amplitudes themselves must satisfy the cubic identity in (7.62). Note that if the $S(ij)$ are just numbers, this cubic identity is trivial. However, as we have pointed out, if some or all the scattering particles have equal masses (if for instance they belong to some degenerate internal symmetry multiplet) then the constraint (7.58) permits $S(ij)$ to be a non-trivial matrix connecting all degenerate-mass particles. In that case, this cubic identity (7.62) is a serious constraint on the two-body S-matrix itself.

The same argument can be repeated to four-body and higher S-matrices to show that they too factorise into products of two-body S-matrices. No independent quartic or higher identities emerge, however. It may be checked that particle displacement and the cubic identity (7.62) are sufficient for factorisation of higher many-body amplitudes. Note the crucial role of the non-trivial nature of Q_N. If we had used the trivial case of $Q_1 = P$, the total momentum operator (which is always conserved in any translationally invariant theory), then $q(k) = k$ and consequently all particles would be displaced by the same amount a. We could not have altered the sequence of collisions. To do so it is necessary that Q_N have a sufficiently high Lorentz tensorial rank so that $q(k)$ is non-linear. For this argument we used only one such Q_N. The fact that our examples of interest permit several (or infinitely many) such conserved charges of increasing complexity only strengthens the argument. Our derivation is undoubtedly somewhat qualitative. We have for instance used one-coordinate wave-functions in (7.59), to represent the overall location of each particle. But neither the explicit form of the momentum distribution $f(k)$, nor of the conserved quantity Q_N has been invoked—only the minimal consequence (which is powerful enough) that $\exp(iaQ_N)$ displaces different particles by different amounts. No doubt the argument can be presented in a formally more rigorous fashion. It should be mentioned that, long before these field theoretical results, similar arguments for factorisation had been developed in the simpler context of δ-function potentials (see for instance McGuire 1964).

Our conclusion, then, is that if a two-dimensional field theory has conserved quantities of sufficient complexity, the general S-matrix is elastic, and factorises into two-body S-matrices which obey the cubic identity (7.62). The task of obtaining the general S-matrix therefore reduces to evaluating just the two-body S-matrix. We shall do this for some models in the next section.

Our discussion has been presented for $(1+1)$-dimensional models. There is not much motivation for looking for $(3+1)$-dimensional models with such remarkable S-matrix properties. For one thing, Coleman and Mandula (1969) long ago showed that in $(3+1)$ dimensions, an S-matrix as severely constrained as the ones we have discussed will be trivial. For another, experimental data in high-energy scattering abounds with prolific inelasticity.

7.5. Some exact S-matrices

In this section we show how exact expressions for the total S-matrix may be derived for a class of two-dimensional theories which possess the requisite conserved quantities to satisfy the preceding requirements. All we need to find are two-body S-matrices. The higher ones are obtained by factorisation. We will take our theories to enjoy some internal $O(N)$ invariance. Many models, including the SG system, the massive Thirring model, the $O(N)$ sigma model and the Gross–Neveu model to be introduced in chapter 9, enjoy such $O(N)$ invariance. We take the two initial particles to belong to a vector multiplet under the $O(N)$ group. They will clearly have a common mass M. Barring fortuitous circumstances, the only other particles with mass M in that theory will also belong to the same multiplet. Hence, by virtue of eq. (7.58), the final state must also consist of only two particles belonging to that vector multiplet. Thus, we are interested in $\langle c(p_1')d(p_2')|S|a(p_1)b(p_2)\rangle$ where p_i and p_i' are two-momenta and a, b, c, d (each varying from 1 to N) refer to internal space labels in that vector multiplet. To obtain this amplitude, we will invoke the cubic identity (7.62) as well as the well-known principles of S-matrix theory. Detailed developments in S-matrix theory will not be needed – only the basic principles like the Lorentz invariance, analyticity, unitarity and crossing symmetry of the S-matrix. We expect the reader to be familiar with these principles (see for example Frautschi 1963).

Lorentz invariance tells us that the S-matrix can, apart from energy–momentum conserving δ-functions, depend on the two-momenta only through Lorentz scalars. Remember that we are in one space dimension, where no spin structure is possible. In $(3+1)$ dimensions, the usual scalar combinations used are

$$s \equiv (p_1+p_2)^2, \qquad t \equiv (p_1-p_1')^2 \qquad \text{and} \qquad u \equiv (p_1-p_2')^2 \tag{7.63}$$

with

$$s+t+u = 4M^2. \tag{7.64}$$

These definitions hold in $(1+1)$ dimensions as well, but unlike $(3+1)$ dimensions, where two of these variables can vary independently in a physical process, in $(1+1)$ dimensions only one of them can vary independently. In the centre-of-mass frame, only forward or backward scattering is possible. Hence, $s = (E_{\text{cm}})^2$, can vary from $4M^2$ to ∞, but either $t = 0$ with $u = 4M^2 - s$, or $u = 0$ with $t = 4M^2 - s$. We can write the scattering amplitude as

$$\langle c(p_1')d(p_2')|S|a(p_1)b(p_2)\rangle = \delta(p_1'-p_1)\delta(p_2'-p_2)[{}_{cd}S_{ab}(s)]. \tag{7.65}$$

This defines the ordering of the labels *cd* with respect to *ab*.

Analyticity tells us that the *S*-matrix ${}_{cd}S_{ab}(s)$ is analytic in the cut *s* plane. It can have bound state poles on the real axis between 0 and $4M^2$, and unitarity cuts from $4M^2$ to ∞ and from $-\infty$ to 0 (fig. 16(a)). It must also obey the reflection principle

$${}_{cd}S^*_{ab}(s) = {}_{cd}S_{ab}(s^*) \tag{7.66}$$

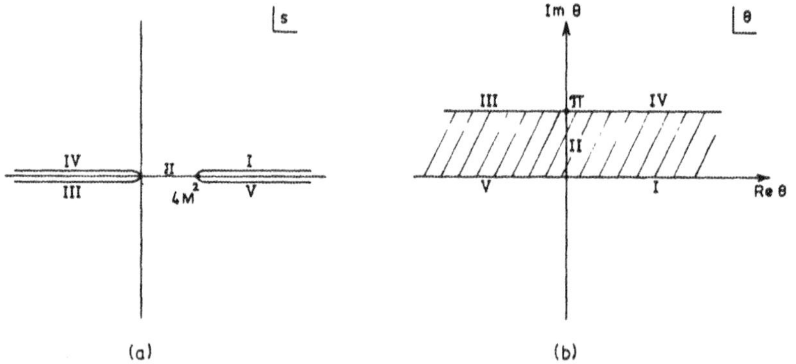

Fig. 16. (a) The cut *s* plane in which the *S*-matrix is analytic. (b) The θ plane. The *s* plane maps into the strip between Im $\theta = 0$ and Im $\theta = \pi$ in the θ plane as shown.

where the star stands for complex conjugation. Now, the infinitely many conservation laws have already ruled out inelastic particle production. Hence the unitarity cuts are purely elastic cuts and are kinematic in origin. (They arise because the momenta are double-valued functions of the energy and hence of *s*.) These cuts can be removed by changing to the rapidity variables θ_i defined in terms of $p_i = \{e_i, k_i\}$ by

$$k_i = M \sinh \theta_i; \qquad e_i = M \cosh \theta_i. \tag{7.67}$$

Then

$$s = 4M^2 \cosh^2(\theta/2) \qquad \text{with } \theta \equiv \theta_1 - \theta_2. \tag{7.68}$$

The *S*-matrix is then a meromorphic function of θ with only bound state poles. Under the change of variables from s to θ, the physical sheet in s maps onto the strip between $\mathrm{Im}\,\theta = 0$ and $\mathrm{Im}\,\theta = \pi$, as shown in fig. 16(b). The reflection condition (7.66) reduces, for real θ, to

$$ {}_{cd}S^*_{ab}(\theta) = {}_{cd}S_{ab}(-\theta). \tag{7.69}$$

The crossing process, $a \leftrightarrow c$ and $s \leftrightarrow u$, under which the S-matrix must enjoy crossing symmetry, reduces for the θ variable to

$$\theta \leftrightarrow \mathrm{i}\pi - \theta. \tag{7.70}$$

These results are easy to check. Next, we exploit the $\mathrm{O}(N)$ invariance of the theory, keeping in mind that the particles form an $\mathrm{O}(N)$ vector. Decomposing ${}_{cd}S_{ab}(\theta)$ into $\mathrm{O}(N)$ invariant combinations we can write, in general,

$$ {}_{cd}S_{ab}(\theta) = \delta_{ab}\,\delta_{cd}\,S_1(\theta) + \delta_{ac}\,\delta_{bd}\,S_2(\theta) + \delta_{ad}\,\delta_{bc}\,S_3(\theta). \tag{7.71}$$

Let us write the crossing and unitarity relations in terms of these S_1, S_2 and S_3. Crossing symmetry, which exchanges $a \leftrightarrow c$ and $\theta \leftrightarrow \mathrm{i}\pi - \theta$ yields

$$S_2(\theta) = S_2(\mathrm{i}\pi - \theta) \tag{7.72}$$

and

$$S_1(\theta) = S_3(\mathrm{i}\pi - \theta). \tag{7.73}$$

The unitarity condition $SS^\dagger = 1$, when written out in full, is

$$\begin{aligned} \sum_{ef} {}_{cd}S_{ef}(\theta)\,{}_{ef}S^\dagger_{ab}(\theta) &= \sum_{ef} {}_{cd}S_{ef}(\theta)\,{}_{ab}S^*_{ef}(\theta) \\ &= \delta_{ac}\,\delta_{bd}. \end{aligned}$$

Upon inserting (7.71) this gives three relations:

$$S_2(\theta)\,S_2(-\theta) + S_3(\theta)\,S_3(-\theta) = 1 \tag{7.74}$$

$$S_2(\theta)\,S_3(-\theta) + S_3(\theta)\,S_2(-\theta) = 0 \tag{7.75}$$

$$\begin{aligned} &N\,S_1(\theta)\,S_1(-\theta) + S_1(\theta)\,S_2(-\theta) + S_2(\theta)\,S_1(-\theta) \\ &\quad + S_1(\theta)\,S_3(-\theta) + S_3(\theta)\,S_1(-\theta) = 0 \end{aligned} \tag{7.76}$$

where the reflection condition (7.69) has been used. Finally the cubic identity (7.62), which the two-particle S-matrix must obey, when expanded in matrix notation states (see fig. 15(a) and (c)),

$$\begin{aligned} &[{}_{b'c'}S_{ef}(\theta')]\,[{}_{a'f}S_{dc}(\theta+\theta')]\,[{}_{de}S_{ab}(\theta)] \\ &= [{}_{b'a'}S_{ed}(\theta)] \times [{}_{c'd'}S_{fa}(\theta+\theta')]\,[{}_{ef}S_{bc}(\theta')]. \end{aligned} \tag{7.77}$$

Here, $\theta \equiv \theta_1 - \theta_2, \theta' \equiv \theta_2 - \theta_3$ and $\theta + \theta' = \theta_1 - \theta_3$ where the θ_i are the rapidities of the three initial particles. Upon inserting (7.71) into this matrix equation, the latter reduces to a set of independent relations for the functions $S_1(\theta)$, $S_2(\theta)$ and $S_3(\theta)$. Obtaining these relations from (7.77) is a matter of straightforward but cumbersome algebra. The results are somewhat different for $N = 2$ from the $N \geqslant 3$ cases. This is because fewer independent three-particle combinations exist when $N = 2$. Let us first take $N \geqslant 3$. The coefficient of $\delta_{a'b}\delta_{b'c}\delta_{c'a}$ on both sides of (7.77) yields

$$S_3(\theta)S_2(\theta+\theta')S_3(\theta') = S_2(\theta)S_3(\theta+\theta')S_3(\theta') + S_3(\theta)S_3(\theta+\theta')S_2(\theta')$$

or, more compactly,

$$S_3S_2S_3 = S_2S_3S_3 + S_3S_3S_2 \tag{7.78}$$

where it is understood that in all terms the arguments of the three S_i are θ, $(\theta+\theta')$ and θ' respectively. In the same fashion, the other independent relations for the $N \geqslant 3$ case contained in (7.77) are

$$S_3S_1S_2 = S_2S_1S_1 + S_3S_2S_1 \tag{7.79}$$

and

$$S_3S_1S_3 = NS_1S_3S_1 + S_1S_3S_2 + S_1S_3S_3 + S_1S_2S_1 + S_2S_3S_1 + S_1S_1S_1 + S_3S_3S_1. \tag{7.80}$$

For $N = 2$, there are altogether only two relations, namely

$$S_3S_2S_3 + S_1S_2S_3 + S_1S_1S_2 = S_2S_1S_3 + S_2S_3S_3 + S_3S_3S_2 \tag{7.81}$$

and

$$S_3S_1S_3 + S_3S_2S_3 = S_3S_3S_1 + S_3S_3S_2 + S_2S_3S_1 + S_2S_3S_3 + 2S_1S_3S_1 + S_1S_3S_2 + S_1S_3S_3 + S_1S_2S_1 + S_1S_1S_1. \tag{7.82}$$

Note that, unlike ${}_{cd}S_{ab}$, these S_1, S_2, S_3 are not matrices, but ordinary functions of the appropriate variables. These cubic identities, along with the unitarity, crossing and analyticity conditions ((7.72)–(7.76)) are sufficient to determine the S-matrix, up to an ambiguity which can also be removed for specific models. We will present the derivation for the $N \geqslant 3$ case. The results for the $N = 2$ case, obtained in the same way, will also be given below. We begin with the cubic conditions, take (7.78) and divide both sides by $S_3 S_3 S_3$ to get

$$\frac{S_2(\theta+\theta')}{S_3(\theta+\theta')}=\frac{S_2(\theta)}{S_3(\theta)}+\frac{S_2(\theta')}{S_3(\theta')}$$

which implies

$$S_2(\theta)/S_3(\theta)=\alpha\theta \tag{7.83}$$

where α is some constant. Next we substitute $S_2(\theta)/\alpha\theta$ for $S_3(\theta)$ in (7.79) and divide both sides by $S_2 S_1 S_1$ to get

$$\alpha\theta=\frac{S_2(\theta')}{S_1(\theta')}-\frac{S_2(\theta+\theta')}{S_1(\theta+\theta')}.$$

This is solved by

$$S_2(\theta)=-(\alpha\theta+\beta)S_1(\theta) \tag{7.84}$$

where β is another constant. Using (7.83)–(7.84) we write (7.80) in terms of the function S_1 alone and equate coefficients of $S_1(\theta)\,S_1(\theta+\theta')\,S_1(\theta')$. This yields the simple relation

$$\beta=(N-2)/2. \tag{7.85}$$

We have exhausted all three cubic identities (7.78)–(7.80) for $N \geqslant 3$. Next, the real-analyticity condition (7.69) requires

$$S_2^*(\theta)/S_3^*(\theta)=S_2(-\theta)/S_3(-\theta).$$

Hence

$$\alpha^*=-\alpha=-\mathrm{i}\lambda, \tag{7.86}$$

where λ is real. The ratio of the crossing relations (7.72) and (7.73) yields

$$\frac{S_2(\mathrm{i}\pi-\theta)}{S_3(\mathrm{i}\pi-\theta)}=\frac{S_2(\theta)}{S_1(\theta)} \quad \text{or} \quad \mathrm{i}\lambda(\mathrm{i}\pi-\theta)=-(\mathrm{i}\lambda\theta+\beta).$$

Therefore

$$\lambda=\beta/\pi=(N-2)/2\pi. \tag{7.87}$$

Hence,

$$S_3(\theta)=\frac{-2\pi\mathrm{i}}{(N-2)\theta}S_2(\theta)$$

and

$$S_1(\theta) = \frac{-2\pi i}{(N-2)(i\pi-\theta)} S_2(\theta). \tag{7.88}$$

Thus, the ratios of S_1, S_2 and S_3 are fully determined. To obtain any of them individually, say $S_2(\theta)$, we resort to the unitarity conditions. The relations (7.75) and (7.76), which involve only the ratios S_i/S_j, are automatically satisfied by the above. The remaining relation (7.74) requires

$$S_2(\theta) S_2(-\theta)(1 + 1/\lambda^2\theta^2) = 1 \tag{7.89}$$

or

$$S_2(\theta) S_2(-\theta) = \theta^2/(\theta^2 + \Delta^2) \tag{7.90}$$

where

$$\Delta \equiv 1/\lambda = 2\pi/(N-2). \tag{7.91}$$

All we need is an $S_2(\theta)$ which satisfies (7.90) along with the crossing relation (7.72). (Recall that so far we have used only the *ratio* of the two crossing relations.) An answer is

$$S_2(\theta) = R(\theta) R(i\pi - \theta) \tag{7.92}$$

where

$$R(\theta) \equiv \Gamma\left(\frac{\Delta - i\theta}{2\pi}\right)\Gamma\left(\frac{\pi - i\theta}{2\pi}\right)\Big/\Gamma\left(\frac{-i\theta}{2\pi}\right)\Gamma\left(\frac{\pi + \Delta - i\theta}{2\pi}\right). \tag{7.93}$$

To remove the mystery out of how (7.93) is arrived at, we note that the simplest solution of (7.90) would be $S_2'(\theta) = \theta/(\theta + i\Delta)$. But this would not satisfy crossing symmetry (7.72). We can try to repair that by using

$$S_2''(\theta) = \frac{\theta}{(\theta + i\Delta)} \frac{(i\pi - \theta)}{(i\pi - \theta + i\Delta)}$$

which is now crossing-symmetric, but no longer satisfies unitarity (7.90). Unitarity is restored by multiplying by another factor, i.e.

$$S_2'''(\theta) = \frac{\theta}{\theta + i\Delta} \frac{(i\pi - \theta)}{(i\pi - \theta + i\Delta)} \frac{(i\pi + \theta + i\Delta)}{(i\pi + \theta)},$$

but now crossing symmetry is again lost. In this manner we can go on to get an infinite product of factors which in the limit should satisfy both unitarity and crossing. The result (7.92)–(7.93) is just a compact representation of this infinite product, in terms of Γ-functions. It can of course be directly verified that (7.92) with (7.93) satisfies both unitarity and crossing symmetry. Note that when $N = 3$, as in the non-linear O(3) model, the result (7.92) reduces to

$$S_2(\theta) = \theta(\mathrm{i}\pi - \theta)/(2\pi\mathrm{i} - \theta)(\mathrm{i}\pi + \theta). \tag{7.94}$$

Notice also that we can obtain another solution by replacing Δ by $-\Delta$ in (7.93), since the condition (7.90) is quadratic in Δ.

For the $N = 2$ case, we use the corresponding cubic identities (7.81)–(7.82) in place of (7.78)–(7.80), and proceed in the same manner. The algebra and the result are a little more complicated, but the principles are the same as in the $N \geqslant 3$ case. We will quote the result, which is (Zamolodchikov and Zamolodchikov 1979),

$$S_3(\theta) = \mathrm{i}\cot(4\pi^2/\gamma')\coth(4\pi\theta/\gamma')\,S_2(\theta);$$
$$S_1(\theta) = S_3(\mathrm{i}\pi - \theta) \tag{7.95}$$

and

$$S_2(\theta) = (2/\pi)\sin(4\pi^2/\gamma')\sinh(4\pi\theta/\gamma')\sinh(4\pi(\mathrm{i}\pi - \theta)/\gamma')\,U(\theta), \tag{7.96}$$

where

$$U(\theta) \equiv \Gamma\left(\frac{8\pi}{\gamma'}\right)\Gamma\left(1 + \frac{\mathrm{i}8\theta}{\gamma'}\right)\Gamma\left(1 - \frac{8\pi + \mathrm{i}8\theta}{\gamma'}\right)\prod_{l=1}^{\infty}\frac{R_l(\theta)\,R_l(\mathrm{i}\pi - \theta)}{R_l(0)\,R_l(\mathrm{i}\pi)} \tag{7.97}$$

$$R_l(\theta) \equiv \Gamma\left(\frac{16\pi l + \mathrm{i}8\theta}{\gamma'}\right)\Gamma\left(1 + \frac{16\pi l + \mathrm{i}8\theta}{\gamma'}\right)\Big/\Gamma\left(\frac{16\pi l + 8\pi + \mathrm{i}8\theta}{\gamma'}\right) \times\Gamma\left(1 + \frac{16\pi l - 8\pi + \mathrm{i}8\theta}{\gamma'}\right). \tag{7.98}$$

In this solution γ' is a real parameter, undetermined so far. This is in contrast to the $N \geqslant 3$ result which had no free parameters. This is related to the fact that the $N = 2$ case had only two independent cubic identities (7.81)–(7.82) in contrast to the three identities (7.78)–(7.80) for $N \geqslant 3$.

Once S_1, S_2 and S_3 are known, we can insert them into (7.71) to get the two-body S-matrix. The general multiparticle S-matrix, involving any number of particles from the $\mathrm{O}(N)$ vector multiplet, may then be written as a suitable sequential product of two-body S-matrices, thanks to factorisation. For instance, the three-body S-matrix may be obtained from the sequence of three two-body S-matrices depicted either in fig. 15(a) or in fig. 15(c). The two possibilities will give the same answer since the cubic identities have been satisfied.

There is, however, a catch. The result (7.95)–(7.98) for $N = 2$, and the result (7.92)–(7.93) for $N \geqslant 3$ along with its counterpart where $\Delta \leftrightarrow -\Delta$, may be called the 'minimal' solutions of the constraints due to the cubic identities and the basic principles of S-matrix theory. They are however not the only solutions. Let us suppose we multiply *all* three functions $S_1(\theta)$, $S_2(\theta)$ and $S_3(\theta)$, by a common factor

$$f(\theta) = \prod_{k=1}^{K} \left(\frac{\sinh\theta + \mathrm{i}\sin\alpha_k}{\sinh\theta - \mathrm{i}\sin\alpha_k} \right), \tag{7.99}$$

where α_k are arbitrary real constants and the number of factors, K, is also arbitrary. Notice that $f(\theta)$ satisfies

$$f(\theta) = f(\mathrm{i}\pi - \theta) \qquad \text{and} \qquad f(\theta) f(-\theta) = 1 .$$

Consequently when $S_2(\theta)$ is multiplied by $f(\theta)$ it continues to satisfy (7.90) and (7.72). The remaining conditions involve only the ratios of the $S_i(\theta)$ and will also remain satisfied. Therefore, in both the $N = 2$ and $N \geqslant 3$ cases, we can obtain a family of solutions by multiplying the 'minimal' solutions by a factor of the form (7.99). Notice that the factor $f(\theta)$ will introduce additional poles at $\theta = i\alpha_k$ in the S-matrix, apart from the poles already present in the minimal solutions. Poles in the S-matrix correspond to bound states provided they lie in the region $0 \leqslant s < 4M^2$, i.e. in the segment of the imaginary θ-axis given by $0 \leqslant Im\theta < \pi$; $Re\theta = 0$. Hence if any of the α_k fall in $0 \leqslant \alpha_k < \pi$, the factor $f(\theta)$ could have the important consequence of indicating further bound states in the system.

The fact that the principles of analiticity, unitarity and crossing determine the S-matrix only up to such ambiguities is not special to our discussion. It is a more general feature, which has been known for some time. These are called CDD ambiguities (Castillejo, Dalitz and Dyson 1956; see also Frautschi 1963). The cubic identities of our systems do not reduce this ambiguity since they involve only the ratios of the $S_i(\theta)$, in each of which the same factor $f(\theta)$ could be introduced.

The presence of such ambiguities (see also Mitra 1977) in our derivation is only to be expected. Recall that nowhere had we written down the Lagrangian or used the full details of the dynamics. All we have used are: (i) the presence of some conservation laws sufficient to force elasticity and factorisation; (ii) the fact that the scattering particles form a vector multiplet in an $O(N)$ invariant theory in two dimensions; and (iii) the general principles of S-matrix theory. There could be several different Lagrangian field theories which satisfy all these conditions including $O(N)$ invariance with a given N. Their dynamics could be different, and it would have been surprising if we had obtained the same S-matrix for all of them. What is remarkable is that we have come this far based on such broad requirements without knowledge of the precise Lagrangian. By the same count, in order to decide what CDD factors of the type (7.99) need to be introduced, if any, we must look closer at the dynamics by starting with the specific Lagrangian of the model in question. The same remark holds for fixing the value of the free parameter γ' in the $N = 2$ case, and for choosing between Δ and $-\Delta$ in the $N \geqslant 3$ case.

To illustrate how this may be done let us consider soliton scattering in the familiar sine–Gordon (SG) theory. As we have already mentioned, this theory gives an infinite number of local conservation laws, more than sufficient to ensure factorisation. Although not obvious from looking at the Lagrangian, this theory also enjoys hidden $O(2)$ invariance. To see this symmetry explicitly, one must invoke the 'disorder' parameter of the theory (Kadanoff and Ceva 1971) and that is beyond the scope of this book. However, we can take advantage of the equivalence of the SG theory to the massive Thirring (MT) model, discussed in section 7.3. The MT model has the well-known $U(1)$ symmetry under $\Psi \to e^{i\alpha}\Psi$. This is essentially $O(2)$ symmetry, with $(\Psi + \Psi^*)/2$ and $(\Psi - \Psi^*)/2i$ acting as an $O(2)$ doublet. Since the fermions and antifermions of the MT model have been identified respectively with the solitons and the antisolitons of the SG model, it is reasonable to suggest that, if A and $\bar{A}$ stand for the soliton and the antisoliton, then

$$A_1 = (A + \bar{A})/2 \qquad \text{and} \qquad A_2 = (A - \bar{A})/2i \tag{7.100}$$

form a doublet under the hidden $O(2)$ symmetry. The decomposition of the S-matrix into S_1, S_2 and S_3 would now be in terms of $O(2)$ tensors. The indices a, b, c and d in (7.71) take the values 1 or 2. Physically more natural definitions correspond to the forward A–$\bar{A}$ scattering (transmission) amplitude $S_T(\theta)$, the backward A–$\bar{A}$ (reflection) amplitude $S_R(\theta)$ and the A–A amplitude $S(\theta)$ for which there is no difference between forward and backward scattering. (The $\bar{A}$–$\bar{A}$ amplitude is also $S(\theta)$ because of the $\phi \leftrightarrow -\phi$ symmetry of the SG theory.) Using the relation (7.100) between $(A, \bar{A})$

and the doublet (A_1, A_2), it is easy to verify that

$$S_{\mathrm{T}}(\theta) = S_1(\theta) + S_2(\theta)$$

$$S_{\mathrm{R}}(\theta) = S_1(\theta) + S_3(\theta)$$

and

$$S(\theta) = S_2(\theta) + S_3(\theta). \tag{7.101}$$

In order to calculate the A–A and A–$\bar{A}$ scattering amplitudes we can therefore use the O(2) result (7.95)–(7.98) for S_1, S_2 and S_3, but an appropriate choice of the constant γ' and the CDD factors needs to be made. To determine them, we consider the poles of the minimal amplitude (7.95)–(7.98). These poles arise from poles of the Γ-functions. There are many such poles, but the only ones that correspond to bound states must fall in the physical region ($0 \leqslant \mathrm{Im}\theta < \pi$; $\mathrm{Re}\theta = 0$). These come from the factors $\Gamma(1 + \mathrm{i}8\theta/\gamma')$ and $\Gamma(1 - (8\pi + \mathrm{i}8\theta)/\gamma')$. These poles are at:

(i) $\theta_n = \mathrm{i}n\gamma'/8 \qquad n = 1, 2, \ldots, < 8\pi/\gamma'$

and (ii) $\theta_n = \mathrm{i}\pi - \mathrm{i}n\gamma'/8, \qquad n = 1, 2, \ldots, < 8\pi/\gamma'. \qquad (7.102)$

The set (ii) corresponds, using (7.68) to,

$$s_n = 4M^2 \cosh^2\left(\frac{\mathrm{i}\pi}{2} - \frac{\mathrm{i}n\gamma'}{16}\right) = 4M^2 \sin^2\left(\frac{n\gamma'}{16}\right);$$

$$n = 1, 2, \ldots, < 8\pi/\gamma'. \tag{7.103}$$

These poles, lying in the range $0 \leqslant s < 4M^2$, give the location of the bound states. The set (i) obviously corresponds to the same bound states, occurring as poles in the u-channel. Now, we have already obtained the bound states of this theory by the WKB method in section 7.1, where we also gave arguments to claim that the bound state spectrum obtained there was exact. That spectrum was (see (7.39)).

$$M_n^2 = 4M^2 \sin^2(n\gamma/16), \qquad n = 1, 2, \ldots, < 8\pi/\gamma \tag{7.104}$$

where γ was the renormalised coupling constant of the SG theory. We see that the spectrum agrees exactly with the bound states (7.103) of our minimal S-matrix provided we set

$$\gamma' = \gamma.$$

This not only helps us fix the parameter γ' in the S-matrix, but also indicates that the minimal solution (7.95)–(7.96) for the O(2) S-matrix may be sufficient for the SG case, without need for additional CDD factors. Further evidence of the correctness of the minimal solution comes from examining the case $\gamma = 8\pi$. Then S_1, S_2 and S_3 in (7.95)–(7.98) greatly simplify to yield, for all θ,

$$S_{\mathrm{T}}(\theta) = S_1(\theta) + S_2(\theta) = 1$$

$$S_{\mathrm{R}}(\theta) = S_1(\theta) + S_3(\theta) = 0$$

and

$$S(\theta) = S_2(\theta) + S_3(\theta) = 1. \tag{7.105}$$

This is just what should happen. The value $\gamma = 8\pi$ corresponds (see section 7.3) to $g = 0$, where g is the MT coupling constant. Consequently we expect solitons and antisolitons to experience no forces. Hence the A–A amplitude S equals unity as does the $\bar{A}$–A transmission amplitude S_{T}, but the $\bar{A}$–A reflection amplitude S_{R} vanishes. Further, as $\gamma \to 8\pi$, $g \to 0$, and our S-matrix may be expanded in powers of $(2g/\pi) = (8\pi/\gamma) - 1$. When this is compared with standard perturbation calculations for the scattering of fermions in the MT model the results agree up to order g^3 (Weisz 1977). These arguments strongly indicate that the minimal solution (7.95)–(7.98) gives the exact S-matrix for the scattering of solitons (and antisolitons) of the SG theory.

Of course, the full S-matrix of this system will also include the scattering of the bound states M_n with one another and with solitons and antisolitons. For these processes, Zamolodchikov (1977b) and Karowski and Thun (1977) offer a derivation capitalising on the fact that any of these bound states M_n is itself a pole in the A–$\bar{A}$ amplitude. Hence the M_n–A amplitude can be extracted as the residue of the corresponding pole in the $A\bar{A}$ subenergy variable, of the $\bar{A}$–A–A amplitude. In the same way, a more complicated amplitude involving several M_n, A and $\bar{A}$ particles is again the residue of a suitable product of poles in some multi-A–$\bar{A}$ amplitude. All multi-A–$\bar{A}$ amplitudes are already known thanks to factorisation, in terms of (7.95)–(7.98), so that the amplitudes involving bound states M_n are also known. For more on how these residues are extracted, see Zamolodchikov and Zamolodchikov (1979). With this added information, the full sine–Gordon S-matrix is known exactly.

In the same way, one can try to resolve CDD ambiguities for other O(N) invariant theories. Minimal solutions seem to work in several cases. For instance, the $N \geqslant 3$ solution (7.92)–(7.93) seems to be correct, with no additional CDD factors, for the non-linear sigma model. Here we have no exact bound state spectrum available for comparison, nor can perturbation theory be used. (Recall from section 3.3 that this model reveals spontaneous symmetry breaking at the classical level. But in quantum theory, continuous symmetry cannot be broken spontaneously in $(1 + 1)$ dimensions (Coleman 1973). The internal direction of the field ϕ will fluctuate violently to restore the symmetry and this will be reflected in the form of infrared divergences in perturbation theory (Ma and Rajaraman 1975).) However, the sigma model can be attacked using a $1/N$ expansion, whose predictions agree with the minimal solution (7.92)–(7.93) (Zamolodchikov and Zamolodchikov 1978a). Similarly the other minimal solution, the one where Δ is replaced by $(-\Delta)$ in (7.93), agrees with the $1/N$ expansion results of the Gross–Neveu model for $N > 4$. (Zamolodchikov and Zamolodchikov 1978b, Witten 1978).

The preceding derivation was for the scattering of a vector multiplet of an $O(N)$ invariant theory. One could try the same principles for other multiplets of other symmetry groups. The basic tensor decomposition (7.71) will be altered, and therefore so will the subsequent equations. Whether the altered equations can still be made to yield a solution will depend on the group and the multiplet of particles. For instance the GN model also yields 'kink' particles in addition to its elementary fermions, which transform as spinors under the $O(N)$ group. Their exact S-matrix has been obtained using these principles by Shankar and Witten (1978a), who also obtain the S-matrix of the supersymmetric sigma model (1978b).

CHAPTER 8

COLLECTIVE COORDINATES AND CANONICAL METHODS

8.1. Treatment of zero-modes: A miniature example

We have already discussed, in section 5.5, the basic physics which gives rise to zero-frequency fluctuation modes in the semiclassical method when continuous symmetries are present. We have also pointed out, at different places in chapters 5 and 6, that the method needs to be modified in the presence of such zero-modes. Although we were able to obtain some results (such as the energy levels in the soliton sectors to $O(\lambda^0)$ where λ is the coupling constant) without paying any special attention to zero-modes, the latter will lead to divergences in higher orders unless treated on a different footing from the other non-zero-frequency modes.

The literature in the field offers a variety of ways for treating zero-modes (see for instance Gervais and Sakita 1975, Rajaraman and Weinberg 1975, Christ and Lee 1975, Faddeev and Korepin 1976, Gervais et al. 1975, Tomboulis 1975, Creutz 1975, Baacke and Rothe 1977, Abbott 1978). Many of the treatments involve using what have come to be known as collective coordinates. (The notion of collective coordinates has been used for several decades in many-body theory; see for instance Bogoliubov and Tyablikov (1949).)

We will devote this chapter to some discussion of the collective coordinate method. Such coordinates can be introduced in either the functional-integral or the canonical-operator formalism. Also, in some cases the collective coordinates can be defined explicitly as functionals of the field variables, while in other cases they have to be introduced implicitly. We shall try to illustrate these different possibilities through examples. In particular, we shall present in sections 8.3 and 8.4 the Christ–Lee canonical Hamiltonian formalism which incorporates collective coordinates. It will give us an opportunity to see how the semiclassical method works in the canonical formalism, complementing the functional-integral approach of chapter 6.

We will begin our discussion, as we have done in other chapters, with a simple example in non-relativistic quantum mechanics where the proper treatment of zero-modes can be demonstrated in miniature. Consider a particle in two dimensions, governed by a Lagrangian

$$L = \tfrac{1}{2}\left((\dot{x}_1)^2 + (\dot{x}_2)^2\right) - V\left(\sqrt{x_1^2 + x_2^2}\right). \tag{8.1}$$

This Lagrangian is invariant under a continuous symmetry; namely, rotations in the (x_1, x_2) plane. Let $V(\sqrt{x_1^2 + x_2^2})$ have a minimum at some non-zero value $\sqrt{x_1^2 + x_2^2} = R \neq 0$. Clearly every point P on the circle of radius R will be a classical static solution. Suppose we naively apply the semiclassical method of chapter 5 to quantise static solutions. The method involves approximating energy eigenfunctions of the quantum system by starting from harmonic oscillator wavefunctions centred around any given static classical solution; in this case, around any given point P on the circle of radius R. The energy levels according to that method would be:

$$E_{\{n_1, n_2\}} = V(R) + \sum_{i=1}^{2} (n_i + \tfrac{1}{2})\hbar\omega_i + \text{higher orders}$$

where ω_1^2 and ω_2^2 are eigenvalues of $(\partial^2 V/\partial x_i \partial x_j)_P$. It is easy to see that, thanks to rotational symmetry, one of the eigenvalues, ω_1^2, will vanish, while the other is given by $\omega_2^2 = V''(R)$. The fact that ω_1 vanishes will cause no problems in the first two terms of $E_{\{n_1, n_2\}}$. One can write

$$E_{\{n_1, n_2\}} = V(R) + (n_2 + \tfrac{1}{2})\hbar\,\omega_2 + \text{higher orders}.$$

But in the higher-order terms, where the frequencies ω_1 and ω_2 will occur in energy denominators as per standard perturbation theory, the vanishing of ω_1 will lead to divergent contributions. This technical difficulty is a reflection of the fact that the underlying assumption behind the quantisation of static solutions is being partially violated. The quantum stationary states of this system are *not* localised around any specific classical static solution, i.e. around any given point P on the circle of radius R. Although the potential may be approximately quadratic near P with respect to radial displacement, it will remain constant with respect to angular displacement because of the symmetry. Consequently, quantum energy eigenstates, while localised radially, will spread in the angular direction throughout the circle. They cannot be associated with any given point on the circle.

This simple model thus contains the ingredients we are interested in, namely zero-mode fluctuations in the semiclassical procedure, caused by a continuous symmetry. A semiclassical method can still be applied, but with some modification to overcome the zero-mode difficulty. The required modification is easy to guess for this simple example. One first changes variables from cartesian to polar coordinates (r, θ). The advantages of the latter are obvious, but let us enumerate them anyway since collective coordinates in field theoretical examples will be chosen with similar criteria in mind: (i) The action of the symmetry is simplified – rotations displace only θ, while the coordinate r, which is orthogonal to θ, is unaffected. (ii) Because of rotational symmetry, the potential $V = V(r)$ is independent of θ. The only θ dependence is in the kinetic energy and that too is only quadratic:

$$L = \tfrac{1}{2}\dot{r}^2 + \tfrac{1}{2}r^2\,\dot{\theta}^2 - V(r). \tag{8.2}$$

(iii) The canonical momentum conjugate to θ is $l = r^2\,\dot{\theta}$. It is conserved, by virtue of the symmetry. (iv) Since the θ dependence is so simple, it can be treated *exactly* without resorting to semiclassical approximations. That will leave behind a residual one-dimensional problem in the variable r, with some effective potential. The semiclassical method of earlier chapters could then be applied only to this residual problem in the variable r, which will no longer enjoy (or, rather, suffer from) any continuous symmetry; the original rotational symmetry leaves r unaffected. Hence there will be no zero-modes or associated difficulties left in implementing the semiclassical method. This is the basic idea behind the modified semiclassical procedure, in the presence of continuous symmetries and zero-modes. With appropriate generalisation, the same procedure is used in field theoretical examples, with collective coordinates playing the same role as the variable θ here.

Solving the θ dependence exactly for the example (8.2) is a simple matter in both the Hamiltonian and the path-integral formalisms. The Hamiltonian yielded by (8.2) is simply

$$H = \tfrac{1}{2}p_r^2 + l^2/2r^2 + V(r); \qquad p_r = \dot{r}. \tag{8.3}$$

Notice that the coordinate θ does not occur at all in H. Furthermore, in the quantum version, since the angular momentum is conserved and hence commutes with H, we can look for energy eigenvalues in any given l sector. In that sector the residual one-dimensional problem (8.3) in the variable r experiences an effective potential $V_{\text{eff}}(r) \equiv V(r) + l^2/2r^2$, i.e. it has an extra centrifugal term. This extra term in the residual radial problem is the net effect of the zero-mode angular fluctuations, when treated properly as above.

Let us apply the semiclassical static-solution method to the radial problem in (8.3) for any given l. A static classical solution $r(t) = \tilde{R}$ will satisfy

$$\left(\frac{\mathrm{d}V_{\mathrm{eff}}}{\mathrm{d}r}\right)_{\tilde{R}} = \left(\frac{\mathrm{d}V}{\mathrm{d}r}\right)_{\tilde{R}} - \frac{l^2}{\tilde{R}^3} = 0. \tag{8.4}$$

The corresponding quantum levels in the harmonic oscillator approximation of chapter 5 will be

$$\begin{aligned} E_{n,l} &= V_{\mathrm{eff}}(\tilde{R}) + \tfrac{1}{2}(n+\tfrac{1}{2})\hbar\tilde{\omega} \\ &= V(\tilde{R}) + \tfrac{1}{2}(n+\tfrac{1}{2})\hbar\tilde{\omega} + l^2/2\tilde{R}^2 \end{aligned} \tag{8.5}$$

where

$$\tilde{\omega}^2 = (\mathrm{d}^2 V_{\mathrm{eff}}/\mathrm{d}r^2)_{\tilde{R}}. \tag{8.6}$$

Notice that while $r(t) = \tilde{R}$ is a static solution of the residual radial problem (8.3) it also corresponds to a time-dependent periodic solution of the parent system (8.2) in the following sense. The equations of motion flowing from (8.2) are

$$\ddot{r} - r\dot{\theta}^2 + \mathrm{d}V/\mathrm{d}r = 0 \tag{8.7a}$$

and

$$(\mathrm{d}/\mathrm{d}t)(r^2\dot{\theta}) = 0. \tag{8.7b}$$

If we set $r(t) = \tilde{R}$ and $\theta = vt$, with $v = l/\tilde{R}^2$, then eq. (8.7b) is automatically solved, while (8.7a) reduces to (8.4). Therefore the semiclassical result (8.5) can also be viewed as an expansion about a time-dependent, circularly rotating solution with angular momentum l. This is consistent with the intuitive argument given earlier. The quantum wavefunction will not stay localised around any given static classical solution on the plane, but will tend to spread in the angular direction, because of rotational symmetry. Hence a rotating time-dependent classical solution, which samples all points related by this symmetry, will be a better classical limit of quantum stationary states. This is what our modified semiclassical method has yielded.

It is worth pointing out, briefly, how the same result can be obtained in the path-integral formalism, since we have seen that the latter offers a straightforward generalisation to field theory. We are interested in energy levels for any given angular momentum l. These are given by the poles of

$$\mathrm{Tr}\left(\frac{\delta(L-l)}{E-H}\right)=\frac{1}{2\pi}\int_0^{2\pi}\mathrm{d}\alpha\,\mathrm{e}^{-il\alpha}\,\mathrm{Tr}\left(\frac{\mathrm{e}^{\mathrm{i}L\alpha}}{E-H}\right)$$

$$=\frac{\mathrm{i}}{2\pi}\int_0^{2\pi}\mathrm{d}\alpha\,\mathrm{e}^{-il\alpha}\int_0^{\infty}\mathrm{d}T\,\mathrm{e}^{\mathrm{i}ET}\int r_0\,\mathrm{d}r_0\int\mathrm{d}\theta_0$$

$$\langle r_0,\theta_0|\,\mathrm{e}^{-\mathrm{i}HT+\mathrm{i}L\alpha}|r_0,\theta_0\rangle, \tag{8.8}$$

where L is the angular momentum operator, $\hbar=1$, and the trace has been taken in the polar coordinate basis $|r_0,\theta_0\rangle$. This can be written as

$$\mathrm{Tr}\left(\frac{\delta(L-l)}{E-H}\right)=\mathrm{i}\int_0^{\infty}\mathrm{d}T\,\mathrm{e}^{\mathrm{i}ET}\,G_l(T),$$

with

$$G_l(T)\equiv\int\frac{\mathrm{d}\alpha}{2\pi}\int\mathrm{d}r_0\int\mathrm{d}\theta_0\,r_0\,\mathrm{e}^{-il\alpha}\langle r_0,\theta_0+\alpha|\,\mathrm{e}^{-\mathrm{i}HT}|r_0,\theta_0\rangle, \tag{8.9}$$

where the operator $\mathrm{e}^{\mathrm{i}L\alpha}$ has been used to rotate the final state. Let us write a path integral for the amplitude $\langle r_0,\theta_0+\alpha|\,\mathrm{e}^{-\mathrm{i}HT}|r_0,\theta_0\rangle$ directly in polar coordinates, using the polar Lagrangian (8.2)

$$G_l(T)=\int\bar{\mathscr{D}}[r(t)]\,r(t)\exp\left(\mathrm{i}\int_0^T(\tfrac{1}{2}\dot{r}^2-V(r))\mathrm{d}t\right)$$

$$\times\int\mathrm{d}\alpha\int\frac{\mathrm{d}\theta_0}{2\pi}\int\mathscr{D}[\theta(t)]\exp\left(\mathrm{i}\int_0^T\tfrac{1}{2}r^2\dot{\theta}^2-l\dot{\theta})\mathrm{d}t\right). \tag{8.10}$$

Here, all paths go from (r_0,θ_0) at $t=0$ to $(r_0,\theta_0+\alpha)$ at $t=T$. Integration over the endpoint value r_0 is included in the notation $\bar{\mathscr{D}}[r(t)]$. Finally, since $\alpha=\theta(T)-\theta(0)$ for all paths, we have replaced α by $\int_0^T\dot{\theta}\,\mathrm{d}t$ in the exponent.

Notice that the exponent in the angular integral is just quadratic in $\dot{\theta}$. Integration over α and θ_0 amounts to integrating over end values of the paths $\theta(t)$ in addition to intermediate values. Thus, the angular integration can be done exactly to yield (apart from constants which can be absorbed into the measure)

$$\prod_t \left[\frac{1}{r(t)} \exp\left(\frac{-\mathrm{i} l^2 \mathrm{d}t}{2r^2(t)} \right) \right],$$

where Π_t stands for a product over each instant t. When this is inserted into (8.10) we are left with

$$G_l(T) = \int \mathscr{D}[r(t)] \exp\left[\mathrm{i} \int_0^T \mathrm{d}t \left(\frac{\dot{r}^2}{2} - V(r) - \frac{l^2}{2r^2} \right) \right]. \tag{8.11}$$

Thus we have, once again, a radial problem with an effective potential $V(r) + l^2/2r^2$. The semiclassical methods of chapters 5 and 6 can now be applied to this radial functional integral with no fear of zero-modes since there is no further rotational symmetry left. (There is an added technical problem in applying the semiclassical method to the radial variable because the range of $r(t)$ is from 0 to ∞ rather than $-\infty$ to $+\infty$. This leads to the so-called Langer modification. We shall not discuss this technicality here since it is unrelated to our main aim of handling the zero-mode, which has already been achieved by the angular integration. Interested readers may see Langer (1937) or Morse and Feshbach (1953).)

A final point, before we leave this simple example, concerns the well-known ordering ambiguity in going from classical to quantum Hamiltonians. In the above discussion, we transformed variables to polar coordinates right at the classical level. Classically, the Lagrangians (8.1) and (8.2) are completely equivalent. Then we quantised the system in the path-integral formalism directly in polar coordinates, which yielded an effective radial potential $V(r) + l^2/2r^2$. Instead suppose we quantise the system in cartesian coordinates. The Lagrangian (8.1) will lead to a Schrödinger equation (with $\hbar = 1$):

$$\left[-\frac{1}{2}\left(\frac{\partial^2}{\partial x_1^2} + \frac{\partial^2}{\partial x_2^2} \right) + V\left(\sqrt{x_1^2 + x_2^2}\right) \right] \psi(x_1, x_2) = E\psi(x_1, x_2). \tag{8.12}$$

When this differential equation is converted to polar form using $\psi(r, \theta) = [u_l(r)/\sqrt{r}]\, \mathrm{e}^{\mathrm{i}l\theta}$, then it is easy to check that $u_l(r)$ obeys

$$\left(-\frac{1}{2}\frac{\partial^2}{\partial r^2} + V(r) + \frac{(l^2 - \frac{1}{4})}{2r^2} \right) u_l(r) = E u_l(r). \tag{8.13}$$

The centrifugal term now contains an extra item $-1/8r^2$, compared with what the polar path integral yielded in (8.11). The same extra factor $-1/8r^2$ could also have been obtained in the path-integral formalism, provided we had started out with the path integral in cartesian coordinates and *then* changed variables to polar coordinates, paying due care to retain the proper order of infinitesimals (for details see Edwards and Gulyaev 1964, Rajaraman and Weinberg 1975). More generally, suppose we replace the Hamiltonian H in (8.3) by

$$\tilde{H} = \tfrac{1}{2}p_r^2 + l^2/2r^2 + V(r) + [p_r, f(r)] \tag{8.14}$$

where $f(r)$ is some function of r. Classically, $\tilde{H}$ is identical to H, since p_r commutes with r. In quantum theory, where $[r, p_r] = i\hbar$, $\tilde{H}$ differs from H by $-i\hbar f'(r)$. Such ordering ambiguities indicate that there can be several quantum Hamiltonians which correspond to the same classical limit. The correct choice, for a given physical system for which the quantum theory is being proposed, is of course decided by which quantum Hamiltonian agrees with the experimental results on that system. Often, the correct form turns out to be the 'standard cartesian form'

$$H = -\frac{1}{2}\sum_i \frac{\partial^2}{\partial x_i^2} + V[\{x_i\}] \tag{8.15}$$

in terms of some cartesian coordinates $\{x_i\}$. In that case, if the same Hamiltonian needs to be described in terms of some other coordinates $\{\xi_i\}$, then the Laplacian $\sum_i \partial^2/\partial x_i^2$ in the Hamiltonian above is rewritten in terms of $\{\xi_i\}$. This yields a unique quantum Hamiltonian in terms of $\{\xi_i\}$. The conversion from (8.12) to (8.13) is an example of this procedure (see also Salomonson 1977).

This completes our discussion of how problems due to a continuous symmetry and resultant zero-mode are handled for the simple example in (8.1). Relevant features of this example, even though obvious, have been emphasised because in the more complicated context of field theories we shall use similar criteria in constructing a proper treatment of zero-modes. The role of the variable θ will be played there by collective coordinates. For the example (8.1), elementary notions of geometry made it possible to immediately identify r and θ as the proper coordinates for handling the zero-mode problem. Further (r, θ) are well known explicitly in terms of the original coordinates x_1 and x_2. In some field theoretical cases, the requisite collective coordinate can be again be written explicitly as functionals of the parent field variables: an example is given in the next section. More generally, collective coordinates have to be introduced implicitly as discussed in sections 8.3 and 8.4.

8.2. An example of an explicit collective coordinate

We now discuss a field theoretical method (Rajaraman and Weinberg 1975) whereby some zero-modes can be treated by a straightforward generalisation of the procedure used in the last section, i.e. by choosing a new set of coordinates, explicitly known in terms of the parent field variables. To illustrate the method, consider a *complex* scalar field $\phi(\boldsymbol{x},t)$ governed by a Lagrangian density

$$\mathscr{L}(\boldsymbol{x}, t) = \tfrac{1}{2}(\partial_\mu \phi)^* (\partial^\mu \phi) - U(|\phi|). \tag{8.16}$$

The form of $U(|\phi|)$ and the space dimensionality can be left arbitrary, as long as $U(|\phi|)$ leaves the theory renormalisable and permits the requisite classical solutions used below. This Lagrangian is clearly invariant under the *global* U(1) symmetry

$$\phi(\boldsymbol{x}, t) \rightarrow e^{i\alpha}\phi(\boldsymbol{x}, t) \tag{8.17}$$

where α is independent of space–time. The classical Euler–Lagrange equation is

$$\Box\phi + \frac{\partial U(|\phi|)}{\partial|\phi|}\frac{\phi}{|\phi|} = 0. \tag{8.18}$$

Given any classical solution $\phi_{\text{cl}}(\boldsymbol{x}, t)$ of (8.18), whether static or time-dependent, the function $e^{i\alpha}\phi_{\text{cl}}(\boldsymbol{x}, t)$ will also be a solution with the same energy for any real α, because of the U(1) symmetry. Consequently there will be zero-mode fluctuations around any given solution, with the associated difficulties discussed earlier. (Of course, the system (8.16) is also invariant under spatial translations and rotations. This will give rise to some more zero-modes. We shall return to consider these modes later. For the moment, let us consider only the zero-mode associated with the U(1) symmetry.)

To overcome these difficulties, we first change variables to polar *fields* $\rho(\boldsymbol{x}, t)$ and $\theta(\boldsymbol{x}, t)$ as defined by

$$\phi(\boldsymbol{x}, t) = \rho(\boldsymbol{x}, t)\exp[i\theta(\boldsymbol{x}, t)]. \tag{8.19}$$

In terms of these polar fields, the Lagrangian (8.16) and field equation (8.18) reduce respectively to

$$\mathscr{L}(\boldsymbol{x}, t) = \tfrac{1}{2}(\partial_\mu\rho)^2 + \tfrac{1}{2}\rho^2(\partial_\mu\theta)^2 - U(\rho) \tag{8.20}$$

and

$$\Box\rho - \rho(\partial_\mu\theta)^2 + \mathrm{d}U/\mathrm{d}\rho = 0 \tag{8.21a}$$

$$\partial_\mu(\rho^2\partial^\mu\theta) = 0. \tag{8.21b}$$

Associated with the one-parameter U(1) symmetry, there will be a conserved charge

$$Q = \frac{\mathrm{i}}{2}\int\left(\phi^*\frac{\partial}{\partial t}\phi - \phi\frac{\partial}{\partial t}\phi^*\right)\mathrm{d}\boldsymbol{x}$$
$$= -\int\rho^2\dot{\theta}\,\mathrm{d}\boldsymbol{x}. \tag{8.22}$$

The conservation of Q is also evident from (8.21b). As is well known, the operator Q in quantum theory will generate the symmetry transformation (8.17), which in terms of polar fields reads as

$$\rho(\boldsymbol{x}, t) \rightarrow \rho(\boldsymbol{x}, t) \tag{8.23a}$$

$$\theta(\boldsymbol{x}, t) \rightarrow \theta(\boldsymbol{x}, t) + \alpha. \tag{8.23b}$$

Since Q is conserved, and hence commutes with the Hamiltonian, we can look for energy eigenvalues in any given sector of charge q. These are given by poles of

$$\mathrm{Tr}_q\left(\frac{1}{E-H}\right) = \mathrm{i}\int_0^\infty \mathrm{d}T\,\mathrm{e}^{\mathrm{i}ET}G_q(T) \tag{8.24}$$

where

$$G_q(T) \equiv \mathrm{Tr}\,[\mathrm{e}^{-\mathrm{i}HT}\delta(Q-q)]$$
$$= \int\frac{\mathrm{d}\alpha}{2\pi}\mathrm{e}^{-\mathrm{i}q\alpha}\int\mathscr{D}\,[\rho_0(\boldsymbol{x})]\rho_0(\boldsymbol{x})\int\mathscr{D}\,[\theta_0(\boldsymbol{x})]$$
$$\times\,\langle\rho_0(\boldsymbol{x}), \theta_0(\boldsymbol{x})+\alpha\,|\,\mathrm{e}^{-\mathrm{i}HT}\,|\,\rho_0(\boldsymbol{x}), \theta_0(\boldsymbol{x})\rangle. \tag{8.25}$$

In the last step, the trace has been taken in field eigenstates written in polar form, and the operator $\mathrm{e}^{\mathrm{i}Q\alpha}$ has been used to transform the final state as per (8.23). Finally, the transition amplitude in (8.25) can be written as a field theoretical path integral. Hence

$$G_q(T) = \int\frac{\mathrm{d}\alpha}{2\pi}\mathrm{e}^{-\mathrm{i}q\alpha}\int\bar{\mathscr{D}}\,[\rho(\boldsymbol{x}, t)]\rho(\boldsymbol{x}, t)\int\bar{\mathscr{D}}[\theta(\boldsymbol{x}, t)]\exp\{\mathrm{i}S[\rho, \theta]\} \tag{8.26}$$

where, (i) all fields go from $(\rho_0(\boldsymbol{x}), \theta_0(\boldsymbol{x}))$ at $t = 0$ to $(\rho_0(\boldsymbol{x}), \theta_0(\boldsymbol{x}) + \alpha)$ at time T, (ii) the integration over $\rho_0(x)$, $\theta_0(x)$ is included, and (iii) $S[\rho, \theta]$ is the action in polar fields given by

$$S[\rho, \theta] = \int_0^T \mathrm{d}t \int \mathrm{d}\boldsymbol{x} \left[\tfrac{1}{2}(\partial_\mu \rho)^2 + \tfrac{1}{2}\rho^2(\partial_\mu \theta)^2 - U(\rho)\right]. \tag{8.27}$$

Up to this stage, the discussion has been a very straightforward generalisation of what was done in the miniature problem in the last section, with Q here playing the same role as L there. There is however an important difference. In section 8.1, $\theta(t)$ was a single dynamical variable, which was changed to $\theta(t) + \alpha$ by the action of the one-parameter rotation in the plane. Now, $\theta(\boldsymbol{x}, t)$ is a field, i.e. an infinite number of variables, one at each $\boldsymbol{x}$. Meanwhile the symmetry group is still a one-parameter U(1) group. Therefore, one is motivated to find some appropriate set of coordinates to describe $\theta(\boldsymbol{x}, t)$ such that only one of them is affected by the U(1) transformation. This is achieved by using Fourier coefficients. Let

$$\theta(\boldsymbol{x}, t) = \sum_{\boldsymbol{k}_n} b_{\boldsymbol{k}_n}(t) \exp(\mathrm{i}\boldsymbol{k}_n \cdot \boldsymbol{x}) \tag{8.28}$$

where, for simplicity, we put the system in a large box, to keep the Fourier wavenumbers $\boldsymbol{k}_n$ discrete. Under the action of the transformation (8.23), $b_0(t) \to b_0(t) + \alpha$, while all the other Fourier coefficients with non-zero wavenumbers are unaltered. In other words, if we write (8.28) in the form

$$\theta(\boldsymbol{x}, t) \equiv b_0(t) + \tilde{\theta}(\boldsymbol{x}, t),$$

where

$$\tilde{\theta}(\bar{x}, t) \equiv \sum_{\boldsymbol{k}_n \neq 0} b_{\boldsymbol{k}_n}(t) \exp(\mathrm{i}\boldsymbol{k}_n \cdot \boldsymbol{x}), \tag{8.29}$$

then under the transformation (8.23),

$$\begin{aligned} &b_0(t) \to b_0(t) + \alpha \\ &\tilde{\theta}(\boldsymbol{x}, t) \to \tilde{\theta}(\boldsymbol{x}, t) \\ &\rho(\boldsymbol{x}, t) \to \rho(\boldsymbol{x}, t). \end{aligned} \tag{8.30}$$

It should be clear by now that it is the variable $b_0(t)$ that plays the same role as $\theta(t)$ did in section 8.1. It, and it alone, is altered by the U(1) symmetry transformation. The remaining (infinitely many) variables $\tilde{\theta}(\boldsymbol{x}, t)$ and $\rho(\boldsymbol{x}, t)$ are unaffected and collectively play the same role as did $r(t)$ in section 8.1. The variable b_0 may be called the 'collective coordinate' associated with the

U(1) symmetry in question. It is a collective coordinate in the sense that it depends on the original field $\phi(\boldsymbol{x}, t)$ at all $\boldsymbol{x}$. In fact b_0 can be explicitly written in terms of $\phi(\boldsymbol{x}, t)$ by inverting the Fourier series (8.28):

$$\begin{aligned} b_0(t) &= (\mathrm{Vol})^{-1} \int \theta(\boldsymbol{x}, t) \mathrm{d}\boldsymbol{x} \\ &= (\mathrm{Vol})^{-1} \int \mathrm{d}\boldsymbol{x}\, \mathrm{Im}(\ln \phi(\boldsymbol{x}, t)). \end{aligned} \tag{8.31}$$

The action $S[\rho, \theta]$ in (8.27), when written in terms of $b_0(t)$ and the 'remaining variables' $\tilde{\theta}(\boldsymbol{x}, t)$ and $\rho(\boldsymbol{x}, t)$, becomes

$$S[b_0, \rho, \tilde{\theta}] = \int_0^T \mathrm{d}t \{\tfrac{1}{2}[\dot{b}_0(t)]^2 A(t) + \dot{b}_0(t) B(t)\} + S'[\rho, \tilde{\theta}], \tag{8.32}$$

where

$$S'[\rho, \tilde{\theta}] \equiv \int_0^T \mathrm{d}t \int \mathrm{d}\boldsymbol{x} \{\tfrac{1}{2}(\partial_\mu \rho)^2 - U(\rho) + \tfrac{1}{2}\rho^2 (\partial_\mu \tilde{\theta})^2] \tag{8.33}$$

$$A(t) \equiv \int \rho^2(\boldsymbol{x}, t) \mathrm{d}\boldsymbol{x} \tag{8.34a}$$

and

$$B(t) \equiv \int \rho^2(\boldsymbol{x}, t) \dot{\tilde{\theta}}(\boldsymbol{x}, t) \mathrm{d}\boldsymbol{x}. \tag{8.34b}$$

Let us rewrite the functional integral (8.26) in terms of b_0, ρ and $\tilde{\theta}$ as

$$\begin{aligned} G_q(T) = &\int \mathscr{D}[\rho(\boldsymbol{x}, t)] \rho(\boldsymbol{x}, t) \int \mathscr{D}[\tilde{\theta}(\boldsymbol{x}, t)] J \exp(\mathrm{i}S'[\rho, \tilde{\theta}]) \\ &\times \int \mathrm{d}b_0(0) \int \frac{\mathrm{d}\alpha}{2\pi} \int \mathscr{D}[b_0(t)] \exp\left[\mathrm{i}\left(\int [\tfrac{1}{2} A(t) \dot{b}_0^2(t) \right.\right. \\ &\left.\left. + B(t) \dot{b}_0(t)] \,\mathrm{d}t \right) - \mathrm{i}q\alpha \right], \end{aligned} \tag{8.35}$$

where J is the Jacobian introduced by the change of variables. All paths $b_0(t)$ will run from $b_0(0)$ to $b_0(T) = b_0(0) + \alpha$. Hence $\alpha = \int_0^T \dot{b}_0 \,\mathrm{d}t$. We notice that the integral over the paths $b_0(t)$, including the end points, is identical in form to the angular integration in (8.10). The integrand is again a quadratic in $\dot{b}_0$, and integration can be done exactly, just as we did in going from (8.10) to (8.11). The result is clearly

$$\begin{aligned} G_q(T) = &\int \mathscr{D}[\rho] \rho \int \mathscr{D}[\tilde{\theta}] \frac{J}{\sqrt{A(t)}} \exp\left(\mathrm{i}S'[\rho, \tilde{\theta}] \right. \\ &\left. - \mathrm{i} \int \frac{[q - B(t)]^2}{2A(t)} \mathrm{d}t \right). \end{aligned} \tag{8.36}$$

Thus, we are left with a functional integral over the 'remaining variables' $\rho(\boldsymbol{x}, t)$ and $\tilde{\theta}(\boldsymbol{x}, t)$, with an effective action in the exponent

$$S_{\text{eff}}[\rho, \tilde{\theta}] = S'[\rho, \tilde{\theta}] - \int \mathrm{d}t \frac{[q - B(t)]^2}{2A(t)} \equiv \int \mathrm{d}t \, L_{\text{eff}}[\rho, \tilde{\theta}], \tag{8.37}$$

where

$$L_{\text{eff}}[\rho, \tilde{\theta}] \equiv \int \mathrm{d}\boldsymbol{x}[\tfrac{1}{2}(\partial_\mu \rho)^2 - U(\rho) + \tfrac{1}{2}\rho^2(\partial_\mu \tilde{\theta})^2] - \frac{(q - \int \rho^2 \dot{\tilde{\theta}} \, \mathrm{d}\boldsymbol{x})^2}{2 \int \rho^2 \, \mathrm{d}\boldsymbol{x}}. \tag{8.38}$$

The semiclassical method may now be applied to the functional integral in (8.36). Since the $U(1)$ transformation leaves $\rho(\boldsymbol{x}, t)$ and $\tilde{\theta}(\boldsymbol{x}, t)$ unaffected, the zero-mode problem associated with $U(1)$ symmetry will not appear any longer. This mode has effectively been integrated, through the separation and integration over the variable $b_0(t)$.

In principle, we have completed our task of handling the zero-mode associated with $U(1)$ symmetry, through a suitable change of variables. In practice, evaluating the functional integral (8.36) is a difficult task, even in the stationary-phase approximation, given the complexity of the integrand in (8.36). However, the leading contributions to the energy levels will come, as we have pointed out in chapters 5 and 6, from the classical solutions themselves, i.e. from the extrema of the exponent $S_{\text{eff}}[\rho, \tilde{\theta}]$. It is only the quantum corrections due to fluctuations that will be very difficult to evaluate from the functional integral (8.36). Let us see if we can find some extrema of $S_{\text{eff}}[\rho, \tilde{\theta}]$ by solving

$$(\delta/\delta\rho(\boldsymbol{x}, t)) S_{\text{eff}}[\rho, \tilde{\theta}] = 0, \tag{8.39a}$$

$$(\delta/\delta\tilde{\theta}(\boldsymbol{x}, t)) S_{\text{eff}}[\rho, \tilde{\theta}] = 0. \tag{8.39b}$$

The action $S_{\text{eff}}[\rho, \tilde{\theta}]$ given in (8.37)–(8.38) is complicated enough that even solving the classical equations (8.39a, b) appears difficult. They are clearly integro-differential, non-linear equations for ρ and $\tilde{\theta}$. Nevertheless, it may be possible to find one non-trivial solution, which furthermore has some nice physical interpretation, through the following procedure.

It can be checked that $\tilde{\theta}(\boldsymbol{x}, t) = 0$ is a solution of (8.39b) provided ρ is time-independent. In that case (8.39a) reduces to

$$-\nabla^2 \rho(\boldsymbol{x}) + \frac{\mathrm{d}U}{\mathrm{d}\rho} - \frac{q^2}{[\int \rho^2 \mathrm{d}\boldsymbol{x}]^2} \rho(\boldsymbol{x}) = 0. \tag{8.40}$$

This equation is still in integro-differential form. However, its solution may be related to certain solutions of the parent field equations (8.18) or (8.21). Consider a periodic solution of the field equations of the form

$$\phi(\boldsymbol{x}, t) = \rho_\nu(\boldsymbol{x})e^{-i\nu t}. \tag{8.41}$$

From (8.21a) we see that $\rho_\nu(\boldsymbol{x})$ must obey

$$-\nabla^2 \rho_\nu(\boldsymbol{x}) + \left(\frac{dU(\rho)}{d\rho}\right)_{\rho_\nu} - \nu^2 \rho_\nu(\boldsymbol{x}) = 0. \tag{8.42}$$

Then, it is clear that a solution $\rho_\nu(\boldsymbol{x})$ of (8.42) will also solve (8.40) provided the charge q in (8.40) is related to the frequency ν by

$$q = \nu \int \rho_\nu^2(\boldsymbol{x})d\boldsymbol{x}. \tag{8.43}$$

In other words, by solving (8.42) for a range of ν, we would have solved (8.40) for a range of q. More importantly, we see that a time-independent solution (with $\tilde{\theta}(\boldsymbol{x}, t) = 0$) of the effective problem (8.39) is related to a rotating solution (8.41) of the parent field equations (8.21). The rotation is taking place in the internal space (the phase of ϕ) in which the U(1) symmetry in question operates. Moreover, the relation (8.43) tells us that q, which was introduced in (8.25) as the charge of the quantum levels, must equal the charge of the classical solution (8.41). In short, our semiclassical method associates energy levels in a given charge sector with classical solutions carrying the same charge. This, in intuitive terms, is very satisfying. It is also evident that our entire derivation has closely paralleled the simple example discussed in the last section. Equations (8.40)–(8.43) correspond to the special case $\tilde{\theta}(\boldsymbol{x}, t) = 0$. Of course, eq. (8.39) will in general also have solutions where $\tilde{\theta}(\boldsymbol{x}, t)$ is non-trivial. These will be very difficult to evaluate, but, for a given q, solutions with $\tilde{\theta}(\boldsymbol{x}, t) = 0$ will carry the lowest energy (see Lee 1976).

The preceding discussion assumes that localised finite-energy solutions of (8.42) exist and can be obtained. Equation (8.42) is however nowhere nearly as complicated as (8.39) or (8.40). It is just a static non-linear wave equation for the real field $\rho(\boldsymbol{x})$, with a potential $U_{\text{eff}}(\rho) = U(\rho) - \frac{1}{2}\nu^2\rho^2$. Recall that we have studied such equations to some extent in chapters 2 and 3. Without loss of generality, we can take $U(|\phi|)$ to vanish at its absolute minima. In order that the charge (8.43) of the solution be finite, the field $\rho(\boldsymbol{x}) = |\phi(\boldsymbol{x})|$ must vanish at the spatial boundary. At the same time, in order that the energy of the solution

$$E = \int [\tfrac{1}{2}|\dot{\phi}|^2 + \tfrac{1}{2}|\nabla\phi|^2 + U(|\phi|)]d\boldsymbol{x}$$

be finite, $U(|\phi|)$ must also vanish (i.e. attain its absolute minimum value) at the spatial boundary. Hence, $U(|\phi|)$ must have its absolute minimum at $|\phi| = 0$. That is, the charge symmetry under question must *not* be spontaneously broken. This is consistent with the requirement that the charge operator be finite. All this also implies that the soliton solution to (8.42) that we are seeking be *non-topological*, in the language of chapters 2 and 3. The existence, stability and other features of charged non-topological solitons of such systems have been discussed with great thoroughness by Lee and co-workers (Lee 1976, Friedberg et al. 1976, a, b and c, Friedberg and Lee 1977). As an example, consider a (1 + 1)-dimensional example with

$$U(|\phi|) = a|\phi|^2 - b|\phi|^4 + c|\phi|^6 \tag{8.44}$$

where a, b and c are real positive parameters, satisfying $b^2 < 4ac$. In (1 + 1) dimensions, this theory is renormalisable. At the same time, if $b^2 < 4ac$, $U(|\phi|)$ will have a non-degenerate absolute minimum at $|\phi| = 0$. Equation (8.42) reduces to

$$d^2\rho/dx^2 = (d/d\rho)U_{eff}(\rho), \tag{8.45}$$

with

$$U_{eff} = U(\rho) - \tfrac{1}{2}\rho^2 v^2 = (a - \tfrac{1}{2}v^2)\rho^2 - b\rho^4 + c\rho^6. \tag{8.46}$$

For v larger than some v_{min} but less than $\sqrt{2a}$, $U_{eff}(\rho)$ will develop a lower minimum at some non-zero ρ (see fig. 17). In that case, a non-topological soliton solution to (8.45) will exist. To see this, we invoke the mechanical analogy used in chapter 2. Equation (8.45) is Newton's equation for a particle in a potential $[-U_{eff}]$, with x playing the role of time. It is clear from fig. 17 that the analogue particle can start at $\rho = 0$ with arbitrarily small velocity at large negative time ($x = -\infty$), go to $\rho = \rho_0$ at $x = 0$, and return to $\rho = 0$ at $x = +\infty$. The explicit solution $\rho(x)$ can be obtained by integrating (8.45) as discussed in chapter 2. Notice that this is a type C solution, in the language of that chapter.

The reader may be concerned by the remark in chapter 2 that static non-topological solitons cannot exist in (1 + 1) dimensions for a single real scalar field. Yet (8.45) seems to have yielded one. The reason is that (8.45) is not really the static equation of a *bona fide* real scalar field. The parent field $\phi(x, t)$ is a complex field and the solution we are discussing, in having the form (8.41), is really time dependent. Only the modulus $\rho(x)$ is static and obeys (8.45).

Since charged scalar solutions are time dependent, the virial theorem of section 3.2 does not apply. Non-topological soliton solutions of (8.42) can

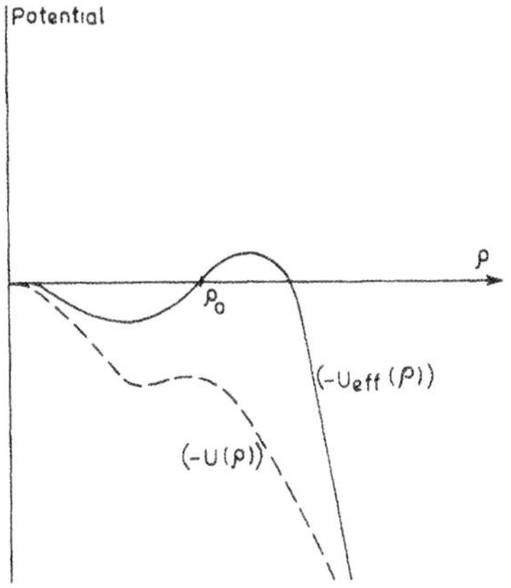

Fig. 17. A schematic plot of $-U(\rho)$ and $-U_{\text{eff}}(\rho)$, for v some $v_{\min}$. Here $U(\rho)$, given in eq. (8.44), is the original potential function in the Lagrangian of the field theory, while $U_{\text{eff}}(\rho) = U(\rho) - \frac{1}{2} v^2 \rho^2$ is the effective potential in the differential equation (8.45). Some typical values of the parameters a, b and c have been used. The analogue particle associated with (8.45) is described by a coordinate $\rho(x)$ and feels a potential $[-U_{\text{eff}}(\rho)]$

exist in more than two dimensions as well, for appropriate choices of $U(\rho)$, although it is difficult to obtain them in exact analytical form. But a detailed discussion of the existence of such solutions and their approximate shapes is given in the papers of Lee and co-workers cited above. These gentlemen also study non-topological solitons of gauge field systems in $(3+1)$ dimensions.

Returning to our semiclassical method, two technical remarks may be made:

(i) We pointed out, in the context of the simple example in section 8.1, that if that system were first quantised in cartesian coordinates and then transformed to polar coordinates, the factor $l^2/2r^2$ in the effective potential is altered to $(l^2 - \frac{1}{4})/2r^2$. A similar effect, in a more complicated way, would also arise in the present case, if we had written the functional integral for $G_q(T)$ in terms of the 'cartesian' fields $\text{Re}(\phi)$ and $\text{Im}(\phi)$, and then changed over to ρ and θ (see Rajaraman and Weinberg 1975 for details). These differences will not affect energy levels to leading order, but only the quantum corrections.

(ii) If the system had included some more fields in addition to $\phi(x, t)$, the action and the field equations would be more complicated. But the basic procedure for handling the U(1) zero-mode would remain the same. The added fields would be unaffected by the symmetry transformation (8.17), and would join $\rho(x, t)$ and $\tilde{\theta}(x, t)$ as part of the 'remaining' variables. The collective coordinate b_0 can still be defined by (8.31) and integrated out exactly. In some of the papers cited above, an extra real scalar field $\sigma(x, t)$ is included in addition to the complex field $\phi(x, t)$. We have suppressed such complications in our discussion.

In conclusion, the method described above for handling the U(1) zero-mode is nice in the sense that it offers a direct generalisation of the simple physical ideas introduced in section 8.1. The collective coordinate $b_0(t)$ could be explicitly defined as a functional of $\phi(x, t)$, and exactly integrated over in the functional integral. The residual problem carried an added 'centrifugal' term, associated with *internal* rotations (the last term in (8.38)). The classical solutions involved were simple rotating solutions, carrying the same charge q as the quantum states. This method can also be generalised to some cases of non-abelian global internal symmetry (see Rajaraman and Weinberg 1975).

The great disadvantage of this method, apart from the cumbersomeness of the functional integral (8.36), is that it cannot be applied to all types of zero-modes, in particular to translation modes. Recall that even for the model considered here, additional zero-modes will arise from translation symmetry, which we have not worried about. The difficulty is that one does not know (or at least, we do not know) how to *explicitly* define a translational collective coordinate, i.e. define $X = X[\phi(x)]$ such that by varying X one can differentiate a given configuration $\phi(x)$ from its translated partners, without altering the 'shape' of the field. Collective coordinates for translational symmetries have to be introduced implicitly. This is described in the following sections.

8.3. Implicit collective coordinates (classical)

Christ and Lee (1975) developed a comprehensive method for quantising solitons using a canonical Hamiltonian procedure. Their work complements and consolidates the results we had described in earlier chapters using a functional-integral approach. It also incorporates collective coordinates – as many as are necessary to replace zero-modes resulting from all the continuous symmetries of the model in question. The collective coordinates in this method are not explicitly written as functionals of the parent field variables. They are introduced implicitly, but still lead to a systematic and elegant semiclassical expansion.

To keep things simple, we will present the Christ-Lee method as applied to models with only one zero-mode, i.e. real scalar field theories in $(1+1)$ dimensions, where the only continuous symmetry leading to a zero-mode is due to spatial translations. The method outlined below for treating the translation mode was also independently proposed by Tomboulis (1975) in a slightly different language. Consider the standard Lagrangian for a two-dimensional real scalar field written in the form (see eq. (6.109)–(6.112))

$$L = \frac{1}{g^2}\int [\tfrac{1}{2}(\dot{\phi}^2) - \tfrac{1}{2}(\phi')^2 - U(\phi)]\,\mathrm{d}x \tag{8.47}$$

in terms of a suitably rescaled field $\phi(x, t)$. Dots and primes refer as usual to derivatives with respect to t and x respectively. The advantage of writing the Lagrangian in the form (8.47) is that in subsequent calculations, powers of the coupling parameter g can easily be kept track of, since g^2 occurs solely as an overall factor in (8.47). Also, as pointed out at the end of chapter 6, powers of $\hbar$ and g^2 go hand in hand. So, we need not keep track of the former, and set $\hbar = 1$.

First, let us recall some salient features of the system (8.47) based on the discussion of earlier chapters. As long as $U(\phi)$ has degenerate absolute minima, a classical static soliton $\phi_{\mathrm{cl}}(x) \equiv \sigma(x - X)$ will exist, satisfying

$$\sigma'' + \mathrm{d}U(\sigma)/\mathrm{d}\sigma = 0 \tag{8.48}$$

and

$$\tfrac{1}{2}(\sigma')^2 = U(\sigma). \tag{8.49}$$

Its classical energy (the classical soliton mass) will be

$$\begin{aligned} M_{\mathrm{cl}} \equiv V[\sigma] &\equiv \frac{1}{g^2}\int [\tfrac{1}{2}(\sigma')^2 + U(\sigma)]\,\mathrm{d}x \\ &= \frac{1}{g^2}\int (\sigma')^2\,\mathrm{d}x \equiv A/g^2. \end{aligned} \tag{8.50}$$

By translation invariance $\sigma(x - X)$ will be a classical static solution for all real X, but its energy A/g^2 will be independent of X. The inadequate procedure in chapter 5 for quantising this soliton amounts to the expansion

$$\phi(x, t) = \sigma(x - X) + g\sum_{0}^{\infty} c_n(t)\,\eta_n(x - X), \tag{8.51}$$

where $\eta_n(x - X)$ are the set of orthonormal eigenfunctions in the equation

$$\left[-\frac{\mathrm{d}^2}{\mathrm{d}x^2} + \left(\frac{\mathrm{d}^2 U}{\mathrm{d}\phi^2}\right)_{\sigma(x-X)}\right]\eta_n(x - X) = \omega_n^2\,\eta_n(x - X). \tag{8.52}$$

Remember that we are explicitly depicting all g^2 dependence. The quantities $\sigma(x-X)$, $\eta_n(x-X)$, ω_n^2, A and $c_n(t)$ are all independent of g, i.e. of order g^0. Thus, when the Lagrangian (8.47) is written in terms of the coordinates $c_n(t)$ instead of $\phi(x,t)$, it becomes

$$L = -\frac{A}{g^2} + \frac{1}{2}\sum_0^\infty [(\dot{c}_n)^2 - \omega_n^2 c_n^2] + O(g). \tag{8.53}$$

To order g^0, this is just a collection of harmonic oscillators, and the energy levels in the soliton sector will be

$$E_{\{N_n\}} = \frac{A}{g^2} + \sum_{n=0}^\infty (N_n + \tfrac{1}{2})\omega_n + O(g). \tag{8.54}$$

This is the central result of chapter 5. The inadequacy of this procedure is that the normalised $n = 0$ mode, given by

$$\eta_0(x-X) = (1/\sqrt{A})\sigma'(x-X), \tag{8.55}$$

carries zero frequency, leading to potential divergences in higher-order terms in equations like (8.54). So far, all we have done is to summarise, for completeness, what is already present in earlier chapters. Our task now is to tackle this zero-mode problem within the canonical Hamiltonian formalism. This is done by elevating the variable X, which indicated the location of the soliton, into a dynamical coordinate replacing the zero-mode coordinate c_0. That is, instead of (8.51), we expand $\phi(x,t)$ in the form

$$\phi(x,t) = \sigma(x - X(t)) + g\sum_{n=1}^\infty q_n(t)\eta_n(x - X(t)). \tag{8.56}$$

The functions σ and η_n are the *same* as in (8.51), but notice that the $n = 0$ term has been omitted in (8.56). $X(t)$ is called the collective coordinate associated with the translational symmetry. The strategy is to use $X(t)$ and $q_n(t)$ $(n > 0)$ as the coordinates of the system, instead of the $c_n(t)$ in (8.51). Clearly,

$$\begin{aligned}\dot{\phi}(x,t) &\equiv (\partial/\partial t)\phi(x,t) \\ &= -\left(\sigma' + g\sum_1^\infty q_n\eta_n'\right)\dot{X} + g\sum_1^\infty \dot{q}_n\eta_n.\end{aligned} \tag{8.57}$$

Let us use the compact notation u_n, $n = 0, 1, \ldots, \infty$, to describe the new coordinates. That is, let

$$u_0(t) = X(t) \quad \text{and} \quad u_n(t) = q_n(t) \quad \text{for } n > 0. \tag{8.58}$$

Then the kinetic energy term in the Lagrangian (8.47) takes the form

$$K \equiv \frac{1}{g^2}\int \mathrm{d}x\,(\tfrac{1}{2}\dot{\phi}^2)$$

$$= \tfrac{1}{2}\sum_{i,j=0}^{\infty} \dot{u}_i D_{ij} \dot{u}_j, \tag{8.59}$$

where D_{ij} is a symmetric matrix whose elements (using (8.57)) are

$$D_{00} = \frac{1}{g^2}\int \mathrm{d}x \left(\sigma' + g\sum_1^{\infty} q_m \eta'_m\right)^2 \tag{8.60a}$$

$$D_{0n}\,(n \neq 0) = -\frac{1}{g^2}\int \mathrm{d}x \left(\sigma' + g\sum_1^{\infty} q_m \eta'_m\right)\left(g\eta_n\right)$$

$$= -\int \mathrm{d}x \left(\sum_1^{\infty} q_m \eta'_m\right)\eta_n \tag{8.60b}$$

$$D_{nm}\,(n, m \neq 0) = \delta_{nm}. \tag{8.60c}$$

We have used the orthonormality of the η_n, and also the orthogonality of $\eta_n\,(n \neq 0)$ to η_0 and therefore to σ' by virtue of (8.55). The functions σ', η'_n, and η_n are all functions of $(x - X(t))$. But since the elements D_{ij} in (8.60) are translationally invariant integrals over x, the D_{ij} will *not* depend on $X(t)$. They are functions only of $\{q_n\}$. Similarly, the potential energy term in the Lagrangian is also translationally invariant, and will not depend on $X(t)$. We can write it as

$$V[\phi] = \int \mathrm{d}x\,[\tfrac{1}{2}(\phi')^2 + U(\phi)] = V(\{q_n\})$$

$$= \frac{A}{g^2} + \tfrac{1}{2}\sum_{n=1}^{\infty} q_n^2 \omega_n^2 + \mathrm{O}(g), \tag{8.61}$$

where (8.56), (8.48) and (8.52) have been used. The $\mathrm{O}(1/g)$ term will be absent because $\sigma(x - X(t))$ satisfies (8.48). The full Lagrangian, in terms of the variables $u_i\,(i = 0, 1, \ldots, \infty)$, is therefore

$$L = \tfrac{1}{2}\sum_{i,j=0}^{\infty} \dot{u}_i D_{ij}(\{q_n\})\dot{u}_j - V(\{q_n\}). \tag{8.62}$$

To get the Hamiltonian, let us find the canonical momenta $\pi_i\,(i = 0, 1, \ldots, \infty)$, conjugate to the coordinates u_i. These are

$$\pi_i = \partial L/\partial \dot{u}_i = D_{ij}\dot{u}_j. \tag{8.63}$$

The classical Hamiltonian is therefore

$$H = \sum_{0}^{\infty} \pi_i \dot{u}_i - L = \sum_{i,j=0}^{\infty} \tfrac{1}{2}\pi_i (D^{-1})_{ij} \pi_j + V(\{q_n\}). \tag{8.64}$$

It is a matter of straightforward algebra to check, given the matrix elements D_{ij} in (8.60), that

$$(D^{-1})_{00} = 1/D$$

$$(D^{-1})_{0n} = -D_{0n}/D; \qquad \mathrm{n} \neq 0$$

and

$$(D^{-1})_{nm} = \delta_{nm} + D_{0n}D_{0m}/D; \qquad \mathrm{n, m} \neq 0, \tag{8.65}$$

where D is the determinant of the matrix D_{ij}, given by

$$D = \mathrm{Det}\ D_{ij} = D_{00} - \sum_{1}^{\infty} D_{0n}^2. \tag{8.66}$$

Inserting (8.60), we get

$$\begin{aligned} D &= \frac{1}{g^2}\int \mathrm{d}x\left(\sigma' + g\sum_{1}^{\infty} q_m \eta_m'\right)^2 - \sum_{n=1}^{\infty}\left[\int \eta_n \left(\sum_{1}^{\infty} q_m \eta_m'\right)\mathrm{d}x\right]^2 \\ &= \frac{A}{g^2} + \frac{2}{g}\int \sigma'\left(\sum_{1}^{\infty} q_m \eta_m'\right)\mathrm{d}x + \int \mathrm{d}x\left(\sum_{1}^{\infty} q_m \eta_m'\right)^2 \\ &\quad - \sum_{n=1}^{\infty}\left[\int \mathrm{d}x\, \eta_n \left(\sum_{1}^{\infty} q_m \eta_m'\right)\right]^2. \end{aligned} \tag{8.67}$$

If we expand the function $(\sum_1^{\infty} q_m \eta_m'(x - X))$ in terms of the complete basis set $\eta_n(x - X)$, including $\eta_0(x - X) = A^{-1/2}\sigma'(x - X)$, then (8.67) reduces to

$$D = \frac{A}{g^2} + \frac{2\sqrt{A}}{g}\alpha + \alpha^2 = \left(\frac{\sqrt{A}}{g} + \alpha\right)^2, \tag{8.68}$$

where

$$\alpha \equiv \int \eta_0 (x - X)\left(\sum_{1}^{\infty} q_m \eta_m'(x - X)\right)\mathrm{d}x. \tag{8.69}$$

The matrix elements $(D^{-1})_{ij}$, using the determinant D from (8.68), may be inserted into (8.64) to get the classical Hamiltonian. The following features of this procedure may be noted.

(i) The zero-mode variable is absent for the simple reason that it was excluded from the start, by fiat, in the decomposition (8.56). But we must make sure that the coordinate $X(t)$, which took the place of the zero-mode, does not lead to similar divergence problems in higher orders.

(ii) As pointed out earlier, the matrix elements D_{ij} as well as the potential energy V are translationally invariant integrals, and therefore independent of $X(t)$. Thus, $u_0(t) = X(t)$ never appears in the Hamiltonian (8.64). Only its conjugate momentum π_0 is present. This is an exact result, not restricted to any given order in g^2. Consequently, the conjugate momentum π_0 must be conserved in time. In fact, π_0 is equal to the conserved total field momentum P. We have

$$
\begin{aligned}
P &= -\frac{1}{g^2}\int \frac{\partial\phi}{\partial t}\frac{\partial\phi}{\partial x}\,\mathrm{d}x \\
&= \frac{1}{g^2}\int\left[\left(\sigma' + g\sum_1^\infty q_n\eta_n'\right)\dot{X} - g\sum_1^\infty \dot{q}_n\eta_n\right]\left(\sigma_0' + g\sum_1^\infty q_m\eta_m'\right)\mathrm{d}x \\
&= D_{00}\dot{X} + \sum_{n=1}^\infty D_{0n}\dot{q}_n = \pi_0,
\end{aligned}
\tag{8.70}
$$

using (8.60) and (8.63). Since the collective coordinate $X(t)$ denotes the spatial location of the field, it is most satisfying that the momentum conjugate to it turns out to be the total momentum of the system.

(iii) Using $\pi_0 = P$ and (8.65), the classical Hamiltonian (8.64) can be written as

$$
H = \frac{P^2}{2D} - \frac{P}{D}\sum_{n=1}^\infty D_{0n}\pi_n + \frac{1}{2}\sum_{n,m=1}^\infty\left(\delta_{mn} + \frac{D_{0n}D_{0m}}{D}\right)\pi_n\pi_m + V(\{q_n\}).
\tag{8.71}
$$

(iv) The collective coordinate $X(t)$ was defined implicitly through (8.56), and not explicitly as a functional of $\phi(x,t)$. This has, however, not come in the way of writing the Lagrangian in terms of $X(t)$ and the 'remaining' coordinates $q_n(t), n > 0$.

Our discussion has so far remained at a classical level. All we have done is change variables from $\phi(x,t)$ to $\{X(t), q_n(t)\}$, thereby getting rid of the zero-mode. In the next section we shall quantise the system in terms of these variables to obtain the energy levels in the soliton-sector.

8.4. Soliton quantisation by canonical operator methods

We proceed to quantise the one-dimensional scalar field theory of the preceding section in terms of the coordinates $u_n = (X; q_n, n > 0)$ and their canonical momenta π_n (P; $\pi_n, n > 0$). Quantisation is achieved by converting the u_n and π_n into operators, obeying the canonical equal-time commutation rules (with $\hbar = 1$)

$$[u_n(t), \pi_m(t)] = \mathrm{i}\delta_{nm}$$
$$[\pi_n, \pi_m] = [u_n, u_m] = 0. \tag{8.72}$$

Recall that in terms of the original field variables $\phi(x,t)$ in (8.47), the canonical commutator would be the familiar object

$$[\phi(x,t), \pi(y,t)] = [\phi(x,t), (1/g^2)\dot{\phi}(y,t)] = \mathrm{i}\delta(x-y). \tag{8.73}$$

It may be verified that the two sets of commutators (8.72) and (8.73) are consistent. Both $\phi(x,t)$ and $\dot{\phi}(y,t)$ are available in terms of $u_n(t)$ and $\pi_n(t)$ upon using eq. (8.56), (8.57), and the inverse of (8.63). Insertion of the commutator (8.72) leads, after some simple algebra, to (8.73). This consistency is important, since the variables $\phi(x,t)$ will be used as usual in the vacuum sector, while we will be using the variables u_n in the soliton sector. The classical Hamiltonian in terms of the u_n and π_n is given in (8.64), or equivalently in (8.71). In quantum theory, the ordering ambiguity in this Hamiltonian has to be resolved. Remember that in the kinetic terms in (8.64), the factors $(D^{-1})_{ij}$ depend on q_n (although not on X), and will not commute with the π_n with $n > 0$. These factors need to be ordered in some specific sequence. Recall our brief introduction to the ordering problem at the end of section 8.1. We shall resolve it along the lines suggested there. We shall choose the quantum Hamiltonian to have the 'cartesian' form (in the sense used in section 8.1) in terms of the original field variables $\phi(x,t)$. That is, we specify the quantum Hamiltonian in the Schrödinger representation to be

$$H = \frac{g^2}{2}\int(\pi(x))^2\,\mathrm{d}x + V[\phi(x)] \tag{8.74}$$

where $\pi(x)$ is the momentum operator conjugate to $\phi(x)$ and, consistent with the commutation rule (8.73), is defined in the Schrödinger representation as the functional derivative

$$\pi(x) = -\mathrm{i}\delta/\delta[\phi(x)]. \tag{8.75}$$

This specifies the quantum Hamiltonian. Remember also that this is the Hamiltonian used in standard calculations in the vacuum sector. Change of variables from $\phi(x)$ to the u_n can now be executed unambiguously, by converting the (functional) differential operator in (8.75) into an operator involving the u_n, by standard rules of calculus. Recall that if there had been a finite number of variables $\alpha_i, i = 1, \ldots, N$, then the Laplacian $\sum_1^N \partial^2/\partial\alpha_i^2$, when written in terms of a different set of coordinates β_i, has the form

$$\sum_{i=1}^{N} \frac{\partial^2}{\partial\alpha_i^2} = \sum_{i,j=1}^{N} \frac{1}{\sqrt{B}} \frac{\partial}{\partial\beta_i} (B^{-1})_{ij} \sqrt{B} \frac{\partial}{\partial\beta_j}, \tag{8.76}$$

where

$$B_{ij} \equiv \sum_{k=1}^{N} \frac{\partial\alpha_k}{\partial\beta_i} \frac{\partial\alpha_k}{\partial\beta_j} \tag{8.77a}$$

and

$$B = \text{Det } B_{ij}. \tag{8.77b}$$

For the present case, we have to generalise this relation to an infinite number of variables, with $\phi(x)$ for all x taking the place of α_i, and the u_n taking the place of the β_i. The matrix B_{ij} in the present case is

$$B_{ij} = \int \mathrm{d}x \frac{\partial\phi(x)}{\partial u_i} \frac{\partial\phi(x)}{\partial u_j}. \tag{8.78}$$

These derivatives are immediately obtained from (8.56), (8.58) and yield the simple result

$$B_{ij} = D_{ij} \tag{8.79}$$

where D_{ij} is the matrix defined in (8.60). Thus the Hamiltonian (8.74) reduces to

$$H = -\tfrac{1}{2} \sum_{n,m=0}^{\infty} \frac{1}{\sqrt{D}} \frac{\partial}{\partial u_n} (D^{-1})_{nm} \sqrt{D} \frac{\partial}{\partial u_m} + V(\{u_n\}). \tag{8.80}$$

We set, consistent with the commutator (8.72),

$$-\mathrm{i}\partial/\partial u_n = \pi_n. \tag{8.81}$$

Then, separating the π_n and u_n into $(P;\ \pi_n, n > 0)$ and $(X;\ q_n, n > 0)$ respectively and using (8.65) for $(D^{-1})_{nm}$, we have

$$H = \frac{1}{2\sqrt{D}} \left[\frac{P^2}{\sqrt{D}} - \sum_{n=1}^{\infty} P\left(\frac{D_{0n}}{\sqrt{D}} \pi_n + \pi_n \frac{D_{0n}}{\sqrt{D}} \right) + \sum_{n,m=1}^{\infty} \pi_n \left(\delta_{nm}\sqrt{D} + \frac{D_{0n}D_{0m}}{\sqrt{D}} \right) \pi_m \right] + V[\{q_n\}]. \tag{8.82}$$

Remember that P commutes with all the other terms since they do not depend on X. Equation (8.82) gives the ordered quantum Hamiltonian in terms of the desired variables. While ordering has been achieved through a specific choice, no approximation, semiclassical or otherwise, has been made so far. Clearly, in the classical limit, where ordering does not matter, (8.82) reduces to the classical Hamiltonian (8.71). Further, recall that P, the momentum conjugate to the collective coordinate X, is also the conserved total momentum, and commutes with H. We can therefore choose simultaneous eigenstates of H and P. Let us begin with the eigenvalue $P = 0$, i.e. consider the energy levels in the zero-momentum frame. Then the Hamiltonian reduces to

$$H_{(P=0)} = \frac{1}{2\sqrt{D}} \sum_{n,m=1}^{\infty} \pi_n \left(\delta_{mn}\sqrt{D} + \frac{D_{0n}D_{0m}}{\sqrt{D}} \right) \pi_m + V[\{q_n\}]. \quad (8.83)$$

This is still exact in the $P = 0$ frame. Now begins the semiclassical approximation, which, as explained at the end of chapter 6, is essentially an expansion in powers of g. Consider the g dependence of the different terms in (8.83). The leading term in D (eq. (8.68)) is A/g^2, where A, given in (8.50), is a c-number and will commute with the π_n. The $D_{0n}\,(n \neq 0)$ are independent of g. The potential $V(\{q_n\})$ has been expanded in (8.61). Using all this we can identify the leading $[\mathrm{O}(1/g^2)$ and $\mathrm{O}(g^0)]$ terms of H. Clearly

$$H_{(P=0)} = H_0 + H_1 \quad (8.84)$$

where

$$H_0 = \frac{A}{g^2} + \frac{1}{2}\sum_{n=1}^{\infty} (\pi_n^2 + q_n^2\omega_n^2) \quad (8.85)$$

and H_1, containing all the remaining terms in $H_{(P=0)}$, will be of order g or higher. The semiclassical method can be considered as standard perturbation theory in powers of H_1, using H_0 as the unperturbed Hamiltonian.

To order g^0, H_1 may be neglected in which case the energy levels are eigenvalues of H_0. Clearly then, the levels are,

$$E_{\{N_n\}} = \frac{A}{g^2} + \sum_{n=1}^{\infty} (N_n + \tfrac{1}{2})\omega_n + \mathrm{O}(g), \quad (8.86)$$

which agrees precisely with the result, to $\mathrm{O}(g^0)$, of the method of chapter 5 (put $\omega_0 = 0$ in eq. (8.54)) where no special effort was made to handle the zero-mode. Such agreement was promised in chapter 5. But whereas the procedure of chapter 5 would have led to problems with zero energy denominators in higher orders because the $\omega_0 = 0$ mode was present in the formalism, there will be no such problems here. Higher-order corrections may be found by applying standard perturbation theory in powers of H_1. In practice this is a little tedious, given the somewhat complicated expression for $H_1(\pi_n, q_n)$ contained in (8.83)–(8.85), but there will be no zero-mode problems because the unperturbed Hamiltonian H_0 (eq. (8.85)) contains only *non-zero* frequencies. The only impact of the coordinate X, which replaced the zero-mode, is to introduce momentum dependence into the Hamiltonian, as depicted in (8.82).

Notice that although this collective coordinate method is different from the methods in sections 8.1 and 8.2, the resulting physics is similar. There, an effective Hamiltonian or Lagrangian could be found (eq. (8.3) in section 8.1, and eq. (8.38) for the problem in section 8.2) in which the collective coordinate (θ in section 8.1 and b_0 in section 8.2) did not appear. The sole impact of the collective coordinate was that the effective Lagrangians contained terms which depended on the corresponding conjugate momentum (l and q respectively) in a quadratic form. In the same way, the Hamiltonian (8.82) does not depend on X but only on the remaining coordinates q_n. The impact of X is to add to the Hamiltonian terms of quadratic form in the conjugate momentum P.

We have discussed the energy eigenvalues in the $P = 0$ frame. To obtain the energy spectrum for some $P \neq 0$, there is in principle no need to go back to the full Hamiltonian (8.82). One can simply invoke Lorentz invariance, i.e. given any energy level $E_{\{N_n\}}$ in the rest frame, there will be a level $\sqrt{P^2 + E^2_{\{N_n\}}}$, for any non-zero momentum P.

What we have described is only a simplified version of the Christ–Lee formalism which, in full, is applicable in an arbitrary number of dimensions and can accomodate as many collective coordinates as are necessary to handle all the continuous symmetries, spatial and internal, of the problem. The treatment in the more general case proceeds along the same lines as our simplified version. Given a system of fields $\phi_1(\boldsymbol{x}, t)$, $\phi_2(\boldsymbol{x}, t), \ldots$ in an arbitrary number of dimensions (collectively denoted by $\boldsymbol{\phi}(\boldsymbol{x}, t)$), subject to a variety of continuous symmetries, a general classical static soliton would have the form $\sigma(\boldsymbol{x}, X_1, X_2, \ldots, X_N)$, where $X_1, \ldots, X_N$ are a maximal number of arbitrary parameters. Each solution belongs to a family of solutions obtained by varying the X_i. This whole family has the same energy since the X_i are such that a change in any of the X_i is the result of the action of some continuous symmetry of the Hamiltonian. To remove the zero-mode problems associated with such variations, one converts all these X_i into dynamical variables. That is, one expands

$$\phi(x, t) = \sigma(\boldsymbol{x}, X_1(t), X_2(t), \ldots, X_N(t)) + g\sum_n q_n(t)\eta_n(\boldsymbol{x}, X_1(t), \ldots, X_N(t)) \tag{8.87}$$

where the $\eta_n(\boldsymbol{x}, X_1 \ldots X_N)$ are a complete orthonormal set subject to the constraint that they are *all* orthogonal to *all* the zero-modes. That is, they satisfy

$$\int \mathrm{d}\boldsymbol{x}\, \eta_n(\boldsymbol{x}, X_1 \ldots X_N)\cdot\eta_m(\boldsymbol{x}, X_1, \ldots, X_N) = \delta_{mn} \tag{8.88}$$

and

$$\int \frac{\mathrm{d}\sigma}{\mathrm{d}X_i}\cdot\eta_n(\boldsymbol{x}, X_1 \ldots X_N)\mathrm{d}\boldsymbol{x} = 0$$

for all $i = 1 \ldots N$ and all n. (8.89)

(In (8.88) and (8.89), inner products between the vectors η and σ in field space are understood, and these vectors are not to be confused with the space coordinate $\boldsymbol{x}$). The dynamical variables are taken to be the X_i and the q_n. Given the expansion (8.87), we have

$$\frac{\partial\phi(\boldsymbol{x}, t)}{\partial t} = \sum_{i=1}^{N}\left(\frac{\partial\sigma}{\partial X_i} + g\sum_n q_n \frac{\partial}{\partial X_i}\eta_n\right)\frac{\mathrm{d}X_i}{\mathrm{d}t} + g\sum_n \frac{\mathrm{d}q_n}{\mathrm{d}t}\eta_n. \tag{8.90}$$

which is the generalisation of (8.57). The Lagrangian is then written in terms of the X_i, q_n, $\dot{X}_i$ and $\dot{q}_n$, and the rest of the steps outlined earlier are repeated with a straightforward generalisation. For more details of the general formalism see Christ and Lee (1975), who also present several applications to quantise static as well as time-dependent soliton solutions.

Finally, the implicit collective coordinate method described in the last two sections in the canonical Hamiltonian formalism can also be presented in the functional integral formalism, where these collective coordinates are introduced through an adaptation of the Faddeev–Popov method (Faddeev and Popov 1967). The physical content is of course the same in both formalisms. In particular, Sakita, Gervais and Jevicki have developed this procedure in a series of papers where not only the quantum energy of the solitons, but the scattering of solitons is also discussed (apart from the papers by the authors already cited, see also Gervais and Jevicki 1976a, 1976b and Jevicki 1976). We shall be content with giving, in Appendix B, a sketch of how an implicit collective coordinate is introduced in the functional integral method, for the simple case of a one-dimensional Euclidean problem.

Other papers dealing with zero-modes and/or collective coordinates include Matveev (1977), Babelon (1977), Chang (1978) and Khrustalev et al. (1980).

CHAPTER 9

SEMICLASSICAL METHODS FOR FERMI FIELDS

9.1. Introduction: Grassmann fields

Up to this point, our discussion of the semiclassical method has been limited to Bose fields. An attempt to extend these methods to cover Fermi fields may appear, at the outset, unwarranted. From the physical point of view, while classical Bose fields are found in nature (e.g. electromagnetic waves, gravitational field, etc.), classical Fermi fields are not, at least not in the same sense. That is to say, a collection of a very large number of bosons present coherently in more or less the same quantum state can be described by classical field theory to a good approximation. The same cannot happen for fermions since the Pauli exclusion principle forbids more than one fermion per state. From the mathematical point of view, classical Bose fields are described by real (or complex) numbers at each space–time point. The calculus of such variables is all too familiar. In particular the functional integral in boson quantum field theory, which is an integral over classical Bose configurations, is just a limit of the familiar c-number multiple integral. In contrast, while a 'classical' limit of Fermi fields can be defined, we shall see that it yields anticommuting fields. The calculus of such objects is not only comparatively unfamiliar, but may be expected to be quite different from that of ordinary commuting variables.

Given these profound differences between Bose and Fermi fields in terms of their classical limit, one may wonder whether functional integrals in general and the semiclassical method in particular, can be extended to systems which include Fermi fields. Interestingly enough, it is possible to do both. Furthermore, many of the results and intervening steps are remarkably similar to those for Bose fields, except for crucial minus signs. This chapter is devoted to an introduction to such techniques for Fermi fields.

The first step is to identify the classical limit of quantum Fermi fields, although the 'classical' Fermi field so obtained will not have the same physical significance, or experimental accessibility, as classical Bose fields. We shall limit ourselves to the spin -$\frac{1}{2}$ Dirac field. Recall that a quantum Bose field, such as the real scalar field $\Phi(\boldsymbol{x},t)$, obeys the equal-time commutation rules:

$$[\Phi(\boldsymbol{x},t),\ (\partial\Phi/\partial t)(\boldsymbol{y},t)] = \mathrm{i}\,\hbar\delta(\boldsymbol{x}-\boldsymbol{y}), \tag{9.1a}$$

$$[\Phi(\boldsymbol{x},t),\ \Phi(\boldsymbol{y},t)] = [(\partial\Phi/\partial t)(\boldsymbol{x},t),(\partial\Phi/\partial t)(\boldsymbol{y},t)] = 0. \tag{9.1b}$$

The non-vanishing of the commutator (9.1a) is what makes the theory quantum mechanical. The classical limit corresponds to $\hbar \to 0$, when the field and its derivatives all commute with one another and behave as ordinary c-numbers at each $(\boldsymbol{x},t)$. In contrast the Dirac field $\Psi_\alpha(\boldsymbol{x}, t)$, whose canonical momentum is $\mathrm{i}\Psi_\alpha^\dagger(x, t)$, obeys the anticommutation relation

$$\{\Psi_\alpha(x, t), \Psi_\beta^\dagger(y, t)\}_+ = \hbar\,\delta_{\alpha\beta}\delta(\boldsymbol{x}-\boldsymbol{y}) \tag{9.2a}$$

with

$$\{\Psi_\alpha(\boldsymbol{x}, t), \Psi_\beta(y, t)\}_+ = \{\Psi_\alpha^\dagger(x, t), \Psi_\beta^\dagger(y, t)\}_+ = 0. \tag{9.2b}$$

The 'classical limit' can again be defined by $\hbar \to 0$, when all the anticommutators vanish. In this limit the Ψ and $\Psi^\dagger$ are no longer non-trivial quantum operators, but they are not ordinary c-number fields either, because they anticommute with one another instead of commuting. They may be called Grassmann fields. As a prelude to developing the fermionic functional integral using such Grassmann fields, let us list some basic properties of Grassmann numbers. What follows will be a capsule summary of some features of Grassmann algebras, limited to our needs. For more details, see for instance Berezin (1966).

Consider a collection of n objects $\{a_1, \ldots, a_n\}$ obeying

$$\{a_i, a_j\}_+ = 0 \qquad \text{for all } i, j. \tag{9.3}$$

Clearly $\{a_i, a_i\}_+ = 0$ implies $a_i^2 = 0$. Therefore $a_i^3 = (a_i^2)a_i = 0$, $a_i^4 = 0$, and so on for all higher powers. Hence, given a single Grassmann number a_i, one can construct only two linearly independent functions of it, namely, $(a_i)^0 \equiv 1$,and a_i. In the same way, given the n independent numbers a_1, $\ldots, a_n$, they generate the following 2^n linearly independent functions:

$$1,$$
$$a_1, a_2, \ldots, a_n$$
$$a_1a_2,\ a_1a_3, \ldots, a_2a_3, \ldots, a_{n-1}a_n$$

$a_1a_2a_3, a_1a_2a_4, \ldots.$

$\ldots.$

$\ldots.$

$a_1a_2a_3 \ldots. a_n.$

Together, these 2^n independent functions span a 2^n-dimensional linear space called a Grassmann algebra G_n. The Grassmann numbers a_i may be called the generators of this algebra. A typical element of the algebra can be written as

$$f(a_1, \ldots, a_n) = f^0 + \sum_i f_i^{(1)} a_i + \sum_{i_1, i_2} f_{i_1 i_2}^{(2)} a_{i_1} a_{i_2} + \ldots$$

$$+ \sum_{i_1, \ldots, i_n} f_{i_1 \ldots i_n}^{(n)} a_{i_1} \ldots a_{i_n} \tag{9.4}$$

where f^0, $f_i^{(1)} f_{i_1 i_2}^{(2)}$ etc. are ordinary (commuting) c-numbers. Notice that because of the antisymmetry of the product $a_1 \ldots a_r$ under the exchange of any two a's, the above decomposition of f is not unique. One can always add a term symmetric in two of the i's to any $f_{i_1 \ldots i_r}^{(r)}$ without altering $f(a_1, \ldots, a_n)$. It is useful to consider the purely 'even' and purely 'odd' subsets of G_n. An element of the even (odd) subset will involve only even (odd) products of the a_i:

$$f^{\text{even}} = f^0 + \sum_{i_1, i_2} f_{i_1 i_2}^{(2)} a_{i_1} a_{i_2} + \ldots \tag{9.5a}$$

$$f^{\text{odd}} = \sum_{i_1} f_{i_1}^{(1)} a_{i_1} + \sum_{i_1, i_2, i_3} f_{i_1 i_2 i_3}^{(3)} a_{i_1} a_{i_2} a_{i_3} + \ldots \tag{9.5b}$$

Clearly, the even subset will commute with all elements of the algebra G_n, while members of the odd subset will anticommute among each other. This is the reflection, in miniature, of the fact that an even number of fermions form a boson, while an odd number of fermions form a fermion.

A calculus, both differential and integral, has been developed for these Grassmann numbers. Before introducing it, we must emphasise that the basic generators of the algebra, $\{a_1, \ldots, a_n\}$, are discrete objects. There is no question of their 'continuously varying their values' in any sense of the term, unlike, say, a real variable x which can take any value between $-\infty$ to $+\infty$. Even though the concept of derivatives and integrals will be developed for Grassmann numbers, with a notation similar to that in ordinary calculus, these derivatives and integrals do not have the familiar interpretation in terms of infinitesimals.

For instance, the derivative is formally defined as

$$\partial a_i/\partial a_j = \delta_{ij}, \qquad \text{the Kronecker symbol.} \tag{9.6}$$

This is just a definition of the symbol on the left-hand side and is not to be viewed as the ratio of two infinitesimal increments. Despite this, one can develop a differential calculus as close as possible in spirit to the rules of ordinary differential calculus. For instance, the derivative of a product is given by

$$\frac{\partial}{\partial a_p}(a_{i_1}a_{i_2}\ldots a_{i_r}) = \delta_{pi_1}(a_{i_2}a_{i_3}\ldots a_{i_r}) - \delta_{pi_2}(a_{i_1}a_{i_3}\ldots a_{i_r}) + \ldots + (-1)^{r-1}\delta_{pi_r}(a_{i_1}\ldots a_{i_{r-1}}). \tag{9.7}$$

That is, we merely commute a_p until it appears on the left and then use (9.6). If a_p does not appear at all in the product $a_{i_1}\ldots a_{i_r}$ then the derivative is zero. It is clear that (9.6) gives the left derivative, which must be distinguished from the right derivative given by

$$(a_{i_1}\ldots a_{i_r})\frac{\overleftarrow{\partial}}{\partial a_p} = \delta_{pi_r}(a_{i_1}\ldots a_{i_{r-1}}) - \delta_{pi_{r-1}}(a_{i_1}\ldots a_{i_{r-2}}a_{i_r}) + \ldots + (-1)^{r-1}\delta_{pi_1}(a_{i_2}\ldots a_{i_r}). \tag{9.8}$$

Proceeding in this manner, a consistent set of rules of differential calculus can be set up. We will however turn our attention to integration, which is what we will need for our fermionic functional integrals.

Let us begin with a single Grassmann number a. We would like to define the object $\int \mathrm{d}a f(a) \equiv I(f)$ where $I(f)$ is a c-number, for any function $f(a)$ belonging to the algebra G_1. Recall our remark that a is not to be viewed as a continuous variable in the sense that some real variable x is. Correspondingly, the integral does not stand for the 'area under the curve $f(a)$', nor is there any meaning to giving upper and lower limits to the integral. Rather, the symbol $\int \mathrm{d}a f(a)$ is to be thought of as just a functional, which associates a c-number $I(f)$ with any element $f(a)$ of the Grassmann algebra G_1. We will however require it to obey two basic properties enjoyed by the usual c-number integral $\int_{-\infty}^{\infty} f(x)$, namely, linearity and translation invariance. That is, we shall require:

$$\text{(i)}\quad \int \mathrm{d}a f(a) = \int \mathrm{d}a f(a+b), \tag{9.9a}$$

where b is any other arbitrary anticommuting number, and

$$\text{(ii)}\quad \int \mathrm{d}a[\alpha f(a) + \beta g(a)] = \alpha\int \mathrm{d}a f(a) + \beta\int \mathrm{d}a\, g(a), \tag{9.9b}$$

where α and β are arbitrary c-numbers. We have already observed that for a single Grassmann variable a, a general function $f(a)$ has the form $(f^{(0)}+f^{(1)}a)$ where $f^{(0)}$ and $f^{(1)}$ are c-numbers. We shall define its integral as

$$I(f) \equiv \int \mathrm{d}a\,(f^{(0)}+f^{(1)}a) \equiv f^{(1)}. \tag{9.10}$$

This satisfies (9.9) and amounts to setting, for the two independent functions 1 and a,

$$\int \mathrm{d}a\,1 = 0 \qquad \text{and} \qquad \int \mathrm{d}a\,a = 1. \tag{9.11}$$

(Equation (9.9a) is satisfied because $\int \mathrm{d}a(a+b) = \int \mathrm{d}a\,a + [\int \mathrm{d}a\,1]b = \int \mathrm{d}a\,a$).

Next, take G_2 generated by a_1 and a_2. We would like to define the double integral $\int \mathrm{d}a_1 \int \mathrm{d}a_2 f(a_1, a_2)$. First we demand that the $\mathrm{d}a_i$ also anti-commute, i.e.,

$$\{\mathrm{d}a_i, \mathrm{d}a_j\}_+ = \{\mathrm{d}a_i, a_j\}_+ = 0. \tag{9.12}$$

In G_2 there are only four independent functions, 1, a_1, a_2 and a_1a_2. Consistent with (9.11), we can define

$$\int \mathrm{d}a_1 \int \mathrm{d}a_2\, 1 = \int \mathrm{d}a_1 [\int \mathrm{d}a_2\, 1] = 0,$$

$$\int \mathrm{d}a_1 \int \mathrm{d}a_2\, a_1 = -\int \mathrm{d}a_2 [\int \mathrm{d}a_1\, a_1] = -\int \mathrm{d}a_2\, 1 = 0,$$

$$\int \mathrm{d}a_1 \int \mathrm{d}a_2\, a_2 = \int \mathrm{d}a_1\, 1 = 0,$$

and

$$\int \mathrm{d}a_1 \int \mathrm{d}a_2\, a_1 a_2 = -\int \mathrm{d}a_1 [\int \mathrm{d}a_2\, a_2] a_1 = -\int \mathrm{d}a_1\, a_1 = -1. \tag{9.13}$$

The integral of an arbitrary function $f(a_1, a_2)$ is then obtained by linearity. These concepts are trivially generalised to G_n. An arbitrary function $f(a_1, \ldots, a_n)$ can be written as an expansion (9.4). In the integral

$$\int \mathrm{d}a_1 \int \mathrm{d}a_2 \ldots \int \mathrm{d}a_n f(a_1, \ldots, a_n),$$

it is clear that only the last term in (9.4) will contribute. All earlier terms will have at least one $\mathrm{d}a_i$ unpaired with the corresponding a_i. Since $\int \mathrm{d}a_i 1 = 0$, the integral of such terms will vanish. The last term is integrated by permuting the product $a_{i_1} \ldots a_{i_n}$ into the form $a_n a_{n-1} \ldots a_1$ with the appropriate signature. Thus

$$\int \mathrm{d}a_1 \int \mathrm{d}a_2 \ldots \int \mathrm{d}a_n (a_{i_1} a_{i_2} \ldots a_{i_n}) = (-1)^P, \tag{9.14}$$

where $(-1)^P$ is the signature of the permutation

$$\begin{pmatrix} i_1, i_2, \ldots, i_n \\ n, n-1, \ldots, 1 \end{pmatrix}.$$

Next, consider the behaviour of the integral under a linear 'change of variables'. Consider G_n generated by $(a_1, \ldots, a_n)$. Let $(b_1, \ldots, b_n)$ be defined by

$$b_i = B_{ij} a_j, \tag{9.15}$$

where B_{ij} is a non-singular *c-number* matrix. Clearly, the set $(b_1 \ldots b_n)$ can be considered an equally good set of generators of the algebra G_n as $(a_1, \ldots, a_n)$. The b_i obey the same anticommutation rule

$$\{b_i, b_j\}_+ = 0$$

and any function $f(a_1, \ldots, a_n)$ which by (9.4) can be written as an nth-order polynomial, can be written equally well as a polynomial in $(b_1, \ldots, b_n)$. Now, the only independent non-zero n-tuple integral in G_n is

$$\int da_1 \ldots \int da_n (a_n a_{n-1} \ldots a_1) = 1. \tag{9.16a}$$

Since the algebra could have been introduced equally well using the generators $(b_1 \ldots, b_n)$, we should also have

$$\int db_1 \ldots \int db_n (b_n b_{n-1} \ldots b_1) = 1. \tag{9.16b}$$

But, using (9.15),

$$b_n b_{n-1} \ldots b_1 = (\text{Det}\, B) a_n a_{n-1} \ldots a_1. \tag{9.17}$$

Hence (9.16a) and (9.16b) will be mutually consistent only if we introduce a Jacobian as follows:

$$db_1 \ldots db_n = (1/\text{Det}\, B) da_1 \ldots da_n. \tag{9.18}$$

Notice that in a similar transformation for c-number integration the determinant would have come in the numerator.

An important integral for our purposes is that of the exponential. Clearly, $e^a = 1 + a$, since higher powers vanish. Hence, for example,

$$\int da_1 \int da_2 \, e^{-\lambda a_1 a_2} = \int da_1 \int da_2 (1 - \lambda a_1 a_2) = \lambda, \tag{9.19}$$

using (9.13). This can be generalised to a multiple integral with a bilinear form in the exponential. Consider the case of $2n$ generators $(a_1, \ldots, a_n, \bar{a}_1, \ldots, \bar{a}_n)$ where, to comply eventually with the usual notation for creation and destruction operators, we have paired the generators through symbols $a_1, \bar{a}_1; a_2, \bar{a}_2, \ldots$ etc. But, although a_1 and $\bar{a}_1$ may share some common symbols they are meant to be entirely independent Grassmann generators. $\bar{a}_1$ is not the conjugate of a_1 in any sense. Consider the integral

$$I = \int (\mathrm{d}\bar{a}_1\, \mathrm{d}a_1\, \mathrm{d}\bar{a}_2\, \mathrm{d}a_2 \ldots \mathrm{d}\bar{a}_n\, \mathrm{d}a_n) \exp\left(- \sum_{i,j=1}^{n} \bar{a}_i A_{ij} a_j \right), \tag{9.20}$$

where A_{ij} is some $n \times n$ c-number matrix, diagonalisable by the transformation

$$(B)_{ki} A_{ij} (B^{-1})_{jl} = \lambda_k \delta_{kl}, \tag{9.21}$$

where λ_k are the eigenvalues of A. We define new Grassmann generators $(b_1, \bar{b}_1, b_2, \bar{b}_2, \ldots b_n, \bar{b}_n)$ by

$$b_i \equiv B_{ij} a_j \tag{9.22a}$$

and independently $\bar{b}_k \equiv \left((B^{\mathrm{T}})^{-1} \right)_{ki} \bar{a}_i$, (9.22b)

where B^{T} is the transpose of B. Now change variables to the set $(b_i, \bar{b}_i)$. Clearly, using (9.22) in (9.18),

$$\mathrm{d}\bar{b}_1\, \mathrm{d}b_1 \ldots \mathrm{d}\bar{b}_n\, \mathrm{d}b_n = \frac{1}{\mathrm{Det}\, B\ \mathrm{Det}[(B^{\mathrm{T}})^{-1}]} \mathrm{d}\bar{a}_1\, \mathrm{d}a_1 \ldots \mathrm{d}\bar{a}_n\, \mathrm{d}a_n$$

$$= \mathrm{d}\bar{a}_1\, \mathrm{d}a_1 \ldots \mathrm{d}\bar{a}_n\, \mathrm{d}a_n. \tag{9.23a}$$

At the same time, using (9.21),

$$\bar{a}_i A_{ij} a_j = \bar{b}_k B_{ki} A_{ij} B_{jl}^{-1} b_l$$

$$= \sum_k \lambda_k \bar{b}_k b_k. \tag{9.23b}$$

Hence

$$I = \int (\mathrm{d}\bar{b}_1\, \mathrm{d}b_1 \ldots \mathrm{d}\bar{b}_n\, \mathrm{d}\bar{b}_n) \exp\left(- \sum_k \lambda_k \bar{b}_k b_k \right)$$

$$= \prod_k \left(\int \mathrm{d}\bar{b}_k\, \mathrm{d}b_k \exp(-\lambda_k \bar{b}_k b_k) \right)$$

$$= \prod_k \lambda_k, \qquad \text{using (9.19)}$$

$$= \mathrm{Det}\, A. \tag{9.24}$$

Compare this with the usual c-number Gaussian integral in n variables (such as in eq. 6.13) whose value would be proportional to $(1/\text{Det}\, A)^{1/2}$. The fact that Det A occurs in (9.24) rather than its square root is simply because we have $2n$ variables here, and the matrix A considered as a $2n \times 2n$ matrix would consist of two off-diagonal $n \times n$ block matrices A_{ij}. However, the fact that Det A occurs in the numerator instead of the denominator is because of the Grassmann nature of the integral. This difference will prove to be crucial in our subsequent analysis.

Our next step is to introduce Grassmann (i.e. anticommuting) fields. We have emphasised that the Grassmann numbers $(a_1, \ldots, a_n)$ generating the algebra G_n are best thought of as some discrete entities, and not as variables taking on a range of values. The same restriction does not hold for general elements of the algebra G_n. An arbitrary element f has c-number coefficients $f^{(0)}$, $f_i^{(1)}, \ldots$ etc. in (9.4). These numbers can certainly be varied continuously. They could also depend on other c-number variables such as space and time coordinates. In that case, the elements of G_n can also be varied in space–time. In particular, consider the linear function

$$f(\boldsymbol{x}, t) = \sum_i f_i^{(1)}(\boldsymbol{x}, t)\, a_i. \tag{9.25}$$

This is our anticommuting field. It is a function of space–time, but anticommutes as per

$$\{ f(\boldsymbol{x}, t), f(\boldsymbol{x}', t') \}_+ = 0. \tag{9.26}$$

Note that one can differentiate or integrate $f(\boldsymbol{x}, t)$ with respect to $\boldsymbol{x}$ or t. These operations will obey the usual c-number calculus, since only the c-number coefficients $f_i^{(1)}(\boldsymbol{x}, t)$ are affected. Such operations are to be distinguished from derivatives with respect to, or integrations over, the Grassmann numbers a_i. To obtain our 'classical' Dirac field, consider a Grassmann algebra with an infinite number of generators which we will choose to write in a paired notation $(a_1, \bar{a}_1, a_2, \bar{a}_2, \ldots, a_n, \bar{a}_n; n \to \infty)$. It is again understood that a_i and $\bar{a}_i$ are quite independent Grassmann generators. Consider also any complete set of orthonormal c-number functions $\phi_i(\boldsymbol{x}, t)$ in space–time, satisfying

$$\int \mathrm{d}\boldsymbol{x} \int \mathrm{d}t\, \phi_i^*(\boldsymbol{x}, t)\, \phi_j(\boldsymbol{x}, t) = \delta_{ij}. \tag{9.27}$$

Define

$$\Psi(\boldsymbol{x}, t) \equiv \sum_{i=1}^{\infty} \phi_i(\boldsymbol{x}, t)\, a_i$$

and

$$\Psi^{\dagger}(x, t) \equiv \sum_{i=1}^{\infty} \phi_i^{*}(x, t)\bar{a}_i. \tag{9.28}$$

These are essentially our classical Dirac fields. [The actual Dirac field will have additional complications, suppressed in our notation purely for simplicity. In the usual notation in quantum Dirac field theory, one has the decomposition (Bjorken and Drell 1965)

$$\Psi_{\alpha}(x; t) = \sum_{p,s} \phi_{ps\alpha}^{(+)}(x, t)b_{ps} + \phi_{ps\alpha}^{(-)}(x, t)d_{ps}^{\dagger}, \tag{9.29}$$

where $\phi_{ps\alpha}^{\pm}(x, t)$ are spinors, α is the spinor component index, s is the spin label and $\pm$ distinguish positive- and negative-frequency subsets of the complete set of functions. In (9.28), the spin label s, along with p, has been absorbed in i, the spinor index α suppressed, and $\phi_{ps}^{\pm}$ unified into the set ϕ_i. Thus, the set a_i in (9.28) stand collectively for the classical limits of the fermion destruction operators b_{ps} as well as the anti-fermion creation operators $d_{ps}^{\dagger}$. Similarly the $\bar{a}_i$ are the classical limits of the full set $(b_{ps}^{\dagger}, d_{ps})$. Summation over the spin components is understood in the normalisation (9.27). The fact that the $\phi_i(x, t)$ are actually spinors will not invalidate our earlier arguments since the spinor is just a collection of c-numbers.]

Finally we are ready to construct the basic functional integral we need in Dirac field theory. Consider the integral

$$G \equiv \int \mathscr{D}[\Psi^{\dagger}(x, t)]\mathscr{D}[\Psi(x, t)] \exp\{\mathrm{i}\int \mathrm{d}x\,\mathrm{d}t\,[\Psi^{\dagger}(x, t)A\Psi(x, t)]\} \tag{9.30}$$

where A is a differential operator in space–time, containing Dirac γ-matrices. Remember that Ψ is a spinor although spin indices are suppressed here. Expand the Ψ and $\Psi^{\dagger}$ as per (9.28), as linear combinations of the a_i and $\bar{a}_i$ respectively. The integration measure is defined as

$$\int \mathscr{D}[\Psi^{\dagger}(x, t)]\mathscr{D}[\Psi(x, t)] \equiv \int \prod_j (\mathrm{i}\,\mathrm{d}\bar{a}_j\,\mathrm{d}a_j), \tag{9.31}$$

where the i in (9.31) is not an index but $\sqrt{-1}$. The operator A acts essentially on the complete set of orthonormal c-number functions ϕ_i in (9.28). If

$$A\phi_i(x, t) = A_{ij}\phi_j(x, t),$$

then the integrand in (9.30) becomes

$$\exp(\mathrm{i}\int \mathrm{d}x \int \mathrm{d}t\, \Psi^{+} A \Psi) = \exp(\mathrm{i}\, \bar{a}_j A_{jk} a_k). \tag{9.32}$$

Inserting (9.31) and (9.32) into (9.30), and using the result (9.24), we get

$$\begin{aligned} G &= \int \prod_j (\mathrm{i}\, \mathrm{d}\bar{a}_j \, \mathrm{d}a_j) \exp(\mathrm{i}\, \bar{a}_j A_{jk} a_k) \\ &= \operatorname{Det} A. \end{aligned} \tag{9.33}$$

The factors of i in the measure (9.31) ensure that all factors of i cancel each other in (9.33). Notice that the determinant of the differential operator occurs in the numerator, unlike the corresponding result for Bose fields.

Notice also that once the result (9.33) has been derived, the Grassmannian aspects of the problem play no further role. The answer in (9.33) depends only on the differential operator A, which is defined on the space of ordinary c-number spinor functions. To evaluate this answer for different choices of A, we only need to solve the usual c-number eigenvalue equation for A.

In the exponent in (9.30) we have considered only bilinear forms in (Ψ, Ψ^{+}). Problems such as the two-dimensional Thirring model will also introduce quartic forms in (Ψ, Ψ^{+}), but these can be converted, by a small sleight of hand, into bilinear forms (see section 9.5). We therefore have collected the requisite mathematical machinery for discussing functional integrals for Fermi fields.

9.2. Functional integral for the free Dirac field

Recall the functional integral (6.34) for evaluating $\mathrm{Tr}[\mathrm{e}^{-\mathrm{i}HT}]$ of a real scalar (Bose) field system. For a complex scalar field $\phi(x, t)$, where ϕ and ϕ^* can be considered as two independent variables instead of Re ϕ and Im ϕ, the equation (6.34) is trivially generalised to

$$\mathrm{Tr}[\mathrm{e}^{-\mathrm{i}HT}] = \int \mathscr{D}[\phi^*(x, t)]\, \mathscr{D}[\phi(x, t)] \exp(\mathrm{i}\, S[\phi, \phi^*]) \tag{9.34}$$

where $S[\phi, \phi^*]$ is the action functional. We now assert that a very similar functional-integral representation holds for the Dirac field, i.e.

$$\mathrm{Tr}[\mathrm{e}^{-\mathrm{i}HT}] = K \int \mathscr{D}[\Psi^{+}(x, t)]\, \mathscr{D}[\Psi(x, t)] \exp(\mathrm{i}\, S[\Psi, \Psi^{+}], \tag{9.35}$$

where K is a normalisation constant, and Ψ, Ψ^{+} are 'classical' Grassmann fields.

Let us begin by proving this assertion for a *free* Dirac quantum field where the eigenvalues of the Hamiltonian are already well known through canonical operator methods. Our discussion is an adaptation of Appendix A of Dashen et al. (1975b). The action for a free Dirac system in $(3+1)$ dimensions is

$$S_0[\Psi, \Psi^{\dagger}] = \int_0^t \mathrm{d}t \int \mathrm{d}x \, \bar{\Psi}(\mathrm{i}\not{\partial} - m)\Psi$$

$$= \int_0^T \mathrm{d}t \int \mathrm{d}x \left\{ \Psi^{\dagger}(x, t)\left(\mathrm{i}\frac{\partial}{\partial t} + \mathrm{i}\boldsymbol{\alpha} \cdot \boldsymbol{\nabla} - \beta m \right)\Psi(x, t) \right\}, \tag{9.36}$$

where we use the standard notation for Dirac matrices (Bjorken and Drell 1964), and spinor indices are again suppressed. $\bar{\Psi} \equiv \psi^{\dagger}\beta, \not{\partial} \equiv \partial_{\mu}\gamma^{\mu}, \gamma^0 \equiv \beta$ and $\gamma^i \equiv \beta\alpha^i$. Here, Ψ and $\Psi^{\dagger}$ can be treated as Grassmann fields. On applying the Grassmannian differentiation rule (9.7) as trivially generalised to fields, the condition

$$(\delta/\delta\Psi^{\dagger}(x, t))S_0[\Psi, \Psi^{\dagger}] = 0 \tag{9.37}$$

will yield the Dirac equation

$$(\mathrm{i}\partial/\partial t + \mathrm{i}\boldsymbol{\alpha} . \boldsymbol{\nabla} - \beta m)\Psi(x, t) = 0. \tag{9.38}$$

(A minor point: In the study of the non-second-quantised Dirac equation, standard textbooks usually treat the 'Dirac wavefunction' as a c-number spinor field, without bothering about Grassmann fields. The reason for this will be explained below.)

When (9.36) is inserted into (9.35), we have

$$\mathrm{Tr}[\mathrm{e}^{-\mathrm{i}H_0 T}] = K \int \mathcal{D}[\Psi^{\dagger}] \, \mathcal{D}[\Psi] \exp\left[\mathrm{i} \int_0^T \mathrm{d}t \int \mathrm{d}x (\Psi^{\dagger}(x, t) A \, \Psi(x, t)) \right], \tag{9.39}$$

where H_0 = the free Dirac Hamiltonian and A is the operator

$$A \equiv \mathrm{i}\,\partial/\partial t + \mathrm{i}\,\boldsymbol{\alpha} \cdot \boldsymbol{\nabla} - \beta m. \tag{9.40}$$

This integral has the form (9.30), and the answer from (9.33) is

$$\mathrm{Tr}\left[\mathrm{e}^{-\mathrm{i}H_0 T} \right] = K \text{ Det } A = K \prod_j (\lambda_j), \tag{9.41}$$

where λ_j are the eigenvalues of the operator A. The eigenvalue equation is

$$(\mathrm{i}\partial/\partial t + \mathrm{i}\boldsymbol{\alpha} \cdot \boldsymbol{\nabla} - \beta m)\psi(x, t) = \lambda\psi(x, t), \tag{9.42}$$

where ψ is a four-component spinor with the spin index suppressed. In contrast to (9.38), the spinor function here (denoted by ψ instead of Ψ) is a c-number object and not a Grassmann field. As emphasised earlier, the operator A acts basically in the space of c-number spinors, and its eigenvalue equation will be couched in this space. When one studies the Dirac equation in text books as a prelude to a second-quantised Dirac field theory, one is actually studying such an eigenvalue equation.

The solutions of (9.42) are well known (Bjorken and Drell 1964). One writes

$$\psi(\boldsymbol{x}, t) = L^{-3/2} \exp[\mathrm{i}(\boldsymbol{k}\cdot\boldsymbol{x} - wt)]\, u(w, \boldsymbol{k}), \tag{9.43}$$

where $u(w, \boldsymbol{k})$ is a spinor. Inserting this into (9.42) yields the eigenvalue condition

$$\mathrm{Det}\, | w - \boldsymbol{k}\cdot\boldsymbol{\alpha} - m\beta - \lambda | = 0, \tag{9.44}$$

where this determinant is in the 4×4 spinor space. This yields eigenvalues

$$\lambda = w \pm \varepsilon_k; \qquad \varepsilon_k = \sqrt{k^2 + m^2}. \tag{9.45}$$

As is well known, the energies ε_k are the eigenvalues of the single-particle free Dirac Hamiltonian $h_0 \equiv -\mathrm{i}\boldsymbol{\alpha}\cdot\boldsymbol{\nabla} + \beta m$; $A = \mathrm{i}\,\partial/\partial t - h_0$. (This h_0 is not to be confused with H_0, the Hamiltonian of the full field theory.) For each $(w, \boldsymbol{k})$ there are four independent spinors, two degenerate ones with $\lambda = w + \varepsilon_k$ and two with $\lambda = w - \varepsilon_k$. Hence, going back to (9.41),

$$\mathrm{Tr}[\mathrm{e}^{-\mathrm{i}H_0 T}] = K \prod_j \lambda_j = K \prod_{\boldsymbol{k}, w} (w + \varepsilon_k)^2 (w - \varepsilon_k)^2. \tag{9.46}$$

The allowed values of $\boldsymbol{k}$, given as usual by periodic boundary conditions in space, are $\boldsymbol{k}_N = 2\pi \boldsymbol{N}/L$, $L \to \infty$. To get the allowed values of w we must decide on the boundary conditions on the c-number functions at $t = 0$ and $t = T$. In the boson case, the natural boundary condition for a trace was seen to be the periodic one $\phi(\boldsymbol{x}, 0) = \phi(\boldsymbol{x}, T)$. For the Fermi case, we shall see that the correct answer (i.e. the same answer as given by canonical operator methods) is obtained by requiring anti-periodicity in T, i.e. $\psi(\boldsymbol{x}, 0) = -\psi(\boldsymbol{x}, T)$. Then the allowed values of w are

$$w_n T = (2n + 1)\pi; \qquad n = -\infty \ldots -1, 0, 1, 2, \ldots, \infty. \tag{9.47}$$

Hence

$$\mathrm{Tr}\,[\mathrm{e}^{-\mathrm{i}H_0 T}] = K \prod_{\boldsymbol{k}} \left(\prod_n (w_n + \varepsilon_{\boldsymbol{k}})^2 (\varepsilon_{\boldsymbol{k}} - w_n)^2 \right). \tag{9.48}$$

But

$$\prod_{n=-\infty}^{\infty} (\varepsilon_k - w_n)^2 = \prod_{n=-\infty}^{\infty} (\varepsilon_k + w_{-n})^2 = \prod_{n=-\infty}^{\infty} (\varepsilon_k + w_n)^2.$$

Hence

$$\begin{aligned}\mathrm{Tr}[\mathrm{e}^{-\mathrm{i}H_0T}] &= K\prod_k\Big(\prod_n (w_n+\varepsilon_k)^4\Big)\\ &= K\prod_k\left[\prod_n\left(\frac{(2n+1)\pi}{T}\right)^4\prod_n\left(1+\frac{\varepsilon_k T}{(2n+1)\pi}\right)^4\right].\end{aligned} \tag{9.49}$$

We choose the normalisation constant, K, such that

$$K\prod_k\prod_n\left(\frac{(2n+1)\pi}{T}\right)^4 = \prod_k (2^4). \tag{9.50a}$$

Note that the factors $(2n+1)\pi/T$ are purely kinematical. They do not depend on the dynamics as represented here by the mass m, contained in ε_k. Also, (see eq. (1.431.3) in Gradshteyn and Ryzhik 1965),

$$\prod_{n=-\infty}^{\infty}\left(1+\frac{\varepsilon_k T}{(2n+1)\pi}\right) = \cos\tfrac{1}{2}\varepsilon_k T. \tag{9.50b}$$

Inserting (9.50) into (9.49),

$$\begin{aligned}\mathrm{Tr}[\mathrm{e}^{-\mathrm{i}H_0T}] &= \prod_k (2\cos\tfrac{1}{2}\varepsilon_k T)^4\\ &= \prod_k \exp(2\mathrm{i}\varepsilon_k T)\prod_k [1+\exp(-\mathrm{i}\varepsilon_k T)]^4\\ &= \sum_{\{n_k\}} \exp\left[-\mathrm{i}T\Big(\sum_k(-2\varepsilon_k + n_k\varepsilon_k)\Big)\right]C(\{n_k\}).\end{aligned} \tag{9.51}$$

The notation in (9.51) is that for each k there is an integer n_k, $0 \leqslant n_k \leqslant 4$. The sum $\sum_{\{n_k\}}$ is over all possible sets of integers $\{n_k\} = (n_{k_1}, n_{k_2}, \ldots)$ subject to $0 \leqslant n_{k_i} \leqslant 4$. The factor $C(\{n_k\})$ is the combinatoric degeneracy factor

$$C(\{n_k\}) = \prod_k \frac{4!}{n_k!\,(4-n_k)!}.$$

Clearly, (9.51) yields energy levels

$$E_{\{n_k\}} = \sum_k (-2\varepsilon_k + n_k\varepsilon_k) \tag{9.52}$$

with degeneracy $C(\{n_k\})$.

This is precisely what we expect for a free Dirac system. The lowest-energy (vacuum) state corresponds to all n_k being zero, and its energy is

$$E_0 = -2\sum_k \varepsilon_k.$$

This has the familiar interpretation as the filled 'negative-energy sea'. In this sea, there are two negative-energy fermions for each $\boldsymbol{k}$, corresponding to helicity ± 1, each with energy $(-\varepsilon_k) = -\sqrt{k^2+m^2}$. If necessary, this E_0 can be removed by normal-ordering, or equivalently by adding the constant E_0 to the Lagrangian. The excited states, obtained when some n_k are non-zero, can be degenerate. One can create upto four particles for each momentum $\boldsymbol{k}$, corresponding to a fermion and/or an antifermion with helicity $+1$ and/or -1. Each particle has the same energy ε_k, and when n_k of them $(0 \leqslant n_k \leqslant 4)$ are present in the $\boldsymbol{k}$-momentum state, the degeneracy factor is $4!/[(n_k)!\,(4-n_k)!]$. In short, we have shown explicitly that the functional-integral representation (9.35) in terms of Grassmann fields is correct for the free Dirac field. We have suppressed factors of $\hbar$ so far. When these are inserted, (9.35) is changed to

$$\mathrm{Tr}[\mathrm{e}^{-\mathrm{i}HT/\hbar}] = K \int \mathscr{D}[\Psi^\dagger]\mathscr{D}[\Psi]\exp(\mathrm{i}S[\Psi,\Psi^\dagger]/\hbar), \tag{9.53a}$$

where the action S is kept the same as before. But we will alter the measure (9.31) correspondingly to

$$\int \mathscr{D}[\Psi^\dagger]\mathscr{D}[\Psi] = \int \prod_j (\mathrm{i}\hbar\, \mathrm{d}\bar{a}_j\, \mathrm{d}a_j) \tag{9.53b}$$

so that the functional integral still equals K Det A with the same operator A as before. Since now $\mathrm{Tr}[\mathrm{e}^{-\mathrm{i}H_0T/\hbar}] = K$ Det A, the energy levels (9.52) will be multiplied by $\hbar$. In this notation, the mass of the fermion is $\hbar m$, where m is the parameter in the operator A (eq. (9.40)). We will continue to suppress $\hbar$ by and large, but it will be resurrected whenever necessary.

9.3. Solitons in the presence of Fermi fields

The functional-integral representation (9.35) for $\mathrm{Tr}[\mathrm{e}^{-\mathrm{i}HT}]$, which we have just verified for the free Dirac field, also holds for the interacting Dirac field. Consider for example the Dirac field interacting with a static external scalar field $\phi(\boldsymbol{x})$ through the interaction Lagrangian

$$L_{\text{int}} = g \int dx\, \bar{\Psi}(x,t)\Psi(x,t)\phi(x), \tag{9.54}$$

where g is a coupling constant. We know what to expect on physical grounds for the spectrum of this system. The energy eigenstates should again consist of an arbitrary number of independent fermions and antifermions superposed on the vacuum, just as in (9.51)–(9.52). But the energy of these particles should not be the pure kinetic energy $\varepsilon_k = \sqrt{k^2+m^2}$, but should instead be eigenvalues of the single-particle Dirac Hamiltonian in the presence of the static external field, which is

$$h_\phi \equiv -\mathrm{i}\,\boldsymbol{\alpha}\cdot\boldsymbol{\nabla} + \beta m - g\beta\,\phi(x). \tag{9.55}$$

We can see that the functional integral (9.35) yields precisely this result. The action here is

$$S = \int dx \int_0^T dt\, \Psi^\dagger(x,t)\left(\mathrm{i}\frac{\partial}{\partial t} + \mathrm{i}\,\boldsymbol{\alpha}\cdot\boldsymbol{\nabla} - \beta(m - g\phi(x))\right)\Psi(x,t)$$

$$= \int dx \int_0^T dt\, \Psi^\dagger(x,t)\, A_\phi \Psi(x,t), \tag{9.56}$$

where

$$A_\phi \equiv \mathrm{i}\,\partial/\partial t + \mathrm{i}\,\boldsymbol{\alpha}\cdot\boldsymbol{\nabla} - \beta(m - g\phi(x))$$
$$= \mathrm{i}\,\partial/\partial t - h_\phi. \tag{9.57}$$

The action is again bilinear in the Fermi field, and differs from the free Dirac action (9.36) only in the replacement of the operator A by the operator A_ϕ. Upon inserting this action (9.56) into the functional integral (9.35), we will get, analogous to (9.41),

$$\mathrm{Tr}[e^{-\mathrm{i}HT}] = K\ \mathrm{Det}\,[A_\phi]. \tag{9.58}$$

Evaluation of Det $[A_\phi]$ again proceeds along the same lines as in the free Dirac case. We need not repeat all the steps. Suffice it to note that A_ϕ differs from A only in that it contains the interacting single-particle Hamiltonian h_ϕ, instead of the free Hamiltonian h_0. Consequently, in the final result, the eigenvalues of h_ϕ will occur in the place of ε_k, the eigenvalues of h_0. The eigenvalues of h_ϕ are given by

$$h_\phi f(x) \equiv [-\mathrm{i}\boldsymbol{\alpha}\cdot\boldsymbol{\nabla} + \beta(m - g\phi(x))]f(x) = \varepsilon f(x). \tag{9.59}$$

Just as in the free Dirac case, the eigenvalues of h_ϕ will occur in pairs $\varepsilon = \pm\varepsilon_r$, $(r = 0, 1, \ldots, \infty)$ because of charge conjugation. That is, given an eigenfunction $f_r(x)$ with eigenvalue $\varepsilon_r > 0$, the function $Cf_r^*(x)$ will be an eigenfunction with eigenvalue $-\varepsilon_r$, where $C \equiv \beta\alpha_2$ is the charge conjugation matrix in the representation

$$\alpha_i = \begin{pmatrix} 0 & \sigma_i \\ \sigma_i & 0 \end{pmatrix}; \qquad \beta = \begin{pmatrix} I & 0 \\ 0 & -I \end{pmatrix}$$

(see Bjorken and Drell 1964). However, unlike the free Dirac case, where for each k and w we had two degenerate functions of different helicity, the same degeneracy may not be true in general for the ε_r in the presence of the external field ϕ. Thus, (9.48) will be replaced here by

$$\mathrm{Tr}[\mathrm{e}^{-\mathrm{i}HT}] = K \prod_{n,r} (w_n + \varepsilon_r)(w_n - \varepsilon_r). \tag{9.60}$$

Correspondingly, (9.51) will be replaced by

$$\begin{aligned} \mathrm{Tr}[\mathrm{e}^{-\mathrm{i}HT}] &= K \;\mathrm{Det}\, A_\phi \\ &= \sum_{\{n_r\}} \tilde{C}(\{n_r\}) \exp\left[-\mathrm{i}T\left(\sum_r (-\varepsilon_r + n_r\varepsilon_r)\right)\right], \end{aligned} \tag{9.61}$$

where now $0 \leqslant n_r \leqslant 2$ and

$$\tilde{C}(\{n_r\}) \equiv \prod_r \left(\frac{2!}{n_r!\,(2-n_r)!}\right).$$

Here, each $\varepsilon_r = \varepsilon_r[\phi]$ is a functional of $\phi(x)$ by virtue of eq. (9.59). The energy levels are

$$E_{\{n_r\}} = \sum_{r=0}^{\infty} (-\varepsilon_r + n_r\varepsilon_r). \tag{9.62}$$

When all $n_r = 0$, we again have the vacuum of this system with the filled Fermi sea of energy $(\sum_{r=0}^{\infty}(-\varepsilon_r))$. When some $n_r \neq 0$, we have a fermion and/or an antifermion, each created with energy $+\varepsilon_r$. This is the result we anticipate intuitively. The constraint $0 \leqslant n_r \leqslant 2$ implies that for each positive ε_r we can have at most one fermion and one antifermion. The loss of helicity degeneracy compared with the free Dirac problem will be compensated by the fact that the label r here will take on, loosely speaking, twice as many values as the momentum label k in (9.51). Note that we have tacitly assumed that none of the ε_r vanish. Special cases where one of the ε_r vanishes will be dealt with separately in section 9.4.

We shall not present at this stage the corresponding calculation for a time-dependent external field $\phi(x, t)$. In any case, there are no energy levels in such a case because energy is not a conserved quantity. We shall, however, have occasion to evaluate K Det A_ϕ for a time-dependent ϕ later in this section.

Next, let ϕ be a quantum field instead of an external field, with an action of the form

$$S = \int \mathrm{d}x\, \mathrm{d}t[\tfrac{1}{2}(\partial_\mu\phi)^2 - U(\phi) + \bar{\Psi}(\mathrm{i}\not{\partial} - m)\Psi + g\bar{\Psi}\Psi\phi]$$

$$= S_\phi + S_\Psi + S_{\text{int}}, \qquad \text{in obvious notation.} \tag{9.63}$$

Given each classical configuration $\phi(x, t)$, it may be thought of as an external field as far as the quantum Fermi system is concerned. The quantum nature of the ϕ field is then incorporated, using the ideas already developed in chapter 6, by integrating over all configurations $\phi(\mathbf{x}, t)$. In other words,

$$\mathrm{Tr}[\mathrm{e}^{-\mathrm{i}HT}] = K \int \mathscr{D}[\phi]\mathrm{e}^{\mathrm{i}S_\phi} \int \mathscr{D}[\Psi^\dagger]\mathscr{D}[\Psi]\mathrm{e}^{\mathrm{i}(S_\Psi + S_{\text{int}})}. \tag{9.64}$$

If the scalar field ϕ were replaced by a vector gauge field A_μ, with an interaction term $g\bar{\Psi}\gamma_\mu A^\mu\Psi$, the latter is still bilinear in the Fermi field. Once again we can write

$$\mathrm{Tr}[\mathrm{e}^{-\mathrm{i}HT}] = K \int \mathscr{D}[A_\mu]\mathrm{e}^{\mathrm{i}S_A} \int \mathscr{D}[\Psi^\dagger]\mathscr{D}[\Psi]\mathrm{e}^{\mathrm{i}(S_\Psi + S_{\text{int}})}. \tag{9.65}$$

If A_μ is a gauge field, gauge-fixing terms will have to be added (see for instance Coleman 1975a and chapter 10 of this book), but this is a complication already present even in the absence of Fermi fields. Clearly, this procedure also holds if the Fermi field couples to a pseudoscalar field through a Yukawa coupling. If there are several species of Fermi fields, we simply have a multiple integral over all the species. Thus, we have essentially covered all cases of interest in $(3+1)$ dimensions. Notice that, in all these cases, the action involves the Fermi field only bilinearly so that our basic Grassmann functional integral (9.30) will suffice as a mathematical tool. Remember that in $(3+1)$ dimensions, the Fermi field cannot self-interact through quartic or higher polynomial interactions, because such theories will not be renormalisable. (In $(1+1)$ dimensions, four-fermion interactions are renormalisable. We shall discuss this possibility separately in section 9.5.)

Now that we have enlarged the functional integral for $\mathrm{Tr}[\mathrm{e}^{-\mathrm{i}HT}]$ to include Fermi fields, the next question is whether the semiclassical method can also be extended to such cases. For the purely bosonic case, the semiclassical methods arose from stationary-phase approximations to the functional integral. For Grassmann fields we did not introduce any corresponding approximation technique. It is not needed, however, because the action is bilinear in $(\Psi, \Psi^\dagger)$ and the fermion functional integral can formally be evaluated exactly by using (9.33). When this is done, we are left in (9.64) or (9.65) with a purely bosonic functional integral, on which the usual stationary-phase approximation may be employed. Based on this idea, Dashen et al. (1974b) extended the functional-integral semiclassical method to include fermions. [Other techniques for obtaining semiclassical results can also be extended to fermions. For instance, Friedberg and Lee (1977) extended the canonical operator methods (introduced in chapter 8 for Bose fields) to cover Fermi fields. In yet another approach, Jackiw and Rebbi (1976c) employed the Goldstone–Jackiw method (described in chapter 5) of sandwiching the operator field equations between a postulated set of soliton states. As with Bose fields, these different techniques yield equivalent results. We will continue to work with the functional-integral method.]

Let us take the example of the system (9.63); the cases of vector fields or pseudoscalar fields can be handled in a similar way. The functional integral (9.64) has the form

$$\begin{aligned}\mathrm{Tr}[\mathrm{e}^{-\mathrm{i}HT}] &= K \int \mathscr{D}[\phi]\mathrm{e}^{\mathrm{i}S_\phi} \int \mathscr{D}[\Psi^\dagger]\mathscr{D}[\Psi] \\ &\qquad \exp\left(\mathrm{i}\int_0^T \mathrm{d}t \int \mathrm{d}x\, \Psi^\dagger A_\phi \Psi\right) \\ &= K \int \mathscr{D}[\phi]\mathrm{e}^{\mathrm{i}S_\phi}\mathrm{Det}[A_\phi] \\ &\equiv \int \mathscr{D}[\phi]\mathrm{e}^{\mathrm{i}S_{\mathrm{eff}}[\phi]},\end{aligned} \tag{9.66}$$

where now

$$A_\phi = \mathrm{i}\,\partial/\partial t + \mathrm{i}\boldsymbol{\alpha}\cdot\nabla - \beta(m - g\phi(\boldsymbol{x}, t)) \tag{9.67}$$

and

$$S_{\mathrm{eff}}[\phi] \equiv S_\phi[\phi] - \mathrm{i}\ln\{K\,\mathrm{Det}[A_\phi]\}. \tag{9.68}$$

We are thus left with a bosonic functional integral where the contribution of the Fermi fields has been entirely absorbed in the effective action $S_{\mathrm{eff}}[\phi]$. We can now apply the familiar stationary-phase approximation to (9.66). Let $\phi_0(\boldsymbol{x}, t)$ be a non-trivial localised solution of the equation

$$\frac{\delta S_{\mathrm{eff}}[\phi]}{\delta\phi(\boldsymbol{x}, t)} = \frac{\delta}{\delta\phi(\boldsymbol{x}, t)}[S_\phi - \mathrm{i}\ln(K\,\mathrm{Det}\,A_\phi)] = 0. \tag{9.69}$$

Then we can expand $S_{\text{eff}}[\phi]$ about $\phi_0(\boldsymbol{x}, t)$ and write (9.66) in the stationary-phase approximation as

$$\begin{aligned}\operatorname{Tr}[\mathrm{e}^{-\mathrm{i}HT}] &\simeq \exp(\mathrm{i}S_{\text{eff}}[\phi_0])\int \mathscr{D}[y(\boldsymbol{x}, t)] \\ &\quad\times \exp\left[\frac{\mathrm{i}}{2}\int \mathrm{d}\boldsymbol{x}\int \mathrm{d}t\, y\left(\frac{\delta^2 S_{\text{eff}}}{\delta\phi\,\delta\phi}\right)_{\phi_0} y\right] \\ &\equiv \exp(\mathrm{i}S_{\text{eff}}[\phi_0])\Delta, \end{aligned} \tag{9.70}$$

where $y(\boldsymbol{x}, t) \equiv \phi(\boldsymbol{x}, t) - \phi_0(\boldsymbol{x}, t)$.

Notice that, in contrast to the purely bosonic system, the term $S_{\text{eff}}[\phi_0]$ here is not an entirely classical contribution. Both the functional $S_{\text{eff}}[\phi]$ as well as the solution $\phi_0(\boldsymbol{x}, t)$ carry quantum effects of the Fermi field through the operator A_ϕ on which they depend. Further quantum effects due to the fluctuations of the Bose field ϕ about the solution ϕ_0 will be introduced by the factor Δ. This factor also depends on the Fermi field implicitly through S_{eff} and ϕ_0.

Formally, (9.70) gives us the answer for the spectrum of the soliton sector associated with $\phi_0(\boldsymbol{x}, t)$, in terms of determinants of linear operators. Once $\operatorname{Det} A_\phi$ is evaluated, $S_{\text{eff}}[\phi]$ is known and $\phi_0(\boldsymbol{x}, t)$ can be obtained from (9.69). The factor Δ is just a gaussian integral and proportional to $[\operatorname{Det}(\delta^2 S_{\text{eff}}/\delta\phi\,\delta\phi)_{\phi_0}]^{-1/2}$. In practice, however, these calculations are very difficult. Because of the second term in (9.68), $S_{\text{eff}}[\phi]$ is in general a non-polynomial non-local functional, and particularly complicated when the time T is finite. To simplify matters, let us make a slightly less satisfactory stationary-phase approximation, but one that will be seen to yield the same energy levels as (9.70) to order $\hbar$. Notice that if we had retained $\hbar$-dependence in our derivation, (9.66) would be altered to

$$\begin{aligned}\operatorname{Tr}[\mathrm{e}^{-\mathrm{i}HT/\hbar}] &= \int \mathscr{D}[\phi]\mathrm{e}^{\mathrm{i}S_\phi/\hbar} K \operatorname{Det}[A_\phi] \\ &= \int \mathscr{D}[\phi]\,\mathrm{e}^{\mathrm{i}S_{\text{eff}}[\phi]/\hbar}. \end{aligned} \tag{9.71}$$

Instead of expanding this integrand about an extremum of S_{eff}, let us expand it about an extremum of S_ϕ, on the grounds that in the small-$\hbar$ limit, $\mathrm{e}^{\mathrm{i}S_\phi/\hbar}$ oscillates rapidly while $K \operatorname{Det}[A_\phi]$ is comparatively slowly varying. (See the prototype equation (6.13) as an approximation to (6.11).) In other words, let $\phi_c(\boldsymbol{x}, t)$ be a soliton solution of

$$\left(\frac{\delta S_\phi[\phi]}{\delta\phi}\right)_{\phi_c} = -\Box\phi_c - \left(\frac{\mathrm{d}U}{\mathrm{d}\phi}\right)_{\phi_c} = 0. \tag{9.72}$$

We approximate (9.71) by

$$\mathrm{Tr}[\mathrm{e}^{-\mathrm{i}HT/\hbar}] = \exp(\mathrm{i}S_\phi[\phi_c]/\hbar)K\,\mathrm{Det}[A_{\phi c}]\Delta_0, \tag{9.73}$$

where

$$\Delta_0 \equiv \int \mathscr{D}[y(\boldsymbol{x},t)]\exp\left(\frac{\mathrm{i}}{2}\int \mathrm{d}\boldsymbol{x}\int \mathrm{d}t\; y\left(\frac{\delta^2 S_\phi}{\delta\phi\delta\phi}\right)_{\phi_c} y\right). \tag{9.74}$$

The result (9.73) is easier to evaluate and interpret than (9.70). Notice that ϕ_c which obeys (9.72) is just a classical soliton that the Bose system would have had in the absence of the Fermi field. Let us, for simplicity, first take ϕ_c to be a static solution $\phi_c(\boldsymbol{x})$. Then

$$\begin{aligned} S_\phi[\phi_c] &= -\int_0^T \mathrm{d}t \int \mathrm{d}\boldsymbol{x}[\tfrac{1}{2}(\nabla\phi_c)^2 + U(\phi_c)] \\ &= -E_{cl}[\phi_c]T, \end{aligned} \tag{9.75}$$

where $E_{cl}[\phi_c]$ is the classical mass of the soliton. The operator $A_{\phi c}$ in (9.67), for a time-independent field $\phi_c(\boldsymbol{x})$, is just the Dirac operator in a static external field. We can use (9.61) to write

$$K\,\mathrm{Det}[A_{\phi c}] = \sum_{\{n_r\}} \tilde{C}(\{n_r\})\exp\left[-\mathrm{i}T\left(\sum_r(-\varepsilon_r + n_r\varepsilon_r)\right)\right], \tag{9.76}$$

where $\varepsilon_r = \varepsilon_r[\phi_c]$ are the eigenvalues of the one-particle Dirac Hamiltonian (9.59) in the presence of $\phi_c(\boldsymbol{x})$. Finally, Δ_0 is the familiar boson fluctuation factor discussed in chapter 6. We can write it as (see eq. (6.39)–(6.43))

$$\Delta_0 = \sum_{\{N_p\}} \exp\left[-\mathrm{i}T\left(\sum_{p=0}^{\infty}(N_p + \tfrac{1}{2})\omega_p\right)\right], \qquad N_p = 0, 1, \ldots, \infty \tag{9.77}$$

where $\omega_p^2 = \omega_p^2[\phi_c]$ are the eigenvalues of $[-\nabla^2 + (\mathrm{d}^2U/\mathrm{d}\phi^2)_{\phi_c(x)}]$. Inserting (9.75)–(9.77) in (9.73), we get

$$\begin{aligned} \mathrm{Tr}[\mathrm{e}^{-\mathrm{i}HT/\hbar}] = \sum_{\{n_r, N_p\}} \tilde{C}(\{n_r\})\exp\Bigg[-\mathrm{i}T\Bigg(&\frac{E_{cl}[\phi_c]}{\hbar} \\ &+ \sum_r(-\varepsilon_r + n_r\varepsilon_r) + \sum_p((N_p + \tfrac{1}{2})\omega_p)\Bigg)\Bigg] \end{aligned} \tag{9.78}$$

with $0 \leqslant n_r \leqslant 2$ and $0 \leqslant N_p \leqslant \infty$. The energy levels given by (9.78) are

$$E_{\{n_r, N_p\}} = E_{\text{cl}}[\phi_c] + \hbar\left(\sum_{r=0}^{\infty}(-\varepsilon_r + n_r\varepsilon_r)\right) + \hbar\left(\sum_{p=0}^{\infty}(N_p + \tfrac{1}{2})\omega_p\right). \quad (9.79)$$

The quantum corrections are divergent. The divergence can be removed, just as in the purely bosonic problems in chapters 5 and 6, by adding the usual normal-ordering and renormalisation counter terms to the action. The nature of the counter terms will depend on the specific model and the space dimensionality. In general, however, the counter terms, being quantum contributions, will be $O(\hbar)$ to leading order. We can write all the counter terms involving Bose and Fermi fields collectively in the form $\hbar S_{\text{ct}}$ and add it to the action. In our stationary-phase approximation about a static configuration $\phi_c(\boldsymbol{x})$, this will clearly introduce a factor into (9.73) of the form

$$\exp(\mathrm{i}S_{\text{ct}}[\phi_c]) = \exp(\mathrm{i}TE_{\text{ct}}[\phi_c]). \quad (9.80)$$

The result (9.79) will correspondingly be modified to

$$E_{\{n_r, N_p\}} = E_{\text{cl}}[\phi_c] + \hbar\left(\sum_r n_r\varepsilon_r + \sum_p N_p\omega_p\right)$$
$$+ \hbar\left(\sum_r(-\varepsilon_r) + \sum_p \tfrac{1}{2}\omega_p - E_{\text{ct}}\right). \quad (9.81)$$

We can expect that in a renormalisable theory the same counter terms that are added in normal pertubation theory will suffice to remove all divergences in the soliton sectors as well, order by order. Accordingly, the potentially divergent parts in the last term in brackets in (9.81) should get cancelled by E_{ct}. These energy levels can be interpreted in the same way as those in chapter 5. They span the soliton sector of states associated with $\phi_c(\boldsymbol{x})$. To start with, let us suppose that none of the ω_p or the ε_r are zero. (This supposition is not valid in general; some of the interpretation given below would have to be modified accordingly – see section 9.4.) Then the lowest level in the set (9.81) is

$$E_{\{N_p = n_r = 0\}} \equiv M_{\text{sol}}$$
$$= E_{\text{cl}}[\phi_c] + \hbar\left(\sum_r(-\varepsilon_r) + \tfrac{1}{2}\sum_p \omega_p - E_{\text{ct}}\right). \quad (9.82)$$

This is the mass of the basic quantum soliton particle in the presence of the Dirac field. On comparing this result with the corresponding problem in the absence of the Fermi field, we see that the solution ϕ_c, its classical energy $E_{cl}[\phi_c]$, as well as the bosonic part of the zero-point energy $\sum_p \frac{1}{2}\hbar\omega_p$, remain the same. This is because of the nature of the approximation (9.73). But the quantum soliton mass now contains an added contribution due to the zero-point energy of the Fermi field in the form of the energy of the filled Fermi sea. The higher levels in (9.81) correspond to excitations which can now arise in both the boson and fermion fluctuation modes. The only difference is that, unlike the boson modes, any given fermion mode can be excited at most twice ($0 \leqslant n_r \leqslant 2$), corresponding to the creation of one fermion and one antifermion in that mode, consistent with the Pauli principle. Notice that the ε_r are eigenvalues of the one-particle Dirac Hamiltonian in the presence of $\phi_c(x)$ (eq. (9.59)). Thus, to order $\hbar$, the soliton acts as a 'background field' for the fermions. The quantum nature of the ϕ field will be felt by the fermions only in higher orders. In the next section, we shall illustrate this whole procedure using a specific model, for which $\phi_c(x)$, $E_{cl}[\phi_c]$, ω_p, ε_r and E_{ct} are all explicitly available.

The soliton sector, in the presence of Fermi fields, can also be explored in the canonical operator formalism as was done for the case of pure Bose fields. If none of the ω_p or ε_r were zero, one would expand the field operators as

$$\Phi(x, t) = \phi_c(x) + \sum_p \left(a_p \frac{e^{-i\omega_p t}}{\sqrt{2\omega_p}} \eta_p(x) + a_p^\dagger \frac{e^{i\omega_p t}}{\sqrt{2\omega_p}} \eta_p^*(x) \right), \tag{9.83a}$$

where $\eta_p(x)$ are the normal modes of boson fluctuations about $\phi_c(x)$, and similarly,

$$\Psi(x, t) = \sum_r [b_r e^{-i\varepsilon_r t} f_r^{(+)}(x) + d_r^\dagger e^{i\varepsilon_r t} f_r^{(-)}(x)], \tag{9.83b}$$

where $f_r^{(\pm)}(x)$ and $\pm\varepsilon_r = \pm\varepsilon_r[\phi_c]$ are respectively the normalised positive- and negative-energy eigenfunctions and eigenvalues of the single-particle Dirac Hamiltonian (9.59) in the presence of ϕ_c. Starting from the basic anticommutation rules (9.2) for the Fermi field, and using the usual normalisation for the $f_r^{(\pm)}(x)$, one can deduce the familiar anticommutation rules

$$\{b_r, b_{r'}^\dagger\}_+ = \{d_r, d_{r'}^\dagger\}_+ = \delta_{rr'}$$

with all other anticommutators vanishing.

The lowest energy state in the soliton sector, the basic quantum soliton, corresponds, in order $\hbar$, to having none of the boson and fermion modes excited. Clearly this state $|\text{sol}\rangle$ will satisfy

$$a_p|\text{sol}\rangle = b_r|\text{sol}\rangle = d_r|\text{sol}\rangle = 0. \tag{9.84}$$

The normal ordered fermion current j_μ (see Bjorken and Drell 1965) is

$$j_\mu = \tfrac{1}{2}[\Psi^\dagger, \gamma_0\gamma_\mu\Psi].$$

The associated charge is

$$Q = \tfrac{1}{2}\int dx(\Psi^\dagger\Psi - \Psi\Psi^\dagger) = \sum_r (b_r^\dagger b_r - d_r^\dagger d_r). \tag{9.85}$$

Clearly, in view of (9.84),

$$Q|\text{sol}\rangle = 0. \tag{9.86}$$

Fermionic excitations are obtained by operating on $|\text{sol}\rangle$ with operators $b_r^\dagger$ or $d_r^\dagger$. For instance, the state $b_{r_0}^\dagger|\text{sol}\rangle$ will carry a fermion number equal to unity. If the mode r_0 is discrete, this will be the bound soliton state with fermion number one. Similarly $d_{r_0}^\dagger|\text{sol}\rangle$ will carry fermion number equal to -1.

It should be emphasised that in (9.83)–(9.86) we have tacitly assumed that none of the ε_r or ω_p vanish. We did so for simplicity, to get a preliminary feel for soliton sectors. In actual fact, while most of the ε_r and ω_p are generally non-zero, a few of them will vanish. This will of course depend on the specific model. We already know that some of the ω_p will vanish because of continuous symmetries like translational invariance. The treatment and consequence of such boson zero-modes have been discussed in the preceding chapter. They will affect the energy levels only in the next order ($O(\hbar^2)$). In the presence of Fermi fields, we have the additional possibility that some of the ε_r may also vanish. The impact of such a zero-energy fermion mode will be discussed in the next section.

The steps leading from (9.73) to (9.81) are valid for a static solution $\phi_c(x)$. In many cases, no static soliton solution may exist. See for instance the virial theorem discussed in chapter 3. In such cases (and also in cases when static solutions do exist), one can still perform a semiclassical expansion about periodic time-dependent solutions of (9.72). For this, we must adopt the full WKB method developed in chapter 6. One works with

$$G(E) \equiv i\int_0^\infty dT e^{iET}\,\text{Tr}[e^{-iHT}] = \text{Tr}[1/(E-H)].$$

The energy levels are the poles of $G(E)$. Once $[\text{Tr}e^{-iHT}]$ is known, the remaining steps proceed exactly as in chapter 6. The approximation (9.73) for $\mathrm{Tr}[e^{-iHT}]$ still holds when ϕ_c is a time-dependent solution $\phi_c(\boldsymbol{x}, t)$ with period T. But the factors (K Det A_{ϕ_c}) and Δ_0 need to be evaluated for the time-dependent function $\phi_c(\boldsymbol{x}, t)$. Δ_0 is a purely bosonic factor and has been already discussed in chapter 6. When ϕ_c is periodic, we saw that the factor $(\omega_p T)$ for each boson mode is replaced by the stability angle v_p. We will now show that a similar replacement takes place for the fermion modes contained in K Det A_{ϕ_c} (Dashen et al. 1975b). The eigenvalue equation for A_{ϕ_c} is, now,

$$[\mathrm{i}\partial/\partial t + \mathrm{i}\boldsymbol{\alpha}\cdot\boldsymbol{\nabla} - \beta(m - g\phi_c(\boldsymbol{x}, t))]\,\psi(\boldsymbol{x}, t) = \lambda\psi(\boldsymbol{x}, t). \tag{9.87}$$

Since ϕ_c is time-dependent, we can no longer write $\Psi(\boldsymbol{x}, t)$ in the form $e^{-i\varepsilon t}f(\boldsymbol{x})$. But, consider the Dirac equation:

$$[\mathrm{i}\partial/\partial t + \mathrm{i}\boldsymbol{\alpha}\cdot\boldsymbol{\nabla} - \beta(m - g\phi_c(\boldsymbol{x}, t))]\,f(\boldsymbol{x}, t) = 0. \tag{9.88}$$

Since the operator in (9.88) is periodic with period T, one can find solutions $f_r(\boldsymbol{x}, t)$ obeying

$$f_r(\boldsymbol{x}, t+T) = e^{-i\alpha_r} f_r(\boldsymbol{x}, t) \qquad \text{for all } \boldsymbol{x}, t.$$

These α_r are called 'Floquet indices'. By comparing the above equation with (6.95), we see that Floquet indices are analogues of the stability angles. Clearly,

$$\psi_{n,r}(\boldsymbol{x}, t) \equiv \exp[-\mathrm{i}(w_n - \alpha_r/T)t]\,f_r(\boldsymbol{x}, t), \tag{9.89}$$

with

$$w_n = (2n+1)\pi/T$$

are eigenfunctions of A_{ϕ_c} satisfying (9.87) with eigenvalues

$$\lambda_{n,r} = w_n - \alpha_r/T.$$

Furthermore, the $\psi_{n,r}(\boldsymbol{x}, t)$ obey the anti-periodicity condition $\psi_{n,r}(\boldsymbol{x}, t+T) = -\psi_{n,r}(\boldsymbol{x}, t)$ as required earlier. Finally, the charge conjugation argument given before again shows that the α's will occur in pairs $\pm\alpha_r$. Hence, analogously to (9.60)–(9.61),

$$K \text{ Det } A_{\phi_c} = \prod_{n,r}(w_n - \alpha_r/T)(w_n + \alpha_r/T), \qquad \text{with all } \alpha_r \text{ positive},$$

$$= \sum_{\{n_r\}} \tilde{C}(\{n_r\}) \exp\left[-\mathrm{i}\left(\sum_r (n_r\alpha_r - \alpha_r)\right)\right]. \tag{9.90}$$

By comparing with (9.76), we see that in going from a static $\phi_c(\boldsymbol{x})$ to a periodic $\phi_c(x, t)$, the Floquet indices α_r take the place of $\varepsilon_r T$. (Notice that had ϕ_c been static we could write $f_r(x, t) = e^{-i\varepsilon_r t} f_r(x)$, and the α_r would trivially reduce to $\varepsilon_r T$.) Instead of computing ε_r, we must now find the α_r by solving the time-dependent Dirac equation. Given the α_r, the result (9.90) can be inserted in (9.73) and the WKB method of chapter 6 executed.

Our results were obtained from the approximation (9.73), wherein we had expanded the functional integral (9.71) about a stationary phase point of $e^{iS_\phi/\hbar}$ rather than that of the full integrand $e^{iS_{\text{eff}}/\hbar}$. This we did because it made subsequent analysis simpler, and on the grounds that in the small-$\hbar$ limit $e^{iS_\phi/\hbar}$ is the most rapidly oscillating part of the integrand. The 'full' stationary-phase approximation, where one expands about an extremum of S_{eff}, was given in (9.70), but it is, as we have said earlier, harder to execute. Even finding the extremum $\phi_0(x, t)$ by solving (9.69) is much harder than solving (9.72). The culprit is the extra term $-i\hbar \ln(K \operatorname{Det} A_\phi)$ which will be a non-local non-polynomial functional of ϕ. However the two approximations (9.70) and (9.73) will yield the same energy levels (9.79) to order $\hbar$. To see this, note that $S_{\text{eff}} - S_\phi = -i\hbar \ln(K \operatorname{Det} A_\phi)$. Therefore, the solutions $\phi_0(x, t)$ and $\phi_c(x, t)$, which are extrema of S_{eff} and S_ϕ respectively, will also differ by $\phi_0 - \phi_c \equiv \delta\phi = O(\hbar)$. This is assuming that, given a ϕ_0, the corresponding ϕ_c can be chosen as the limit of ϕ_0 as $\hbar \to 0$. Now, (9.70) can be written, when $\hbar$-dependence and the counter terms are included, as

$$\operatorname{Tr}[e^{-iHT/\hbar}] = \exp\{(i/\hbar)[S_\phi[\phi_0] - i\hbar \ln(K \operatorname{Det} A_{\phi_0}) - i\hbar \ln\Delta + \hbar S_{ct}[\phi_0]]\}. \tag{9.91}$$

But

$$\begin{aligned} S_\phi[\phi_0] &= S_\phi[\phi_c] + (\delta S_\phi/\delta\phi)_{\phi_c}\delta\phi + O(\Delta\phi^2) \\ &= S_\phi[\phi_c] + O(\hbar^2), \qquad \text{using (9.72).} \end{aligned} \tag{9.92}$$

Similarly, the remaining terms in the exponent in (9.91) are already of one higher order in $\hbar$. The error introduced in replacing ϕ_0 by ϕ_c, and $(\partial^2 S_{\text{eff}}/\partial\phi\delta\phi)_{\phi_0}$ by $(\partial^2 S_\phi/\delta\phi\delta\phi)_{\phi_c}$ in these terms will be of still higher order. These replacements reduce (9.70) to (9.73). Therefore, the improved approximation (9.70) will give the same energy levels (9.81) to order $\hbar$, but will yield some higher-order corrections as well. Despite its complexity some authors have successfully used essentially the approximation (9.70) for specific models in the $T \to \infty$ limit. (Dashen et al. 1975b, Campbell and Liao 1976).

Another procedure, popular in this field, uses a classical solution which extremises neither $S_\phi[\phi]$ nor the full $S_{\text{eff}}[\phi]$, but something in between. Let us describe this method for static solutions, in which case the extremisation can be done amongst static configurations $\phi(x)$. For such configurations, the integrand in the basic functional integral (9.66) can be written as,

$$\exp[\mathrm{i}(S_\phi - \mathrm{i}\ln(K\,\mathrm{Det}\,A_\phi))]$$
$$= \sum_{\{n_r\}} \tilde{C}\Big(\{n_r\}\Big)\exp\Big[-\mathrm{i}T\Big(E_{\mathrm{cl}}[\phi] + \sum_r n_r\varepsilon_r[\phi] + \sum_r -\varepsilon_r[\phi]\Big)\Big]. \tag{9.93}$$

We have again set $\hbar = 1$ and used (9.61). Now, if we extremise the entire exponent on the left-hand side and proceed with the semiclassical method, we will get (9.70). If we extremise only S_ϕ, or equivalently $E_{\mathrm{cl}}[\phi]$, we get the approximation (9.73). Let us do something in between, and extremise $E_{\mathrm{cl}}[\phi] + \sum_r n_r\varepsilon_r[\phi]$. This of course depends on the choice of the fermion excitation numbers $\{n_r\}$. This procedure tries to optimise the extremisation depending on the occupation numbers of the state we are interested in. The energy of the Fermi sea, $(\sum_r - \varepsilon_r[\phi])$ is not included in the extremisation, but is instead evaluated at the extremum. We know of no satisfactory justification for doing this, other than that the procedure is an improvement over merely extremising $E_{\mathrm{cl}}[\phi]$ alone. Remember, furthermore, that most of the energy of the Fermi sea (all its divergent parts) will be cancelled by counter terms, when the latter are added.

The extremisation condition now is

$$\frac{\delta}{\delta\phi(x)}\Big(E_{\mathrm{cl}}[\phi] + \sum_r n_r\varepsilon_r[\phi]\Big)$$
$$= -\nabla^2\phi + \frac{\mathrm{d}U}{\mathrm{d}\phi} + \sum_r n_r\frac{\delta}{\delta\phi}\Big(\varepsilon_r[\phi]\Big) = 0. \tag{9.94}$$

One can rewrite (9.94) as follows. $\varepsilon_r[\phi]$ are eigenvalues of the one-particle Dirac Hamiltonian, and are given by (9.59). If ϕ is changed to $\phi + \delta\phi$, f will change to $f + \delta f$ and ε to $\varepsilon + \delta\varepsilon$. To first order in infinitesimals, (9.59) will yield

$$\{-\mathrm{i}\boldsymbol{\alpha}\cdot\nabla + \beta(m - g\phi)\}\,\delta f - g\,\delta\phi\,\beta f$$
$$= \delta\varepsilon f + \varepsilon\,\delta f. \tag{9.95}$$

By taking an inner product with f^{+} and using the normalisation

$\int dx f^\dagger f = 1$, this gives

$$g\int f^\dagger(x)\beta f(x)\delta\phi(x)dx = \delta\varepsilon$$

or

$$\delta\varepsilon[\phi]/\delta\phi(x) = gf^\dagger(x)\beta f(x) = g\bar{f}(x)f(x). \tag{9.96}$$

(Remember, $f(x)$ is a spinor, with spin indices suppressed and spin summation understood in the above equations; $\bar{f} \equiv f^\dagger\beta$.) Inserting (9.96) into (9.94) we have

$$-\nabla^2\phi + \frac{dU}{d\phi} = -g\sum_r n_r \bar{f}_r f_r, \tag{9.97}$$

where each f_r obeys (9.59). We thus get a set of coupled equations (9.59) and (9.97) to solve self-consistently for ϕ and the f_r. In the simplest case, when only one fermion mode r_0 is excited once, $n_r = \delta_{rr_0}$. Then the equations (9.59) and (9.97) reduce to the pair:

$$[-i\boldsymbol{\alpha}\cdot\nabla + \beta(m - g\phi(x))]f_{r_0}(x) = \varepsilon_{r_0}f_{r_0} \tag{9.98a}$$

$$-\nabla^2\phi + dU/d\phi = -g\bar{f}_{r_0}f_{r_0}. \tag{9.98b}$$

The earlier approximation (9.72) amounts, for a static solution, to setting the right-hand side of (9.98b) equal to zero, thereby decoupling the classical Bose field from any feedback from the Fermi field. In (9.98), the classical Bose field and the Dirac wavefunction f_{r_0} are coupled, and that is the price we have to pay for the improvement in the approximation (9.94). Notice that (9.98) looks just like the Euler–Lagrange equations we would get for a static $\phi(x)$ starting from the Lagrangian (9.63) provided we inserted the Dirac wavefunction of the occupied state ($\psi_{r_0}(x,t) = f_{r_0}(x)e^{-i\varepsilon_{r_0}t}$) in the place of the Fermi field $\Psi(x,t)$. This gives the coupled equations (9.98) some intuitive appeal. Indeed, field equations with such a replacement of the Fermi field by a *c*-number Dirac wavefunction have often been introduced, *ab initio*, in semiclassical calculations in the literature, particularly in constructing models for hadrons. Our discussion shows the relevance of such equations in the context of systematic semiclassical approximations of the parent quantum field theory. We can also see the limitations on the relevance of such equations. One gets these two coupled equations (9.98) from the condition (9.94) only when a single fermion mode is occupied. When N different modes are occupied, we get $N+1$ coupled equations– eq. (9.97) plus the N equations of the form (9.59), one for each f_r. Also, the parent approximation (9.94) from which these equations flow does not extremise the full $S_{\text{eff}}[\phi]$. The energy of the Fermi sea has been left out of the extremisation.

Despite these limitations, eqs. (9.98) do offer an improvement over the uncoupled equation (9.72). Although more difficult to solve than (9.72), they are still tractable. Such equations have been solved, either exactly or numerically, for a variety of models. Apart from the papers cited already, see Bardeen et al. (1975; this is the famous SLAC Bag), Lee and Wick (1974) and Chang et al. (1975). Once a solution ϕ_c to the coupled equations (9.98) is found, the rest of the semiclassical method proceeds as before, upon expanding the functional integral (9.66) about ϕ_c.

9.4. The possibility of solitons of charge $\frac{1}{2}$

Let us illustrate the semiclassical method (9.72)–(9.81) with a concrete example. The example we choose will also yield a zero-energy fermion mode and give us an opportunity to discuss the implications of such a mode. The model will be a $(1+1)$-dimensional theory, in which the kink system discussed in chapters 2 and 5 couples to fermions through a Lagrangian density:

$$\mathscr{L}(x,t) = \tfrac{1}{2}(\partial_\mu \phi)^2 - \tfrac{1}{2}(\phi^2 - 1)^2 + \bar{\Psi}(\mathrm{i}\partial\!\!\!/)\Psi + g\bar{\Psi}\Psi\phi. \tag{9.99}$$

Compared with the Lagrangian (5.36) of the kink system, here we have, for simplicity, set the kink parameters $\lambda = m^2 = 2$. The soliton sector of this system (9.99) was discussed by Dashen et al. (1974b), who also pointed out the existence of the zero-energy fermion mode. The implications of such a mode were explored further and related to the possibility of solitons with charge $\pm\frac{1}{2}$ by Jackiw and Rebbi (1976c).

The methods of the preceding section, adapted to $(1+1)$ dimensions, can be applied to this model. All we need to remember is that Ψ will now be a two-component spinor, and there will only be two Dirac matrices. We will use the representation

$$\beta = \sigma_1 \qquad \text{and} \qquad \alpha_1 = \sigma_2. \tag{9.100}$$

(Equivalently, $\gamma^0 = \beta = \sigma_1$ and $\gamma^1 = \beta\alpha_1 = \mathrm{i}\sigma_3$ in $\mathrm{i}\partial\!\!\!/ \equiv \mathrm{i}(\partial_t\gamma^0 + \partial_x\gamma^1)$.) Before proceeding to the soliton sector, remember that in this model, the vacuum has an expectation value $\langle \phi \rangle = \pm 1$. In such a vacuum, the elementary fermion will have a mass $m_F = |g\langle \phi \rangle|$ to lowest order, due to the last term in (9.99) even though the Lagrangian has no explicit fermion mass term. This mass can always be taken to be positive regardless of the sign of $\langle \phi \rangle$ or g, since the sign of the mass term can always be reversed by

$\Psi \to \gamma_5 \Psi \equiv \gamma^0 \gamma^1 \Psi$. With no loss of generality we can take $g > 0$, $\langle \phi \rangle = +1$ and $m_F = g$.

Keeping all this in mind, let us consider the soliton sector and use the approximation (9.73). This leads to the result (9.81) for the energy levels. All we need to do is evaluate the quantities appearing in (9.81), namely, ϕ_c, $E_{cl}[\phi_c]$, $\omega_p[\phi_c]$, $\varepsilon_r[\phi_c]$ and $E_{ct}[\phi_c]$, for the model (9.99). A static classical solution $\phi_c(x)$ must obey (9.72) which now reduces to

$$d^2\phi_c/dx^2 - 2\phi_c(\phi_c^2 - 1) = 0. \tag{9.101}$$

The solution to this is just the old kink function. Putting $\lambda = m^2 = 2, a = 0$ in (5.45)–(5.46) we have

$$\phi_c(x) = \tanh x \tag{9.102a}$$

with

$$E_{cl}[\phi_c] = 4/3. \tag{9.102b}$$

(The antikink is also a solution. A sector of states built around the antikink can be discussed in the same way.) The quantities ω_p in (9.81) are just the stability frequencies of the kink and have already been given in (5.50). The energies $\varepsilon_r[\phi_c]$ of the fermion modes satisfy, on applying (9.59),

$$\begin{aligned} h_{\phi_c} f_r(x) &= (-i\alpha_1 d/dx - g\beta\phi_c(x)) f_r(x) \\ &= \varepsilon_r f_r(x). \end{aligned} \tag{9.103}$$

If we write the two-component spinor in the form

$$f_r(x) = \begin{pmatrix} u_r(x) \\ v_r(x) \end{pmatrix},$$

then in the representation (9.100), (9.103) reduces to the coupled equations:

$$(-d/dx - g\phi_c(x))\, v_r(x) = \varepsilon_r u_r(x), \tag{9.104a}$$

$$(d/dx - g\phi_c(x))\, u_r(x) = \varepsilon_r v_r(x), \tag{9.104b}$$

which yield

$$\begin{aligned} &\left(-\frac{d^2}{dx^2} + g^2\phi_c^2 + g\frac{d\phi_c}{dx}\right) u_r(x) \\ &\quad = (-d^2/dx^2 + g^2\phi_c^2 + g(1-\phi_c^2))\, u_r(x) = \varepsilon_r^2 u_r(x) \end{aligned} \tag{9.105a}$$

and

$$(-d^2/dx^2 + g^2\phi_c^2 - g(1-\phi_c^2))\, v_r(x) = \varepsilon_r^2 v_r(x), \tag{9.105b}$$

where we have used the fact (see eq. (2.20)) that $d\phi_c/dx = \sqrt{2U(\phi_c)} = 1 - \phi_c^2$. Equation (9.105) is exactly solvable in terms of elementary functions. It is essentially just a pair of one-dimensional Schrödinger equations in a 'potential' $V(x) = g^2\phi_c^2 \pm g(1-\phi_c^2) = g^2 \tanh^2 x \pm g \operatorname{sech}^2 x$. As $x \to \pm\infty$, $V(x) \to g^2$. Hence the eigenvalues ε_r^2 will consist of a continuum $\varepsilon_r^2 = g^2 + k_r^2$ for $\varepsilon_r^2 > g^2$. Below g^2, the discrete bound levels can be exactly obtained (see Morse and Feshbach 1953, Dashen et al. 1974b) as

$$\varepsilon_r^2 = 2rg - r^2; \qquad r = 0, 1, \ldots, < g. \tag{9.106}$$

We notice that the lowest level is $\varepsilon_0 = 0$. Apart from this, for every non-zero eigenvalue ε_r^2 of (9.105), we have a pair of eigenvalues $\pm\varepsilon_r$ of the Dirac Hamiltonian in (9.103), in accordance with the general argument given earlier based on charge conjugation. That the continuum modes begin at $\varepsilon_r = g$ is also consistent with the fact that the elementary fermion has mass g in this model.

The last ingredient needed in (9.81) is $E_{ct}[\phi_c]$. Apart from the counter terms $(D + \frac{1}{2}\partial m^2 \phi^2)$ already introduced in chapter 5 (eq. (5.66)–(5.67)), we will have additional counter terms due to the presence of the Fermi field in the Lagrangian (9.99). These again have the form $\bar{D} + \frac{1}{2}\overline{\partial m^2}\,\phi^2$. Here $\bar{D}$ is the energy of the fermion vacuum, i.e. $\bar{D} = \sum_r^{\infty}(-\varepsilon_r[\phi = 1])$. The term $\frac{1}{2}\overline{\partial m^2}\,\phi^2$ comes from the boson self-energy graph due to the fermion loop, i.e.

$$\overline{\partial m^2} = g^2 \mathrm{i}\int \frac{d^2 k}{(2\pi)^2}\, \mathrm{Tr}\left(\frac{1}{(\not{k} - g)(-\not{k} - g)}\right). \tag{9.107}$$

Altogether, the counter terms to be added to the lagrangian density (9.99) are

$$\mathscr{L}_{ct} = D + \bar{D} + \tfrac{1}{2}(\partial m^2 + \overline{\partial m^2})\phi^2. \tag{9.108}$$

As a result

$$E_{ct}[\phi_c] = \int dx\,[D + \bar{D} + \tfrac{1}{2}(\partial m^2 + \overline{\partial m^2})(\phi_c^2(x) - 1)], \tag{9.109}$$

where the contribution of the vacuum value $\phi = 1$ has been subtracted just as we did in (5.68). Thus, we have accumulated all the inputs needed to calculate the energy levels (9.81) in the soliton sector associated with the kink function $\phi_c(x)$. Evaluating the energies $E_{\{n_r, N_p\}}$ is now just a matter of straightforward algebra. We showed in chapter 5 that the counter terms

$\int dx\,(D+\frac{1}{2}\partial m^2(\phi_c^2-1))$ cancelled the divergence in $\sum_p(\frac{1}{2}\omega_p)$ (see eq. (5.69)). In the same way it can be checked that $\int dx\,(\bar{D}+\frac{1}{2}\overline{\partial m^2}\,(\phi_c^2-1))$ will cancel the divergence in $\sum_{r=0}^{\infty}(-\varepsilon_r[\phi_c])$, leaving behind a finite answer for the energy level $E_{\{n_r,N_p\}}$ for any finite excitation numbers $\{n_r, N_p\}$.

Instead of performing a semiclassical expansion about the kink function, one can also try to expand about an extremum of the full effective action $S_{\text{eff}}[\phi]$, including the negative-energy-sea contribution. This is the approximation (9.70), and is as we have stated, much harder to execute. Campbell and Liao (1976) show how this may be done analytically for the model (9.99). Interested readers are referred to their paper, which also includes a generalisation to a chirally invariant model, where the fermion field couples to both a scalar and a pseudoscalar field.

Returning to our calculation of the kink sector for the model (9.99), we had so far paid no special attention to the fact that the lowest fermion mode in (9.106) has zero energy. The energy spectrum *per se* does not have to be modified to order $\hbar$ in the presence of such a zero-mode. One can simply insert $(\varepsilon_{r=0})=0$ in (9.81). But, as Jackiw and Rebbi (1976c) argue, other properties of the states associated with these levels are significantly affected. To see this, let us explore the zero-mode of (9.103) a little further. For $\varepsilon_0=0$, the functions $u_0(x)$ and $v_0(x)$ in (9.104) decouple, and are easily integrated to yield

$$u_0(x)=u_0\exp\Big[g\int_0^x\phi_c(x')\,dx'\Big]=u_0(\cosh x)^g$$

and

$$v_0(x)=v_0\exp\Big[-g\int_0^x\phi_c(x')\,dx'\Big]=v_0(\cosh x)^{-g}. \tag{9.110}$$

We have taken g to be positive. Clearly $v_0(x)$ is normalisable, but $u_0(x)$ is not. Therefore we have to set $u_0=0$, and this gives a single non-degenerate solution for $\varepsilon_0=0$:

$$f_0(x)=v_0\begin{pmatrix}0\\(\cosh x)^{-g}\end{pmatrix}. \tag{9.111}$$

Notice that $\bar{f}_0 f_0=f_0\,\sigma_1 f_0=0$. Thus, the solutions $\phi_c(x)=\tanh x$ and $f(x)=f_0(x)$ in fact together satisfy the coupled equations (9.98). The right-hand side of (9.98b) simply vanishes. Therefore, if only this discrete mode $r=0$ were occupied $(n_r=\delta_{r0})$, then we are actually working with the improved stationary-phase approximation (9.94) and not just with (9.72).

More importantly, since $f_0(x)$ is non-degenerate, its charge conjugate $Cf_0^*(x)$, which must also be a zero-energy solution, must equal f_0 itself, apart from a constant factor. That is, $f_0(x)$ must be self-charge-conjugate, if we appropriately choose the phase of the charge conjugation matrix C. This can be verified. In (1 + 1) dimensions, with the representation (9.100), we can use $C = (-\sigma_3)$ as the charge conjugation matrix. Given any solution $f_r(x)$ of (9.103) with energy ε_r, it is easy to check that $f^c \equiv -\sigma_3 f_r^*$ is also a solution, but with energy $-\varepsilon_r$. For the $\varepsilon = 0$ solution (9.111), which is non-degenerate, one can explicitly see that $Cf_0 = -\sigma_3 f_0 = f_0$.

Now, if there had been no self-charge-conjugate, zero-energy fermion mode, then we could expand the quantum Fermi field as per eq. (9.83b). The 'basic soliton state', i.e. the ground state in the soliton sector, would then be the non-degenerate state satisfying (9.84), with charge $Q = 0$. However, in the presence of the self-conjugate zero-mode, clearly (9.83) is modified to

$$\Psi(x,t) = b_0 f_0(x) + \sum_{r \geqslant 1} \left(b_r e^{-i\varepsilon_r t} f_r^{(+)}(x) + d_r^\dagger e^{i\varepsilon_r t} f_r^{(-)}(x)\right). \tag{9.112}$$

Starting from the basic anticommutator (9.2) for the fields, one can check that b_0 obeys the same standard anticommutation rules as b_r and d_r;

$$\{b_0, b_0^\dagger\}_+ = 1, \qquad \{b_0, b_r\}_+ = \{b_0, d_r\}_+ = 0, \text{etc.}$$

Then the same charge operator Q reduces, in the place of (9.85), to

$$Q = \tfrac{1}{2}\int (\Psi^\dagger\Psi - \Psi\Psi^\dagger)\,dx$$

$$= b_0^\dagger b_0 - \tfrac{1}{2} + \sum_{r \geqslant 1} (b_r^\dagger b_r - d_r^\dagger d_r). \tag{9.113}$$

The state where no fermion mode is excited, now labelled $|\text{sol} -\rangle$, will satisfy analogously to (9.84),

$$b_0|\text{sol} -\rangle = b_r|\text{sol} -\rangle = d_r|\text{sol} -\rangle = 0. \tag{9.114}$$

(For simplicity, we shall consider only fermion modes. Boson excitations can be incorporated as before into the discussion.) With all $n_r = 0$, and no boson excitations ($N_p = 0$), this state $|\text{sol} -\rangle$ will have the same energy M_{sol} given in (9.82). The energy levels to $O(\hbar)$ will not be affected by the zero-mode as we had earlier stated. But the charge of this state will now be, by virtue of (9.113) and (9.114),

$$Q|\text{sol} -\rangle = -\tfrac{1}{2}|\text{sol} -\rangle. \tag{4.115}$$

Contrast this with the 'basic soliton state' we introduced earlier assuming that none of the ε_r vanished, which had $Q = 0$. Next, excite the zero-mode once, i.e., consider

$$|\,\text{sol}\,+\rangle \equiv b_0^\dagger\,|\,\text{sol}\,-\rangle. \tag{9.116a}$$

Clearly

$$\begin{aligned} b_0\,|\,\text{sol}\,+\rangle &= b_0 b_0^\dagger\,|\,\text{sol}\,-\rangle \\ &= (1 - b_0^\dagger b_0)\,|\,\text{sol}\,-\rangle = |\,\text{sol}\,-\rangle. \end{aligned} \tag{9.116b}$$

Since only the zero-energy mode has been excited ($n_r = \delta_{r0}$, $N_p = 0$, $\varepsilon_0 = 0$ in (9.81)), the state $|\,\text{sol}\,+\rangle$ will have the same energy M_{sol} as $|\,\text{sol}\,-\rangle$. But

$$\begin{aligned} Q\,|\,\text{sol}\,+\rangle &= Q b_0^\dagger\,|\,\text{sol}\,-\rangle = +\tfrac{1}{2} b_0^\dagger\,|\,\text{sol}\,-\rangle \\ &= \tfrac{1}{2}\,|\,\text{sol}\,+\rangle, \end{aligned} \tag{9.117}$$

upon using the anticommutation rules for the b's and d's. We therefore have two degenerate lowest-energy states, $|\,\text{sol}\,\pm\rangle$ in the kink sector, with charges $\pm\frac{1}{2}$ respectively. Note that, unlike the non-zero energy modes, which come in pairs $\pm\varepsilon_r$ leading to $0 \leqslant n_r \leqslant 2$ in (9.76), the zero-mode is non-degenerate. It will occur only once in the eigenvalue decomposition (9.60) of Det A_{ϕ_c}. Hence for $r = 0$, $0 \leqslant n_0 \leqslant 1$, i.e. this mode can be excited only once. Thus $|\,\text{sol}\,-\rangle$ and $|\,\text{sol}\,+\rangle$ are the only two independent degenerate lowest-energy states in the kink sector. Furthermore, we can see that the states $|\,\text{sol}\,\pm\rangle$ will be charge conjugates of each other. The field theoretical charge conjugation operator $\mathscr{C}$ transforms fields by

$$\begin{aligned} \mathscr{C}\,\Psi(x,t)\,\mathscr{C}^{-1} &= C\Psi^{+}(x,t), \\ \mathscr{C}\,\Phi(x,t)\,\mathscr{C}^{-1} &= \Phi(x,t), \end{aligned} \tag{9.118}$$

where C is the charge conjugation matrix introduced earlier (see Bjorken and Drell 1965; in our notation, the matrix C differs from theirs by a factor γ^0). This reduces, given the expansion (9.112), to

$$\mathscr{C}\,b_r\,\mathscr{C}^{-1} = d_r; \qquad \mathscr{C}\,b_r^\dagger\,\mathscr{C}^{-1} = d_r^\dagger, \qquad \text{for } r \geqslant 1$$

and

$$\mathscr{C}\,b_0\,\mathscr{C}^{-1} = b_0^\dagger. \tag{9.119}$$

Now, $\mathscr{C}$ is unitary and a symmetry of the Hamiltonian of the model (9.99). Also, $|\,\text{sol}\,+\rangle$ is the only other independent state with the same energy as $|\,\text{sol}\,-\rangle$. Hence,

$$\mathscr{C}\,|\,\text{sol}\,-\rangle = \alpha\,|\,\text{sol}\,-\rangle + \beta\,|\,\text{sol}\,+\rangle \qquad \text{with } |\alpha|^2 + |\beta|^2 = 1. \tag{9.120}$$

But

$$\langle \text{sol} - |\mathscr{C}|\text{sol} - \rangle = \langle \text{sol} + |b_0^+\mathscr{C}|\text{sol} - \rangle, \qquad \text{using (9.116b)}$$
$$= \langle \text{sol} + |\mathscr{C}b_0|\text{sol} - \rangle, \qquad \text{using (9.119)}$$
$$= 0, \qquad \text{using (9.114).}$$

Thus, $\alpha = 0$, $\beta = 1$ in (9.120), and

$$\mathscr{C}|\text{sol} - \rangle = |\text{sol} + \rangle. \tag{9.121}$$

In summary, we have two degenerate lowest-energy states $|\text{sol} \pm \rangle$ in the kink sector, which carry charge $\pm\frac{1}{2}$, and are charge conjugates of each other. Each will be a bound quantum soliton particle. Fermionic excitations of each of these two states can be obtained as before by applying the remaining creation operators $b_r^\dagger$ and $d_r^\dagger$; $r \geqslant 1$. For instance

$$(b_{r_1}^\dagger b_{r_2}^\dagger \ldots b_{r_n}^\dagger d_{\rho_1}^\dagger d_{\rho_2}^\dagger \ldots d_{\rho_m}^\dagger)|\text{sol} - \rangle,$$

with distinct r_i and distinct ρ_j, will yield a state with charge $(n - m - \frac{1}{2})$, whose conjugate will be the state

$$(d_{r_1}^\dagger d_{r_2}^\dagger \ldots d_{r_n}^\dagger b_{\rho_1}^\dagger \ldots b_{\rho_m}^\dagger)|\text{sol} + \rangle,$$

with charge $(m - n + \frac{1}{2})$. Both will have equal energy, equal to

$$M_{\text{sol}} + \sum_{i=1}^{n} \varepsilon_{r_i} + \sum_{j=1}^{m} \varepsilon_{\rho_j},$$

to order $\hbar$. It is clear that all the energy eigenstates in the kink sector consist of such pairs of charge conjugate states, the states within each pair having equal energy but opposite charges. Further, all these states have half-integral charge. This property will clearly be unaffected when boson excitations are incorporated.

An identical discussion can be carried out separately in the antikink sector, built around the classical antikink solution of (9.101).

Remember that the model (9.99) also includes the vacuum sectors built around $\phi = \pm 1$. In these sectors, fermionic excitations over the vacuum are the usual elementary fermions and antifermions, all carrying integral charges, where the same charge operator $Q = \frac{1}{2}\int [\Psi^\dagger, \Psi]\,dx$ is being used. Of course, we could add by hand a constant $\frac{1}{2}$ to the operator Q. This would render the charges in the kink sector integral, but now the charges in the vacuum sector would be half-integral! Therefore the difference, by a half-integer, between charges in the vacuum sector on the one hand, and charges in the kink sector on the other, seems to be an intrinsic feature brought about by the presence of the self-conjugate zero-energy mode (9.111). If only we could do experiments on all sectors of this two-dimensional model, this difference would be a measurable quantity. For instance if the system (9.99) were imbedded in an external electromagnetic field (by adding a term $\frac{1}{2}eA_\mu[\bar{\Psi}, \gamma_\mu \Psi]$ to the Lagrangian density (9.99)), then the eQ measures the electric charge as probed by the field A_μ. Particles from the kink sector should reveal half-integral electric charges, while those in the vacuum sector would reveal integral electric charges.

9.5. Further comments

We have just studied the implications of the zero-energy fermion mode present in the kink sector of the model (9.99). The presence of such a zero-mode is not special to the kink solution, or to the model (9.99). It is clear from (9.110) that the crucial feature of the kink solution, which made the normalizable zero-mode $f_0(x)$ possible, is that $\phi_c(+\infty)$ and $\phi_c(-\infty)$ had non-zero values of opposite signs. Any other topological solution with the same property would also yield such a zero-mode. Nor is this phenomenon limited to $(1+1)$ dimensions. Recall from chapter 3 that the non-abelian Higgs theory in $(3+1)$ dimensions (eq. (3.42)) yielded the 'tHooft–Polyakov monopole solution. This is also a static solution carrying a topological index. Jackiw and Rebbi (1976c) couple this system (3.42) to massless fermions. In the monopole sector of the enlarged theory, the relevant fermion modes will, as per our general method, be eigenfunctions of the one-particle Dirac Hamiltonion in the background field of the classical monopole solution. Jackiw and Rebbi take the fermions to be iso-spinors under the internal SU(2) group of the theory and find that there is again a non-degenerate zero-energy fermion mode. Similarly consider the Yang–Mills system in Euclidean four dimensions, discussed in chapter 4. It yielded the classical instanton solution carrying a topological index. 'tHooft (1976a) has shown that the massless Dirac equation in Euclidean four dimensions, in the background field of the Yang–Mills instanton, again yields a zero eigenvalue. We shall discuss this in chapter 11. The presence of fermionic zero-modes in these different examples, all involving some topological boson solutions, is clearly not coincidental. We understand that this phenomenon is related to basic theorems in mathematics (Atiyah and Singer 1968, Atiyah et al. 1976; see also Kiskis 1977) which lie, however, far beyond the scope of our presentation.

All the stable static soliton solutions of classical boson field equations discussed in the book so far have carried some non-zero topological index. This is because, at least for the classes of Lagrangians we have considered, static solutions seem to need a conserved topological index to keep them stable against collapsing into the trivial solution. Therefore, when these solutions are coupled to fermions, the presence of fermionic zero-modes becomes less an exception than the rule. However, time-dependent solutions of the various models we have studied need not carry any topological index. Recall that the doublet solutions of the sine–Gordon model satisfy $\phi(x,t) \to 0$ as $x \to \pm\infty$, and are non-topological. See also the approximate, time-dependent solutions given by Friedberg et al. (1976a) for three-dimensional scalar field models, where again $\phi(\mathbf{x},t) \to 0$ as $\mathbf{x} \to \infty$. In these cases, when the Dirac field is added through Yukawa coupling there need in general be no zero-modes of fermions. (In the context of time-dependent solutions $\phi(\mathbf{x},t)$, by the term fermionic zero-modes, we mean cases where the Floquet indices vanish.) The discussion of section (9.3) can therefore be adapted without modification. See Friedberg and Lee (1977) for detailed calculations using time-dependent solutions in $(3+1)$ dimensions.

We have been discussing models in which the Lagrangian involves the Dirac field in a bilinear form. All cases of interest in $(3+1)$ dimensions come under this category, because interactions involving a product of four Fermi fields or more are not renormalisable. This can be checked by the usual argument of counting powers of momenta in Feynman graphs. Coming to $(1+1)$ dimensions, four-fermion interactions can be renormalisable. A prototype example is the model

$$\mathscr{L}_1(x,t) = \bar{\Psi}(\mathrm{i}\partial\!\!\!/ - m)\Psi + \tfrac{1}{2}g^2(\bar{\Psi}\Psi)^2. \tag{9.122}$$

Now, compare this with an alternative model

$$\mathscr{L}_2(x,t) = \bar{\Psi}(\mathrm{i}\partial\!\!\!/ - m)\Psi + g\phi\bar{\Psi}\Psi - \tfrac{1}{2}\phi^2, \tag{9.123}$$

where $\phi(x,t)$ is a scalar field. It is easy to see that the models (9.122) and (9.123) are equivalent. This can be shown by either canonical or functional-integral methods. The Euler–Lagrange equations arising from $\mathscr{L}_2(x,t)$ are

$$\phi(x,t) = g\bar{\Psi}(x,t)\Psi(x,t) \tag{9.124a}$$

$$(\mathrm{i}\partial\!\!\!/ - m)\Psi = -g\phi\Psi. \tag{9.124b}$$

Equation (9.124a) is not a genuine 'equation of motion' for the field $\phi(x,t)$ since no time derivative of ϕ is involved. This is because the Lagrangian

(9.123) contained no kinetic terms for the field ϕ. The field ϕ is not an independent degree of freedom in the system (9.123), and eq. (9.124a) is a constraint equation which fixes ϕ, given Ψ and $\bar{\Psi}$ at each (x, t). The standard theory (Dirac 1950, Mukunda and Sudarshan 1974) of constraints, of which (9.124a) is an elementary example, permits this constraint to be inserted into the Lagrangian. When ϕ is eliminated from the Lagrangian (9.123) by inserting (9.124a), we obtain the four-fermion interaction model (9.122). This equivalence can also be shown using functional integrals. On applying the functional-integral formula (9.64) to the model (9.123), we have

$$\begin{aligned}\operatorname{Tr}[e^{-iHT}] &= \int \mathscr{D}[\phi]\mathscr{D}[\Psi^{\dagger}]\mathscr{D}[\Psi]\, e^{iS_{\Psi}} \exp(i\textstyle\int dx\, dt\, (g\phi\bar{\Psi}\Psi \\ &\quad -\tfrac{1}{2}\phi^{2})) \\ &= \int \mathscr{D}[\Psi^{\dagger}]\mathscr{D}[\Psi]\, e^{iS_{\Psi}} \int \mathscr{D}[\phi] \exp(i\textstyle\int dx\, dt \\ &\quad [-\tfrac{1}{2}(\phi - g\bar{\Psi}\Psi)^{2} + \tfrac{1}{2}g^{2}(\bar{\Psi}\Psi)^{2}]),\end{aligned} \tag{9.125}$$

where $S_{\Psi} = \int dx\, dt\, \bar{\Psi}(i\not{\partial} - m)\Psi$ refers to the free-fermion terms in the action. Now, upon changing variables from ϕ to $\tilde{\phi} = \phi - g\bar{\Psi}\Psi$, the boson integration can be performed exactly (the exponent is a pure quadratic), giving some constant A independent of dynamics. Then (9.125) reduces to

$$\operatorname{Tr}[e^{-iHT}] = A \int \mathscr{D}[\Psi^{\dagger}]\mathscr{D}[\Psi] \exp(iS_{\Psi} + i\textstyle\int \tfrac{1}{2}g^{2}(\bar{\Psi}\Psi)^{2}\, dx\, dt). \tag{9.126}$$

This is just the functional-integral representation we would have used for the four-fermion model (9.122), apart from the constant A which can be absorbed into the normalisation. (Although we have not gone into that aspect, these functional integrals can be used not only for finding energy levels, but to obtain all Green's functions of the theory, by adding source terms (see for example Abers and Lee 1973).) Thus, we again find that the models (9.122) and (9.123) are equivalent.

Having transformed the four-fermion interaction model (9.122) into (9.123), which has only bilinear terms in $(\Psi, \Psi^{\dagger})$, we can apply the methods developed in the preceding sections to the latter. The procedure can be trivially extended to other renormalisable four-fermion interactions, such as $(\bar{\Psi}\gamma_5\Psi)^2$. All we need to do is replace the role of the scalar field ϕ by a pseudo-scalar field.

Using this technique, Dashen et al. (1975b) calculate the spectrum of the Gross–Neveu model. This model is described by

$$\mathscr{L}(x, t) = \sum_{j=1}^{N} (\bar{\Psi}_j i\not{\partial}\Psi_j) + \tfrac{1}{2}g^{2}\Big(\sum_{j=1}^{N} \bar{\Psi}_j\Psi_j\Big)^{2}, \tag{9.127}$$

where $j = 1, \ldots, N$ is an index of an internal symmetry $U(N)$, with respect to which $\mathscr{L}(x,t)$ is obviously invariant. Gross and Neveu (1974) studied this theory in the $N \to \infty$ limit and found that it yields spontaneous symmetry breaking. The composite field $\bar{\Psi}_j \Psi_j$ develops a non-zero vacuum expectation value. This opens up the possibility of topological soliton sectors, and Dashen et al. computed the energy spectrum in the different sectors using the semiclassical method described above. By a trivial generalisation of (9.123), the system (9.127) is equivalent to

$$\mathscr{L}'(x,t) = \sum_{j=1}^{N} \bar{\Psi}_j (\mathrm{i}\not{\partial} + g\phi)\Psi_j - \tfrac{1}{2}\phi^2.$$

We know how to treat this Lagrangian. The fact that a multiplet of N fermion species is involved instead of just one is easily incorporated into the functional-integral formalism by using a multiple integral over all the Fermi fields. That is, (9.66) is replaced by

$$\begin{aligned}\mathrm{Tr}[\mathrm{e}^{-\mathrm{i}HT}] &= \int \mathscr{D}[\phi]\mathrm{e}^{\mathrm{i}S_\phi} \int \prod_j (K\mathscr{D}[\Psi_j^\dagger]\mathscr{D}[\Psi_j]) \\ &\quad \exp\left(\mathrm{i}\int \mathrm{d}t\,\mathrm{d}x \sum_j \Psi_j^\dagger A_\phi \Psi_j\right) \\ &= \int \mathscr{D}[\phi] \exp(\mathrm{i}S_{\mathrm{eff}}[\phi]),\end{aligned} \tag{9.128}$$

where

$$S_\phi \equiv \int \mathrm{d}x\,\mathrm{d}t\,(-\tfrac{1}{2}\phi^2)$$

$$S_{\mathrm{eff}}[\phi] \equiv S_\phi - \mathrm{i}N \ln(K\,\mathrm{Det}\,A_\phi)$$

and A_ϕ is the same as in (9.67).

The rest of the semiclassical calculation proceeds as in section 9.3. Dashen et al. succeed in employing, as $T \to \infty$, the 'full stationary phase' approximation (9.69)–(9.70), where the classical solution used is an extremum of the entire effective action including the energy of the Fermi sea. This is, as we have pointed out earlier, a difficult technical feat, and readers are referred to their paper for details. The principles of quantisation used are however the same as those described in section 9.3. Indeed, we had abstracted many of those principles from their paper.

As a final comment, note that with the extension of our methods to include Dirac fields we have the basic machinery for constructing approximate models of hadrons using semiclassical field theory. The study of non-perturbative soliton sectors through semiclassical methods is interesting enough as an important theoretical development. But, as mentioned in the Introduction, an added impetus for studying such methods came from the possibility of explaining the structure and spectrum of hadrons, since these are known to be extended particles, and are believed to be described by some relativistic quantum field theory. But while the various soliton solutions and associated quantum states discussed in chapters 2–7 were useful in illustrating the methodology, they were not realistic models for hadrons. Most of the examples we dealt with in these chapters were two-dimensional scalar-field models. We did study a $(3+1)$-dimensional gauge theoretical example in section 3.3 but it carried a monopole charge – an interesting feature, but one not known to be shared by real hadrons. Most importantly, the systems discussed in earlier chapters lacked the ingredient of Fermi fields, whereas a crucial component of hadrons is the fermionic quark.

Now that we have incorporated Fermi fields into the method, we have the requisite theoretical framework for semiclassical models of real hadrons. Hadrons are widely believed to consist of quarks held together by gauge bosons. As a starting approximation, one can use scalar bosons in the place of vector-gauge bosons. Then we have typically a model of the form (9.63). (If one wanted to be more realistic, gauge fields could be used. While being technically more complicated, this involves the same principles, augmented by gauge-fixing requirements.) Given a model of the form (9.63) or (9.65), the rest of our semiclassical method follows. In particular, we shall be interested mostly in a small number of low-lying fermionic excitations. Baryons, to a good approximation, are bound states of three quarks, while mesons are bound states of a quark–antiquark pair. Thus, one can realistically attempt the intermediate approximation outlined at the end of section 9.3, where the classical solution obeys the coupled equations (9.97) and (9.59). Furthermore, to leading order in the semiclassical expansion, we have seen that the energy, form factors etc., of the quantum extended particle are given by the corresponding properties of the classical solution itself. Therefore, in the leading semiclassical approximation, for what it is worth, one merely solves the classical coupled equations of the type (9.98). This has been done in several papers, notably the one on the SLAC bag (Bardeen et al. 1975), where several hadronic properties, including their quantum numbers, form factors etc., are calculated from such classical solutions. Our discussion in this chapter provides the theoretical framework underlying such calculations.

CHAPTER 10

INSTANTONS IN QUANTUM THEORY

10.1. Instantons and the periodic potential

In chapter 4, we presented several examples of instanton solutions. These were finite-action classical solutions of Euclidean equations of motion, and the examples given ranged from non-relativistic one-particle systems, to relativistic field systems including gauge theories. The motivation for studying instantons was far from clear from our discussion at the classical level, since they are solutions of Euclidean equations where the time variable is essentially imaginary. We will now show that, despite their Euclidean origins, instantons can play an important role in the physics of the corresponding quantum theories.

As we did while quantising solitons, let us again begin with non-relativistic quantum mechanics and introduce to the reader instanton effects in this familiar context. For example, Polyakov (1977b) and Gildener and Patrascioiu (1977) had applied instanton methods to a particle in a double-well potential. Another popular example is the periodic-potential problem (Coleman 1979). We shall use the latter example for illustration since it offers a closer analogy to instanton physics in gauge theories.

Consider a unit-mass particle in one dimension, under a potential $V(q)$ satisfying

$$V(q) = V(q+2\pi) \tag{10.1}$$

for all q in the range $[-\infty, \infty]$. Further, let the minima of $V(q)$ lie at $q = 2N\pi$. Without loss of generality, let $V(2N\pi) = 0$. As a concrete example one can bear in mind the specific case

$$V(q) = 1 - \cos q \tag{10.2}$$

but our considerations will hold for any smooth bounded periodic potential with an absolute minimum within each period (fig. 18).

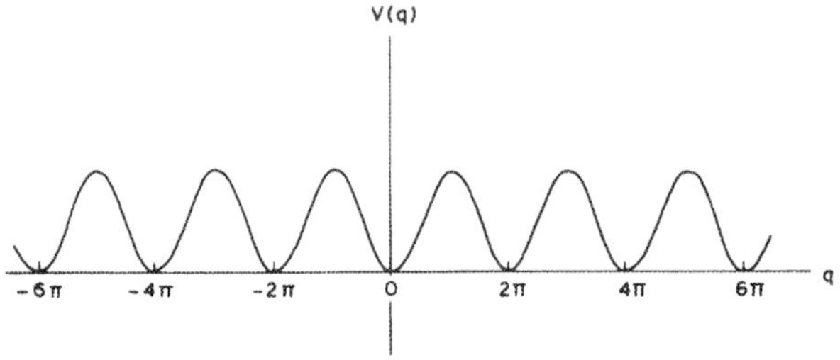

Fig. 18. A periodic potential $V(q)$ of period 2π, with its minima at $q = 2N\pi$.

The periodic-potential problem is a very familiar one, particularly in solid-state physics where it is a model for the behaviour of electrons in a one-dimensional lattice (Kittel 1959). The following properties, obtainable by standard methods in quantum mechanics without recourse to instantons, are well known.

(i) Expand $V(q)$ around the minimum at $q = 0$. Let

$$V(q) = \tfrac{1}{2}\omega^2 q^2 + \sum_{r=3}^{\infty} \lambda_r \frac{q^r}{r!}. \tag{10.3}$$

We shall discuss the low-lying levels of this system under weak-coupling conditions $\lambda_r \ll \omega^2$. This is also the 'tight-binding' approximation for the electrons in the lattice. In this approximation, tunnelling of low-energy wavefunctions from one potential well in fig. (18) to the neighbouring wells will be small. If we were to neglect tunnelling altogether, we could construct a ground state wavefunction $u_0(q)$ in the well around $q = 0$, with some energy E_0. To lowest order in λ_r in (10.3), $E_0 = \frac{1}{2}\hbar\omega$. Since the problem is periodic, we could also construct a similar ground state $u_0(q - 2N\pi)$ around any of the other minima $q = 2N\pi$, again with energy E_0. In the absence of tunnelling, then, we would have an infinitely degenerate ground state with energy $E_0 \simeq \frac{1}{2}\hbar\omega$.

(ii) In actuality, tunnelling effects, although small, will be present. The single level E_0, which was infinitely degenerate in the absence of tunnelling, will split into a band. The associated energy eigenfunctions will, to leading order, be linear combinations of the unperturbed wave packets in the separate wells.

$$\phi_\theta(q) = \sum_{N=-\infty}^{\infty} e^{iN\theta}\, u_0(q - 2N\pi) \tag{10.4}$$

where θ parametrises the states in the band. The coefficients $e^{iN\theta}$ are so chosen that the states (10.4) are eigenfunctions of the unitary operator effecting the symmetry $q \to q + 2\pi$, under which the Hamiltonian is invariant. Under this symmetry

$$\phi_\theta(q + 2\pi) = e^{i\theta}\, \phi_\theta(q). \tag{10.5}$$

(iii) Bloch's theorem (Bloch 1928, Floquet 1883) states that energy eigenfunctions in a periodic potential must obey

$$\phi_k(q) = e^{ikq}\, v_k(q), \tag{10.6}$$

where k is a wavenumber, and $v_k(q)$ is periodic under lattice displacement, in our case under $q \to q + 2\pi$. Under $q \to q + 2\pi$, (10.6) gives

$$\phi_k(q + 2\pi) = e^{i2\pi k}\, \phi_k(q), \tag{10.7}$$

which agrees with (10.5) under the identification $2\pi k = \theta$.

(iv) The energy of the levels in this lowest band are given by (see Appendix J of Kittel (1959)),

$$E_k \simeq E_0 - \alpha(\cos ka), \tag{10.8}$$

where α is a constant, and a is the lattice size. Using for our case $a = 2\pi$, $k = \theta/2\pi$ and $E_0 \simeq \frac{1}{2}\hbar\omega$, we can write (10.8) as

$$E_\theta \simeq \tfrac{1}{2}\hbar\omega - \alpha\cos\theta. \tag{10.9}$$

The system also includes higher-energy bands, but we will be interested only in the lowest band arising from the splitting of the ground state.

Let us now proceed to show how these well known results may also be derived using instantons and the Euclidean functional-integral methods of chapter 6. At the end of the discussion it will be clear as to why Euclidean path integrals (and hence instantons) are more useful than the real-time path integral for studying such tunnelling effects by semiclassical methods.

Consider the Euclidean amplitude defined in eq. (6.30)–(6.31) for the transition from $q = 0$ at $\tau' = (-\tau/2)$ to $q = 2\pi$ at $\tau' = (+\tau/2)$:

$$\begin{aligned} K_E(2\pi, \tau/2; 0, -\tau/2) &\equiv \langle 2\pi | e^{-H\tau/\hbar} | 0 \rangle \\ &= \int \mathscr{D}_E[q(\tau')] \exp\{-(1/\hbar) S_E[q(\tau')]\} \\ &= \sum_n \langle 2\pi | \phi_n \rangle \langle \phi_n | 0 \rangle e^{-E_n\tau/\hbar}. \end{aligned} \tag{10.10}$$

Here $|\phi_n\rangle$ are energy eigenstates, $|0\rangle$ and $|2\pi\rangle$ are position (q) eigenstates, $S_E[q(\tau')]$ is the Euclidean action, and all paths go from $q(-\tau/2)=0$ to $q(\tau/2)=2\pi$. We are interested in the lowest-energy band of our periodic-potential problem. The sum $\sum_n$ in (10.10) will be dominated by this band in the large τ limit. We shall therefore work in this limit. It is to be taken as understood that the arguments given below are meant to hold only in the $\tau\to\infty$ limit, even though we shall display some τ-dependence explicitly.

We shall evaluate the functional integral in (10.10) in the semiclassical Gaussian approximation, by expanding $S_E[q(\tau')]$ about its extrema. These extrema obey

$$\frac{\delta S_E}{\delta q(\tau')} = -\frac{d^2 q}{d\tau'^2} + \frac{dV}{dq} = 0 \tag{10.11}$$

This is just the classical Euclidean equation of motion. Its finite-action solutions (satisfying $q(-\tau/2)=0$, $q(\tau/2)=2\pi$ with $\tau\to\infty$) are the classical instantons of this system. The anti-instanton solution will satisfy $q(-\tau/2)=2\pi$, $q(+\tau/2)=0$, with $\tau\to\infty$.

Now, in chapter 4, we have pointed out that finite-action solutions of (10.11) are the same as finite-energy static solitons of the corresponding $(1+1)$-dimensional scalar field theory with a potential $V(\phi)$. In chapter 2, we have discussed the existence and properties of such static solitons. Combining this discussion in chapters 2 and 4, we know that a finite-action instanton solution to (10.11) will exist for any periodic potential of the form depicted in fig. 18. Let this solution, $q_{cl}(\tau')$, have an action $S_E[q_{cl}(\tau')]=S_0$. This solution has been schematically sketched in fig. 19(a). We shall not be interested in its precise shape, which will depend on the details of $V(q)$, except that it will asymptotically tend exponentially fast to zero as $\tau'\to-\infty$ and to 2π as $\tau'\to\infty$. It will carry a topological index

$$Q=(1/2\pi)[q(\infty)-q(-\infty)]=1.$$

(We know that these features will exist from the discussion in chapter 2. For the specific example of $V(q)$ in (10.2), this instanton will be the same function as the sine–Gordon soliton.)

Let us evaluate the contribution from the neighbourhood of the single-instanton solution $q_{cl}(\tau')$ to the functional integral in (10.10). Using the Gaussian approximation in chapter 6, this contribution is

$$e^{-S_0/\hbar}B_E(\tau)\{\mathrm{Det}[-d^2/d\tau'^2+V''(q_{cl})]\}^{-1/2}, \tag{10.12}$$

where $B_E(\tau)$ is the measure factor in the functional integral.

At this stage, two related points must be made, which arises when τ, the

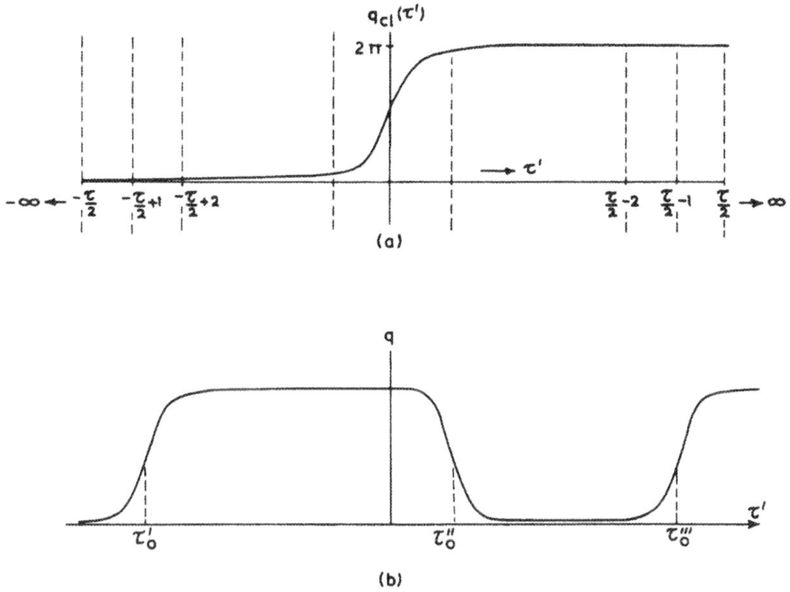

Fig. 19. (a) Schematic sketch of a one-instanton solution $q_{cl}(\tau')$. Its detailed shape will depend on the choice of $V(q)$. The range of τ' is $[-\tau/2, \tau/2]$, with $\tau \to \infty$. The dashed vertical lines divide this interval into segments $[-\tau/2, -\tau/2+1]$, $[-\tau/2+1, -\tau/2+2] \ldots [\tau/2-1, \tau/2]$. In most of these segments, $V''[q_{cl}(\tau')] \simeq V''(0) = V''(2\pi)$. (b) Sketch of a two-instanton, one-anti-instanton solution. The solution is exact when their locations τ_0', τ_0'' and τ_0''' are infinitely far apart.

total range of τ', tends to infinity. On the one hand, the operator

$$-d^2/d\tau'^2 + V''(q_{cl})$$

will have a zero eigenvalue because the action is translationally invariant in the τ' variable. Formally, this will cause a divergence in (10.12). On the other hand, due to the same translational invariance, given $q_{cl}(\tau')$, $q_{cl}(\tau' - \tau_0)$ will also be an instanton solution for any τ_0 with the same action S_0. Therefore each of the translated partners of $q_{cl}(\tau')$ will make the same contribution as (10.12). This cumulative contribution of all single instantons would then require that we multiply (10.12) by $\int_{-\tau/2}^{\tau/2} d\tau_0 = \tau$. These two features, namely the existence of a zero-mode and the need to integrate over the corresponding collective coordinate τ_0, are, as we have discussed in chapter 8, reflections of the same phenomenon. The correct way to handle both features, as mentioned there, is to replace the zero-mode coordinate by the collective coordinate through an appropriate change of variables. In Appendix B, we show how this may be done for the present example, in the functional-integral formalism. The resulting contribution of all single-instantons to (10.10) is (see eq. (B.14)),

$$\langle 2\pi | e^{-H\tau/\hbar} | 0 \rangle_{1,0} = e^{-S_0/\hbar} J \tau B_E(\tau) \{ \mathrm{Det}' [-d^2/d\tau'^2 + V''(q_{cl})] \}^{-1/2}. \tag{10.13}$$

Here, Det′ stands for the determinant with the zero-mode removed, J is the Jacobian factor in eq. (B.13) and the subscript (1, 0) stands for the fact that it is the contribution of one instanton and no anti-instantons. Further contributions are forthcoming. The next step is to notice that the instanton of this problem has 'finite size'. That is, the instanton solution $q_{cl}(\tau' - \tau_0)$ is very near the trivial solutions, either $q = 0$ or $q = 2\pi$, for most values of τ' (see fig. 19(a)). This is a general feature not dependent on the detailed functional form of $V(q)$ as long as $V''(0) = \omega^2 \neq 0$. It is only in some finite region of τ' that the solution significantly differs from 0 or 2π. The action is also localised within this region, except for an exponential tail. The size of this region, i.e. the 'size of the instanton', depends on the parameters of $V(q)$ such as ω^2, but not on τ, the overall range of τ'. Now, the total range of τ' is $[-\frac{1}{2}\tau, \frac{1}{2}\tau]$, with $\tau \to \infty$. This range may be divided into subintervals $[-\frac{1}{2}\tau, -\frac{1}{2}\tau + 1], [-\frac{1}{2}\tau + 1, -\frac{1}{2}\tau + 2], \ldots\ldots, [\frac{1}{2}\tau - 1, \frac{1}{2}\tau]$. The operator $e^{-H\tau/\hbar}$, and the path integral in (10.10) may be written as products of the corresponding quantitities in the subintervals. In most of the subintervals, $q_{cl} \simeq 0$ or 2π, $V''(q_{cl}) \simeq (d^2V/dq^2)_{0,2\pi} = \omega^2$, and the operator $-\partial^2_{\tau'} + V''(q_{cl}) \simeq -\partial^2_{\tau'} + \omega^2$. This will fail only in the finite interval whose size is independent of τ and where q_{cl} differs substantially from 0 or 2π. We can therefore write

$$\{ \mathrm{Det}' [-\partial^2_{\tau'} + V''(q_{cl})] \}^{-1/2} \equiv \{ \mathrm{Det}(-\partial^2_{\tau'} + \omega^2) \}^{-1/2} K. \tag{10.14}$$

This equation merely defines K, but the thrust of the argument above is that K is a constant independent of τ as $\tau \to \infty$. Hence (10.13) becomes

$$\langle 2\pi | e^{-H\tau/\hbar} | 0 \rangle_{(1,0)} = e^{-S_0/\hbar} J K \tau B_E(\tau) [\mathrm{Det}(-\partial^2_{\tau'} + \omega^2)]^{-1/2}. \tag{10.15}$$

We already know the value of $B_E(\tau)[\mathrm{Det}(-\partial^2_{\tau'} + \omega^2)]^{-1/2}$. It is just the value of the Gaussian functional integral, which is the analytic continuation to $\tau = iT$ of the harmonic oscillator path integral. The latter has been given in eq. (6.19) in the form

$$B'(T)\{\operatorname{Det}(-\partial^2/\partial t^2-\omega^2)\}^{-1/2},$$

whose value (see (6.21) and (6.24)) was

$$\left(\frac{1}{2\pi\mathrm{i}\hbar T}\,\frac{\omega T}{\sin\omega T}\right)^{1/2}.$$

Continuing to $\tau = \mathrm{i}T$,

$$B_{\mathrm{E}}(\tau)[\operatorname{Det}(-\partial^2_{\tau'}+\omega^2)]^{-1/2}=\left(\frac{1}{2\pi\hbar}\,\frac{\omega}{\sinh\omega\tau}\right)^{1/2}$$

$$\xrightarrow[\tau\to\infty]{}\left(\frac{\omega}{\pi\hbar}\right)^{1/2}\mathrm{e}^{-\omega\tau/2}. \tag{10.16}$$

Hence

$$\lim_{\tau\to\infty}\langle 2\pi|\mathrm{e}^{-H\tau/\hbar}|0\rangle_{(1,0)}=\mathrm{e}^{-S_0/\hbar}JK\tau\left(\frac{\omega}{\pi\hbar}\right)^{1/2}\mathrm{e}^{-\omega\tau/2}. \tag{10.17}$$

This is the contribution of single instantons.

Next, consider a configuration consisting approximately of an instanton located at τ_0', followed by an anti-instanton at $\tau_0''\gg\tau_0'$, followed again by an instanton at $\tau_0'''\gg\tau_0''$ (fig. 19(b)). The configuration will have the correct boundary conditions for our problem, namely $q(-\infty)=0$, $q(\infty)=2\pi$. As long as the instantons are separated by a finite distance (along the τ' axis), this configuration will not in general be an exact classical solution. But as the separations $\tau_0''-\tau_0'$ and $\tau_0'''-\tau_0''$ increase, such a configuration will solve (10.11) to better and better accuracy since the overlap between the instantons becomes arbitrarily small. In fact, this remark holds for any set of n_1 instantons and n_2 anti-instantons all widely separated from one another. As long as $n_1-n_2=1$, the boundary conditions will be satisfied. Since we are working in the $\tau\to\infty$ limit where τ is the range of the variable τ', configurations of such instantons with arbitrarily large separations are permitted. Indeed, it is clear that instantons with mutual separation greater than R will occupy infinitely more phase space than those with separation less than R, for any R however large.

Strictly speaking, for any finite separation, however large, such multi-instanton–antiinstanton configurations will not be stationary points of the action. However, the first derivative $\delta S[q]/\delta q(\tau')$ will be very small and will tend to zero as the mutual separation increases. Consequently a Gaussian approximation of the functional integrand about such a configuration, where we would drop this first derivative, will be increasingly well justified as the separation increases. Therefore we must include in our semiclassical expansion contributions from such widely separated instantons and anti-instantons as well (see also Bogomol'nyi 1980).

The contribution of such asymptotically exact multi-instanton solutions to $\langle 2\pi|e^{-H\tau/\hbar}|0\rangle$ is easy to obtain within our approximation, given the single-instanton contribution (10.17). Since the n_1 instantons and n_2 anti-instantons are taken to be widely separated, each instanton (or anti-instanton) will give an additive contribution S_0 to the action. Each will give a factor K to the determinant since $V''(q_{cl})$ will differ from ω^2 to the same extent in the vicinity of each. Finally, subject to the constraint that they be widely separated (i.e. separation much greater than instanton size) each instanton (or anti-instanton) can be located anywhere on the τ'-axis. Hence, integration over its location will produce a factor $J\tau$. In allowing each instanton to be anywhere on the τ'-axis, we are also including configurations where they will not be widely separated. However, the phase space for configurations where some instantons are near to each other will be infinitesimally small compared with that of configurations where they are all far apart. This is a purely kinematic phase-space statement in the $\tau \to \infty$ limit, given instantons of finite size. That the dynamics does not reverse this phase-space argument needs to be checked. We shall do so later (see comment (iv) in the discussion below).

Therefore, altogether, the contribution of n_1 instantons and n_2 anti-instantons will be

$$\lim_{\tau\to\infty}\langle 2\pi|e^{-H\tau/\hbar}|0\rangle_{(n_1,n_2)}$$
$$= [\delta_{n_1-n_2,1}]\frac{\exp[-(n_1+n_2)S_0/\hbar]}{n_1!\,n_2!}(JK\tau)^{n_1+n_2}e^{-\omega T/2}\left(\frac{\omega}{\pi\hbar}\right)^{1/2} \tag{10.18}$$

where $\delta_{n_1-n_2,1}$ is the Kronecker symbol. The factors $(1/n_1!)\,(1/n_2!)$ arise because of indistinguishability. We have allowed each instanton to be located freely over the entire τ'-axis. We shall clearly be over-counting configurations where instantons merely exchange positions–hence the $1/n_1!$. The same argument holds for anti-instantons. Note, incidentally, that in our problem instantons and anti-instantons need not alternate in sequence. Since not only $q = 0$ or 2π, but all $q = 2N\pi$ are zeroes of $V(q)$, one can have finite-action configurations where the n_1 instantons and n_2 anti-instantons are arranged in arbitrary sequence. The only requirement, because of boundary conditions, is $n_1 - n_2 = 1$. This is to be contrasted with the double-well potential problem, which has only two minima, and consequently instantons would have to alternate with anti-instantons in order to maintain finite action.

Returning to our problem, the total contribution in the semiclassical approximation comes from summing (10.18) over n_1 and n_2.

$$\lim_{\tau\to\infty} \langle 2\pi | e^{-H\tau/\hbar} | 0 \rangle = \left(\frac{\omega}{\pi\hbar}\right)^{1/2} e^{-\omega\tau/2} \sum_{n_1, n_2} \frac{(JK\tau e^{-S_0/\hbar})^{n_1+n_2}}{n_1! n_2!} [\delta_{n_1-n_2,1}]. \tag{10.19}$$

The sums over n_1 and n_2 can be decoupled using the identity

$$\int_0^{2\pi} \frac{d\theta}{2\pi} \exp[-i\theta(n_1 - n_2 - 1)] = \delta_{n_1-n_2,1}. \tag{10.20}$$

With this decoupling the sums over n_1 and n_2 are trivial, giving

$$\lim_{\tau\to\infty} \langle 2\pi | e^{-H\tau/\hbar} | 0 \rangle = \lim_{\tau\to\infty} \sum_n \langle 2\pi | \phi_n \rangle \langle \phi_n | 0 \rangle e^{-E_n\tau/\hbar}$$

$$= \int_0^{2\pi} \frac{d\theta}{2\pi} e^{i\theta} \left(\frac{\omega}{\pi\hbar}\right)^{1/2} \exp(e^{-S_0/\hbar} 2JK\tau \cos\theta - \tfrac{1}{2}\omega\tau). \tag{10.21}$$

This is our main result. It may be noticed that the well-known features of the lowest-energy band have been reproduced. The low-lying energy levels, which will dominate the left-hand side of (10.21) as $\tau \to \infty$, can be seen from the right-hand side to form a continuous band parametrised by θ, $0 \leqslant \theta \leqslant 2\pi$. The energy of the band levels is clearly

$$E_\theta = \tfrac{1}{2}\hbar\omega - \hbar(2JK e^{-S_0/\hbar} \cos\theta). \tag{10.22}$$

This agrees with the expression (10.9), with the constant α appearing there equal to $2JK\hbar e^{-S_0/\hbar}$ in the semiclassical approximation. 2α represents the band-width. Given an explicit expression for $V(q)$, we can in principle find the instanton solution $q_{cl}(\tau')$, its action $S_0[q_{cl}]$, and the Jacobian $J = \sqrt{S_0}$. The constant K can also be obtained from (10.14), using the eigenvalues of $-\partial_{\tau'}^2 + V''(q_{cl})$. (For the sine–Gordon potential (10.2), for example, these eigenvalues are just the squares of the stability frequencies of the SG soliton, used in section 7.1.) Thus, our semiclassical derivation also permits the evaluation of the band-width, in principle.

Turning to the eigenfunctions $|\phi_\theta\rangle$ associated with the energy levels in (10.22), we see from the expressions equated in (10.21) that

$$\langle 2\pi|\phi_\theta\rangle\langle\phi_\theta|0\rangle = (\mathrm{e}^{\mathrm{i}\theta}/2\pi)\,(\omega/\pi\hbar)^{1/2}. \tag{10.23}$$

Now, if we had calculated the transition amplitude $\langle 2N\pi|\mathrm{e}^{-H\tau/\hbar}|0\rangle$, it is clear that the preceding analysis would go through, with the sole difference that $n_1 - n_2 = N$, instead of 1. This would lead to the replacement of $\mathrm{e}^{\mathrm{i}\theta}$ in (10.21) by $\mathrm{e}^{\mathrm{i}N\theta}$. Hence

$$\langle 2N\pi|\phi_\theta\rangle\langle\phi_\theta|0\rangle = (\mathrm{e}^{\mathrm{i}N\theta}/2\pi)\,(\omega/\pi\hbar)^{1/2}.$$

From this it is clear that

$$\langle 2N\pi|\phi_\theta\rangle = \mathrm{e}^{\mathrm{i}N\theta}\langle 0|\phi_\theta\rangle. \tag{10.24}$$

This is consistent with the property of Bloch waves, given in (10.5) or (10.7). The relation (10.24) gives the quasi-periodicity (10.5) of the wavefunctions, at the points $q = 2N\pi$. A similar relation connecting $\phi_\theta(q)$ to $\phi_\theta(q+2N\pi)$ for arbitrary q can also be obtained by similar semiclassical methods, if one begins with $\langle q+2N\pi|\mathrm{e}^{-H\tau/\hbar}|q\rangle$.

We have thus rederived, using instanton methods, the basic properties listed earlier of the lowest-energy band in a periodic potential. The motivation is the same as in earlier chapters. Similar features exist for field theories, particularly gauge theories, which are most conveniently investigated using Euclidean functional integrals and instantons. With appropriate generalisation, our instanton exercise in the periodic potential problem can be carried over to these more complicated theories. Before going on to gauge theories, the following comments on the preceding derivation may be helpful.

(i) We have employed only finite-action classical solutions, namely, the instantons and anti-instantons. Infinite-action solutions are also formally extrema of the Euclidean action. But their contribution will vanish in the semiclassical expansion. Notice (eq. (10.17)) that the contribution is proportional to $\exp[-S_{\mathrm{cl}}]$.

(ii) In as much as we are studying tunnelling effects, one might ask why we did not use the real-time amplitude $\langle 2\pi|\mathrm{e}^{-\mathrm{i}HT/\hbar}|0\rangle$ for the transition from the bottom of one potential well to the next. Indeed the spectral decomposition of this real-time amplitude does contain, in principle, all the information we seek, and we have used such amplitudes in earlier chapters. But it is not convenient for semiclassical approximations to the quantities we seek now. We are interested in tunnelling effects on low energy levels (i.e. the splitting of the ground state due to tunnelling). The corresponding semiclassical approximation would involve, as evident from chapter 6, low-energy classical orbits. But no low-energy classical paths connecting $q = 0$ and $q = 2\pi$ are permitted, thanks to the potential barrier in between. By contrast, the Euclidean path integral can be expanded around solutions of the Euclidean equation of motion (10.11). This Euclidean equation is identical to the real-time Newton's equation for an associated problem whose potential is $[-V(q)]$. In $[-V(q)]$, the points $q = 0$ and $q = 2\pi$ are maxima instead of minima, and a low-energy solution does exist. These are the instantons of the original problem. This is why, in studying tunnelling effects on low-lying levels through the semiclassical method, Euclidean path integrals and hence instantons come into play.

(iii) Our calculation may be called the 'dilute instanton-gas approximation'. The instantons contributing to (10.21) have been treated in a manner similar to a one-dimensional gas of particles and antiparticles spread along the τ'-axis. The summation over n_1, n_2 in (10.19) is very similar to what occurs in the partition function of a very dilute gas of particles and antiparticles in one dimension, with the constraint that the total 'particle number', $(n_1 - n_2)$, be unity.

(iv) An important assumption has been the diluteness of the instanton gas, i.e. that the instantons and anti-instantons contributing to (10.21) are widely separated. A crucial ingredient for this is obviously that the instantons, to start with, have finite size. For a typical periodic potential, $V(q)$ in fig. (18) with $V''(0) \neq 0$, this will be true. In some field theories, such as the two-dimensional abelian Higgs model discussed in the next section, the instantons again have finite size, so that our derivation can be applied there with suitable generalisation. But, for Yang–Mills theory in four dimensions, which is scale-invariant, instantons come in all sizes. (Recall the arbitrary scale parameter λ_1 in the Yang–Mills instantons derived in chapter 4.) For this system, therefore, the dilute instanton-gas approximation runs into difficulty, as we shall see in section 10.5. Returning to our periodic potential problem, although finiteness of instanton size permits dilute configurations in the $\tau \to \infty$ limit, we must ensure that they do make the dominant contribution to (10.21). The n-dependence of the instanton contribution in (10.19) is of the form $(JK\tau e^{-S_0/\hbar})^n/n!$ which reaches a maximum when $n \simeq JK\tau e^{-S_0/\hbar}$. The corresponding 'instanton-density' is $n/\tau = (JKe^{-S_0/\hbar})$. For small $\hbar$, this indeed corresponds to low density.

(v) In (10.22), it is the second term that represents the effect of tunnelling. Through the factor $e^{-S_0/\hbar}$, this term reveals the expected exponential dependence on $1/\hbar$ as well as on the 'barrier strength'. Recall, from eq. (2.20) of chapter 2 as applied to the present case, that the instanton solution $q_{cl}(\tau')$ will obey

$$\tfrac{1}{2}(\dot{q}_{cl})^2 = V(q_{cl}), \qquad \text{where } \dot{q}_{cl} \equiv dq_{cl}/d\tau'.$$

Hence

$$S_0 = \int_{-\infty}^{\infty} (\tfrac{1}{2}\dot{q}_{cl}^2 + V)\,d\tau' = \int_{-\infty}^{\infty} (\dot{q}_{cl})^2\,d\tau'$$
$$= \int_0^{2\pi} \dot{q}_{cl}\,dq = \int_0^{2\pi} \sqrt{2V(q)}\,dq. \tag{10.25}$$

This quantity represents the strength of the potential barrier between 0 and 2π. This dependence of the tunnelling amplitude on

$$\exp\left(-\frac{1}{\hbar}\int_0^{2\pi} \sqrt{2V}\,dq\right)$$

is in accord with conventional WKB treatments (Landau and Lifshitz 1958).

(vi) We have pointed the similarity between instantons of this problem and static solitons of the corresponding $(1+1)$-dimensional scalar field theory. It is well known that in $(1+1)$-dimensional systems, discrete symmetry will be restored at any temperature, however small, by the formation of solitons and antisolitons ('domain formation'). For a concrete example in the statistical mechanics of a quantum field theory, see Dashen et al. (1975c). In analogous fashion, instantons here have restored the $q \to q+2\pi$ symmetry ('for any $\hbar$, however small'). In the absence of tunnelling, the ground state wavefunction ϕ_0 would be located in any *one* of the potential wells, violating symmetry of ϕ_0 under $q \to q+2\pi$. Such spontaneous symmetry breaking cannot really happen in a system with just one degree of freedom. The reason of course, is that tunnelling will necessarily take place, as represented here by instanton effects. In the resulting energy band (20.22), the real ground state corresponds to $\theta = 0$. Insert $\theta = 0$ in the corresponding wavefunction (10.5), which makes it completely symmetric under $q \to q+2\pi$. Thus, instantons have restored the symmetry of the ground state.

(vii) Contrast the periodic potential problem (PPP) we have just studied, with a closely related problem, where the particle lies on a unit circle described by $0 \leqslant q \leqslant 2\pi$, but with the same Lagrangian $L = \tfrac{1}{2}\dot{q}^2 - V(q)$.

The particle on a circle (POC) will have the same classical equation of motion and the same differential Schrodinger equation as PPP. But the quantum energy levels will be different. This is because the coordinate space has been reduced from the entire real line to just the circle $0 \leqslant q \leqslant 2\pi$, and accordingly the boundary conditions change.

In POC, the points $q = 0$ and $q = 2\pi$ are physically the same. The natural boundary condition one would think of is that all wavefunctions should satisfy $\phi(2\pi) = \phi(0)$. In particular, this amounts to choosing, for the energy eigenfunctions, only those Bloch waves from the set (10.5) for which $\theta = 0$. Corresponding to each band of levels in PPP we would just have one level (the $\theta = 0$ member) for POC. The ground state of POC, for instance, will be the $\theta = 0$ member of the set (10.22) with energy,

$$E_0 \simeq \tfrac{1}{2}\hbar\omega - \hbar(2JK\,\mathrm{e}^{-S_0/\hbar}). \tag{10.26}$$

This result can also be derived *ab initio* using instantons. Since the classical equation for PPP and POC are the same, the same instanton solutions are valid for both. (In POC, a multi-instanton solution is to be interpreted as the particle traversing the circle several times as τ' goes from $-\infty$ to $+\infty$.) We can repeat the same instanton-gas calculation as we did for PPP, with the sole difference that the constraint $n_1 - n_2 = 1$ imposed in (10.19) can be dropped. Any solution with n_1 instantons and n_2 anti-instantons starting from $q = 0$ will return to $q = (n_1 - n_2)\,2\pi$, which is the same physical point for all n_1 and n_2. If we therefore drop the factor $\delta_{n_1 - n_2,\,1}$ from (10.19), the exponential series in n_1 and n_2 are already decoupled, without recourse to (10.20). The result will clearly be

$$\lim_{\tau \to \infty} \langle q = 2\pi|\mathrm{e}^{-H\tau/\hbar}|q = 0\rangle = (\omega/\pi\hbar)^{1/2}\exp(-\tfrac{1}{2}\omega\tau + \mathrm{e}^{-S_0/\hbar}2JK\tau). \tag{10.27}$$

Thus, the energy level E_0 in (10.26) is reproduced. Instanton-based derivations have connotations of tunnelling. For POC, it may appear strange to be talking of tunnelling from the same physical point $q = 0$ to itself. However, the notion of tunnelling has meaning even for POC. In the path integral for the transition amplitude from $q = 0$ back to the same physical point, there are additive contributions from classes of paths which wind zero times, once, twice, etc., around the circle. For instance, in paths that wind once the paricle in going from 0 to 2π definitely overcomes a potential barrier. In multiple traverses, it overcomes the barrier several times. The corresponding contributions to the transition amplitude can certainly be viewed as tunnelling effects.

Equation (10.26) gives, for the ground state of POC, the $\theta = 0$ member of the energy band of PPP. There is a modification by which, without changing the classical physics of POC in any way, we can alter its quantum ground state energy to agree with any $\theta \neq 0$ member of the PPP band. Change the Lagrangian of POC to

$$L\left(q, \frac{dq}{dt}\right) = \frac{1}{2}\left(\frac{dq}{dt}\right)^2 - V(q) - \frac{\theta}{2\pi}\frac{dq}{dt}, \tag{10.28}$$

where θ is a real constant. The extra term added in (10.28) is a total time derivative. It will not affect the classical Euler–Lagrange equations. But the quantum levels will change, as one may anticipate from the change of canonical momentum from $p = \dot{q}$ to $p = \dot{q} - \theta/2\pi$. For different choices of θ in (10.28) one gets different quantum theories, but each can be considered an equally valid quantum version of the parent classical POC system. How does the additional term in (10.28) affect our instanton calculation for POC? The classical instanton solutions would remain the same, but the Euclidean action of a single instanton will be altered to

$$S_\theta = \int_{-\infty}^{\infty} \left[\frac{1}{2}\left(\frac{dq}{d\tau'}\right)^2 + V(q) + \frac{i\theta}{2\pi}\frac{dq}{d\tau'}\right] d\tau'$$

$$= S_0 + \frac{i\theta}{2\pi}\int_0^{2\pi} dq = S_0 + i\theta. \tag{10.29}$$

Since the anti-instanton flows from 2π to 0, its action will be altered to $S_0 - i\theta$. It is easy to trace the resulting change in the algebra, and check that (10.26) will now be replaced by

$$E_\theta = \tfrac{1}{2}\hbar\omega - 2JK\hbar e^{-S_0/\hbar}\cos\theta.$$

This equation looks identical to the result (10.22) for PPP, but whereas in PPP all values of θ exist forming an energy band, in POC only one value of θ exists in a given theory. This is the value of the parameter θ in the Lagrangian (10.28). Instead of a whole band in PPP, there will be only one level in POC even though we may alter this level by varying the parameter θ. This reduction in the number of levels from a band in PPP to a single level in POC is consistent with the fact that the former has a larger configuration space by an infinite factor.

Clearly, the level E_θ in POC corresponds to the particular Bloch function

ϕ_θ with that value of θ. This function obeys $\phi_\theta(2\pi) = e^{i\theta}\phi_\theta(0)$, i.e. the boundary condition is only quasi-periodic (see Bernard and Weinberg 1977, Rothe and Swieca 1978, 1979). It may appear strange that even though the points 0 and 2π are physically the same, yet $\phi_\theta(2\pi) \neq \phi_\theta(0)$. However, as Rothe and Swieca point out, it is perfectly possible to construct a consistent quantum theory for POC with only quasi-periodic boundary conditions. The physical identity of $q = 0$ and $q = 2\pi$ will still be reflected by the fact that matrix elements of physical operators will be the same for $\phi_\theta(2\pi)$ and $\phi_\theta(0)$. The phase factor $e^{i\theta}$ will cancel out in the matrix element.

We have belaboured these features of POC, and its differences from PPP, because an analogous comparison arises while studying topological vacua in gauge theories (Bernard and Weinberg 1977; Wadia and Yoneya 1977). Although the vacuum tunnelling phenomenon in gauge theories bears some analogy to PPP, it is even closer in spirit to POC. This will be discussed in the subsequent sections.

(viii) In PPP, the discrete symmetry of the Hamiltonian under $q \to q + 2\pi$, although broken by the individual 'classical vacua' at $q = 2N\pi$, is restored to the quantum ground state because of tunnelling. In the functional-integral formalism this happens through instantons. One may wonder what prevents the same thing from happening in field theories where some discrete symmetry is known to be spontaneously broken by the vacuum state. Consider for example the familiar $(1+1)$-dimensional sine–Gordon model. Its potential $U(\phi) = 1 - \cos\phi$ has the same features as the PPP potential in fig. 18. To investigate whether tunnelling between neighbouring minima takes place in the model, suppose we compute the amplitude $\langle \phi(x) = 2\pi | e^{-H\tau} | \phi(x) = 0 \rangle$ by generalising the semiclassical method of this section to field theory. We would use the field theoretical Euclidean functional integral, with the Euclidean action

$$S_E = \int d^2x \left(\sum_{\mu=1}^{2} \tfrac{1}{2}(\partial_\mu\phi)(\partial_\mu\phi) + 1 - \cos\phi \right),$$

where $\mu = 1, 2$ corresponds to Euclidean space–time ($x_1 = x$, $x_2 = it$). In the semiclassical method, we would expand fields about the solutions of the Euclidean field equation which is

$$\nabla^2\phi = \sin\phi, \qquad \text{where} \quad \nabla^2 = \frac{\partial^2}{\partial x_1^2} + \frac{\partial^2}{\partial x_2^2}. \tag{10.30}$$

In accordance with the remark in chapter 4, this Euclidean field equation is just the static field equation of the same model, but in $(2+1)$ dimensions.

Finite-action instantons of this Euclidean equation (10.30) would just be the finite-energy static solitons of the (2 + 1)-dimensional model. But the Derrick–Hobart virial theorem (section 3.2) precisely forbids static finite-energy solitons for this model in two space dimensions. Therefore (10.30) has no finite-action instantons. Infinite-action classical solutions do exist for (10.30) but, as pointed out earlier, an infinite-action solution will give no contribution to the semiclassical tunnelling amplitude. In short we see, at least in the semiclassical approximation, that in systems like the SG model in (1 + 1) dimensions, tunnelling between classical minima does not occur because finite-action instantons are absent. The same remark holds for the kink system discussed in chapter 5, where $U(\phi) = \frac{1}{4}\lambda(\phi^2 - 1)^2$. In these systems separate vacua can be built around each potential minimum which will not tunnel into one another. Symmetry will be broken spontaneously.

In order to obtain non-zero tunnelling effects in a field theory through semiclassical methods, we must find models that permit finite-action instantons. In the following sections, we shall consider certain gauge theories which have this feature.

10.2. Topological vacua in the abelian Higgs model

We shall now generalise the method outlined in the preceding section to a gauge field theory. Consider the so-called abelian Higgs model. This system consists of the electromagnetic field $A_\mu(\mathbf{x}, t)$ coupled to a complex scalar field $\phi(\mathbf{x}, t)$ through the Lagrangian density,

$$\mathscr{L}(\mathbf{x}, t) = -\tfrac{1}{4}F_{\mu\nu}F^{\mu\nu} + \tfrac{1}{2}(D_\mu\phi)^* (D^\mu\phi) - \tfrac{1}{4}\lambda(|\phi|^2 - F^2)^2 \qquad (10.31)$$

where $F_{\mu\nu} \equiv \partial_\mu A - \partial_\nu A_\mu$, $D_\mu\phi \equiv (\partial_\mu - ieA_\mu)\phi$, and F is a real constant. We have already discussed properties of this system at the classical level in sections 3.5 and 4.4. Now we take up the quantum version of this theory (Callan et al. 1976, 1978a, Coleman 1979).

It is well known that according to perturbation theory this system should display the famous Higgs phenomenon (Higgs 1964, 1966, Englert and Brout 1964, Guralnik et al. 1964). In this phenomenon there is spontaneous breaking of the U(1) symmetry of the Lagrangian but, in the place of an associated Goldstone boson, the vector field acquires a mass and a longitudinal degree of freedom, in contrast to the normal photon field which would have been massless and purely transverse. One expects such a behaviour because of the following argument. In perturbation theory, thanks to the last term in (10.31), we would expect ϕ to acquire a symmetry-breaking, non-zero vacuum expectation value

$$\langle \phi \rangle_{\text{vac}} = F e^{i\alpha} \tag{10.32}$$

where the space–time-independent phase angle α could take any value. Conventionally one chooses the sector of states where $\langle \phi \rangle = F$ is real. Now let us write ϕ in terms of polar coordinates:

$$\phi(x,t) = \rho(x,t) e^{i\theta(x,t)}. \tag{10.33}$$

Since within perturbation theory one works with fields near $\phi(x,t) = F \neq 0$, such polar coordinates are well defined. Recall that this Lagrangian is invariant under U(1) gauge transformations of the form

$$\begin{aligned} \phi(x,t) &\rightarrow e^{i\Lambda(x,t)}\phi(x,t) \\ A_\mu(x,t) &\rightarrow A_\mu(x,t) + (1/e)\partial_\mu \Lambda(x,t). \end{aligned} \tag{10.34}$$

We choose Λ to be the negative of the phase angle θ in (10.33), i.e. gauge transform to

$$\begin{aligned} \phi(x,t) &\rightarrow \rho(x,t) = e^{-i\theta(x,t)}\phi(x,t) \\ A_\mu(x,t) &\rightarrow A'_\mu(x,t) = A_\mu - (1/e)\partial_\mu \theta. \end{aligned} \tag{10.35}$$

Effectively we have made the field ϕ real everywhere. Note that $F'_{\mu\nu} \equiv \partial_\mu A'_\nu - \partial_\nu A'_\mu = F_{\mu\nu}$. Since the Lagrangian is gauge-invariant, when written in terms of ρ and A'_μ it becomes

$$\mathcal{L} = -\tfrac{1}{4}F'_{\mu\nu}F'^{\mu\nu} + \tfrac{1}{2}(\partial_\mu - ieA'_\mu)\rho\,(\partial^\mu + ieA'^\mu)\rho - \tfrac{1}{4}\lambda(\rho^2 - F^2)^2$$

Since $\langle \rho \rangle_{\text{vac}} = \langle |\phi| \rangle_{\text{vac}} = F$, we expand $\rho(x,t) = F + \eta(x,t)$; $\partial_\mu \rho = \partial_\mu \eta$. Then

$$\begin{aligned} \mathcal{L} = &-\tfrac{1}{4}F'_{\mu\nu}F'^{\mu\nu} + \tfrac{1}{2}(\partial_\mu\eta)(\partial^\mu\eta) + \tfrac{1}{2}e^2F^2A'_\mu A'^\mu - \lambda F^2\eta^2 \\ &+ \tfrac{1}{2}e^2A'_\mu A'^\mu(2\eta F + \eta^2) - \tfrac{1}{4}\lambda(\eta^4 + 4\eta^3 F). \end{aligned} \tag{10.36}$$

Written in this form, we see that the system consists of a vector field of mass eF, and a real scalar field of mass $\sqrt{2\lambda F^2}$. The missing component of the complex scalar field has been 'eaten up' by the vector field in acquiring a mass and hence a longitudinal component, in addition to its two transverse components.

This qualitative argument represents the conventional wisdom to the effect that the system (10.31) will exhibit the Higgs phenomenon. In (3 + 1) dimensions, we know of no reason to doubt that conclusion. However, in (1 + 1) dimensions, where homotopy considerations are different, we will see that in fact the Higgs phenomenon does not take place. Where does the argument given above go wrong? It does not seem to depend manifestly on the space dimensionality. The trouble is that it is based on perturbation theory around $\phi = F$ and $A_\mu = 0$, whereas non-perturbative mechanisms through instanton effects alter the conclusion in $(1+1)$ dimensions. We devote this section and the next to demonstrating this. In particular, we shall discuss the structure of the vacuum and find that it is quite different from the Higgs vacuum centred around $A_\mu = 0$ and $\phi = F$. Despite the complexity of this model compared with the periodic potential problem (PPP) in the last section, the physics and our presentation of it will be very similar. (Actually, the analogy to a particle on a circle (POC) is closer, but for simplicity let us begin by drawing parallels to PPP.)

Recall that in PPP, the energy had its minima at configurations $q(t) = 2N\pi$. In the absence of tunnelling, one could construct a degenerate family of ground states, each localised around one of these minima. However, tunnelling does take place, splitting these degenerate levels into a band. We calculated the structure of this energy band by evaluating the Euclidean amplitude for transition from one of these minima to another. This amplitude was evaluated using instantons in a semiclassical approximation to the Euclidean path integral. With suitable generalisation, let us follow the same procedure for the two-dimensional abelian Higgs model starting from the parent Lagrangian (10.31).

The first step is to identify the minimum-energy configurations (the 'classical vacua') of this system, so that a suitable vacuum state may be constructed around them. It is clear (see section 3.5) that this system has minimum (zero) energy if and only if

$$\phi(x, t) = F e^{i\alpha(x,t)}, \tag{10.37a}$$

and

$$D_\mu \phi(x, t) = 0 = F_{\mu\nu}(x, t)$$

which is satisfied by

$$A_\mu(x, t) = (1/ie\phi)\partial_\mu \phi = (1/ie)(e^{-i\alpha})\partial_\mu(e^{i\alpha}). \tag{10.37b}$$

Therefore we have a continuous infinity of classical minima, each characterised by given function $e^{i\alpha(x,t)}$. The usual perturbation theory vacuum expectation value $\langle \phi(x, t) \rangle = F$, $\langle A_\mu \rangle = 0$ belongs to this set, but so do other configurations obtained from it by the gauge transformations (10.34). We may call the configurations in (10.37) 'pure gauges'. One can already see how perturbation theoretical predictions (including the occurrence of the Higgs phenomenon) may fail. In that approach, one constructs the vacuum around just one of the classical vacua from the set (10.37), usually around $\phi(x, t) = F$, $A_\mu = 0$. This would be akin to constructing the ground state in PPP just around the $q = 0$ minimum whereas the correct ground state was actually spread into wave packets around all the minima. In the same fashion, we will find that the correct vacuum state here samples all the classical vacua. The analysis has to be done with care, however, since the configurations (10.37) are partly redundant, because of gauge invariance. Let us examine this question.

Firstly, we notice that the classical vacua in (10.37) include static as well as time-dependent configurations, as distinct from familiar examples in particle mechanics (harmonic oscillator, PPP, etc.) where the Lagrangian has the form $L = \frac{1}{2}\dot{q}^2 - V(q)$ and consequently the energy is minimised only by static solutions. However, the time dependence in (10.37) is spurious and is entirely because of the time-dependent gauge freedom contained in (10.34). It can be removed by partially fixing the gauge so that $A_0(x, t) = 0$. Henceforth we shall work in this gauge. The set of classical vacua in (10.37) reduce, in the $A_0 = 0$ gauge, to a time-independent set

$$\phi(x) = F e^{i\alpha(x)}$$

and

$$A_x(x) = \frac{1}{ie}(e^{-i\alpha(x)})\frac{d}{dx}(e^{i\alpha(x)}) = \frac{1}{e}\frac{d\alpha(x)}{dx}. \tag{10.38}$$

(Remember that in (1 + 1) dimensions, with $A_0 = 0$, A_μ has only one component left, namely, A_x.) (10.38) gives the complete set of degenerate static configurations which minimise the classical energy. If there were no further gauge freedom, we could say that the set (10.38) is the analogue of the set of coordinates $q = 2N\pi$ in PPP. However the choice $A_0 = 0$ does not exhaust all gauge freedom; time-independent gauge transformations are still permitted. We must examine the precise content of gauge invariance under such transformations, in the quantum field theoretical context. In the canonical Hamiltonian procedure in the presence of constraints, this freedom due to time-independent gauge transformations can be traced to Gauss's law:

$$\begin{aligned} I(x) &= \text{Div}\, E(x) - j_0(x) \\ &= dE_x/dx - \tfrac{1}{2}ie(\phi^* D_0\phi - \phi(D_0\phi)^*) = 0. \end{aligned} \tag{10.39}$$

Recall that this equation is one of the Euler–Lagrange equations arising from the Lagrangian (10.31). But it involves no time derivatives of the electromagnetic field and is a constraint equation rather than a genuine equation of motion. In the quantised theory, it cannot be considered an operator equation since it conflicts with canonical commutation rules. This difficulty is well known (see for example Bjorken and Drell 1965). Equal-time commutation rules give

$$[E_x(x_1), A_x(x_2)]_{t=0} = \mathrm{i}\delta(x_1 - x_2);$$

$$[\phi, A_\mu]_{t=0} = [D_0\phi, A_\mu]_{t=0} = 0.$$

Hence

$$[I(x_1), A_x(x_2)]_{t=0} = \mathrm{i}(\partial/\partial x_1)[\delta(x_1 - x_2)] \neq 0 \tag{10.40}$$

and $I(x_1)$, as an operator, cannot vanish. One therefore imposes Gauss's law as a constraint on 'physical' states. These are defined as that subset of states which obey

$$I(x)|\Psi\rangle_{\text{phys}} = 0. \tag{10.41}$$

Next, consider the operator

$$U_\Lambda = \exp\left(\frac{\mathrm{i}}{e}\int_{-\infty}^{\infty} \Lambda(x) I(x)\,\mathrm{d}x\right), \tag{10.42}$$

where $\Lambda(x)$ is any c-number function. Then

$$U_\Lambda|\Psi\rangle_{\text{phys}} = |\Psi\rangle_{\text{phys}} \tag{10.43}$$

Now, suppose we restrict ourselves to that subset of functions, labelled $\tilde{\Lambda}(x)$, which satisfies $\tilde{\Lambda}(x) \to 0$ as $x \to \pm\infty$. Then $U_{\tilde{\Lambda}}$ can be written as

$$U_{\tilde{\Lambda}} = \exp\left(\frac{\mathrm{i}}{e}\int_{-\infty}^{\infty} \mathrm{d}x[(\partial_x E_x - \tfrac{1}{2}\mathrm{i}e(\phi^* D_0\phi - \phi(D_0\phi)^*))\tilde{\Lambda}(x)]\right)$$

$$= \exp\left\{\mathrm{i}\int_{-\infty}^{\infty}\left[(-E_x)\left(\frac{\partial_x\tilde{\Lambda}}{e}\right) - (\mathrm{i}\tilde{\Lambda}\phi^*)\left(\frac{D_0\phi}{2}\right)\right.\right.$$

$$\left.\left. + (\mathrm{i}\tilde{\Lambda}\phi)\left(\frac{D_0\phi}{2}\right)^*\right]\mathrm{d}x\right\}$$

$$= \exp\left\{ i \int_{-\infty}^{\infty} \left[\pi_x \left(\frac{\partial_x \tilde{\Lambda}}{e} \right) + \pi_{\phi^*}(-i\tilde{\Lambda}\phi^*) + \pi_\phi(i\tilde{\Lambda}\phi) \right] dx \right\}, \tag{10.44}$$

where an integration by parts has been carried out in the first term using $\tilde{\Lambda}(\pm\infty) = 0$, and where

$$\pi_x \equiv \delta L/\delta(\partial_0 A_x) = -E_x,$$
$$\pi_{\phi^*} \equiv \delta L/\delta(\partial_0 \phi^*) = \tfrac{1}{2}(D_0\phi) \quad \text{and} \quad \pi_\phi = \tfrac{1}{2}(D_0\phi)^*. \tag{10.45}$$

Since the π's are the canonical momenta which displace fields conjugate to them, we see that $U_{\tilde{\Lambda}}$ is just the operator which executes time-independent gauge transformations, in the quantum context, corresponding to

$$\phi \to \phi e^{i\tilde{\Lambda}(x)}, \qquad \phi^* \to \phi^* e^{-i\tilde{\Lambda}(x)}$$

and

$$A_x \to A_x + (1/e)\partial_x \tilde{\Lambda}. \tag{10.46}$$

Equation (10.43) thus tells us that physical states must be invariant under such transformations. This is hardly an unfamiliar result, but in deriving it this way we notice the important restriction that $\tilde{\Lambda}(\pm\infty) = 0$. Gauge transformations certainly exist for a general $\Lambda(x)$ not satisfying $\Lambda(\pm\infty) = 0$, and they do leave the Lagrangian and the Hamiltonian invariant. But Gauss's law does not necessarily force all wavefunctionals to be invariant under them. Gauge transformations by the restricted subset of functions $\tilde{\Lambda}(x)$ may be called, for want of a better name, 'small gauge transformations'. The name implies not that $\tilde{\Lambda}(x)$ is small everywhere, but that $\tilde{\Lambda}(\pm\infty) = 0$. Gauge transformations that are not 'small', will be called 'large'. Eigenstates of the field operators, $|\phi(x); A_x(x)\rangle$, are not physical states since they are not invariant under $U_{\tilde{\Lambda}}$, which transforms them into

$$|e^{i\tilde{\Lambda}}\phi; \quad A_x + (1/e)(d\tilde{\Lambda}/dx)\rangle.$$

However, starting with $|\phi, A_x\rangle$ one can form a gauge-invariant physical state by superposing all small-gauge related states:

$$|\phi, A_x\rangle_{\text{phys}} \equiv \int \mathscr{D}[\tilde{\Lambda}(x)] U_{\tilde{\Lambda}} |\phi, A_x\rangle, \tag{10.47}$$

where the integral is over all $\tilde{\Lambda}(x)$ such that $\tilde{\Lambda}(\pm\infty) = 0$. The state $|\phi, A_x\rangle_{\text{phys}}$ is no longer a field eigenstate, but it is the closest entity we can form in the physical subspace of states. Another way of saying all this is that the set of all field configurations $\{\phi, A_x\}$ divides into equivalence classes, the members within each class being related to one another by small gauge transformations $\tilde{\Lambda}(x)$. As a result of the constraint of Gauss's law, only the classes themselves are coordinates of the system and not the individual configurations. For each class we can have a physical quasi-field-eigenstate $|\phi, A_x\rangle_{\text{phys}}$ as defined in (10.47). Remember that this entire discussion is in the presence of the condition $A_0 = 0$.

Using these principles, let us compare our gauge field theory with PPP. Each equivalence class of field configurations (equivalent only under *small* gauge transformations) is like one value of the position coordinate q in PPP. In particular, given the zero-energy configurations in (10.38), it is each *class* of such configurations that is analogous to one of the points $q = 2N\pi$ in PPP. It is therefore necessary to identify the distinct equivalence classes contained in the set (10.38). We shall now see that this equivalence classification of the set (10.38) turns out to be the same as its homotopy classification. Note that for each function $e^{i\alpha(x)}$, we have one member of (10.38). To begin with, we have placed no restriction on the boundary values of $\alpha(x)$. However, we shall shortly show that if we start with an $\alpha(x) = 0$ configuration, then quantum transitions can take place only to those vacuum configurations that satisfy $e^{i\alpha(\pm\infty)} = 1$. For convenience of presentation, let us anticipate this result and restrict $e^{i\alpha(x)}$ to satisfy $e^{i\alpha(\pm\infty)} = 1$. For such functions, since $e^{i\alpha(\infty)} = e^{i\alpha(-\infty)}$, one-dimensional space (the x-axis) is compactified and rendered equivalent to a circle S_1. Thus, all such functions $e^{i\alpha(x)}$ form mappings of S_1 into the U(1) group. Such mappings, as we know from chapter 3, can be divided into homotopy sectors, characterised by a winding number given by

$$N = \frac{1}{2\pi} \int_{-\infty}^{\infty} \frac{d\alpha}{dx} dx = \frac{1}{2\pi} [\alpha(\infty) - \alpha(-\infty)]. \tag{10.48}$$

[Note that our entire discussion of abelian and non-abelian gauge theories treats *global* gauge transformations (sometimes called gauge transformations of the 'first kind') on a separate footing from space–time-dependent local transformations, on the grounds that the former are genuine symmetries leading to conserved Noether charges. Recall that we took the same stance in section 4.3 while counting instanton degrees of freedom. Similarly, the present analysis is meant to be applied after factoring out the global gauge group. In other words, we can set $\alpha(x = -\infty) = 0$ for all our configurations. But the condition $e^{i\alpha(x = +\infty)} = 1$ still allows $\alpha(\infty) = 2N\pi$ where N is precisely the homotopy index in (10.48).]

Furthermore, using (10.38 and 10.46), under any small gauge transformation $\tilde{\Lambda}(x)$,

$$\alpha(x) \to \alpha'(x) = \alpha(x) + \tilde{\Lambda}(x) \tag{10.49}$$

so that

$$\alpha'(\pm\infty) = \alpha(\pm\infty) + \tilde{\Lambda}(\pm\infty) = \alpha(\pm\infty)$$

and

$$N' = N. \tag{10.50}$$

Thus, under small gauge transformations the classical vacua stay within the same homotopy sector. Conversely, all members of a given homotopy class can be related by small gauge transformations. Therefore, classification of the classical vacua (10.38) based on small-gauge equivalence, becomes the same as the homotopy classification described above. Each class is characterised by its winding number N and plays the same role as one of the minima of the potential ($q = 2\pi N$) in PPP.

The remaining steps are very similar to those in PPP, with appropriate generalisation. Around each homotopy class N of classical vacua we can construct a perturbative vacuum state, i.e. a minimum-energy field-theoretical wavefunctional. Let us call these states $|N\rangle$. They are sometimes referred to as 'topological vacua', since they are distinguished by topological considerations. They are the field-theoretical analogues of the wave packet states $u_0(q-2N\pi)$ in PPP (see eq. 10.4)). The naive perturbation theory vacuum built around $\phi(x) = F$, $A_\mu(x) = 0$ (i.e. $\alpha(x) = 0$, $N = 0$) is just the state $|N = 0\rangle$. The point is that a similar topological vacuum with the same energy can be built in each homotopy sector N. More importantly, as we shall demonstrate next, there will be tunnelling between them. As a result, none of these topological vacua $|N\rangle$ is the real vacuum. The actual low-energy eigenstates will be, to leading order, linear combinations of these perturbative topological vacua and the energy degeneracy will be split.

10.3. Vacuum tunnelling in the abelian Higgs model

Let us now demonstrate that tunnelling does indeed take place in the two-dimensional abelian Higgs model between classical vacua belonging to different homotopy sectors. This we do by evaluating once again the Euclidean transition amplitude connecting two of these classical vacua, which can be written as a Euclidean functional integral as outlined in chapter 6. In the process, we shall also obtain the energy spectrum of the correct vacua of this theory.

If it were not for complications due to gauge invariance, we would write the Euclidean functional integral, giving the Euclidean transition amplitude for this model, as per the principles outlined in chapter 6. We would have

$$G(\tau) \equiv \int \mathscr{D}[\phi(x,\tau')]\,\mathscr{D}[(\phi^*(x,\tau')]\,\mathscr{D}[A_\mu(x,\tau')] \times \exp(-S_E[\phi,\phi^*,A_\mu]), \tag{10.51}$$

where $\hbar = 1$ and the Euclidean action is

$$S_E[\phi,\phi^*,A_\mu] = \int_{-\infty}^{\infty} dx_1 \int_{-\tau/2}^{\tau/2} dx_2\,[\tfrac{1}{4}F_{\mu\nu}F_{\mu\nu} + \tfrac{1}{2}(D_\mu\phi)^*(D_\mu\phi) + \tfrac{1}{4}\lambda(|\phi|^2 - F^2)^2]. \tag{10.52}$$

Here $x_1 \equiv x$, $x_2 \equiv \tau'$, the Euclidean 'time', and μ, ν take values in the range 1, 2. Since we are interested in the vacuum state, we shall take the limit $\tau \to \infty$, just as in PPP. Because of redundancy due to gauge invariance, this functional integral will have to be modified. In the preceding section we studied the impact of gauge equivalence in the language of field-theoretical wavefunctionals $|\Psi(\phi, A_\mu)\rangle$ in the Schrödinger representation. Now we shall present, for pedagogical clarity, the corresponding discussion in terms of the functional integral. Let us begin *ab initio*, without fixing the gauge in any way, with the full gauge freedom under the transformations (10.34). Correspondingly, classical zero-energy configurations are the full set of space–time-dependent 'pure gauges' given in (10.37). Now, consider the possible boundary conditions on the fields in the integral (10.51). Since we are interested in transitions from one classical vacuum to another, the initial and final configurations (at $\tau' = \pm\tau/2$; $\tau \to \infty$) must be of the pure gauge form (10.37). This is analogous to the condition in PPP that all paths go from one of the minima of the potential to another. In addition, in a field-theoretical path integral, boundary conditions at $x \to \pm\infty$ must also be specified. Consistent with our prescription in chapter 6, we shall require that ali fields in (10.51) must approach zero-energy (pure gauge) configurations as $x \to \pm\infty$. Altogether, therefore, throughout the perimeter of Euclidean space–time, the fields in (10.51) must approach a pure gauge form. In two dimensions, this perimeter is topologically equivalent to a circle and can be parametrised by some angle θ. The boundary condition then is that on the perimeter of space–time:

$$\phi(x,\tau') \to F e^{i\alpha(\theta)} \quad \text{and} \quad A_\mu \to (1/ie)e^{-i\alpha(\theta)}\partial_\mu(e^{i\alpha(\theta)}). \tag{10.53}$$

Each set of boundary conditions amounts to specifying a function of the form $e^{i\alpha(\theta)}$, on the circle. Such functions, as before, can be divided into homotopy classes, characterised by the winding number (see eq. (3.96) and (3.99)):

$$Q = \frac{1}{2\pi}\int_0^{2\pi} \frac{d\alpha}{d\theta}\, d\theta = \frac{e}{2\pi}\oint_{S_1} \mathbf{A}\,.\,d\mathbf{l} = \frac{e}{4\pi}\int F_{\mu\nu}\,\varepsilon_{\mu\nu}\, d^2x. \tag{10.54}$$

Note that:

(i) This homotopy classification and the associated index Q must not be confused with the homotopy classification with the associated index N (eq. (10.48)) in the preceding section. Both involve mappings of some circle into the U(1) group. But here, the circle is the boundary of Euclidean space–time in two dimensions, and $e^{i\alpha(\theta)}$ specifies the boundary condition on the fields throughout their history. In the last section, we discussed time-independent pure gauges as a function of space alone, which was compacted into a circle. Q and N are therefore different, but they can be related to one another in a special gauge, as we shall see.

(ii) The index N in (10.48) was defined after the gauge was partially fixed. Recall that we had set $A_0 = 0$ and $e^{i\alpha(x)} = 1$ at $x = \pm\infty$ in the preceding section. Hence N has a gauge-dependent definition. In the present discussion, we have so far left the gauge free, and Q itself is gauge-invariant as evident from the last expression in (10.54) in terms of $F_{\mu\nu}$.

(iii) We encountered this homotopy index Q and the associated boundary condition (10.53) earlier in chapter 4 for the same model. There, the condition arose from the requirement of finite action of Euclidean solutions. Here, we do not demand that all the field configurations being integrated over in (10.51) have finite action: indeed, the opposite is true. It can be shown (see Coleman 1979) that such functional integrals are predominantly over infinite-action fields. In the present discussion we have arrived at the same boundary conditions, not from finite action, but from general requirements on fields at the space–time boundary in any vacuum-to-vacuum functional integral. However, since our boundary conditions are the same as in chapter 4, they are consistent with finite action. Therefore the set of fields permitted will also include the classical finite-action instantons of the system.

At the outset, we have one functional integral of the vacuum-to-vacuum type for any given boundary condition, which is in turn specified by a given function $e^{i\alpha(\theta)}$ belonging to some homotopy sector Q. All fields obeying this boundary condition are to be integrated over. We would appear to have a continuous infinity of such functional integrals, one for each function $e^{i\alpha(\theta)}$. This statement however needs to be re-examined in the face of gauge invariance.

Consider any general gauge transformation from the complete set (10.34). Formally, these transformations hold in the Euclidean metric as well, with the replacement of t by τ'. Let the gauge function $\Lambda(x, \tau')$ have a boundary behaviour:

$$\lim_{r \to \infty} \Lambda(x, \tau') \equiv \Lambda(\theta), \tag{10.55}$$

where (r, θ) are polar coordinates in the (x, τ') plane. It is clear that under this transformation,

$$\alpha(\theta) \to \alpha'(\theta) = \alpha(\theta) + \Lambda(\theta). \tag{10.56}$$

In general, therefore, the boundary function changes under gauge transformation, and we have a different functional integral. But the value of the new functional integral will be the same as the old one. Since the action (10.52) is gauge-invariant and the measure (not specified here in detail) will also be chosen in a gauge-invariant manner, the integral has the same value. Furthermore, although $\alpha(\theta)$ is altered to $\alpha'(\theta)$, the homotopy index Q, being gauge-invariant, is unchanged. This is explicit from the expression (10.54). [Also, $\Lambda(\theta)$, as the boundary value of a gauge transformation throughout space–time, will belong to a $Q = 0$ sector. The argument for this is essentially the same as was used in chapter 4 for the Yang–Mills system. Hence $\alpha(\theta)$ and $\alpha'(\theta)$ belong to the same Q sector.] Conversely any two members $\alpha(\theta)$ and $\alpha'(\theta)$ belonging to the same Q sector can be related by a gauge transformation by choosing $\Lambda(\theta) = \alpha'(\theta) - \alpha(\theta)$.

From these arguments we see that all functional integrals whose boundary functions fall in the same homotopy sector Q are equal in value, related to one another by gauge transformations, and hence gauge-equivalent. Thus the continuous infinity of such integrals reduces, under gauge equivalence, to a discrete infinity of distinct classes, one for each integer Q. Integrals belonging to different Q sectors are of course unrelated by gauge transformations and, in general, distinct. Notice the analogy to PPP. There the corresponding amplitudes, connecting any two potential minima, are $\langle (N + \tilde{Q}) 2\pi | e^{-H\tau} | 2\pi N \rangle$. This amplitude does not depend on N, i.e. on the initial or final configurations, but only on $\tilde{Q}$, i.e. on the difference in boundary values. The homotopy index Q plays the same role in our gauge theory. (To push the analogy closer, we shall show below that in the $A_0 = 0$ gauge, Q can be written as the difference between a final and an initial contribution.)

In short, for each Q we have a distinct amplitude, which may be evaluated using any boundary function $e^{i\alpha(\theta)}$ belonging to that Q sector. Of course, even after choosing a particular function $e^{i\alpha(\theta)}$, there is still some residual gauge freedom. Consider those gauge transformations where the functions $\Lambda(x, \tau')$vanish at infinity, i.e. when $\Lambda(\theta) = 0$. Then, by (10.56), the boundary function $e^{i\alpha(\theta)}$ is unchanged. Consequently, the set of fields being integrated over remains the same. What these gauge transformations do is to merely interchange gauge-related field configurations, each of which makes the same contribution $\exp(-S_E[\phi, A_\mu])$. It is well known that it is desirable to factor out this redundancy by adding a 'gauge-fixing' term. This aspect has been discussed in detail even in the pre-instanton gauge-theory literature (see for instance Coleman 1975a) and we need not elaborate on it here. A popular gauge-fixing choice is to add to the action a gauge-fixing term

$$S_{gf} = -\tfrac{1}{2}\int d^2x\,(\partial_\mu A_\mu)^2. \tag{10.57}$$

This is the Feynman gauge, and removes the residual freedom.

Taking this into account, the functional integral (10.51) has to be modified, in any given Q sector, to the form

$$[G(\tau)]_Q = \int \{\mathscr{D}[\phi]\,\mathscr{D}[\phi^*]\,\mathscr{D}[A_\mu]\}_{\alpha_Q(\theta)} \exp[-(S_E + S_{gf})], \tag{10.58}$$

where all fields obey a boundary condition of the form (10.53) characterised by any function $\alpha_Q(\theta)$ belonging to the homotopy class Q. We can choose any convenient $\alpha_Q(\theta)$; this is equivalent to working in a convenient gauge. To match our discussion in the preceding section, we can choose to work with $A_0(x, \tau') = 0$. Let us take the space–time boundary to be the rectangle ABCD (fig. 20(a)). This is topologically equivalent to a circle and all our arguments hold. The index Q (eq. (10.54) is just the line integral

$$Q = \frac{e}{2\pi}\oint \mathbf{A}\,.\,d\mathbf{l},$$

taken clockwise around this rectangle. Since $A_0 = 0$, the sides AD and BC do not contribute to Q. Hence

$$Q = N_+ - N_- \tag{10.59}$$

where

$$N_\pm = \frac{e}{2\pi}\int_{-\infty}^{\infty} A_x(x, \tau' = \pm\infty)\,dx$$

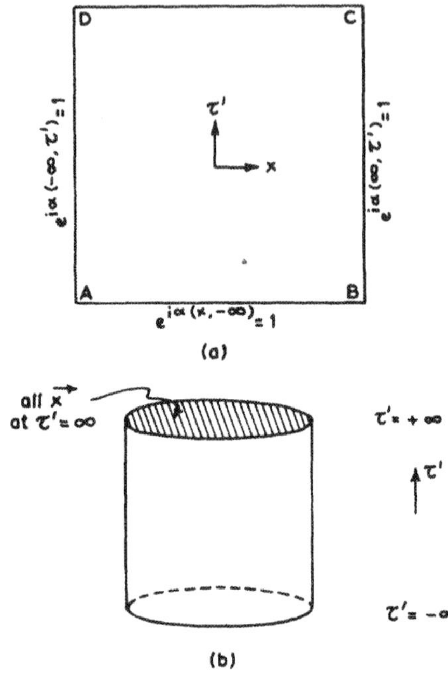

Fig. 20. (a) The rectangle ABCD representing the boundary of two-dimensional Euclidean space–time. (b) The hypercylinder representing the boundary of four-dimensional Euclidean space–time. This will be used in discussing the Yang–Mills system. The flat surfaces of the cylinder at the top and bottom refer to all of three-dimensional $\mathbf{x}$-space, at $\tau' = \pm\infty$ respectively. The curved surface refers to all of time τ', but at spatial infinity $\mathbf{x} \to \infty$.

$$= \frac{1}{2\pi} \int_{-\infty}^{\infty} \mathrm{d}x \left(\frac{\mathrm{d}\alpha}{\mathrm{d}x} \right)_{\tau' = \pm\infty} \tag{10.60}$$

upon using (10.53) on AB and CD. Thus Q becomes, as promised in the $A_0 = 0$ gauge, the difference between contributions at $\tau' = \pm\infty$. The choice $A_0 = 0$ permits only static gauge transformations. We use this time-independent gauge freedom to make A_x (along with A_0) vanish at $\tau' = -\infty$. Thus $\alpha(x, -\infty) = 0$ on the line AB. All gauge freedom has been used up, and now

$$Q = N_+ = \frac{1}{2\pi} \int_{-\infty}^{\infty} \mathrm{d}x \left(\frac{\mathrm{d}\alpha}{\mathrm{d}x} \right)_{\tau' = \infty}. \tag{10.61}$$

Since $A_0(x, \tau') = 0$, we have on AD and BC

$$0 = ieA_0(\pm\infty, \tau') = e^{-i\alpha(\pm\infty, \tau')}(\partial/\partial\tau')e^{i\alpha(\pm\infty, \tau')}.$$

Hence at C and D, the value of $e^{i\alpha}$ must be the same as at B and A respectively. On the line AB, $e^{i\alpha(x)} = 1$. Hence $(e^{i\alpha})_D = (e^{i\alpha})_C = 1$. Therefore in the $A_0 = 0$ gauge, the transition amplitude connects the initial configuration $\alpha(x, -\infty) = 0$ only with those final configurations (at $\tau' = +\infty$) which satisfy $e^{i\alpha(x)} = 1$ at $x = \pm\infty$. Recall that we used this fact as input in the preceding section with the promise of a proof later. Further, in this gauge, eq. (10.61) tells us that $Q = N_+$, where N_+ is just the homotopy index N defined in the last section (eq. (10.48)) as applied to the final configuration at $\tau' = \infty$. The functional integral can therefore be interpreted as giving the Euclidean quantum transition (tunnelling) amplitude, starting from the $A_\mu = 0$ configuration (which belongs to the $N = 0$ sector) at $\tau' = -\infty$, to a configuration in some other homotopy sector $N = Q$ at $\tau' = +\infty$.

To calculate this tunnelling amplitude, we must evaluate the functional integral (10.58). This we do by the semiclassical approximation, just as in PPP. We expand fields about the extrema of the action, which obey the classical Euclidean field equations. The solutions are just the collection of instantons and anti-instantons of this model. Recall that we have already discussed these solutions in chapters 3 and 4. They first arose in chapter 3 as static solitons of this model in (2 + 1) dimensions. Later in chapter 4 we argued that soliton solutions from (2 + 1) dimensions can be used as instantons in (1 + 1) dimensions with the trivial replacement of (x, y) and (A_x, A_y) in the former by $(x, \tau' = it)$ and (A_x, iA_0) respectively in the latter. We did not derive the exact solutions but the work of Taubes (1980) indicates that they exist. We derived their asymptotic behaviour (eq. (3.105)) which showed that these instantons had finite size of order F^{-1}. They had finite action, and a homotopy index given precisely by the Q in (10.54). Finally, since each instanton has $Q = 1$ and finite size (localised action density and localised electromagnetic fields $F_{\mu\nu}$), a collection of widely separated n_1 instantons and n_2 anti-instantons is also a solution in the sector $Q = n_1 - n_2$, exact in the limit of infinite separation. Given all this information, we can perform a dilute instanton-gas expansion just as we did for PPP. The fact that we do not know the exact instanton solution will not matter for our purposes, just as it did not in PPP.

We could evaluate the functional integral (10.58) for any Q. Let us choose $Q = 1$ which is analogous to the amplitude $\langle 2\pi|e^{-H\tau}|0\rangle$ evaluated in PPP. In the semiclassical expansion all solutions with n_1 instantons and n_2

anti-instantons, all widely separated and with $n_1 - n_2 = 1$, will contribute. With obvious generalisations, all the steps in PPP leading from eq. (10.10) to eq. (10.21), can be repeated. We shall omit the steps, but clearly the answer will have the same structure as (10.21) and can be written in the form

$$\lim_{\tau \to \infty} [G(\tau)]_{Q=1} = A \int_0^{2\pi} d\theta e^{i\theta} \exp[(e^{-S_0} 2B \cos\theta - C) L\tau], \qquad (10.62)$$

where S_0 is the value of $(S_E + S_{gf})$ for a single instanton, and $L \to \infty$ is the length of space. In the place of the factor τ in (10.21), we have $L\tau$ here since the location of each instanton can now vary over the (x, τ') plane. The details of the factors A, B and C are not important for our discussion. Using the usual expansion of the transition amplitude $G(\tau)$ in terms of energy eigenstates, (10.62) yields, as $\tau \to \infty$, a band of low-lying energy levels characterised by the continuous variable θ. Each level corresponds to an energy *density*,

$$E_\theta / L = C - e^{-S_0} 2B \cos\theta. \qquad (10.63)$$

We see that this has the same structure as the band (10.22) in PPP. It can be given a similar interpretation. Recall that $[G(\tau)]_{Q=1}$ in (10.62) stands for the (Euclidean) transition amplitude from the $N = 0$ vacuum sector to the $N = 1$ sector. Therefore, given that S_0 is finite, these states in different homotopy sectors *do* tunnel into one another. Consequently the perturbative topological vacua $|N\rangle$ introduced in the last section will mix with one another. The correct energy levels, to leading order, will be linear combinations of them. The correct linear combinations are obtained using the same principles as in PPP. Recall that there the band states had the form (eq. (10.4))

$$|\phi_\theta(q)\rangle = \sum_{N=-\infty}^{\infty} e^{iN\theta} |u_0(q - 2N\pi)\rangle. \qquad (10.64)$$

The coefficients were required to be $e^{iN\theta}$ essentially because of Bloch's theorem. Under the transformation $q \to q + 2\pi$ (effected by the unitary operator $T = e^{-i2\pi\hat{p}}$ where $\hat{p}$ is the momentum), the PPP Hamiltonian is unchanged. Hence $[T, H] = 0$, and energy eigenstates are also eigenstates of T with some unimodular eigenvalue $e^{i\theta}$. The linear combinations (10.64) are chosen so as to satisfy this. Let us apply the corresponding argument for the abelian gauge model.

Consider a time-independent gauge transformation with a gauge function $\Lambda_1(x)$ such that $\Lambda_1(-\infty) = 0$ and $\Lambda_1(+\infty) = -2\pi$. (An example is

$\Lambda_1(x) = -\pi(1+\tanh x)$.) The classical vacua in (10.38), characterised by some $e^{i\alpha(x)}$ will be changed by such a transformation. In particular,

$$\alpha(\pm\infty) \to \alpha(\pm\infty) + \Lambda_1(\pm\infty)$$

and

$$N = (1/2\pi)[\alpha(\infty) - \alpha(-\infty)] \to N - 1. \tag{10.65}$$

(Remember that only small gauge transformations leave N invariant. $\Lambda_1(x)$, however, is a large gauge transformation since $\Lambda_1(\infty) = -2\pi \neq 0$.) The operator T_1 which executes this transformation is constructed by inserting $\Lambda_1(x)$ into the last expression in (10.44):

$$T_1 = \exp\left(\mathrm{i}\int_{-\infty}^{\infty}[\pi_x(\partial_x\Lambda_1/e) + \pi_{\phi^*}(-\mathrm{i}\Lambda_1\phi^*) + \pi_\phi(\mathrm{i}\Lambda_1\phi)]\,\mathrm{d}x\right).$$

We can integrate the first term by parts and reverse the steps in eq. (10.44), but now a surface term will survive since $\Lambda_1(\infty) \neq 0$:

$$\begin{aligned} T_1 &= \exp\left(-\mathrm{i}\,E_x\frac{\Lambda_1(x)}{e}\right)_{x=\infty} \exp\left(\frac{\mathrm{i}}{e}\int_{-\infty}^{\infty}\mathrm{d}x\{\Lambda_1(x) \right. \\ &\qquad \left. \times[\partial_x E_x - \tfrac{1}{2}\mathrm{i}e(\phi^* D_0\phi - \phi(D_0\phi)^*)]\}\right) \\ &= \exp[+(2\pi\mathrm{i}/e)E_x(\infty)], \end{aligned} \tag{10.66}$$

where the second exponential factor is just unity when acting on physical states, thanks to Gauss's law. The quantum version of (10.65) is

$$T_1|N\rangle = |N-1\rangle. \tag{10.67}$$

Clearly T_1 plays the same role here as the operator T in PPP which changed $u_0(q-2\pi N)$ to $u_0(q-2\pi(N-1))$. Further since T_1 is a gauge transformation, even if not a 'small' one, it does leave the Hamiltonian invariant. Hence,

$$[T_1, H] = 0.$$

H and T_1 may be simultaneously diagonalised, and T_1, as explicit in (10.66) is a unitary operator, whose eigenvalues must be of the form $e^{i\theta}$. Putting all this together it is clear that the linear combinations of the perturbative vacua $|N\rangle$ which give the correct vacua as a result of tunnelling will be

$$|\theta\rangle = \sum_{N=-\infty}^{\infty} e^{\mathrm{i}N\theta}|N\rangle, \tag{10.68}$$

satisfying

$$T_1|\theta\rangle = \sum_{-\infty}^{\infty} e^{iN\theta}|N-1\rangle = e^{i\theta}|\theta\rangle. \tag{10.69}$$

We thus get a band of states, parametrised by θ ($0 \leqslant \theta < 2\pi$). These are precisely the states in the spectral decomposition of (10.62) with energies E_θ given in (10.63). Since the states $|\theta\rangle$ are correct energy eigenstates,

$$\langle\theta|e^{-H\tau}|\theta'\rangle = 2\pi\delta(\theta-\theta')e^{-E_\theta\tau}.$$

Hence

$$\begin{aligned}\langle\theta|e^{-H\tau}|\theta\rangle &= 2\pi\delta(0)e^{-E_\theta\tau}\\ &= \sum_{NM} e^{i(N-M)\theta}\langle M|e^{-H\tau}|N\rangle \qquad \text{using (10.68)}\\ &= \sum_N\left(\sum_Q e^{-iQ\theta}\langle N+Q|e^{-H\tau}|N\rangle\right).\end{aligned} \tag{10.70}$$

We saw that the transition amplitude between topological sectors N and $N+Q$, depends only on Q, and is given by the functional integral (10.58). Using $\sum_N = 2\pi\delta(0)$, we thus have

$$\lim_{\tau\to\infty} e^{-E_\theta\tau} = \lim_{\tau\to\infty}\sum_Q e^{-iQ\theta}\int\{\mathscr{D}[A_\mu]\,\mathscr{D}[\phi]\,\mathscr{D}[\phi^*]\}_Q e^{-(S_E+S_{gf})}. \tag{10.71}$$

Since

$$Q = \frac{e}{4\pi}\int d^2x\,\varepsilon_{\mu\nu}F_{\mu\nu}$$

for all the fields being integrated in (10.71), we can bring the factor $e^{-iQ\theta}$ as well as the summation $\sum_Q$ inside the functional integral. Hence

$$\lim_{\tau\to\infty} e^{-E_\theta\tau} = \lim_{\tau\to\infty}\int\mathscr{D}[A_\mu,\phi,\phi^*]_{\text{all }Q}\exp\left[-\left(S_E+S_{gf}+\frac{ie\theta}{4\pi}\int d^2x\,F_{\mu\nu}\varepsilon_{\mu\nu}\right)\right]. \tag{10.72}$$

Here, $\mathscr{D}[A_\mu,\phi,\phi^*]_{\text{all }Q}$ denotes that fields from all Q sectors are to be integrated over. In other words, to obtain E_θ we can simply integrate all field configurations which become pure gauge at infinity without bothering to make a homotopy classification, provided we use a modified θ-dependent action S_θ given by

$$S_\theta = S_{\mathrm{E}} + S_{\mathrm{gf}} + \frac{\mathrm{i}e\theta}{4\pi} \int \mathrm{d}^2x \, \varepsilon_{\mu\nu} F_{\mu\nu}. \tag{10.73}$$

This amounts to adding to the parent Lagrangian density (10.31) in Minkowski space, an extra term:

$$\Delta \mathscr{L}_\theta = (e\theta/4\pi)\, \varepsilon_{\mu\nu} F^{\mu\nu} \tag{10.74}$$

This result offers us an alternate approach to θ-vacua. One can at the outset start with a Lagrangian $\mathscr{L} + \Delta\mathscr{L}_\theta$ and compute the vacuum state using the functional integral. In this approach, θ can be considered as a parameter in the Lagrangian (and the Hamiltonian) and for each θ, we have a different theory with a sector of states whose vacuum is the state $|\theta\rangle$. In this respect the $|\theta\rangle$ states in this gauge theory are different from the states $|\phi_\theta\rangle$ in PPP. In PPP, all the states $|\phi_\theta\rangle$ in the lowest-energy band were part of a single physical spectrum of states. Only the lowest-energy state ($\theta = 0$) was the ground state and higher-θ states were true excited states. In principle, one can excite the particle from the $\theta = 0$ ground state to the higher-θ excited states by, say, shining radiation on it. By contrast, in our gauge theory, each state $|\theta\rangle$ is the vacuum of a separate sector of states. No gauge-invariant operator can connect states from different θ sectors (see below). Thus the gauge theory is closer in spirit to a particle on a circle (called POC at the end of section 10.1). By adding a θ-dependent term to the Lagrangian of such a particle (see eq. (10.28)), we could make its ground state any one but *only one* of the states in the PPP band. For any given θ, we had a complete Hilbert space of states for POC. By altering θ, which is a parameter in POC, we go to a different Lagrangian with its own separate space of states. The situation in our gauge theory is somewhat similar.

Notice that like the extra term in (10.28), the extra term $\Delta\mathscr{L}_\theta$ in (10.74) added to the Lagrangian density, is a total divergence. $\Delta\mathscr{L}_\theta = \partial^\mu((e/2\pi)\varepsilon_{\mu\nu}A^\nu)$. Therefore it will not affect the classical equations or the instanton solutions. The different θ sectors of these quantum theories have the same classical limit. But the extra term $\Delta\mathscr{L}_\theta$ does violate the parity symmetry P enjoyed by the parent Lagrangian (10.31). In $(1+1)$ dimensions, $\frac{1}{2}\varepsilon_{\mu\nu}F^{\mu\nu}$ is just the electric field E_x, which transforms as $E_x \to -E_x$ under parity. The same remark holds for charge conjugation symmetry C. This violation of C and P is one of the important consequences of quantum instanton effects in this model. It is only in the $\theta = 0$ sector, where $\Delta \mathscr{L}_\theta$ vanishes, that this C, P violating term is absent.

The claim that the different θ-vacua belong to separate, physically unconnected sectors can be supported by the following argument. In a gauge theory, one takes all physical observables and the associated operators to be invariant under all gauge transformations. Consider in particular the transformation T_1 (eq. (10.68)). Any physical operator B must commute with T_1. Hence

$$\begin{aligned} 0 &= \langle \theta | [B, T_1] | \theta' \rangle \\ &= \langle \theta | B | \theta' \rangle \ (e^{i\theta'} - e^{i\theta}). \end{aligned} \tag{10.75}$$

Hence

$$\langle \theta | B | \theta' \rangle = 0 \qquad \text{if } \theta \neq \theta'. \tag{10.76}$$

No physical operator can connect the state $|\theta\rangle$ to the state $|\theta'\rangle$ and in this sense they are vacua of separate sectors of states. The reader may be concerned that on the one hand we have claimed that all physical observables commute with the 'large' gauge transformation T_1. At the same time we have distinguished between wave functions $|N\rangle$ and $|N+1\rangle$ which are related precisely by the operator T_1. Here again it will be helpful to draw an analogy to the particle on a circle. Recall that there $q = 0$ and $q = 2\pi$ are physically equivalent in the sense that all observables have the same value at $q = 0$ and $q = 2\pi$. Yet, when we add the θ-dependent term (10.28) to the Lagrangian leading to quasi-periodic boundary conditions, we have $\phi_\theta(2\pi) \neq \phi_\theta(0)$. The status of the $|N\rangle$ vacua is to be viewed on a similar footing.

We emphasise that an interpretation of the θ-vacua and their non-degenerate energies E_θ as being caused by tunnelling between topological vacua $|N\rangle$ is a gauge-dependent interpretation. In defining the $|N\rangle$ states, we had used gauge conditions amounting to $A_0 = 0$ as well as $e^{i\alpha(\pm\infty)} = 1$ for the static classical vacua. We chose this gauge because it offers an attractive analogy to the much simpler quantum mechanics problems PPP and POC. In a different gauge, however, the interpretation would in general be different. This aspect has been lucidly discussed by Bernard and Weinberg (1977). For instance, in a physical gauge, i.e. one where A_μ is uniquely determined in terms of $F_{\mu\nu}$, there will be no residual gauge freedom at all. The classical vacuum will be unique and there is no meaning to our tunnelling picture. Nevertheless, different θ values and θ sectors can exist since we can start with a θ-dependent Lagrangian by adding the term (10.74). The concrete results obtained, namely the existence of the θ-vacua and their energies E_θ, are gauge-invariant results. Our gauge-dependent discussion in section 10.2 notwithstanding, these concrete results were obtained using the functional integral (10.51), which depends only on Q, and is gauge-invariant.

Finally, regardless of which θ sector we are in, the corresponding vacuum $|\theta\rangle$ is quite different from the perturbative vacuum constructed around $\phi = F$, $A_\mu = 0$. Hence the Higgs phenomenon and associated features of the perturbative vacuum need not be present. In the next chapter, we will explicitly see that the θ-vacua of this model permit a long-range Coulomb force, in contrast to the Higgs phenomenon where the A_μ field would acquire a mass and yield only short range forces.

10.4. The Yang–Mills vacuum

We finally turn to instanton effects in quantum non-abelian gauge theories. Let us begin with the structure of the vacuum state in the pure Yang–Mills system which consists of three SU(2) gauge fields $A_\mu^a(x, t)$, self-coupled in (3 + 1) dimensions. Instanton physics in fact originated with this system, first at the classical level (Belavin et al. 1975) and soon thereafter, at the quantum level ('tHooft 1976a, 1976b; Callan et al. 1976, Jackiw and Rebbi 1976d). In the initial stages, our arguments will be essentially the same as in the preceding sections on the abelian Higgs model. We only need to repeat them, in condensed form, in the Yang–Mills context. An important difference will however arise at the later stages, and we shall lay emphasis on that. Our discussion will also hold for self-coupled gauge fields of higher SU(N) groups.

Recall that we had already introduced the Yang–Mills system in chapter 4. We shall use the notation (eq. (4.5)–(4.7)) where A_μ^a and $G_{\mu\nu}^a$ are compactly written in terms of anti-hermitian (2 × 2) matrices A_μ and $G_{\mu\nu}$. The Lagrangian density is

$$\mathscr{L}(\boldsymbol{x}, t) = (1/2g^2)\,\mathrm{Tr}[G_{\mu\nu}G^{\mu\nu}]. \tag{10.77}$$

This is invariant under the SU(2) gauge group whose elements are $e^{-\Lambda(\boldsymbol{x}, t)}$ where, to be consistent in notation, we use anti-hermitian (2 × 2) matrices $\Lambda(\boldsymbol{x}, t) = (\sigma^a/2\mathrm{i})\,\Lambda^a(\boldsymbol{x}, t)$ as generators. Under these transformations (see eq. (4.8))

$$A_\mu \to e^{-\Lambda}(A_\mu + \partial_\mu)\,e^{\Lambda}. \tag{10.78}$$

As in preceding sections, let us work in the gauge $A_0(\boldsymbol{x}, t) = 0$. This condition still permits gauge transformations generated by time-independent matrices $\Lambda(\boldsymbol{x})$. Consequently the set of classical vacua of this system are the static pure gauges satisfying $G_{\mu\nu} = 0$, or equivalently (see eq. (4.12))

$$A_i(\boldsymbol{x}) = \mathrm{e}^{-\alpha(\boldsymbol{x})} \nabla_i \mathrm{e}^{\alpha(\boldsymbol{x})}, \tag{10.79}$$

where $\alpha(\boldsymbol{x})$ is any traceless anti-hermitian (2×2) matrix function. Further, we can restrict ourselves to those $\alpha(\boldsymbol{x})$ which satisfy $\mathrm{e}^{\alpha(\boldsymbol{x})} = 1$ at all points $|\boldsymbol{x}| = \infty$ at spatial infinity. The argument for this is the same as in the abelian Higgs model; we will show later that quantum tunnelling as calculated by the appropriate Euclidean functional integral will connect the $\alpha(\boldsymbol{x}) = 0$ configuration with only those classical vacua which satisfy $[\mathrm{e}^{\alpha(\boldsymbol{x})}]_{\boldsymbol{x}\to\infty} = 1$. Therefore we need to consider only those classical vacua in the set (10.79) which satisfy this condition. Since $[\mathrm{e}^{\alpha(\boldsymbol{x})}]_{\boldsymbol{x}\to\infty} = 1$, three-dimensional space is compacted for purposes of these functions into the surface of a hypersphere S_3. Each such function gives a mapping of this S_3 into the group space of SU(2). Such mappings, as we know from chapter 4, can be classified into homotopy sectors which form the homotopy group $\pi_3(\mathrm{SU}(2)) = Z$. Each sector is characterised by an integer, the Pontryagin index. This index, which we shall now label N, is obtained by a straightforward adaptation of eq. (4.27) to compactified three-dimensional space:

$$N = \frac{1}{24\pi^2} \int \mathrm{d}^3x \, \mathrm{Tr}[(\mathrm{e}^{\alpha}\nabla_i \mathrm{e}^{-\alpha})(\mathrm{e}^{\alpha}\nabla_j \mathrm{e}^{-\alpha})(\mathrm{e}^{\alpha}\nabla_k \mathrm{e}^{-\alpha})]\, \varepsilon_{ijk}. \tag{10.80}$$

In short, our classical vacua (10.79) can be divided into a discrete infinity of sectors, each sector associated with a given homotopy class N of the corresponding gauge group element $\mathrm{e}^{-\alpha(\boldsymbol{x})}$. For example, the case $\alpha(\boldsymbol{x}) = 0$ yielding $A_i(\boldsymbol{x}) = 0$, belongs to the $N = 0$ class. A prototype example of the $N = 1$ class corresponds to

$$A_i^{(1)}(\boldsymbol{x}) = \mathrm{e}^{-\alpha_1} \nabla_i \mathrm{e}^{\alpha_1} \tag{10.81}$$

with

$$\alpha_1(\boldsymbol{x}) = \frac{\mathrm{i}\pi\boldsymbol{\sigma}\cdot\boldsymbol{x}}{(\boldsymbol{x}^2 + a^2)^{1/2}} + \mathrm{i}\pi\sigma_3. \tag{10.82}$$

It can be verified by inserting $\mathrm{e}^{\alpha_1(\boldsymbol{x})}$ into (10.80) that this belongs to the $N = 1$ sector. (The term $\mathrm{i}\pi\sigma_3$ is added to fulfil our demand that $\mathrm{e}^{\alpha(\boldsymbol{x})} \to 1$ as $\boldsymbol{x} \to \infty$. Without this term, it would tend to -1. Actually, our entire discussion on local gauge invariance is modulo overall global gauge transformations. It is therefore sufficient if $e^{\alpha(x)}$ tends to some constant e^{α_0} as $x \to \infty$. For convenience, we have taken $e^{\alpha_0} = 1$. We shall not elaborate on this technicality further. Recall that a corresponding statement was made for the abelian Higgs model.)

The gauge transformation which takes the $N = 0$ configuration $A_i(\mathbf{x}) = 0$, to the $N = 1$ configuration $A_i^{(1)}(\mathbf{x})$ is clearly

$$g_1(x) = e^{-\Lambda_1(x)} \tag{10.83}$$

with $\Lambda_1(x) = \alpha_1(x)$. From general homotopy considerations, $g_1(x)$ will also transform any classical vacuum in the N sector into one in the $(N+1)$ sector.

Around each homotopy class, N, of classical vacua, we can construct a topological vacuum state $|N\rangle$. The states $|N\rangle$ in the different homotopy sectors are not gauge equivalent (Callan et al. 1978a). The reason is the same as in the abelian Higgs model. The extent of gauge equivalence under time-independent transformations (which are the only ones left in the $A_0 = 0$ gauge) is again fixed by Gauss's law. As generalised to Yang–Mills theory, this law states that

$$I(x) \equiv D_i G^{0i} = \partial_i E^i + [A_i, E^i] = 0. \tag{10.84}$$

This is one of the field equations derivable from the Lagrangian (10.77), but it has no time derivatives and is therefore a constraint. Here $E^i \equiv G^{0i}$ is the Yang–Mills 'electric' field in 2×2 matrix form. (Note: the commutator $[A_i, E^i]$ is understood to be in the space of 2×2 matrices. It should not be confused with commutators of operators in the Hilbert space of quantum states.) Just as in electrodynamics, eq. (10.84) cannot be considered as an operator equation in quantum Yang–Mills theory. It has to be imposed as a condition on physical states which are required to satisfy

$$D_i E^i |\Psi\rangle_{\text{phys}} = 0. \tag{10.85}$$

Now, under infinitesimal elements $e^{-\delta\Lambda(x)} \simeq 1 - \delta\Lambda(x)$, the transformation (10.78) reduces to

$$\begin{aligned} A_i &\to A_i + [A_i, \delta\Lambda] + \nabla_i(\delta\Lambda) \\ &= A_i + D_i(\delta\Lambda) \end{aligned} \tag{10.86}$$

where D_i is the covariant derivative. Suppose we consider for a moment only those gauge functions $\tilde{\Lambda}(x)$ which satisfy $\tilde{\Lambda}(x) \to 0$ as $x \to \infty$. In our matrix notation, the canonical momentum conjugate to A_i is $(2/g^2)E_i$. Hence the operator in quantum theory generating the gauge transformations $e^{-\tilde{\Lambda}(x)}$ is

$$U = \exp\left(\frac{i2}{g^2}\int d^3x \operatorname{Tr}((D_i\tilde{\Lambda})E^i)\right)$$

$$= \exp\left(\frac{-2i}{g^2}\int d^3x \operatorname{Tr}(\tilde{\Lambda}D_i E^i)\right), \tag{10.87}$$

where integration by parts has been done using $\tilde{\Lambda}(x \to \infty) = 0$. By virtue of Gauss's law (eq. (10.85)),

$$U_{\tilde{\Lambda}}|\Psi\rangle_{\text{phys}} = |\Psi\rangle_{\text{phys}}. \tag{10.88}$$

Thus Gauss's law forces gauge equivalence only under gauge transformations, which satisfy $\tilde{\Lambda}(x \to \infty) = 0$. By contrast, consider the transformation $g_1(x)$ (eq. (10.83)) which takes the $N = n$ sector into the $N = (n+1)$ sector. As explicitly clear from (10.82), the corresponding gauge function $\Lambda_1(x) = \alpha_1(x)$ does *not* satisfy $\Lambda_1(x \to \infty) = 0$. Therefore, states in different homotopy sectors are not forced to be gauge equivalent by Gauss's law.

We therefore have a distinct topological vacuum $|N\rangle$ in each sector. These are not the true vacua since they will tunnel into one another. We can anticipate this, at least in a semiclassical expansion, since the Yang–Mills system permits finite-action instanton solutions. We had explicitly obtained these solutions in chapter 4. An attractive alternate argument has been given by Jackiw and Rebbi (1976d) to suggest that such tunnelling will take place. Consider the class of field configurations

$$A_i^{(\beta)}(x) = \beta A_i^{(1)}(x) \tag{10.89}$$

where β is a real parameter ($0 \leqslant \beta \leqslant 1$) and $A_i^{(1)}$ is the pure-gauge classical vacuum in the $N = 1$ sector (eq. (10.81)). For $\beta = 0$ and $\beta = 1$, we get the pure gauges $A_i = 0$ and $A_i = A_i^{(1)}$ respectively. They will both yield $G_{\mu\nu} = 0$ and hence zero classical energy. For $0 < \beta < 1$, $A_i^{(\beta)}(x)$ is not a pure gauge. The 'electric' field G^{oi} will still vanish, since $\dot{A}_0^{(\beta)} = 0$ and $A_i^{(\beta)}$ is time-independent. But the 'magnetic' field $B_i = \frac{1}{2}\varepsilon_{ijk}G^{jk}$ will not vanish:

$$\begin{aligned} G_{jk} &= \beta(\partial_j A_k^{(1)} - \partial_k A_j^{(1)}) + \beta^2[A_j^{(1)}, A_k^{(1)}] \\ &= (\beta^2 - \beta)[A_j^{(1)}, A_k^{(1)}] \\ &\neq 0, \qquad \text{for } 0 < \beta < 1. \end{aligned} \tag{10.90}$$

The energy is proportional to $\int \operatorname{Tr}(G_{jk}G_{jk})d^3x$, and will therefore be non-zero for $0 < \beta < 1$, but finite, since $A_j^{(1)}$ falls as $1/|x|$ as $x \to \infty$ (see eq.

(10.82)). Now, the fields $A_i^{(\beta)}$ form a curve in field-space as β varies from 0 to 1. The curve connects two classical vacua from the $N = 0$ and $N = 1$ sectors respectively, but offers a non-zero but finite energy barrier between the two. Consequently we may expect quantum tunnelling between the two sectors. The argument is trivially extended to connect any two sectors.

We shall try to calculate the tunnelling amplitude shortly, using functional integrals. In the meantime, anticipating that some tunnelling does take place, the correct vacua will be linear combinations of the topological vacua and are given by

$$|\theta\rangle = \sum_{N=-\infty}^{\infty} \mathrm{e}^{\mathrm{i}N\theta}|N\rangle. \tag{10.91}$$

The factor $e^{\mathrm{i}N\theta}$ comes from the same requirements as in the abelian model. Consider the unitary operator $U(g_1)$, which executes the gauge transformation g_1 and gives

$$U(g_1)|N\rangle = |N+1\rangle. \tag{10.92}$$

Since $[U(g_1), H] = 0$ and $U(g_1)$ is unitary, the eigenstates of H must satisfy $U(g_1)|\theta\rangle = \mathrm{e}^{-\mathrm{i}\theta}|\theta\rangle$, which is fulfilled by (10.91). Finally, if B is any gauge-invariant operator, $[B, U(g_1)] = 0$. Hence

$$0 = \langle\theta|[B, U(g_1)]|\theta'\rangle = (\mathrm{e}^{-\mathrm{i}\theta'} - \mathrm{e}^{-\mathrm{i}\theta})\langle\theta|B|\theta'\rangle.$$

Therefore

$$\langle\theta|B|\theta'\rangle = 0 \qquad \text{if } \theta \neq \theta'. \tag{10.93}$$

Hence each $|\theta\rangle$ is the vacuum of a separate sector of states, unconnected by any gauge-invariant operators. All the supporting remarks made for the abelian model in section (10.3) are understood to be valid here also.

Next let us turn to the Euclidean functional integral which gives the tunnelling amplitude. As in all earlier models, the fields $A_\mu(\boldsymbol{x}, \tau')$ to be integrated over must approach vacuum (pure gauge) configurations at the boundary of Euclidean space–time. This boundary in this case is a surface S_3. Since a pure gauge field has the form $A_\mu = \mathrm{e}^{-\alpha}\partial_\mu \mathrm{e}^{\alpha}$ where $\mathrm{e}^{-\alpha}$ is an element of SU(2), each given boundary condition is a mapping of $S_3 \to [\mathrm{SU}(2)]$. Such mappings are again classificable into homotopy sectors. The associated Pontryagin index Q has already been obtained in eq. (4.18), (4.26):

$$Q = -\frac{1}{16\pi^2}\int \mathrm{Tr}[G_{\mu\nu}\tilde{G}_{\mu\nu}]\,\mathrm{d}^4x$$

$$= \frac{1}{24\pi^2} \oint_{S_3} d\sigma_\mu \varepsilon_{\mu\nu\rho\sigma} \mathrm{Tr}[A_\nu A_\rho A_\sigma]. \tag{10.94}$$

The Euclidean functional integral in the sector Q will be

$$\lim_{\tau \to \infty} [G(\tau)]_Q = \int \{\mathscr{D}[A_\mu]\}_Q \exp[-S_{\mathrm{Euc}}]. \tag{10.95}$$

The Euclidean action S_{Euc} is given in eq. (4.10). There will also be the usual gauge-fixing terms S_{gf} (see for instance Coleman 1975a) just as in the abelian case, suppressed here for simplicity. The fields A_μ being integrated over must satisfy some boundary value on S_3 which belongs to the Q sector. As in the abelian model, this functional integral is gauge-invariant and depends on the Pontryagin index Q, which is also gauge-invariant.

Next, consider the $A_0 = 0$ gauge, and distort the boundary of space–time into a large closed cylinder (fig. 20(b)). This is the analogue of the rectangle in fig. 20(a). The two flat surfaces of the cylinder stand for all of three-dimensional space, at $\tau' = \pm\infty$ respectively, while the curved surface stands for the boundary of space ($\boldsymbol{x} \to \infty$) at all τ'. In the $A_0 = 0$ gauge, the curved surface will make no contribution to the surface integral for Q in (10.94). Hence

$$Q = N_+ - N_-$$

where

$$N_\pm = \frac{1}{24\pi^2} \int d^3x\, \varepsilon_{ijk} \mathrm{Tr}[A_i A_j A_k]_{\tau' = \pm\infty}. \tag{10.96}$$

Furthermore, one can use the freedom of time-independent gauge transformations to set $\alpha(\boldsymbol{x}) = 0$, $(A_i(\boldsymbol{x}) = 0)$, at $\tau' = -\infty$. Then $N_- = 0$ and $Q = N_+$. The functional integral can be interpreted, in this gauge, as the Euclidean transition amplitude connecting the $N = 0$ sector to the $N = Q$ sector. Finally, consider the curved surface of the cylinder where, in the $A_0 = 0$ gauge,

$$0 = A_0(\boldsymbol{x} \to \infty, \tau') = [e^{-\alpha(\boldsymbol{x},\tau')}(\partial/\partial\tau')e^{\alpha(\boldsymbol{x},\tau')}]_{\boldsymbol{x} \to \infty}. \tag{10.97}$$

Hence $[e^{\alpha(\boldsymbol{x},\tau')}]_{\boldsymbol{x} \to \infty}$ is independent of τ'. Since in the initial configuration at $\tau' = -\infty$ we have chosen $e^{\alpha(\boldsymbol{x}, -\infty)} = 1$, therefore in the final configuration at $\tau' = +\infty$, again $e^{\alpha(\boldsymbol{x} \to \infty, \tau' = \infty)} = 1$. Thus, transition takes place only to those classical vacua where $[e^{\alpha(\boldsymbol{x})}]_{\boldsymbol{x} \to \infty} = 1$. Our homotopy analysis of the classical vacua, which we based on this restriction, is hence justified.

Finally, analogously to (10.72), we can interpret the θ-vacua as caused by a θ-dependent term added to the Lagrangian. We have, as $\tau \to \infty$,

$$\langle \theta | e^{-H\tau} | \theta \rangle = \sum_{N,Q} e^{-iQ\theta} \langle N+Q | e^{-H\tau} | N \rangle$$

$$= 2\pi\delta(0) \sum_Q e^{-iQ\theta} \int \mathscr{D}[A_\mu]_Q e^{-S_{\text{Euc}}}$$

$$= 2\pi\delta(0) \int \mathscr{D}[A_\mu]_{\text{all }Q}$$

$$\times \exp\left(-S_{\text{Euc}} + \frac{i\theta}{16\pi^2} \int \text{Tr}(G_{\mu\nu}\tilde{G}_{\mu\nu}) d^4x \right). \tag{10.98}$$

This amounts to adding to the Minkowskian Lagrangian density (10.77) an extra term

$$\Delta\mathscr{L}_\theta = (\theta/16\pi^2)\,\text{Tr}[G_{\mu\nu}\tilde{G}^{\mu\nu}]. \tag{10.99}$$

As in the abelian model, this term is a total divergence (see (4.21)–(4.24)). It will not affect the classical Yang–Mills equations, but will yield a different quantum theory for each value of θ. The state $|\theta\rangle$ may be considered as the vacuum of the corresponding theory with a Hilbert space of states constructed over it. This is an alternative way of understanding the θ sectors. Notice that $\text{Tr}[G_{\mu\nu}\tilde{G}^{\mu\nu}] = \frac{1}{2}\varepsilon^{\mu\nu\rho\sigma}\,\text{Tr}[G_{\mu\nu}G_{\rho\sigma}] = \text{Tr}[4E_i B_i]$, where E_i and B_i are the non-abelian electric and magnetic fields. Under time-reversal, $E_i \to E_i$ while $B_i \to -B_i$. Under parity, $E_i \to -E_i$ while $B_i \to B_i$. Thus, under either time reversal or parity, the extra term (10.99) changes sign while the original Lagrangian (10.77), proportional to $\text{Tr}(E_i E_i - B_i B_i)$, is invariant. Hence, with the additional term $\Delta\mathscr{L}_\theta$, the system violates P and T symmetry. The exception is the $\theta = 0$ case when the extra term vanishes.

Up to this point, our discussion has closely parallelled that of earlier sections. This is because the four-dimensional Yang–Mills system has the same homotopy group as the two-dimensional abelian Higgs model. By using the compact matrix notation $A_\mu(\boldsymbol{x})$ to represent the Yang–Mills fields, we have also made many of the equations appear almost identical to the abelian case. In fact, our analysis will also hold for self-coupled gauge fields of any higher SU(N) group in (3 + 1) dimensions, with minor modifications. As mentioned in chapter 4, any continuous mapping of S_3 into the SU(N) group may be deformed to a mapping into an SU(2) subgroup of SU(N). Therefore, all we need to do is pick an SU(2) subgroup of SU(N), with the associated gauge fields, and repeat the same analysis. Our discussion therefore has some relevance to the theory of quantum chromodynamics

(QCD). This is an SU(3) gauge theory, with fermions added. The presence of fermions can of course make a difference, some of which we shall discuss in the next chapter. For the present, let us return to the prototype system of SU(2) Yang–Mills fields.

10.5. The infrared problem of large instantons

Our next step, if we are to proceed as in the abelian Higgs model, would be to actually evaluate the tunnelling amplitude $[G(\tau)]_Q$ and the energy of the θ-vacuum, by making a semiclassical dilute instanton-gas approximation to the functional integral (10.95). We obtained, in chapter 4, exact one-instanton solutions to the Yang–Mills system, with $Q = 1$ and finite action. Once again, a collection of instantons and anti-instantons will also become, in the limit of infinite separation, a solution. There are, however, two differences between this system and the abelian Higgs model or PPP. The first difference is not very significant for our purposes. Unlike the abelian Higgs model, the Yang–Mills system yielded exact multi-instanton solutions, which were described in chapter 4. However, although they represent an important result at the classical level, exact multi-instanton solutions will not significantly affect the functional integral. This is because the functional integral, in the instanton-gas approximation, will be dominated by a finite number *density* of instantons and anti-instantons. Recall that in PPP we had explicitly shown this density to be finite and proportional to e^{-S_0}: the same argument holds here. A finite density implies an infinite number of instantons in infinite volume. By comparison, any solution (exact or otherwise) with a finite number of instantons will make negligible contribution. In the analogue-gas language, a finite number of instantons will not carry sufficient entropy. Indeed, even an infinite number (finite density) of instantons alone will not do. Recall from PPP that the dominant contribution also requires a finite density of anti-instantons. Regretably, then, our knowledge of exact multi-instanton solutions is not of much help in evaluating the functional integral. The collection of widely separated single instantons and anti-instantons will suffice.

There is, however, a more significant difference between instantons in PPP or the abelian Higgs model, and Yang–Mills instantons: the former had some fixed finite size, but the latter come in all sizes. Recall that the exact Yang–Mills instanton (eq. (4.74)) involves a size parameter λ_1 that could take any value $0 < \lambda_1 < \infty$. As pointed out in chapter 4, this is related to the scale-invariance of the Yang–Mills Lagrangian. Unfortunately this feature casts some doubt on the entire dilute-gas approximation. Firstly one can wonder if the dilute-gas concept of widely separated instantons makes any sense for instantons which are arbitrarily large. One could try to overcome this problem by assuming that given an instanton of any size λ_1, however large, we can take the Euclidean volume ($V\tau$) to be much larger. Even assuming that this can be done in a mathematically satisfactory way (by ordering the limits $V\tau \to \infty$ and $\lambda_1 \to \infty$), there are further problems if we proceed with a dilute-gas calculation. Recall from chapter 4 that a single Yang–Mills instanton has altogether eight degrees of freedom, i.e. given a specific one-instanton solution, we can obtain others by applying any one of eight continuous symmetries. Four correspond to space–time translations, three to global SU(2) rotations and one to scale change. Each symmetry will lead to a zero-mode in the quantum fluctuations about a given instanton, which is handled by integrating over the corresponding collective coordinate. Consequently, if we go ahead with a dilute-gas approximation exactly as in PPP or the abelian Higgs model, we shall get a set of vacuum levels (analogous to (10.22) or (10.63))

$$E_\theta = -V\cos\theta\, e^{-S_0} \int_0^\infty d\lambda_1\, 2B(\lambda_1) \tag{10.100}$$

where we have dropped the unimportant θ-independent additive constant C compared with (10.63). S_0 is the action of a single instanton, and V is the volume of three-dimensional space. Note that integration over global SU(2) symmetry will yield only a finite constant equal to the group volume. This has been absorbed in $2B(\lambda_1)$, which also contains all other factors from quantum fluctuations about an instanton of size λ_1. Integration over translations yields a divergent factor $V \to \infty$, but this is amenable to the satisfactory interpretation that we expect the vacua to have a finite energy *density* E_θ/V. But if the integration over the instanton size λ_1 (which is also the collective coordinate for scale symmetry) diverges, then no similar interpretation can save the situation. To see if such divergence occurs, let us investigate the factor $B(\lambda_1)$ more carefully than we had done for the abelian model. This factor essentially involves the determinant due to quantum fluctuations. The latter is the exponential of the sum over all non-zero eigenvalues of the logarithm of the second derivative of the action, evaluated about the background field of the instanton. In the Yang–Mills system, as in many other field theories, this sum over frequencies will suffer from ultraviolet divergence. Recall that in evaluating the quantum kink mass in chapter 5, we encountered a similar ultraviolet divergence in the sum over normal-mode frequencies. In a renormalisable theory such divergences are handled by the renormalisation procedure. Obviously we cannot reproduce here the complicated procedure for renormalising non-abelian gauge theories. These have been amply reviewed in the pre-instanton literature (see for instance, Abers and Lee 1973, Coleman 1975a). Nor can we reproduce here 'tHooft's Herculean work ('tHooft 1976b) in which he has evaluated, amongst other things, these quantum fluctuations about the instanton solution after renormalisation (see also Belavin and Polyakov 1977, Ore 1977 and Chadha et al. 1977). It will suffice at our introductory level to recall just the following general features of renormalisation theory:

(i) We are evaluating fluctuations about the instanton solution and not about $A_\mu = 0$. However, ultraviolet divergences are not affected by a smooth background field, and therefore the standard renormalisation procedure may be used.

(ii) This procedure involves a renormalisation mass M on which the renormalised $B(\lambda_1)$ will depend. Since λ_1 is the only dimensional variable in the system aside from M (the coupling constant g is dimensionless) one can see by comparing dimensions on both sides of (10.100) that $B(\lambda_1)$ must have the form $(1/\lambda_1^5)f(\lambda_1 M)$ where f is some function.

(iii) Since g is the only parameter in the Lagrangian, the M dependence due to renormalisation comes entirely through the replacement of g by a 'running coupling constant' $\bar{g}(\lambda_1 M)$ in (10.100). Note that $\bar{g}$, being dimensionless, must depend on M through the combination $(\lambda_1 M)$.

We must therefore extract the g dependence of (10.100) not explicitly shown so far. Firstly, note that in our matrix notation (4.5)–(4.7), the action (4.10) depends on g only through an overall factor $1/g^2$. In that case, the change of variables from a zero-mode to the corresponding collective coordinate will, as shown in Appendix B, lead to a Jacobian factor proportional to $1/g$. In that Appendix, we consider a simple case with only a one-parameter continuous symmetry and hence one zero-mode. In the present Yang–Mills case, there are eight zero-modes to be replaced by the corresponding collective coordinates. The resulting Jacobian will therefore be proportional to $1/g^8$. In addition, the action of the single instanton is $S_0 = 8\pi^2/g^2$. Thus, the g dependence in the unrenormalised energy (eq. (10.100)), not explicitly shown in that equation, is of the form

$$(1/g^8)\mathrm{e}^{-8\pi^2/g^2}. \tag{10.101}$$

In the renormalised expression for the energy, we only need to replace g by $\bar{g}(\lambda_1 M)$ to get the M dependence.

Using all this, the renormalised energies are

$$\frac{E_\theta}{V} = -2B\cos\theta \int_0^\infty \frac{\mathrm{d}\lambda_1}{\lambda_1^5}\frac{\exp[-8\pi^2/(\bar{g}(\lambda_1 M))^2]}{(\bar{g}(\lambda_1 M))^8} \tag{10.102}$$

where we have absorbed all other constants in B. Now, we know from the discovery of asymptotic freedom (Gross and Wilczek 1973, Politzer 1973) that the running coupling constant $\bar{g}(\lambda_1 M)$ of the Yang–Mills theory obeys

$$\frac{d\bar{g}(\lambda_1 M)}{d[\ln(\lambda_1 M)]} = \frac{11}{24\pi^2}[\bar{g}(\lambda_1 M)]^3 + \text{higher orders in } \bar{g}. \tag{10.103}$$

For small $\bar{g}$ therefore,

$$\left(\frac{1}{\bar{g}(\lambda_1 M)}\right)^2 = \frac{1}{g_0^2} - \frac{11}{12\pi^2}\ln(\lambda_1 M) \tag{10.104}$$

where g_0 is a constant. Notice that as the instanton size λ_1 goes to zero, $\bar{g}$ also goes to zero, and the above formula (10.104) is good. Therefore for small λ_1, the integrand in (10.102) behaves as

$$\frac{e^{-8\pi^2/g_0^2}}{g_0^8}\frac{(\lambda_1 M)^{22/3}}{\lambda_1^5}\left(1 - \frac{11g_0^2}{12\pi^2}\ln(\lambda_1 M)\right)^4. \tag{10.105}$$

Clearly this vanishes as $\lambda_1 \to 0$, and there is no small-instanton divergence in the expression (10.102) for the energy density. If we continue to use (without justification) the same integrand for large instantons as well, then the integral (10.102) will have an infrared divergence as $\lambda_1 \to \infty$. Of course, the behaviour of the coupling constant given in (10.103)–(10.104) is valid only for small $\bar{g}$ and hence small λ_1. We do not really have at the time of writing a theory of how $\bar{g}$ will behave for large λ_1. All we can say is that the relation (10.104) drives $\bar{g}$ towards larger values as λ_1 increases, and therefore ceases to be valid. In short, we do not really know at present whether the integral (10.102) is finite or not, and if finite, how to evaluate it.

Given this unhappy situation, one could take one of several approaches (represented in different segments of the literature). An optimistic attitude would be to hope that in the large λ_1 region, the large coupling effects, which we cannot compute analytically now, will somehow render the integral finite. At the present time this is only a hope. One could also try to improve on the calculation by incorporating corrections to the dilute 'ideal' instanton-gas limit. One form of improvement would be to include instanton–instanton interactions, i.e. the fact that at large but finite separations the action of a set of N instantons is not N times the action of a single instanton. Inclusion of such effects would improve the calculation for larger instantons as well as for the higher instanton-density domain. Another possibility is to include solutions other than instantons. In the basic semiclassical approximation, one should use only the finite-action solutions of the Euclidean field equations. These are the instantons. However, this approximation is valid only in the weak-coupling limit, whereas in systems like the Yang–Mills theory, indications are that one must necessarily face strong-coupling domains at large distance scales. The asymptotic freedom formula (10.104) is already driving the effective coupling constant towards stronger values at large distance scales. The formula of course is not valid at large coupling. But there is reason to believe, from the experimental fact that quarks are confined (or at best rarely free, if one incorporates the very few experimental claims of quark sighting; see La Rue et al. 1977) that they are bound by strong forces at large separations. It is now widely believed that hadrons are described by a non-abelian gauge theory, the most popular candidate being quantum chromodynamics (QCD). From this one might expect the Yang–Mills coupling also to become large at large distances. Now, notice that the coupling g^2 plays the same role as temperature in the statistical mechanical analogy, where the Euclidean functional integral is compared to the partition function. In (10.101) g^2 divides the action in the exponential just as the temperature divides the energy in the partition function. Consider a cold gas of some molecules. As the temperature increases, higher-energy configurations come into play. The molecules will dissociate into atoms, which in turn will ionise into further constituents etc., etc. In the same way, in our functional integral at small g^2, instantons which are finite-action classical solutions (minima of the action functional) will be important. In the domain of larger g^2, however, configurations with higher action, including possibly those with infinite action, may become prominent. In this spirit, Callan et al. (1978) have added on the contribution of the 'meron' solutions. These are solutions of the Yang–Mills system with topological charge $Q = 1/2$ (De Alfaro et al. 1977). Since Q is non-integral, they have infinite action, which is why we did not discuss them in chapter 4. But a meron–meron configuration, with finite separation, has finite action, with $Q = 1$. In this sense an instanton can be considered as a bound meron pair. At large g^2 we may expect instantons to 'break up into constituent merons'. A single meron's action happens to diverge logarithmically (see Callan et al. 1978b). Naively we may expect merons not to contribute because the integrand has the form $\exp[-(\text{action})]$ which would behave for merons as $\exp[-(C/g^2)\ln R]$ where R, the radius of Euclidean space, tends to infinity. But their entropy (i.e. the integral over their locations) would yield a factor $R^4 = e^{4\ln R}$ which may offset the suppression factor $\exp[-(C/g^2)\ln R]$ for large enough g^2. This qualitative remark has to be consolidated after including the effect of fluctuations – a formidable task. However, similar phenomena have been demonstrated in the condensed-matter physics of certain two-dimensional systems where the interaction energies again diverge logarithmically with distance.

Meron effects for the Yang–Mills system have been discussed in some detail by Callan et al. (1977a, 1978a), who also studied the non-dilute gas of instantons by including some of their interactions. Such calculations have also been done for the O (3) model and the CP_N model in two dimensions which share similar problems (Fateev et al. 1979a, Berg and Luscher 1979a, b). For these latter models, the infrared divergence of the dilute instanton gas seems to be removed by inclusion of these improvements. Beyond the qualitative remarks given above, we shall not discuss either meron calculations or dense instanton-gas calculations in more detail. They are quite complicated and the results are not completely conclusive, at least for non-abelian gauge theories in four dimensions. Proceeding in yet another direction some people have suggested that for systems at a finite temperature, instanton calculations may be sensible since the temperature provides an infrared cut-off. See Affleck (1980) for such a calculation in the CP_N model. Readers interested in pursuing all these questions are referred to the papers listed above, as a starting point.

In contrast to the efforts outlined above, a more conservative attitude to adopt would be to say that for scale-invariant four-dimensional non-abelain gauge theories, such semiclassical attempts are futile. Semiclassical methods generally require weak coupling, whereas in these gauge theories one is unavoidably faced with large couplings at large distance scales. In the strong-coupling functional integral (which is like a high-temperature partition function), not only instantons or merons, but essentially all field configurations in the functional integral may be *a priori* significant. If such is the situation, then any semiclassical calculation which picks out some subset of (classical) configurations may be meaningless. This attitude is pessimistic from the point of view of semiclassical calculations, but it may well be correct. It should be pointed out in fairness to instantons that this problem of how to handle strong couplings at large distances is not special to such semiclassical methods. The same problem plagues most investigations of non-abelian gauge theories, including attempts to demonstrate quark confinement. Recently Witten (1979a) has suggested that in view of these problems, a better approach may be to use the $1/N$ expansion where N is the index in the group $SU(N)$, and has argued that instantons disappear in the large-N limit.

It should be emphasised that the basic infrared difficulty in instanton calculations exists only for scale-invariant theories where instantons come in all sizes. For systems like PPP, the $(1+1)$-dimensional abelian Higgs model, or the $(2+1)$-dimensional Georgi–Glashow model (studied by Polyakov 1977b), where instantons have finite action and some finite size, there is no reason to dismiss the dilute instanton-gas calculation. We only need to choose the coupling constant at the instanton size scale to be small, and the semiclassical approximation should be good. It is just unfortunate that a physically important model, namely QCD in $(3+1)$ dimensions, does not fall into this category. Here again, it should be pointed out that even if semiclassical calculations using instantons, merons, etc. fail for the Yang–Mills system or QCD, not all the information we have learnt along the way is wrong or useless. Exact one- and multi-instantons solutions certainly exist for these systems. The homotopy classification of gauge transformations and of the classical vacua has nothing to do with strong or weak coupling. Finally the existance of the continuous family of vacua described by an angle θ may also survive the strong-coupling infrared problems. Recall our alternative way of introducing θ-vacua. One can simply start with a θ-dependent family of Lagrangians $\mathscr{L}+\Delta\mathscr{L}_\theta$, all of which are equally good parents of the Yang–Mills field equations. For each Lagrangian with a given θ, a quantum theory can be defined using the corresponding action in the functional integral. We may not be able to evaluate the functional integral using instantons in view of the strong-coupling problem. However, as long as configurations with $\int \mathrm{Tr}(G_{\mu\nu}\tilde{G}_{\mu\nu})\,\mathrm{d}^4x \neq 0$ make some contribution, we can expect the term $\Delta\mathscr{L}_\theta$ to yield a different theory for each θ. This θ-vacuum can be considered as the vacuum of the corresponding Hilbert space of states. This argument for θ-vacua and θ sectors of states is only suggestive, but it is free of the details of instanton calculations.

CHAPTER 11

SOME MORE INSTANTON EFFECTS

11.1. Instantons and confinement

In the last chapter we showed how instantons and the associated homotopy considerations can drastically alter the structure of the vacuum state in some field theories, compared with the perturbative vacuum. It is not surprising that accompanying this there are other instanton-related consequences on the behaviour of such theories. One of these consequences deals with the phenomenon of 'confinement'.

We have already touched on this phenomenon in earlier chapters. In nature, quarks, which are believed to be constituents of hadrons, are rarely, if at all, seen. It is also believed that hadrons are in all likelihood described by a non-abelian gauge theory where quarks couple to the gauge fields. (The most popular candidate is quantum chromodynamics where the gauge group is 'colour' SU(3), taken to be unbroken spontaneously (see section 11.2). In larger 'grand unified' models which attempt to unify strong interactions with electro–weak interactions, this SU(3) group is imbedded in larger gauge groups.) Therefore one of the major tasks of theoretical particle physicists is to see whether it can be proved that in such gauge theories quarks will experience a long-range attractive force, sufficiently strong as to keep them confined in hadrons. It is clear that such a force can arise, if at all, only from non-perturbative effects. In perturbation theory, the forces between quarks due to the exchange of gauge bosons in $(3+1)$ dimensions would fall off as $1/r^2$ just as in usual electrodynamics. As we are familiar from electrodynamics, such a force will not confine the corresponding charges.

One of the motivations behind the development of instanton physics, from the very outset, was the hope that non-perturbative instanton effects might lead to such quark confinement (see Polyakov 1975b). Unfortunately, subsequent investigations revealed that for quantum chromodynamics in (3 + 1) dimensions, instantons are unlikely to yield a confining potential between quarks – at least not in a straightforward instanton-gas calculation. Nevertheless, the intuitive feeling that instantons should produce a tendency towards confinement is not incorrect. Indeed, in lower-dimensional models, it can be shown that instantons do confine, in the sense of producing an attractive interaction energy which continues to grow with separation. Two prominent examples where this can be shown to happen are the two-dimensional abelian Higgs model (Callan et al. 1977b) and the three- (2 + 1)-dimensional non-abelian Higgs model (Polyakov 1977b). We shall present the former example because it is simpler and much of the ground work has already been done in section 10.3.

Recall that for this two-dimensional abelian Higgs model perturbation theory would predict the Higgs phenomenon. In this phenomenon the gauge field has a mass proportional to eF (see eq. (10.36)) and would give rise to a short-range force which falls exponentially fast with a range of order $(eF)^{-1}$. But, we have seen from section 10.3 that the correct vacua are not perturbative. They are the θ-vacua, obtained by incorporating instanton (tunnelling) effects which are non-perturbative. Correspondingly, we would expect that features associated with the Higgs phenomenon will also be absent. In particular, we shall show that the system will support a long-range electric field. When two oppositely charged particles are placed in the θ-vacuum they will feel an attractive potential which grows linearly with their separation. It will cost infinite energy to separate them infinitely far apart, i.e. they will be confined.

Before we demonstrate this, notice that even in the absence of external charges the θ-vacuum provides a constant background field. Consider the expectation value of the Euclidean electric field $\langle F_{12} \rangle_\theta$ in the θ-vacuum. Since the vacuum is translation-invariant,

$$\begin{aligned}\langle F_{12}(x, \tau') \rangle_\theta &= (1/2L\tau) \langle \int \varepsilon_{\mu\nu} F_{\mu\nu} \mathrm{d}^2 x \rangle_\theta \\ &= (2\pi/eL\tau) \langle Q \rangle_\theta \end{aligned} \tag{11.1}$$

where $L\tau \to \infty$ is the volume of Euclidean space–time and Q is the winding number in (10.54). In the Euclidean functional-integral formalism, such an expectation value is given by

$$\langle Q \rangle_\theta = \frac{\int [\mathscr{D} A_\mu][\mathscr{D}\phi][\mathscr{D}\phi^*] \mathrm{e}^{-S} \mathrm{e}^{\mathrm{i}Q\theta} Q}{\int [\mathscr{D} A_\mu][\mathscr{D}\phi][\mathscr{D}\phi^*] \mathrm{e}^{-S} \mathrm{e}^{\mathrm{i}Q\theta}}. \tag{11.2}$$

Recall that the vacuum functional for θ-vacua involves (eq. (10.72)) the sum over field configurations in all Q sectors, weighted by the factor $e^{iQ\theta - S}$; $S \equiv S_E + S_{gf}$. The denominator in (11.2) is for normalisation. Hence,

$$\begin{aligned}
\langle F_{12}(x,\tau') \rangle_\theta &= (2\pi/eL\tau)\,\langle Q \rangle_\theta \\
&= -\frac{2\pi i}{eL\tau}\frac{d}{d\theta}\left[\ln \int [\mathscr{D}A_\mu][\mathscr{D}\phi][\mathscr{D}\phi^*]\, e^{iQ\theta - S}\right] \\
&= -(2\pi i/eL\tau)\,(d/d\theta)\,[-E_\theta\tau], \qquad \text{using (10.72)}, \\
&= (2\pi i/eL\tau)\,(d/d\theta)\,[(C - 2B\,e^{-S_0}\cos\theta)\,L\tau], \\
&\qquad\qquad \text{using (10.63)}, \\
&= (2\pi i/e)\,2B\,e^{-S_0}\sin\theta. \qquad (11.3)
\end{aligned}$$

Thus, we see that the θ-vacuum carries a constant 'background' electric field which is non-zero if $\theta \neq 0$ or π. The imaginary factor i is just an artifact of the Euclidean metric, where $(F_{12})_{\text{Euc}} = i\,(F_{01})_{\text{Min}}$.

Next let us place two point-charges $\pm q$ separated by a large but finite distance $\tilde{L}$ in this θ-vacuum, and compute the resultant change in energy. A prescription for doing this in a gauge-invariant and Lorentz-invariant manner has been proposed by Wilson (1974). Let us explain his prescription as adapted to our problem. Let the two charges $\pm q$ be created at some point P in the Euclidean (x, τ') plane. Let them then separate to a distance $\tilde{L}$, stay static at that separation for Euclidean time $\tilde{\tau}$, and then move together and annihilate at Q. The world lines of the two charges will together form a closed loop (fig. 21). The parameters of the loop, $\tilde{L}$ and $\tilde{\tau}$, should not be confused with L and τ which describe the dimensions of all of Euclidean space–time in which the loop is imbedded. We shall work in the limit where L, τ and $\tilde{\tau}$ all tend to infinity with the understanding that $\tau \gg \tilde{\tau}$. The separation $\tilde{L}$ will be kept large but finite.

Now, the interaction of any external current density j_μ with A_μ is described by the action $S_{\text{int}} = \int d^2x\, j_\mu A_\mu$. In our case, the current density j_μ is caused entirely by the two point charges moving on the perimeter of the loop. Hence $j_\mu d^2x \to q\,dx_\mu$ where dx_μ describes the line-element on the loop. Hence S_{int} reduces to a line integral over the loop, given by

$$S_{\text{int}} = q \oint A_\mu\, dx_\mu. \qquad (11.4)$$

This is the additional action due to the external charges. In the path-integral formalism this will clearly introduce into the vacuum functional integral an extra factor

$$W \equiv \exp(iS_{\text{int}}) = \exp(iq \oint A_\mu\, dx_\mu). \qquad (11.5)$$

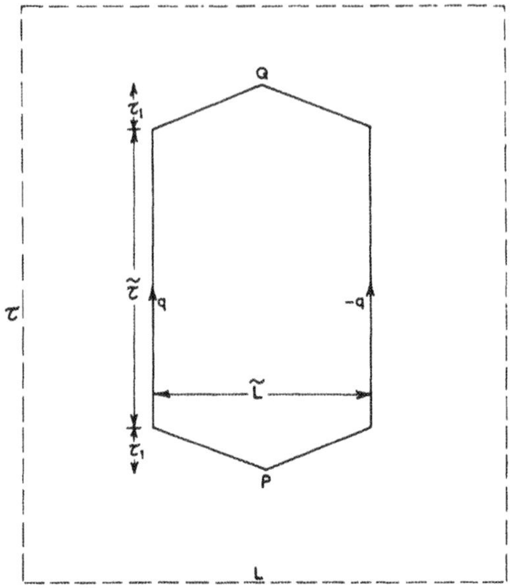

Fig. 21. The Wilson loop. The charges $\pm q$ are created at P, and move apart in (Euclidean) time τ_1 to a separation $\tilde{L}$. Then they stay static for time $\tilde{\tau}$, and finally come together in time τ_1, to annihilate. τ_1 and $\tilde{L}$ are finite while $\tilde{\tau} \to \infty$. The dashed rectangle represents the boundary of space–time.

(Note that even in the Euclidean metric, the exponent in W will still carry a factor i. Recall that $\mathrm{i}\,[A_\mu \mathrm{d}x^\mu]_{\mathrm{Min}} = -\mathrm{i}\,[A_\mu \mathrm{d}x_\mu]_{\mathrm{Euc}}$, where the minus sign has been absorbed in choosing the loop integral to be anticlockwise in fig. 21. Notice that the factor W, which may be referred to as the Wilson loop factor, is both gauge-invariant and Lorentz-invariant.) The vacuum functional integral in the presence of these charges will therefore be

$$\int \mathscr{D}[A_\mu, \phi, \phi^*]_{\mathrm{all}\,Q} \{\mathrm{e}^{-S} \mathrm{e}^{\mathrm{i}Q\theta}\, W\}.$$

Let us normalise this with respect to the usual integral in the absence of these charges and consider

$$\langle W \rangle_\theta = \frac{\int \mathscr{D}[A_\mu, \phi, \phi^*]_{\mathrm{all}\,Q} \{\mathrm{e}^{-S} \mathrm{e}^{\mathrm{i}Q\theta}\, W\}}{\int \mathscr{D}[A_\mu, \phi, \phi^*]_{\mathrm{all}\,Q} \{\mathrm{e}^{-S} \mathrm{e}^{\mathrm{i}Q\theta}\}}. \tag{11.6}$$

The denominator, which is the same as in (11.2), will give $\mathrm{e}^{-E_\theta \tau}$. In the numerator, we have the extra interaction due to external charges. It should yield the additional potential energy $\Delta E_\theta(\tilde{L})$ of the two static charges separated by a distance $\tilde{L}$ for a duration $\tilde{\tau}$. The finite time τ_1 it takes for the charges to separate or recombine (see fig. 21), can be neglected compared with the time $\tilde{\tau}$ for which they are static, since $\tilde{\tau} \to \infty$. Hence, the numerator should yield, as we will verify below, $\exp[-E_\theta \tau - \Delta E_\theta(\tilde{L})\tilde{\tau}]$. Therefore,

$$\lim_{\tilde{\tau} \to \infty} \langle W \rangle_\theta = \frac{\exp[-E_\theta \tau - \Delta E_\theta(\tilde{L})\tilde{\tau}]}{\exp[-E_\theta \tau]} \tag{11.7a}$$

or,

$$\Delta E_\theta(\tilde{L}) = \lim_{\tilde{\tau} \to \infty} -(1/\tilde{\tau}) \ln \langle W \rangle_\theta. \tag{11.7b}$$

This is Wilson's prescription for obtaining the potential energy of the external charge pair as a function of their separation. We proceed to evaluate $\langle W \rangle_\theta$ using the dilute instanton-gas approximation introduced in the last chapter. This was done by Callan et al. (1977b); see also Coleman (1979). We sum over all configurations with n_1 instantons and n_2 anti-instantons, all widely separated. There is no constraint on the integers n_1 or n_2, since we are summing over all $Q = n_1 - n_2$ in (11.6). The denominator in (11.6) is the familiar integral for the vacuum energy. As a sum over instantons, it will clearly have the form (see chapter 10):

$$\begin{aligned} \text{Denominator} &= A e^{-CL\tau} \sum_{n_1, n_2} e^{i(n_1 - n_2)\theta} \frac{(BL\tau e^{-S_0})^{n_1 + n_2}}{n_1 ! n_2 !} \\ &= A e^{-CL\tau} \exp(2BL\tau \cos\theta e^{-S_0}) \end{aligned} \tag{11.8}$$

where the constants A, B and C, in which we continue to remain uninterested, are the same as in chapter 10. This gives the familiar vacuum energy E_θ in (10.63).

In the numerator of (11.6) we have the Wilson factor W whose exponent involves the integral $\oint A_\mu \, dx_\mu$ over the perimeter of the loop in fig. 21. As a result, the numerator factorises into a part inside the loop and a part outside. Any configuration containing n_1 instantons will have:

(a) some number $(n_1)^{\text{in}}$ of instantons well inside the loop (This statement is meaningful since instantons in this model have finite size);

(b) some number $(n_1)^{\text{out}}$ of instantons well outside the loop; and

(c) some instantons overlapping with the perimeter of the loop.

Since we are working in the limit of large $\tilde{L}$, and $\tilde{\tau} \to \infty$, the area-to-perimeter ratio for the loop will be large, hence the fraction of instantons in category (c) will be very small, and we shall neglect them. These remarks also hold for the n_2 anti-instantons. We are summing over not only all n_1 and n_2, but all possible locations of the instantons and anti-instantons for each n_1 and n_2. If we denote this by $\sum_{\{n_1, n_2\}}$, then symbolically,

$$\sum_{\{n_1, n_2\}}^{\infty} = \sum_{\{n_1^{\text{in}}, n_2^{\text{in}}\}}^{\infty} \sum_{\{n_1^{\text{out}}, n_2^{\text{out}}\}}^{\infty}. \tag{11.9}$$

Furthermore

$$W = \exp\left(iq \oint A_\mu \, dx_\mu \right)$$

$$= \exp\left(\frac{2\pi i q}{e} \left[\frac{e}{4\pi} \int_{\text{in}} (d^2x) \varepsilon_{\mu\nu} F_{\mu\nu} \right] \right). \tag{11.10}$$

The term in the square of brackets looks just like the winding number Q in (10.54), but the integral is taken only *inside* the loop. Hence it will yield the winding number $n_1^{\text{in}} - n_2^{\text{in}}$ of the *inside* instantons and give zero for the outside ones. Finally, in the limit $\tilde{\tau} \to \infty$, the area inside the loop is $\tilde{L}\tilde{\tau}$ to leading order in $\tilde{\tau}$. Using all this, an instanton-gas approximation to the numberator of (11.6) will yield

$$\text{Numerator} = A e^{-CL\tau} \left(\sum_{n_1^{\text{in}}, n_2^{\text{in}}} \exp[i(n_1^{\text{in}} - n_2^{\text{in}})(\theta + 2\pi q/e)] \frac{(B\tilde{L}\tilde{\tau} e^{-S_0})^{n_1^{\text{in}} + n_2^{\text{in}}}}{(n_1^{\text{in}})!(n_2^{\text{in}})!} \right)$$

$$\times \left(\sum_{n_1^{\text{out}}, n_2^{\text{out}}} \exp[i(n_1^{\text{out}} - n_2^{\text{out}})\theta] \frac{(B(L\tau - \tilde{L}\tilde{\tau}) e^{-S_0})^{n_1^{\text{out}} + n_2^{\text{out}}}}{(n_1^{\text{out}})!\,(n_2^{\text{out}})!} \right)$$

$$= A e^{-C\tau L} \exp\{2B e^{-S_0}[(L\tau - \tilde{L}\tilde{\tau})\cos\theta + \tilde{L}\tilde{\tau}\cos(\theta + 2\pi q/e)]\} \tag{11.11}$$

Here we have used the fact that the inside area is $\tilde{L}\tilde{\tau}$, while the outside area is $(L\tau - \tilde{L}\tilde{\tau})$. Dividing the expression (11.11) by (11.8), we have for (11.6)

$$\langle W \rangle_\theta = \exp\{2B\, e^{-S_0} \tilde{L}\tilde{\tau}[\cos(\theta + 2\pi q/e) - \cos\theta]\}. \tag{11.12}$$

Inserting this in (11.7),

$$\Delta E_\theta(\tilde{L}) = 2B e^{-S_0} \tilde{L}[\cos\theta - \cos(\theta + 2\pi q/e)]. \tag{11.13}$$

The most important feature of this result is that the potential energy increases linearly with the charge separation $\tilde{L}$, for large $\tilde{L}$. It corresponds to a confining force, which remains constant at large distances. Note also that the energy (11.13) is periodic in the external charge q. For $q = Ne$, where N is any integer, this energy vanishes. The confining force exists only for non-integral external charges (in units of e). A natural explanation is that when $q = Ne$, the system produces N charged pairs $\pm e$ between the external charges $\pm q$ and screens them. When q/e is not an integer, such screening is incomplete since the system can produce only integrally charged particles, leaving behind a confining potential (11.13). This interpretation suggests that the system can yield particles of charge $\pm e$, again in contrast to the perturbative Higgs mechanism result (see eq. (10.36)) where only a neutral scalar field η is left. Indeed, instanton effects seem to make the Higgs model behave similarly to the 'non-Higgs' model. By the latter we mean the model in which the constant F^2 in the parent Lagrangian (10.31) is taken to be negative. That model is just ordinary scalar electrodynamics, which has a unique vacuum, particles of charge $\pm e$, and the standard Coulomb interaction which will also increase linearly in one space dimension. There is however a crucial difference between the linear Coulomb potential in the normal non-Higgs electrodynamics in $(1+1)$ dimensions and the linear potential (11.13) in our model. In the former, we expect a linear potential even classically. The quantum result will survive even as $\hbar \to 0$. By contrast in (11.13) the linear potential is exponentially small in $\hbar$. Note that if we had retained $\hbar$-dependence, (11.13) would have a factor $e^{-S_0/\hbar}$. The confining potential in the abelian Higgs model is therefore very much a non-perturbative quantum effect, caused by instantons.

A similar investigation can also be done for non-abelian theories. The Wilson loop factor for non-abelian theories (Wilson 1974) is just the path-ordered integral $P[\exp(\mathrm{i}\oint A_\mu \mathrm{d}x_\mu)]$ in the matrix notation for the non-abelian gauge fields (see chapter 4 for the definition of 'path-ordered'). Using this, Polyakov (1977b) studied instanton effects on confinement in the $(2+1)$-dimensional non-abelian Higgs model. The model is more complicated than our abelian one and so is the calculation, but again he finds a linear potential due to instanton effects. Recall that in two space dimensions, the usual classical Coulomb potential would increase only logarithmically with distance.

It would have been very nice if instantons had yielded such a confining potential in $(3+1)$ dimensions as well – either for QCD or for the prototype SU(2) Yang–Mills case. Unfortunately they do not seem to, at least not in our simple dilute-gas approximation. Recall from chapter 4 that although Q, the Pontryagin index for the Yang–Mills case, is gauge-invariant, it can be written as the surface integral of a current which is locally gauge-variant.

$$Q = \oint_{S_3} j_\mu \mathrm{d}\sigma^\mu, \qquad \text{where} \quad j_\mu = \frac{1}{24\pi^2} \mathrm{Tr}\,[\varepsilon_{\mu\nu\rho\sigma} A_\nu A_\rho A_\sigma]. \tag{11.14}$$

For instantons far away from the Wilson loop, A_μ will be a pure gauge on the loop and can be gauge-transformed to $A_\mu = 0$ on the loop without altering Q. The entire 'Pontryagin flux' $j_\mu d\sigma^\mu$ could be transformed into a thin tube (essentially a 'string' in four dimensions) emanating from the instanton, and this tube could be chosen to avoid intersecting the loop. The only contribution to the Wilson loop would then come from instantons near the perimeter of the loop. Their contributions would be proportional to the perimeter rather than to $\tilde{L}\tilde{\tau}$ (in the sense of the exponent in (11.12)) and no confining force will be produced (see also section 11.5). This statement of course, can be made only for instantons of some finite size. For the arbitrarily large Yang–Mills instantons with their associated infrared, strong-coupling problems, no clear-cut statement can be made at our level of discussion. Callan et al. (1977a, 1978a) have given arguments to suggest that when meron configurations are also included in the semiclassical calculations, a confining potential will result. We briefly introduced the notion of merons in the last chapter. As mentioned there however, it will be beyond the scope of this book to discuss meron calculations. Interested readers are referred to the work of Callan et al.

11.2. The relevance of massless fermions to QCD

We have studied the vacuum structure of pure Yang–Mills theory, resulting from the tunnelling between topological vacua. When massless spin-1/2 fields are coupled to the Yang–Mills fields, they lead to further modifications in the behaviour of the vacuum state. Apart from leading to interesting theoretical consequences, the system of massless fermions coupled to non-abelian gauge fields has some relevance to hadron physics of the real world. It is worth pointing out this relevance, briefly, before studying the vacuum structure of this system. Recall that we have repeatedly referred to quantum chromodynamics (QCD) as a promising candidate for describing hadron physics. Let us now specify this model in detail (Fritzsch and Gell-Mann 1972, Pati and Salam 1973, Fritzsch et al. 1973). QCD is a non-abelian gauge theory in (3 + 1) dimensions. The gauge group (the so called 'colour' group) is SU(3), and therefore calls for eight gauge fields A_μ^a; $a = 1, \ldots, 8$. These can be represented collectively by an anti-hermitian traceless 3×3 matrix, analogous to what was done for the Yang–Mills system. Let $\lambda^a/2$ be the hermitian traceless generators of SU(3) in the fundamental (3×3) representation. They satisfy the commutation rules

$$[\lambda^a/2, \lambda^b/2] = \mathrm{i} f^{abc} \lambda^c/2 \qquad (11.15)$$

where f^{abc} are the structure constants of the group SU(3), as well as $\mathrm{Tr}\,(\lambda^a \lambda^b) = 2\delta^{ab}$. Then we write, analogously to (4.5),

$$A_\mu(x, t) = \frac{g}{2\mathrm{i}} \sum_a \lambda^a A^a_\mu(x, t), \tag{11.16}$$

where g is the gauge coupling constant. Tensor fields are given formally by the same equation as in the SU(2) case, namely

$$G_{\mu\nu} = \partial_\mu A_\nu - \partial_\nu A_\mu + [A_\mu, A_\nu]. \tag{11.17}$$

In QCD, these gauge fields couple to several species of spin-1/2 quarks. Each species is represented by the Dirac field Ψ^f_α, where f is a 'flavour' index, α is a 'colour' index and the spin index has been suppressed. The flavour index $f = 1, 2, 3, 4$, etc., stands respectively for up, down, strange, charmed, etc. quarks. The total number of flavours will not play a role in our discussion. We shall leave it unspecified, since new flavours are continuing to be discovered. For a given flavour f, Ψ^f_α ($\alpha = 1, 2, 3$), transforms according to the fundamental (triplet) representation of the colour SU(3) group. The quarks have masses m_f. These in general may depend on flavour, but not on colour, which is an exact gauge symmetry of the model. The gauge-invariant Lagrangian density for the model is

$$\mathscr{L}(x, t) = \sum_{f,\,\alpha,\,\beta} \bar{\Psi}^f_\alpha (\mathrm{i}\gamma^\mu D_\mu - m_f)_{\alpha\beta} \Psi^f_\beta + \frac{1}{2g^2} \mathrm{Tr}\,[G_{\mu\nu} G^{\mu\nu}], \tag{11.18}$$

where

$$(D_\mu)_{\alpha\beta} = (\partial_\mu + A_\mu)_{\alpha\beta} \tag{11.19}$$

and 3×3 unit matrices are understood to be present where necessary. We shall not present in detail the cumulative wisdom from experiments which motivates writing down the model (11.18). (See for instance the reviews by Marciano and Pagels (1978) and Novikov et al. (1978).) Very briefly, the three-fold colour symmetry of quarks is motivated by Fermi–Dirac statistics as used in constructing baryons from three quarks. The need to make this a local gauge symmetry arises from the latter's property of asymptotic freedom (quarks appear to be weakly coupled in deep-inelastic lepton scattering off hadrons). Finally, the presence of quark fields of different flavours in (11.18) is called for by the discovery of different families of hadron resonances, interpreted as bound states of quarks with these different flavours.

There is also some motivation for setting the masses of the 'up' and 'down' quarks (m_f; $f = 1, 2$) equal to zero in (11.18), as a starting approximation. Indirect evidence on quark masses suggests that m_1 and m_2 are much smaller than the masses of quarks of other flavours. Moreover, when $m_1 = m_2 = 0$ the model (11.18) enjoys further symmetries beyond colour SU(3). Massless up and down quarks clearly obey the field equations

$$\gamma^\mu D_\mu \Psi^f = \gamma^\mu(\partial_\mu + A_\mu)\Psi^f = 0, \tag{11.20}$$

and

$$\bar{\Psi}^f(-\overleftarrow{\partial}_\mu + A_\mu)\gamma^\mu = 0 \tag{11.21}$$

for $f = 1, 2$. In (11.20)–(11.21) we have suppressed colour indices, in addition to spinor indices. Now consider the doublet

$$\chi \equiv \begin{pmatrix} \Psi^1 \\ \Psi^2 \end{pmatrix}$$

to be an isospinor under an SU(2) group. Then it is easy to check that the six currents $\bar{\chi}(\sigma^i/2)\gamma^\mu(1 \pm \gamma_5)\chi$ are all conserved by virtue of (11.20)–(11.21), where $\sigma^i/2$, $i = 1, 2, 3$, are the usual SU(2) generators acting here on the flavour spinor χ. This indicates that the theory also enjoys chiral SU(2) $\otimes$ SU(2) symmetry. (The SU(2) groups appearing in the last sentence are not to be confused with any SU(2) subgroups of the colour SU(3) gauge group. This chiral symmetry is a global symmetry, operating in the flavour space of up–down quarks, and is independent of the colour group.) Such a chiral SU(2) $\otimes$ SU(2) symmetry is again supported by experiment. Its diagonal SU(2) subgroup, associated with the three vector currents, is just the familiar isotopic spin group, well known to be a fairly good symmetry of strong interactions. The remaining symmetry, associated with the three axial vector currents, is also known to be a good symmetry, but it is manifest in a spontaneously broken form. Associated with this spontaneously broken symmetry, we expect to find an isovector of three pseudoscalar Nambu–Goldstone bosons (Nambu 1960, Bernstein et al. 1960, Goldstone 1961, Goldstone et al. 1962). The three pi-mesons found in nature are identified as these bosons. In addition, chiral SU(2) $\otimes$ SU(2) symmetry has yielded a host of 'soft pion' results, well supported by experiments. (For a survey see Adler and Dashen (1968).) Of course, in reality pions have small but non-zero mass, indicating that chiral SU(2) $\otimes$ SU(2) is not an exact symmetry and that the axial vector currents are only approximately conserved. This feature can be introduced in the model

(11.18) by giving the up–down quarks a small mass. But, to a first approximation, SU(2) $\otimes$ SU(2) can be considered an exact symmetry, and m_1 and m_2 set equal to zero.

Notice however that the choice $m_1 = m_2 = 0$, yields not only the desired chiral SU(2) × SU(2) symmetry, but appears to yield a larger U(2) $\otimes$ U(2) symmetry. In addition to the six isovector currents mentioned above, two isoscalar currents $\bar{\chi}\gamma^\mu(1 \pm \gamma_5)\chi$ also appear to be conserved, formally, by the field equations (11.20)–(11.21). The conservation of the vector current $\bar{\chi}\gamma^\mu\chi$, associated with the conservation of the appropriate quark number, is physically acceptable. But the axial vector current $\bar{\chi}\gamma^\mu\gamma_5\chi$, associated with the U(1) symmetry under $\Psi^f \to e^{i\alpha\gamma_5}\Psi^f$, $f = 1, 2$, creates problems. If this symmetry is not spontaneously broken, then massive hadrons must occur in parity doublets, which they do not. On the other hand, if this symmetry is spontaneously broken, then there should be an associated isoscalar, Lorentz-pseudoscalar Goldstone boson. However, the known meson spectrum offers no suitable candidate for this Goldstone boson. Of course, we do not expect to find this boson to be massless in nature; the same quark-mass correction terms that render chiral SU(2) $\otimes$ SU(2) approximate and give the pi-mesons some mass, will also render the chiral U(1) symmetry approximate. But Weinberg (1975) estimates the resulting mass of this boson to be less than $\sqrt{3}m_\pi$ whereas the lowest isoscalar, $J^P = 0^-$ meson in the particle spectrum is the η-meson, whose mass is nearly $4m_\pi$!

This absence in nature of parity-doubling on the one hand, and of the appropriate Goldstone boson on the other, was viewed as a serious problem with regard to QCD (Fritzsch et al. 1973, Weinberg 1975). A resolution of this problem has been offered by 'tHooft (1976a, 1976b; see also Kogut and Susskind 1975 and Pagels 1976; for a dissenting view, argued in detail, see Crewther 1979). We shall not be able to go into a discussion of the suggested resolution of this U(1) problem, but one feature of the problem, which distinguishes the axial vector U(1) current from the axial vector SU(2) current in QCD, namely that the former suffers from an 'anomaly', will be of relevance to our discussion of vacuum structure.

11.3. Suppression of vacuum tunnelling by massless fermions

The preceding paragraphs are to be viewed as a capsule summary of QCD. with emphasis on the fact that, as a starting approximation, there is good reason to take some of the quarks to be massless. This motivates us to study the system of massless fermions coupled to non-abelian gauge fields. Specifically, we shall concentrate on how the vacuum tunnelling phenomenon, discussed in the last chapter, is affected by the presence of massless fermions. Our discussion will be presented, not in terms of the QCD model (11.18) in its full complexity, but in terms of a simpler miniature model (called a 'baby' version by Coleman (1979), whose analysis we follow). This miniature model will still be a non-abelian gauge theory in (3 + 1) dimensions, but will differ from full QCD in two respects:

(i) The gauge group is chosen to be SU(2) rather than SU(3).

(ii) Quarks with only one flavour will be used. These will be taken to be massless, and to transform according to the fundamental (doublet) representation of the SU(2) gauge group.

The gauge-invariant Lagrangian for this miniature QCD will clearly be (again suppressing spin indices)

$$\mathscr{L}(x, t) = \bar{\Psi}_\alpha (\mathrm{i}\gamma^\mu D_\mu)_{\alpha\beta} \Psi_\beta + (1/2g^2)\,\mathrm{Tr}\,[G_{\mu\nu} G^{\mu\nu}];$$
$$\alpha, \beta = 1, 2, \qquad (11.22)$$

where, formally, $G_{\mu\nu}$ and D_μ are still given by (11.17) and (11.19), but A_μ is now just the SU(2) Yang–Mills field, described by a 2×2 matrix, as per eq. (4.5). One can see that the miniature model (11.22) contains the essential features we are interested in. It has massless fermions, and an axial vector current $\bar{\Psi}_\alpha \gamma^\mu \gamma^5 \Psi_\alpha$ which appears to be conserved by the field equations. That the gauge group is SU(2) rather than SU(3) simplifies matters without losing the richness of the homotopy structure. From Bott's theorem, quoted in chapter 4, we know that the relevant homotopy group $\pi_3(\mathrm{SU}(3))$ for the full QCD is any case related to π_3 of an SU(2) subgroup of SU(3).

We shall study the vacuum structure and associated features of the model (11.22), by essentially combining the techniques of chapters 9 and 10. The Fermi field has no c-number classical limit, nor can it have a vacuum expectation value. The set of 'classical vacua' around which the quantum ground state of (11.22) is to be constructed, will therefore involve only the Yang–Mills field, i.e. its zero-energy pure gauge configurations. These we have already classified into homotopy sectors N; at any given instant of time, and associated with each sector N we can construct a 'topological vacuum' state$|N\rangle$. The Euclidean transition amplitude $\langle N + Q | \mathrm{e}^{-H\tau} | N \rangle$ will once again be given by a Euclidean functional integral in four dimensions, where the boundary conditions on A_μ will be characterised by the gauge-invariant Pontryagin index Q. Up to this point the analysis is unaffected by the fermions. But the value of the amplitude

$\langle N+Q \,|\, e^{-H\tau} | N \rangle$ will of course be altered by the presence of the massless fermions. The field Ψ will be present in the Euclidean action and will have to be integrated over, in addition to the gauge field, in the functional integral for this transition amplitude. The Euclidean action S_E associated with the Lagrangian (11.22) is

$$S_E = S_A + S_\Psi;$$

$$S_A = -\frac{1}{2g^2}\int \mathrm{Tr}\,(G_{\mu\nu}G_{\mu\nu})\mathrm{d}^4x$$

$$S_\Psi = \mathrm{i}\int \bar{\Psi}\gamma_\mu D_\mu \Psi \mathrm{d}^4x \equiv \int \Psi^\dagger M(A_\mu)\Psi \mathrm{d}^4x, \tag{11.23}$$

where

$$M(A_\mu) \equiv \mathrm{i}(\partial_\mu + A_\mu)\gamma_4\gamma_\mu. \tag{11.24}$$

Gauge-fixing terms and colour indices have been suppressed for simplicity; they are understood to be included. The matrices γ_μ, in the Euclidean metric, are taken to be hermitian, satisfying $\{\gamma_\mu, \gamma_\nu\}_+ = \delta_{\mu\nu}$. The matrix $\gamma_5 \equiv \gamma_1\gamma_2\gamma_3\gamma_4$ is also hermitian, with $\gamma_5^2 = 1$. The 'tunnelling amplitude' between the topological vacua $|N\rangle$ and $|N+Q\rangle$ will be given by

$$\lim_{\tau\to\infty} \langle N+Q|e^{-H\tau}|N\rangle$$

$$= \int \mathscr{D}[A_\mu]_Q \int \mathscr{D}[\Psi^\dagger]\mathscr{D}[\Psi]e^{-S_E}, \tag{11.25}$$

where integrations over the Grassmann fields Ψ, $\Psi^\dagger$ have been defined in chapter 9. The dependence of the action on the Fermi field, through S_Ψ, is bilinear and hence the fermion integration can be done exactly. We have, analogously to (9.66),

$$\lim_{\tau\to\infty} \langle N+Q|e^{-H\tau}|N\rangle$$

$$= K_1 \int \mathscr{D}[A_\mu]_Q e^{-S_A}\,\mathrm{Det}\,[M(A_\mu)], \tag{11.26}$$

where K_1 is a constant.

We shall now show that this integral vanishes exactly. The operator $M(A_\mu)$ given in (11.24) is a Dirac operator in the presence of the external field A_μ. We have discussed similar operators in chapter 9 (see for example the one-particle Dirac Hamiltonian in eq. (9.55)). We saw that in some cases where the external field obeyed topologically non-trivial boundary conditions, the one-particle Dirac Hamiltonian had some zero energy eigenvalues. In the present case, where the metric is Euclidean, the operator $M(A_\mu)$ in (11.24) is hermitian and rather like a Dirac Hamiltonian. Furthermore, the field A_μ in the operator $M(A_\mu)$ obeys non-trivial boundary conditions characterised by the index Q in (11.26). Not surprisingly, then, we shall find that $M(A_\mu)$ also has one or more zero eigenvalues, when $Q \neq 0$. As we have mentioned earlier (see section 9.5 and the references cited there), the vanishing of such eigenvalues is related to basic mathematical theorems on spectral flow. It is also related to the occurence here of the well-known phenomenon of current anomalies. We shall use the latter to derive the existence of the zero eigenvalues of $M(A_\mu)$.

Since the Dirac field is massless in (11.22), the field equations $\mathrm{i}(\partial_\mu + \mathrm{A}_\mu)\gamma_\mu \Psi = 0 = \mathrm{i}\bar{\Psi}(-\overleftarrow{\partial}_\mu + \mathrm{A}_\mu)\gamma_\mu$ appear to formally conserve the axial vector current $J_{\mu 5} = \bar{\Psi}(x)\gamma_\mu\gamma_5\Psi(x)$, in addition to the vector current $J_\mu = \bar{\Psi}(x)\gamma_\mu\Psi(x)$. This would imply chiral $\mathrm{U}(1)\otimes\mathrm{U}(1)$ symmetry. However, in actuality the current $J_{\mu 5}$ will not be conserved. The bilinear product of the form $\bar{\Psi}(x)\Psi(y)$ diverges in quantum field theory as $x \to y$. Hence J_μ and $J_{\mu 5}$ which, if defined as above, both contain a product of $\bar{\Psi}$ and Ψ at the same point x, are ill-defined because of this divergence. One has to redefine these currents more carefully, by separating the two points slightly in a gauge-invariant fashion (Schwinger 1951). That is,

$$J_{\mu 5} = \bar{\Psi}(x + \tfrac{1}{2}\varepsilon)\gamma_\mu\gamma_5 P\left[\exp\left(-\int_{x-\varepsilon/2}^{x+\varepsilon/2} (A_\mu \mathrm{d}x'_\mu)\right)\right]\Psi(x - \tfrac{1}{2}\varepsilon) \qquad (11.27)$$

and similarly for J_μ. The limit $\varepsilon \to 0$ is taken at the end. Notice that without the path-ordered exponential over the gauge field inserted the combination (11.27) would not be gauge-invariant under the SU(2) transformations,

$$\Psi_\alpha(x) \to (\mathrm{e}^{-\Lambda(x)})_{\alpha\beta}\Psi_\beta(x)$$

$$A_\mu(x) \to \mathrm{e}^{-\Lambda(x)}(A_\mu + \partial_\mu)\mathrm{e}^{+\Lambda(x)},$$

where $\Lambda(x)$ is the 2×2 anti-hermitian matrix used in (10.78).

Now, when the divergences of J_μ and $J_{\mu 5}$, properly defined as in (11.27), are taken, keeping track of terms singular as $\varepsilon \to 0$, it turns out that while conservation of J_μ can still be retained, the divergence of $J_{\mu 5}$ picks up an anomalous term, given by

$$\partial_\mu J_{\mu 5} = (\mathrm{i}/8\pi^2)\,\mathrm{Tr}\,(G_{\mu\nu}\tilde{G}_{\mu\nu}). \qquad (11.28)$$

We shall not derive this equation, which is a particular case of the phenomenon of current anomalies – a major topic in its own right (Bell and

Jackiw 1969, Adler 1969, Bardeen 1974); see Jackiw (1972) for a review of how such anomalies are derived. In comparing eq. (11.28) with these references, note that the factor of i arises because we are in the Euclidean metric. We shall take (11.28) as given and study its relationship to the vacuum tunnelling phenomenon. In particular, we shall use the anomaly to establish the existence of zero eigenvalues for the massless Dirac operator $M(A_\mu)$ (Schwarz 1977, Brown et al. 1977, Nielsen and Schroer 1977).

Let us approach the massless case by starting with a *massive* quantum Dirac field, in the presence of an external field A_μ characterised by a Pontryagin index Q. The Euclidean action for such a system is

$$S = \mathrm{i}\int \bar{\Psi}(\gamma_\mu D_\mu - m)\Psi \mathrm{d}^4x. \tag{11.29}$$

The presence of the mass term in (11.29) and in the associated field equations will, as is well known and easy to verify, lead to an explicit chiral symmetry-breaking term $-2m\bar{\Psi}\gamma_5\Psi$ in $\partial_\mu J_{\mu 5}$, in addition to the anomalous term. Note also that $\mathrm{Tr}[G_{\mu\nu}\tilde{G}_{\mu\nu}]$, for the external field A_μ, is just proportional to the Pontryagin index density $Q(x)$ defined in eq. (4.18). Hence (11.28) is altered for the system (11.29) to

$$\partial_\mu J_{\mu 5} = -2m\bar{\Psi}\gamma_5\Psi - 2\mathrm{i}Q(x). \tag{11.30}$$

We integrate (11.30) over the Euclidean volume, and take the Euclidean vacuum expectation value. The left-hand side yields

$$\int \langle \partial_\mu J_{\mu 5} \rangle_{\mathrm{Euc}} \mathrm{d}^4x = \oint_{S_3} \mathrm{d}\sigma_\mu \langle J_{\mu 5} \rangle_{\mathrm{Euc}} = 0, \tag{11.31}$$

where the surface integral at infinity will vanish because the Dirac field is massive (short-ranged) and A_μ is just an external field in (11.29). The right-hand side of (11.30) will therefore yield

$$0 = -2m\int \langle \bar{\Psi}\gamma_5\Psi \rangle_{\mathrm{Euc}} \mathrm{d}^4x - 2\mathrm{i}\int Q(x)\mathrm{d}^4x$$

or

$$2m\int \langle \bar{\Psi}\gamma_5\Psi \rangle_{\mathrm{Euc}} \mathrm{d}^4x = -2\mathrm{i}Q. \tag{11.32}$$

Let us evaluate the vacuum expectation value $\int \langle \bar{\Psi}\gamma_5\Psi \rangle_{\mathrm{Euc}} \mathrm{d}^4x$ using the Euclidean functional integral for the system (11.29). We have

$$\langle \int \bar{\Psi}(y)\gamma_5\Psi(y)\mathrm{d}^4y \rangle_{\mathrm{Euc}} = \frac{\int \mathscr{D}[\Psi^\dagger(x)]\mathscr{D}[\Psi(x)]\{\mathrm{e}^{-S}\int \mathrm{d}^4y[\bar{\Psi}(y)\gamma_5\Psi(y)]\}}{\int \mathscr{D}[\Psi^\dagger(x)]\mathscr{D}[\Psi(x)]\mathrm{e}^{-S}}, \tag{11.33}$$

where the Euclidean action S is given in (11.29). Now, let $f_r(x)$ be the spinor eigenfunctions of the hermitian operator $(iD_\mu\gamma_\mu)$, with eigenvalues λ_r respectively, and normalised such that $\int \bar{f}_{r'}(x)f_r(x)d^4x = \delta_{r'r}$. Of course, $f_r(x)$ will also be eigenfunctions of $(iD_\mu\gamma_\mu - im)$, satisfying

$$(iD_\mu\gamma_\mu - im)f_r(x) = (\lambda_r - im)f_r(x). \tag{11.34}$$

We expand the Grassmann field $\Psi(x)$, as explained in chapter 9, in the form

$$\Psi(x) = \sum_r a_r f_r(x); \qquad \bar{\Psi}(x) = \sum_r \bar{a}_r \bar{f}_r(x), \tag{11.35}$$

so that

$$S = \int \bar{\Psi}(x)(iD_\mu\gamma_\mu - im)\Psi(x)d^4x$$
$$= \sum_r \bar{a}_r a_r(\lambda_r - im). \tag{11.36}$$

Hence the denominator of (11.33) becomes, using the measure (9.31),

$$\int \prod_r (d\bar{a}_r da_r)\exp\left(-\sum_r \bar{a}_r a_r(\lambda_r - im)\right) = \prod_r (\lambda_r - im). \tag{11.37}$$

(We have omitted factors of i compared with the measure in (9.31). These will in any case cancel in the numerator and denominator of (11.33). We can also choose to define the Euclidean measure without the factors of i.) The numerator of (11.33) equals, upon using the expansion (11.35),

$$\int \prod_r (d\bar{a}_r da_r)\prod_r \exp[-\bar{a}_r a_r(\lambda_r - im)]\left(\sum_{ss'} \bar{a}_{s'} a_s \int \bar{f}_{s'}(y)\gamma_5 f_s(y)d^4y\right)$$
$$= \int \prod_r (d\bar{a}_r da_r[1 - \bar{a}_r a_r(\lambda_r - im)])\left(\sum_{ss'} \bar{a}_{s'} a_s \int \bar{f}_{s'}(y)\gamma_5 f_s(y)d^4y\right). \tag{11.38}$$

Recall the rules of Grassmanian integration from chapter 9. When the integrand of (11.38) is expanded as a sum of products of the a's and $\bar{a}$'s, only those terms will give a non-zero integral which contain each a and $\bar{a}$ once and only once. This rule when applied here requires that $s = s'$, and reduces (11.38) to

$$\sum_s \int d^4y(\bar{f}_s(y)\gamma_5 f_s(y)) \prod_{r \neq s} (\lambda_r - im). \tag{11.39}$$

Dividing (11.39) by (11.37), we have for (11.33)

$$\langle \int \bar{\Psi}(y)\gamma_5 \Psi(y) \mathrm{d}^4 y \rangle_{\mathrm{Euc}} = \sum_s \frac{\int \mathrm{d}^4 y \bar{f}_s(y)\gamma_5 f_s(y)}{(\lambda_s - \mathrm{i}m)}. \tag{11.40}$$

Remember that f_r satisfy

$$(\mathrm{i}D_\mu \gamma_\mu) f_r(y) = \lambda_r f_r(y)$$

with

$$\int \bar{f}_{r'} f_r \mathrm{d}^4 y = \delta_{r'r}. \tag{11.41}$$

Since $(\mathrm{i}D_\mu\gamma_\mu)$ anticommutes with γ_5, the function $\gamma_5 f_r$ will also be an eigenfunction of $(\mathrm{i}D_\mu\gamma_\mu)$, but with eigenvalue $(-\lambda_r)$. Hence, by the orthogonality condition (11.41),

$$\int \bar{f}_s \gamma_5 f_s \mathrm{d}^4 y = 0 \qquad \text{if} \qquad \lambda_s \neq 0. \tag{11.42}$$

If any λ_s vanishes, then f_s and $\gamma_5 f_s$ are degenerate, and one can always choose f_s to have definite chirality, i.e. $\gamma_5 f_s = +f_s$ or $-f_s$. Hence for each zero eigenvalue λ_s, $\int \bar{f}_s \gamma_5 f_s \mathrm{d}^4 y = \pm 1$. Altogether (11.40) reduces to

$$\langle \int \bar{\Psi}\gamma_5 \Psi \mathrm{d}^4 y \rangle_{\mathrm{Euc}} = (1/-\mathrm{i}m)(n_+ - n_-), \tag{11.43}$$

where $n_+(n_-)$ is the number of positive (negative) chirality zero-λ eigenfunctions. Inserting (11.43) into (11.32), we get

$$Q = n_- - n_+. \tag{11.44}$$

We have thus obtained, starting from the axial current divergence equation (11.30), a simple relation between the number of zero eigenvalues of the operator $(\mathrm{i}D_\mu\gamma_\mu)$ and the Pontryagin index Q of the field A_μ. The relation was derived starting from a massive Dirac field, but (11.44) has no m dependence and the $m \to 0$ limit can be taken with no damage. In particular, when Q is a non-zero integer, n_+ and n_- cannot both vanish. Hence there must be at least one zero eigenvalue.

Now, let us go back to the functional integral (11.26) for any given non-zero integer Q. Each configuration A_μ in the integral carries the same Pontryagin index Q, and we have shown that $\mathrm{i}D_\mu\gamma_\mu$ must have at least one zero eigenvalue, for each such A_μ. Clearly the same holds for the operator $M(A_\mu) = \gamma_4(\mathrm{i}D_\mu\gamma_\mu)$ used in (11.23)–(11.26). Hence Det $[M(A_\mu)] = 0$ for each A_μ in the functional integral (11.26) and therefore that integral vanishes exactly! This in turn implies that the tunnelling amplitude

$$\lim_{\tau \to \infty} \langle N + Q | \mathrm{e}^{-H\tau} | N \rangle$$

vanishes. The vacuum tunnelling phenomenon we described in the last chapter for the pure Yang–Mills system is suppressed by the addition of massless fermions.

It should be pointed out that this phenomenon was first discovered by 'tHooft (1976a), who explicitly obtained a zero-eigenvalue mode of the operator $\mathrm{i}(\partial_\mu + A_\mu)\gamma_\mu$ for the special case where the field configuration A_μ was the one-instanton solution. The eigenfunction he obtained had the form

$$f_0(x - X, \rho) = \frac{\rho}{(\rho^2 + (x - X)^2)^{3/2}} u_0 \tag{11.45}$$

where u_0 is some space-independent spinor, X and ρ respectively are the location and size of the one-instanton solution. The reader can verify that (11.45), with an appropriate choice of u_0, indeed satisfies

$$(\partial_\mu + A_\mu)\gamma_\mu f_0 = 0.$$

when the one-instanton solution in eq. (4.74) is inserted for A_μ. From this, clearly, $\mathrm{Det}[M(A_\mu)] = 0$ for the one-instanton configuration, and hence the functional integral (11.26) will vanish in a semiclassical dilute instanton-gas expansion. The result (11.44) derived earlier is, however, more general. It implies that $\mathrm{Det}\, M[A_\mu] = 0$ for every configuration A_μ in (11.26), and hence that the functional integral vanishes exactly. The result is therefore free from any dilute instanton-gas approximation, with its associated difficulties described in section 10.5.

The connection between the axial-current anomaly and the suppression of tunnelling can also be shown in a different way. Let us return for a moment to Minkowski space, where the anomaly equation (11.28) changes to

$$\partial^\mu J_{\mu 5} = -(1/8\pi^2)\,\mathrm{Tr}[\tilde{G}_{\mu\nu} G^{\mu\nu}] \equiv 2Q_\mathrm{M}(\boldsymbol{x}, t). \tag{11.46}$$

Here $Q_\mathrm{M}(\boldsymbol{x}, t)$ is the Minkowskian analogue of the Pontryagin density function $Q(x)$. We saw in eq. (4.23)–(4.24) that $Q(x)$ can be written as a pure divergence. That construction can be trivially adopted to Minkowski space, to yield

$$\partial^\mu J_{\mu 5} = 2Q_\mathrm{M}(\boldsymbol{x}, t) = 2\partial^\mu k_\mu, \tag{11.47}$$

where

$$k_\mu \equiv -(1/8\pi^2)\,\varepsilon_{\mu\nu\rho\sigma}\,\mathrm{Tr}[A^\nu(\partial^\rho A^\sigma + \tfrac{2}{3} A^\rho A^\sigma)]. \tag{11.48}$$

Although $J_{\mu 5}$ is not conserved because of the anomaly, the modified current

$$\tilde{J}_{\mu 5} \equiv J_{\mu 5} - 2k_\mu, \tag{11.49}$$

is clearly conserved by virtue of (11.47). The corresponding charge

$$\tilde{Q}_5 \equiv \int \tilde{J}_{05}\, \mathrm{d}^3 x \tag{11.50}$$

is a constant in time, and must therefore commute with the Hamiltonian:

$$[\tilde{Q}_5, H] = 0. \tag{11.51}$$

Notice that while $J_{\mu 5}$ is, by the construction (11.27), gauge-invariant, the current $\tilde{J}_{\mu 5}$ is not. This is because the 'topological current' k_μ is, as we have remarked in the Euclidean context in chapter 4, not gauge-invariant. Hence $\tilde{Q}_5 = Q_5 - 2\int k_0\, \mathrm{d}^3 x$ is a conserved, but gauge-variant charge. In particular, consider g_1, the prototype example of a 'large' gauge transformation, defined in eq. (10.83). This is the transformation that changes the topological vacuum $|N\rangle$ into $|N+1\rangle$. Under this transformation, $\tilde{Q}_5$ will be altered because of the change in $\int k_0 \mathrm{d}^3 x$. This change can be obtained by inserting the transformed gauge field (combine (10.78), (10.82) and (10.83)) into the expression (11.48) for k_μ. Straightforward algebra yields, under the gauge transformation g_1,

$$\int k_0 \mathrm{d}^3 x \rightarrow \{\int k_0 \mathrm{d}^3 x\} + 1 \tag{11.52}$$

and hence

$$U(g_1)\tilde{Q}_5 U(g_1)^{-1} = \tilde{Q}_5 - 2.$$

(Recall that for a static pure gauge configuration, the term $\int k_0 \mathrm{d}^3 x$ is just the topological charge N (see (10.96)) which we know is altered to $N+1$ by the action of g_1, consistent with (11.52). For an arbitrary A_μ, $\int k_0 \mathrm{d}^3 x$ is no longer an integer-valued topological charge, but under the action of g_1 this term still changes by unity, giving rise to (11.52).)

Now, the topological vacuum $|0\rangle$ corresponds to the $N = 0$ sector of the gauge field. It is, of course, also the fermionic vacuum. Hence,

$$\tilde{Q}_5|0\rangle = 0.$$

Then

$$\begin{aligned} \tilde{Q}_5|1\rangle &= \tilde{Q}_5\, U(g_1)|0\rangle \\ &= \{U(g_1)\tilde{Q}_5 + 2U(g_1)\}|0\rangle, \qquad \text{by (11.52)} \\ &= 2|1\rangle. \end{aligned}$$

Proceeding similarly, we obtain

$$\tilde{Q}_5|N\rangle = 2N|N\rangle. \tag{11.53}$$

But $[\tilde{Q}_5, e^{-H\tau}] = 0$, by (11.51); hence

$$\begin{aligned} 0 &= \langle M|[\tilde{Q}_5, e^{-H\tau}]|N\rangle \\ &= 2(M-N)\langle M|e^{-H\tau}|N\rangle. \end{aligned}$$

Therefore

$$\langle M|e^{-H\tau}|N\rangle \propto \delta_{MN}. \tag{11.54}$$

Thus, we once again have the result that the tunnelling amplitude vanishes when $M \neq N$. The physics behind what is happening is simply this: The state $|M\rangle$ is related to $|N\rangle$ by the gauge transformation g_1 acting $(M-N)$ times. Under this transformation $\tilde{Q}_5$ must change, because of (11.52). But $\tilde{Q}_5$ is conserved. Hence there can be no transition from $|N\rangle$ to any other $|M\rangle$.

The situation then is that the topological vacua $|N\rangle$ are (i) the lowest energy states in the corresponding homotopy sectors, (ii) energetically degenerate with one another (since large gauge transformations also commute with H), and (iii) furthermore do *not* tunnel into one another when massless fermions are coupled. One can ask whether there is any need for constructing the θ-vacua in this case, or whether the states $|N\rangle$ can serve as the correct vacua of this theory.

Callan et al. (1976) argue that the θ-vacua still need to be constructed as the correct vacua because of the cluster decomposition property – a requirement on the vacuum of a field theory. Let $D(x)$ and $B(x)$ be two local operators at, say, $t = 0$. Then the cluster decomposition property of the vacuum state requires that

$$\langle \text{vac}|D(x)B(y)|\text{vac}\rangle \xrightarrow[|x-y|\to\infty]{} \langle \text{vac}|D(x)|\text{vac}\rangle \langle \text{vac}|B(y)|\text{vac}\rangle. \tag{11.55}$$

We shall see that the $|N\rangle$ states violate this for operators of non-zero chirality. An operator $B(x)$ can be said to have chirality c if

$$[\tilde{Q}_5, B(x)] = cB(x). \tag{11.56}$$

Clearly $B^+(x)$ will have chirality $(-c)$. For instance the operator

$$\phi(x) \equiv \bar{\Psi}(x)(1-\gamma_5)\Psi(x)$$

has chirality $c = 2$, as can be checked using the canonical equal-time commutation rules. Now consider the matrix element of any operator $B(\boldsymbol{x})$ with *non-zero* chirality $2c$, between topological vacua $|N\rangle$ and $|M\rangle$. We have

$$2c\langle M|B(\boldsymbol{x})|N\rangle = \langle M|[\tilde{Q}_5, B(\boldsymbol{x})]|N\rangle$$
$$= 2(M-N)\langle M|B(\boldsymbol{x})|N\rangle.$$

Therefore

$$\langle M|B(\boldsymbol{x})|N\rangle \propto \delta_{c,M-N}. \tag{11.57}$$

In particular,

$$\langle N|B(\boldsymbol{x})|N\rangle = 0, \qquad \text{since } c \neq 0. \tag{11.58}$$

Cluster decomposition, if obeyed by the $|N\rangle$ vacua, would require that

$$\lim_{|\boldsymbol{x}-\boldsymbol{y}|\to\infty} \langle N|B^{+}(\boldsymbol{x})B(\boldsymbol{y})|N\rangle$$
$$= \langle N|B^{+}(\boldsymbol{x})|N\rangle\langle N|B(\boldsymbol{y})|N\rangle = 0, \qquad \text{by (11.58)}.$$

But this is not true, since, when expanded in terms of intermediate states, $\langle N|B^{\dagger}(\boldsymbol{x})B(\boldsymbol{y})|N\rangle$ will have a non-zero contribution $\langle N|B^{\dagger}(\boldsymbol{x})|N+c\rangle\langle N+c|B(\boldsymbol{y})|N\rangle$. This contribution, by translation invariance of each matrix element, will equal $\langle N|B^{\dagger}(0)|N+c\rangle\langle N+c|B(0)|N\rangle$ and will remain non-zero as $|\boldsymbol{x}-\boldsymbol{y}|\to\infty$, violating cluster decomposition. The basic reason is that, while the off-diagonal matrix element $\langle N+c|B|N\rangle \neq 0$, the diagonal element $\langle N|B|N\rangle = 0$. By contrast, consider the matrix element between $|\theta\rangle$ states.

$$\langle\theta|B|\theta'\rangle = \sum_{N,M} \exp[\mathrm{i}(N\theta - M\theta')]\langle N|B|M\rangle$$
$$= \sum_{M} \exp[\mathrm{i}M(\theta-\theta') + \mathrm{i}c\theta]\langle M+c|B|M\rangle$$
$$\propto \delta(\theta-\theta'), \tag{11.59}$$

since, for a gauge-invariant operator B, $\langle M+c|B|M\rangle$ will not depend on M. Hence, when expanded in intermediate states,

$$\langle\theta|B^{\dagger}(\boldsymbol{x})B(\boldsymbol{y})|\theta\rangle \xrightarrow[|\boldsymbol{x}-\boldsymbol{y}|\to\infty]{} \int \mathrm{d}\theta'\langle\theta|B^{\dagger}(\boldsymbol{x})|\theta'\rangle\langle\theta'|B(\boldsymbol{y})|\theta\rangle$$
$$= \langle\theta|B^{\dagger}(\boldsymbol{x})|\theta\rangle\langle\theta|B(\boldsymbol{y})|\theta\rangle, \tag{11.60}$$

consistent with the cluster decomposition property (11.55).

In short, even though massless fermions suppress tunnelling between the topological ground states $|N\rangle$, the correct vacua are their linear combinations represented by the $|\theta\rangle$ states. Of course, the fact that the $|N\rangle$ states do not tunnel should result in the $|\theta\rangle$ states continuing to be energetically degenerate. This is indeed so. We have,

$$\langle\theta|e^{-H\tau}|\theta'\rangle = \sum_{N,M} \langle N|e^{-H\tau}|M\rangle \exp[i(N\theta - M\theta')]$$

$$= \sum_{N} \langle N|e^{-H\tau}|N\rangle \exp[iN(\theta-\theta')], \qquad (11.61)$$

using (11.54). But since the different $|N\rangle$ states are related by large gauge transformations, which commute with H, $\langle N|e^{-H\tau}|N\rangle$ is independent of N. Hence (11.61) leads to

$$\langle\theta|e^{-H\tau}|\theta'\rangle = e^{-E_0\tau}\sum_{N} \exp[iN(\theta-\theta')]$$

$$= 2\pi\delta(\theta-\theta')e^{-E_0\tau}. \qquad (11.62)$$

Thus, the $|\theta\rangle$ vacua all have the same θ-independent energy E_0. A functional-integral representation can be written for (11.62) analogously to (10.98), with the only difference that the action will include Fermi fields as well, which must also be integrated over:

$$G_\theta = \lim_{\tau\to\infty} \langle\theta|e^{-H\tau}|\theta\rangle$$

$$= 2\pi\delta(0)\sum_{Q} e^{-iQ\theta}\int \mathscr{D}[\Psi^\dagger]\,\mathscr{D}[\Psi]\,\mathscr{D}[A_\mu]_Q\, e^{-S}$$

$$= 2\pi\delta(0)\int \mathscr{D}[\Psi^\dagger]\,\mathscr{D}[\Psi]\,\mathscr{D}[A_\mu]_{\mathrm{all}\,Q}$$

$$\exp\left(-S + \frac{i\theta}{16\pi^2}\int \mathrm{Tr}[\tilde{G}_{\mu\nu}G_{\mu\nu}]\,d^4x\right). \qquad (11.63)$$

Therefore

$$\langle\theta|\int \mathrm{Tr}(\tilde{G}_{\mu\nu}G_{\mu\nu})d^4x|\theta\rangle = -16\pi^2 i\frac{\partial}{\partial\theta}(\ln G_\theta) = 0, \qquad (11.64)$$

since G_θ, by (11.62), is independent of θ.

That all the θ-vacua have the same energy, and all carry a vanishing expectation value of $\mathrm{Tr}(G_{\mu\nu}\tilde{G}_{\mu\nu})$ is a reflection of the fact that in the presence of massless fermions, the different θ sectors can be obtained from one another by chiral rotation. The current associated with this transformation $\Psi \to e^{i\alpha\gamma_5}\Psi$ is the axial vector current $J_{\mu 5}$ defined earlier. Although it is naively conserved by the Lagrangian (11.22), we have seen that the axial current actually carries an anomaly (eq. (11.28)). As a result, under the chiral rotation mentioned above, the Euclidean action will change (as per the usual steps leading to Noether's theorem) by

$$\Delta S = \alpha \int d^4x\, \partial_\mu J_{\mu 5} = \frac{i\alpha}{8\pi^2} \int d^4x\, \mathrm{Tr}[G_{\mu\nu}\tilde{G}_{\mu\nu}].$$

Compare this with the θ-dependent part of the effective action in (11.63). It is evident that chiral rotation alters θ to $\theta - 2\alpha$. Therefore the different θ sectors are related to one another by just a chiral rotation, which, for massless Fermi fields, is a mere redefinition of the fields. Hence, the equivalence of chirally invariant results in the different θ sectors. Contrast this with the pure gauge field system discussed in section 10.4. There the energy E_θ (eq. (10.100)), and $\langle \mathrm{Tr}\, G_{\mu\nu}\tilde{G}_{\mu\nu} \rangle_\theta$ (obtained by taking the logarithmic derivative of (10.98) with respect to θ) are θ-dependent, showing that the different θ sectors are inequivalent. The addition of massless Fermi fields restores their equivalence.

11.4. Instantons and Borel summability

An important feature of instantons, as we have seen, is that they lead to non-perturbative results for physical quantities. This aspect of instantons is also reflected in their relevance to the Borel summability of perturbation expansions. We devote this section to pointing out this relevance.

Consider a quantum system characterised by some expansion parameter α. Let $G(\alpha)$ be some physical quantity in that system, such as its vacuum energy or one of its Green's functions. We suppress, for simplicity, the dependence of $G(\alpha)$ on variables other than α. Perturbation theory would yield, typically, an infinite series for $G(\alpha)$ of the form

$$G(\alpha) = \sum_{n=0}^{\infty} a_n \alpha^n. \tag{11.65}$$

Let the theory be renormalisable, so that ultraviolet divergences in each order are absorbed by using renormalised parameters. We can then take each term in the series (11.65) to be finite for any finite α. That still leaves open the question of whether the infinite series converges to a finite sum for the values of α of interest. There is also the distressing possibility that the series may not converge for any α, however small. Consider for example, quantum electrodynamics, where α stands for the fine-structure constant $e^2/4\pi$. Perturbative predictions when calculated to low orders agree spectacularly with experiment. Nevertheless, as Dyson (1952) has argued, perturbation theory for QED may actually diverge when one goes to high orders. Dyson pointed out that when α is negative (of however small magnitude), the perturbative vacuum will be unstable. It can decay into electron–positron pairs, and lower the energy by separating the pairs far apart. (Remember that for negative α unlike charges will repel.) As a result, the vacuum energy will have an imaginary part for all negative α and a branch cut along the negative α-axis, with a singularity at $\alpha = 0$. This implies a zero radius of convergence for perturbative expansions in the theory! In that case the agreement with experiment of low-order QED calculations has to be attributed to perturbation series being just an asymptotic expansion for the correct answer.

Similar divergence of perturbation series has also been discovered subsequently for other models, by estimating the behaviour of high-order terms in the series (see for example Bender and Wu 1973, 1976, Lipatov 1976, 1977, Brezin et al. 1977a, 1977b, Parisi 1977a, 1977b, 'tHooft 1979). These results underline the possibility that for many field theories of interest perturbation series may not converge even for arbitrarily small α.

Now, given a perturbation expansion such as (11.65) in a field theory, its radius of convergence R is given by

$$1/R = \lim_{n \to \infty} \sup |a_n|^{1/n}.$$

The coefficient a_n, for any n, can in principle be obtained by calculating appropriate Feynman diagrams of that order, duly renormalised. An explicit way to obtain R is by trying to calculate a_n for large n and estimating its behaviour as $n \to \infty$. An alternative way of obtaining information on convergence is by studying the analyticity properties of the Borel function associated with this infinite series, in the Borel summation procedure. This procedure also allows the possibility of 'improving' the convergence of the series, i.e. giving meaning to $G(\alpha)$ for some $|\alpha| > R$, by analytical continuation. The idea, briefly, is as follows. Using $n! = \int_0^\infty e^{-t} t^n dt$, one can write (11.65) in the form

$$G(\alpha) = \sum_{n=0}^{\infty} \frac{a_n}{n!} \int_0^{\infty} e^{-t} (t\alpha)^n dt. \tag{11.66}$$

Suppose, with the intention of analytically continuing this function, we reverse the order of summation and integration, and consider the function

$$\tilde{G}(\alpha) = \int_0^\infty e^{-t}\left(\sum_{n=0}^\infty \frac{a_n (t\alpha)^n}{n!}\right) dt$$

$$\equiv \int_0^\infty e^{-t} F(t\alpha)\, dt \tag{11.67}$$

where

$$F(z) \equiv \sum_{n=0}^\infty \frac{a_n}{n!} z^n. \tag{11.68}$$

$F(z)$ may be called the Borel function associated with the original series. The key point is that the series (11.68) enjoys a better convergence than the original series (11.65) thanks to the extra factor $1/n!$ in each term. Let R, the radius of convergence of the original series (11.65), be finite and *non-zero*. Then $\tilde{R}$, the radius of convergence of the Borel series (11.68), is given by

$$\frac{1}{\tilde{R}} = \limsup_{n \to \infty} \left|\frac{a_n}{n!}\right|^{1/n} = \frac{1}{R} \lim_{n \to \infty} (n!)^{-1/n}$$

$$= 0, \qquad \text{since } R > 0. \tag{11.69}$$

In that case, the series (11.68) is convergent for all z, and $F(z)$ is analytic in the entire z-plane. $F(t\alpha)$, for all positive t, can be obtained by summing (11.68) and inserted into the integrand in (11.67). Then, for those values of α for which the integral exists, $\tilde{G}(\alpha)$ is well defined. It is called the Borel sum of the parent series (11.65). When $|\alpha| < R$, it can be shown (see for instance Whittaker and Watson 1920) that $\tilde{G}(\alpha)$ is the same as $G(\alpha)$, i.e. that the interchange of summation and integration used in going from (11.66) to (11.67) is justified. For those α outside the circle of convergence ($|\alpha| > R$), for which the integral (11.67) exists, $\tilde{G}(\alpha)$ provides an analytical continuation of $G(\alpha)$.

Of course, this procedure requires that $F(t\alpha)$ be non-singular for all $0 \leqslant t < \infty$, for any given α. When $R > 0$, this is satisfied by virtue of (11.69). The Borel series will have an infinite radius of convergence and $F(z)$ will be analytic throughout the z-plane. When $R = 0$, this may not hold true. Conversely, suppose we are given the physical quantity $G(\alpha)$ directly in the integral form (11.67) with an $F(\alpha t)$ given in closed form. Then we may be able to directly examine the analyticity properties of the function $F(z)$. If we find that $F(z)$ has singularities at some finite z, then the Borel series (11.68) will have a finite radius of convergence $\tilde{R}$, and, by virtue of (11.69), we can conclude that R, the radius of convergence of the original perturbation expansion for $G(\alpha)$, will be zero.

Let us apply these ideas to quantum field theory, following 'tHooft (1979). We shall work in the Euclidean metric. Let $\phi(x)$ represent the field in D Euclidean dimensions. (It need not be a scalar field; $\phi(x)$ could stand, symbolically, for scalar or vector fields.) Let $\langle \phi(x_1) \ldots \phi(x_n) \rangle$ be any Euclidean Green's function of the theory. It will also be a function of the coupling parameter, α, of the theory. For definiteness, we shall take α to be positive in the physical region. (For QED, $\alpha = e^2/4\pi > 0$; for QCD, $\alpha = g^2$; for the sine–Gordon system (eq. (7.1)) $\alpha = \lambda/m^2 > 0$ etc.). The Green's function is given in terms of functional integrals as

$$\langle \phi(x_1) \ldots \phi(x_n) \rangle = \frac{\int \mathscr{D}[\phi(x)] e^{-(\text{action})} \phi(x_1) \ldots \phi(x_n)}{\int \mathscr{D}[\phi(x)] e^{-(\text{action})}}. \qquad (11.70)$$

By using a suitably rescaled field $\phi(x)$, we can write the Euclidean action in the form $(1/\alpha)S[\phi]$, where the α-dependence is factored out and $S[\phi]$ is positive. This can be done for most models of interest as pointed out in chapter 6. Let us first consider the numerator and denominator of (11.70) separately. The numerator may be written as

$$N(\alpha) = \int \mathscr{D}[\phi(x)] \exp\left(-\frac{1}{\alpha} S[\phi] \right) (\phi \ldots . \phi) \qquad (11.71)$$

where, for simplicity, we have suppressed dependence on the coordinates $x_1, \ldots, x_n$. From the fact that we may have rescaled the field by some power of α in order to cast the action in the form $(1/\alpha)S[\phi]$, we may expect infinitely many factors of α in the measure $\mathscr{D}[\phi(x)]$ compared with the unscaled field. But these factors will also exist in the denominator of (11.70), and may be cancelled. Now, since α and $S[\phi]$ are positive, we have the identity:

$$\alpha \int_0^\infty dt \, \delta(\alpha t - S[\phi(x)]) = 1. \qquad (11.72)$$

Inserting this into (11.71), we have

$$N(\alpha) = \alpha \int_0^\infty dt \int \mathscr{D}[\phi(x)] \, \delta(\alpha t - S[\phi]) \exp\left(-\frac{1}{\alpha} S[\phi] \right) (\phi \ldots \phi)$$

$$\equiv \alpha \int_0^\infty dt \, F_N(\alpha t) e^{-t}, \qquad (11.73)$$

where

$$F_N(z) = \int \mathscr{D}[\phi(x)]\,\delta(z - S[\phi])\,(\phi \ldots \phi). \tag{11.74}$$

Thus, we have cast $N(\alpha)$ in the Borel integral form apart from an overall factor α, which will not matter. A similar factor will again arise in the denominator of (11.70). Let us cancel it out.

Let us look for singularities of $F_N(z)$ as defined by the functional integral in (11.74). The functional integral is just a multiple integral over the field at each x. The multiple integral of a δ-function is defined by

$$\int \mathrm{d}y_1 \ldots \mathrm{d}y_n\,\delta[z - f(y_1, \ldots, y_n)] = \oint_{f(y)=z} \mathrm{d}\sigma\,|\nabla f|^{-1}, \tag{11.75}$$

where the right-hand side is an integral over the surface $f(y) = z$ in y-space.

$$|\nabla f| \equiv \left[\sum_i \left(\frac{\partial f}{\partial y_i}\right)^2\right]^{1/2}$$

Consequently, the integral (11.75) will be singular for some $z = \bar{z}$, if $|\nabla f|$ vanished anywhere on the surface $f(y) = \bar{z}$. But $|\nabla f| = 0$ implies $\partial f/\partial y_i = 0$ for all i. Thus, the integral (11.75) will be singular at $\bar{z}$, if the function $f(y_1 \ldots y_n)$ had a stationary point on the surface $f(y) = \bar{z}$. Generalising this to the functional integral (11.74), we see that the Borel function $F_N(z)$ will be singular at $\bar{z}$ if, on the surface $S[\phi] = \bar{z}$ (in the space of functions $\phi(x)$), there is a point satisfying

$$\delta S[\phi]/\delta\phi(x) = 0 \qquad \text{for all } x. \tag{11.76}$$

This, however, is just the Euclidean field equation, whose finite-action solutions are instantons. Therefore, if the field theory in question contains finite-action instantons, then the Borel function $F_N(z)$ in (11.74) will have singularities at those finite values of z which equal the action of those instantons. From this we can conclude, using arguments given earlier, that a perturbation expansion of $N(\alpha)$ in powers of α will have zero radius of convergence. The same analysis can be repeated for the denominator of (11.70) – call it $D(\alpha)$. The Borel function $F_D(z)$ associated with $D(\alpha)$ will also contain singularities due to the same instantons, at the same values of z. Hence $D(\alpha)$ will also have zero radius of convergence if expanded as a power series in α. Then, barring fortuitous circumstances where the singularities at $\alpha = 0$ of $N(\alpha)$ and $D(\alpha)$ cancel each other exactly, the power series for the ratio $N(\alpha)/D(\alpha)$, which gives the Green's function $\langle \phi(x_1) \ldots \phi(x_n) \rangle$ in (11.70), will also have zero radius of convergence.

In short, when finite-action instantons are permitted by a field theory, perturbation expansions in powers of α of its Green's functions will not converge for any α. We have obtained this result without explicitly calculating the large-order coefficients of the perturbation series. Notice also that even though instantons have entered into the discussion we have not used any semiclassical approximations in the derivation.

Of course, our semiclassical instanton calculations in earlier sections support this general result. For instance, the Yang–Mills system contains finite-action instantons. Correspondingly we found that the vacuum energy for this system had an essential singularity when the coupling parameter g^2 is zero (see e.g. (10.102)–(10.105)). That was only a semiclassical result but, for what it is worth, it indicates that a perturbation expansion of this energy would have zero radius of convergence. Similar statements also hold for the periodic potential model, the two-dimensional abelian Higgs model etc.

It is also instructive to consider models which permit no real instanton solutions with positive action but do permit complex solutions with negative action. For example, consider a real scalar theory in four dimensions, with Euclidean action (see 'tHooft 1979 and Fainberg and Iofa 1980)

$$\frac{1}{\alpha}\int d^4x\left(\frac{1}{2}(\partial_\mu\phi)^2+\frac{\phi^4}{4!}\right)\equiv\frac{1}{\alpha}S[\phi]. \tag{11.77}$$

The Euclidean field equation

$$\partial_\mu\partial_\mu\phi-\phi^3/6=0 \tag{11.78}$$

can have no real finite-action solutions, because of the virial theorem described in chapter 3. But it does have a solution if ϕ is extended to complex values. The solution is

$$\phi(x)=i\rho\sqrt{48}/(x^2+\rho^2), \tag{11.79}$$

where ρ is an arbitrary scale parameter (the action (11.77) is scale-invariant). This solution has negative action

$$S[\phi]=-16\pi^2. \tag{11.80}$$

Now, if we wish to study $F_N(z)$ as defined in (11.74) for complex z, we shall have to extend the real field ϕ to complex values in order that $\delta(z-S[\phi])$ may be non-zero. (If the original field ϕ had itself been a complex (charged) field, with a positive action, then both $\operatorname{Re}\phi$ and $\operatorname{Im}\phi$ will have to be extended to complex values, in order to obtain complex action.) In that case, the above solution will produce a singularity at $z = -16\pi^2$. Once again the existence of this singularity means that the Borel series has only a finite radius of convergence, and that the parent perturbation series has zero radius of convergence. But, since the singularity in $F_N(z)$ is only at a negative z, it may still be possible to evaluate $\tilde{G}(\alpha)$ in (11.67) for positive α, provided there are no other singularities, unrelated to instantons, in $F_N(z)$ for positive z. In that case, we can consider the perturbation series to be Borel-summable, in a generalised sense of the term. The original perturbation series of course is meaningless for any α, since its radius of convergence is zero. But $\tilde{G}(\alpha)$, as evaluated through (11.67), if the integral exists for some $\alpha > 0$, may give a sensible value of the physical quantity in question. A prototype example of this kind is the power series

$$\sum_n (-1)^n n!\,\alpha^n. \tag{11.81}$$

Its radius of convergence is of course zero. The corresponding Borel function is

$$F(z) = \sum_0^\infty (-1)^n z^n = \frac{1}{1+z}. \tag{11.82}$$

This certainly has a singularity, but at $z = -1$. Hence, for positive $\alpha > 0$, the integral

$$\tilde{G}(\alpha) = \int_0^\infty \mathrm{d}t\, \mathrm{e}^{-t}\left(\frac{1}{1+\alpha t}\right) \tag{11.83}$$

will still exist. The function $\tilde{G}(\alpha)$ can be called the Borel sum of the original series (11.81), but it is in no sense an analytical continuation of that series because the latter does not exist in any finite domain in the α-plane. If a series such as (11.81) emerges in the perturbation expansion of a physical quantity, we could take the attitude that while the perturbation expansion is meaningless for any α, the parent physical quantity may still be well defined in terms of a Borel integral such as (11.83).

By contrast, when the instanton is a genuine one, i.e. having finite positive action as in all our examples of chapter 4 and 10, then $F(z)$ will be singular for some positive z. This singularity will fall, for physical values of $\alpha > 0$, within the range of integration of the Borel integral. In that case, the perturbation expansion is not really Borel-summable. But 'tHooft (1979) argues that even in such cases there may be hope. Even when the singularity falls at some positive t in the integral (11.67), the latter may be interpreted through appropriate analytical continuation, provided the singularity is integrable. For example we saw that the Yang–Mills vacuum energy, in the dilute instanton-gas approximation, had a coupling constant ($g^2 = \alpha$) dependence of the form

$$G(\alpha) = (1/\alpha^4)e^{-8\pi^2/\alpha}[1 + O(\alpha)]. \tag{11.84}$$

It is easy to check that, formally,

$$(1/\alpha^4)e^{-8\pi^2/\alpha} = \int_0^\infty dt\, e^{-t} \frac{\partial^3}{\partial(\alpha t)^3}[\delta(\alpha t - 8\pi^2)]. \tag{11.85}$$

By comparing (11.85) with (11.67), we see that the associated Borel function is

$$F(z) = \frac{\partial^3}{\partial z^3}[\delta(z - 8\pi^2)]. \tag{11.86}$$

This is of course singular. Notice that the singularity occurs at $8\pi^2$, the Yang–Mills instanton's action, in accordance with our general result. This singularity is, however, integrable. We know how to integrate over a δ-function or its derivatives. Therefore it is possible that even though instantons produce singularities in the Borel function on the positive real axis, the function may still be integrable, and may yield an evaluation of the physical quantity $G(\alpha)$ as a function of $\alpha > 0$. These are open questions. Furthermore, although we have shown that instantons yield singularities in the Borel function, we have not shown that these are the only singularities. In general, other singularities, unrelated to instantons, may also exist in $F(z)$. All these must be unearthed before the possibility of Borel summation can be conclusively determined. For more on these questions, readers are referred to the lectures of 'tHooft (1979) and to Crutchfield (1979). For an extension of some of these ideas to include Fermi fields, see Parisi (1977b).

11.5. Further developments

We have described in detail the basic non-perturbative impact of instantons on the vacuum structure of the associated quantum theory, and a few of the resulting consequences. Of course further work has been done in this general area, both in terms of refining or generalising these calculations, and towards searching for some more physical consequences of instantons. In this concluding section we make a brief mention of some of these developments, along with references.

One set of calculations evaluates the heavy-quark potential caused by instantons in QCD, by using the Wilson loop function. We introduced this function in section 11.1 in the context of the two-dimensional abelian Higgs model, and found that it leads to confinement of charges in that model in the presence of a dilute instanton gas. We also pointed out that a similar calculation in (3 + 1)-dimensional QCD will, unfortunately, not lead to quark confinement. Nevertheless, instantons do make a significant contribution to the interaction between heavy quarks, even if this contribution does not grow at large distances and result in confinement. Callan et al. (1978a) computed a spin-independent instanton-induced heavy quark potential, using the non-abelian Wilson loop function. The procedure is similar to what we used in section 11.1, with the loop corresponding to the Euclidean world lines of a $q\bar{q}$ pair. These authors found a potential which behaved at short distances as

$$V(R) \underset{R\to 0}{\sim} 11.27 R^2 \int_0^{\lambda_c} \frac{d\lambda_1}{\lambda_1^4} D(\lambda_1), \tag{11.87}$$

where R is the quark separation, λ_1 is the instanton size, and $D(\lambda_1)$ is the 'density function' for instantons of size λ_1, given by

$$D(\lambda_1) \simeq 0.10 \left(\frac{8\pi^2}{\bar{g}^2(\lambda_1)} \right)^6 \exp\left(-\frac{8\pi^2}{\bar{g}^2(\lambda_1)} \right). \tag{11.88}$$

Notice that compared with the Yang–Mills density function in eq. (10.102), the above function carries a power $(\bar{g})^{-12}$ rather than $(\bar{g})^{-8}$. That is because eq. (11.88) corresponds to the SU(3) gauge group associated with QCD. The same SU(2) Yang–Mills instanton can be used, as pointed out in chapter 4, but now this instanton will carry four more zero-modes, due to transformations by the additional generators of SU(3). (SU(3) has actually five more generators than does SU(2), but rotation by one of them leaves the SU(2) instanton invariant and yields no zero-modes.) Although this potential grows quadratically at short distances, Callan et al. found that at large separation, it levels off to a constant value:

$$V(R) \underset{R\to \infty}{\sim} (a + b/R). \tag{11.89}$$

Thus, the dilute instanton gas in QCD does not give a confining potential. The constant a is interpreted as a mass-renormalisation and the constant b as a renormalisation of the Coulomb potential. Subsequently Callan et al. (1978b) extracted additional spin-dependent terms (see also Wilczek and Zee 1978). Such a spin-dependent potential will clearly be of value in phenomenological models of quarkonium ($q\bar{q}$) states. A nice feature they found was that the spin-dependent terms are obtainable from the spin-independent term (11.87) by taking appropriate derivatives with respect to R. Consequently, the spin-dependent part depends on lower powers of R. It turns out to be constant at short distances, and falls off as $R \to \infty$.

Notice that in (11.87), a cut-off λ_c has been introduced on the range of instanton sizes λ_1. This is a manifestation of the infrared problem pointed out in section 10.5. Instantons in Yang–Mills theory (and hence also in QCD) come in all sizes varying from 0 to ∞. To get a concrete result, Callan et al. therefore introduced a cut-off λ_c, which they took to be the inverse of the typical hadron mass scale, where the effective coupling $\bar{g}(\lambda_c)$ is also small. The reliability of such a procedure, the sensitivity of the results to the precise value of λ_c and possible alternative ways of handling the large-instanton problem are still somewhat controversial matters. See for instance Appelquist and Shankar (1978) and McDougall (1980). For some other approaches to the effect of instantons on the dynamics of quarks (including light quarks) and their bound states, see Carlitz and Creamer (1979), Duncan (1978), Caldi (1977), Geshkenbein and Ioffe (1980), and Suzuki (1978).

Another aspect that has received attention is the threat of strong violation of P and T symmetries due to instanton effects, and some plausible way of avoiding it. Recall from section 10.4 that the effective Lagrangian associated with θ-vacua in Yang–Mills theory (eq. (10.98)) carries a θ-dependent extra term and is not invariant under P and T, except for the special case of $\theta = 0$. The same statement holds for the colour gauge group SU(3) and becomes relevant to the real world in the context of QCD. In nature, there *is* violation of both P and T, but it is very small and is attributed in conventional wisdom to weak or superweak interactions. In that case QCD, which refers solely to strong interactions, should conserve P and T, i.e. the vacuum angle θ must vanish. Alternatively, if some $\theta \neq 0$ is to be permitted, the observed minuteness of T violation (as indicated by the upper bound on the neutron electric dipole moment) requires that (Wilczek 1978)

$$\theta \leqslant 10^{-5}.$$

What is puzzling is the mechanism which forces θ to either vanish or be such an unnaturally small number less than 10^{-5}.

One possible explanation is that one of the quarks coupling to the gauge field in QCD is massless. In that case, as pointed out at the end of section 11.3, the different θ sectors are equivalent and can be reduced to $\theta = 0$ by chiral rotation, restoring P and T conservation by QCD. However, indirect evidence on quark masses, through current-algebra estimates, indicates that while 'up' and 'down' quarks have low masses by hadronic scales, they are probably not massless. In view of this, Peccei and Quinn (1977a, b) suggested an alternative mechanism for conserving P and T while permitting all quarks to have a non-zero physical mass. Masses for some quarks are obtained by coupling them to Higgs scalars which are made to acquire a non-zero vacuum expectation value. It was, however, soon realised (Weinberg 1978, Wilczek 1978) that this type of mechanism, by involving a spontaneously broken U(1) symmetry, requires the existence of a new pseudoscalar Goldstone boson. It would have zero bare mass, but, through higher-order interactions, acquire a finite but small physical mass. This (as yet hypothetical) light boson has come to be known as the 'axion'. Wilczek and Weinberg have estimated various properties that this particle would have and suggested experimental processes whereby it may be detected (see also Treiman and Wilczek 1978, Kandaswamy et al. 1978, Goldman and Hoffmann 1978). In particular, a rough estimate of the axion mass yields about 100 keV. Several experimental searches have been made to find this particle, but unfortunately they have so far yielded a negative result (Donnelly et al. 1978, Bechis et al. 1979, Belotti et al. 1978). If this situation persists and the axion turns out not to exist in nature, then some other mechanism would have to be found for suppressing strong P and T violation. Possibilities are being suggested, including some that involve a much heavier axion (Yang 1978, Baluni 1978). As of now, a fully satisfactory explanation of the observed P and T violation is unavailable.

Meanwhile, Callan, Dashen and Gross have been continuing their prodigous efforts to extract as much physics as possible about the structure of real hadrons by applying instanton methods and related techniques to QCD. We have already referred to, and used, their early work while discussing the basics of instanton physics. Their subsequent papers (Callan et al. 1979a, 1979b, 1980) go well beyond the dilute instanton-gas approximation, in the way of refinements and generalisations of the semiclassical method. Among the results claimed are the emergence of a bag-like structure of hadrons, possible quark confinement due to merons, and a sharp rise in the running gauge coupling constant $\bar{g}$ as the distance scale is increased. The last-mentioned sharp rise in $\bar{g}$, bridging the weak- and strong-coupling domains of non-abelian gauge theories, has also been obtained by Creutz (1980) in a computer simulation of such a theory. Callan et al. (1980) view Creutz's result as a vindication of their instanton-based theories.

An opposing viewpoint questions the usefulness of instanton-based results in the context of QCD, particularly because of the strong-coupling, large-instanton problem described in section 10.5. We have already referred to the work of Witten (1979a) which recommends the $1/N$ expansion as a more meaningful approximation to SU(3) gauge theory and, furthermore, argues that instanton effects disappear in the leading $N \to \infty$ limit of this expansion. However, Jevicki (1979), finds some instanton effects persisting even in the $N \to \infty$ limit of the CP_N model; see also Cant (1980), and Neuberger (1980).

Apart from these basic questions, a lot of work has been done on the technical front of instanton physics. In particular, methods have been developed for evaluating instanton determinants, i.e. the determinant of quantum fluctuations (of bosons and fermions) about instanton solutions. Recall that in chapter 5 we displayed in detail the calculation of such a determinant for an illustrative example, when we computed quantum corrections to the kink mass. For subsequent examples, especially gauge-theoretical solutions like the abelian Higgs vortex or the Yang–Mills instanton, we did not evaluate the corresponding determinant of fluctuations. They are contained in factors like B in eq. (10.63) or $B(\lambda_1)$ in eq. (10.100), and their exact value was not needed for our subsequent discussion. If one wishes to study instanton effects in more detail, evaluation of such fluctuation determinants is essential. Following the pioneering work of 'tHooft (1976b), quantum fluctuation corrections in the presence of instantons have been studied by a lot of people for a variety of models. As well as the papers cited already in section 10.5, some more references are Brown and Creamer (1978), Berg and Luscher (1979a, b), Levine (1979), Levine and Yaffe (1979), Bernard (1979) and Osborn and Moody (1980).

As we remarked in chapter 10, the tunnelling effects we have derived in the Euclidean functional-integral formalism should also be obtainable in the real-time formalism, as well as by generalising the ordinary Schrödinger equation and its WKB approximation to field theory. Studies of this nature have been done by Gervais and Sakita (1977), Bitar and Chang (1978a, b), De Vega et al. (1978) and Bender and Rothe (1978).

We introduced the notion of meron solutions in section 10.5. Apart from the work of De Alfaro et al. (1977) and Callan et al. (1978a) cited already, classical meron solutions as well as their role in confinement have been studied by Glimm and Jaffe (1978a, b). Interpolating solutions bridging merons and instantons have been presented by Khare (1980) for σ-models and Baseyan and Matinyan (1980) for Yang–Mills theory.

APPENDIX A

For any classical orbit, the factor Δ_1 in (6.59) is given by

$$\Delta_1 = \int \mathscr{D}[y(t)] \exp\left(\frac{\mathrm{i}}{\hbar}\int_0^T (\tfrac{1}{2}\dot{y}^2 - \tfrac{1}{2}y^2 V''(t))\,\mathrm{d}t\right), \tag{A.1}$$

where

$$V''(t) = V''(q_{\mathrm{cl}}(t)). \tag{A.2}$$

$q_{\mathrm{cl}}(t)$ is a classical periodic path of period T and

$$y(0) = y(T) = 0. \tag{A.3}$$

Here the basic period of the path need not be T, but T/n for any integer n. If (A.1) is written on a time lattice it is just a Gaussian multiple integral. It can be evaluated exactly if we could calculate the determinant of the operator $(-\partial^2/\partial t^2 - V''(t))$ that occurs in the exponent, subject to the boundary conditions (A.3). To find this determinant, we effectively diagonalise it by a change of variables:

$$z(t) \equiv y(t) - \int_0^t \frac{\dot{f}(t')}{f(t')}\, y(t')\,\mathrm{d}t', \tag{A.4}$$

where $f(t)$ is the solution of the zero eigenvalue equation of the operator in question:

$$(\partial^2/\partial t^2 + V''(t))f(t) = 0. \tag{A.5}$$

The inverse of (A.4) is

$$y(t) = f(t)\int_0^t \frac{\dot{z}(t')}{f(t')}\,\mathrm{d}t'. \tag{A.6}$$

The advantage of this change of variables is, as can be checked by direct substitution,

$$\int_0^T (\tfrac{1}{2}\dot{y}^2 - \tfrac{1}{2}V''(t)y^2)\mathrm{d}t = \int_0^T (\tfrac{1}{2}\dot{z}^2)\mathrm{d}t. \tag{A.7}$$

Thus, in terms of $z(t)$, the action behaves like that of a free particle! Now, if time is divided into a lattice $t_0 = 0, t_1, \ldots, t_{N-1}, t_N = T$, as in (6.8), then the path integral involves only $\prod_{i=1}^{N-1} \mathrm{d}y(t_i)$. The end points $y(0)$ and $y(T)$ are fixed by the condition (A.3). It will be useful to add an extra integration over $y_N = y(T)$ by multiplying (A.1) by

$$1 = \int \mathrm{d}y_N \delta(y_N) = \int_{-\infty}^{\infty} \frac{\mathrm{d}\alpha}{2\pi} \int \mathrm{d}y(T) \mathrm{e}^{\mathrm{i}\alpha y(T)}$$

$$= \int \frac{\mathrm{d}\alpha}{2\pi} \int \mathrm{d}y(T) \exp\left(\mathrm{i}\alpha f(T) \int_0^T \frac{\dot{z}(t')}{f(t')}\mathrm{d}t'\right). \tag{A.8}$$

When (A.1) is multiplied by (A.8) and variables changed to $z(t)$, we have

$$\Delta_1 = J \int \frac{\mathrm{d}\alpha}{2\pi} \int \mathrm{d}z(T) \int \mathscr{D}[z(t)] \exp\left[\frac{\mathrm{i}}{\hbar} \int_0^T \left(\frac{\dot{z}^2}{2} + \alpha\hbar \frac{f(T)}{f(t)}\dot{z}\right)\mathrm{d}t\right], \tag{A.9}$$

where J is the Jacobian of the transformation (A.4). Going to the time lattice,

$$\Delta_1 = \lim_{\substack{N\to\infty \\ N\varepsilon = T}} \left\{ J \int \frac{\mathrm{d}\alpha}{2\pi} \int \prod_{i=1}^{N} \left(\frac{\mathrm{d}z_i}{g}\right) \exp\left[\frac{\mathrm{i}}{\hbar} \sum_{i=1}^{N} \left(\frac{(z_i - z_{i-1})^2}{2\varepsilon} + \hbar\alpha f(T) \frac{z_i - z_{i-1}}{f_i}\right)\right]\right\}, \tag{A.10}$$

where $g = (2\pi\mathrm{i}\varepsilon\hbar)^{1/2}$ is the correct measure factor. On changing variables to

$$b_i \equiv z_i - z_{i-1}, \qquad i = 1, \ldots, N, \tag{A.11}$$

this factorises into a product of simple Gaussians in b_i, giving

$$\Delta_1 = \lim_{\substack{N\to\infty \\ N\varepsilon = T}} \left[J \int \frac{\mathrm{d}\alpha}{2\pi} \exp\left(-\mathrm{i}\hbar \frac{\alpha^2}{2} f^2(T) \sum_{i=1}^{N} (\varepsilon/f_i^2)\right)\right]$$

$$= J\int\frac{\mathrm{d}\alpha}{2\pi}\exp\left(-\frac{\mathrm{i}\hbar\alpha^2}{2}f^2(T)\int_0^T\frac{\mathrm{d}t}{f^2(t)}\right)$$

$$= J\left(\frac{-\mathrm{i}}{2\pi\hbar}\right)^{1/2}\left(f^2(T)\int_0^T\frac{\mathrm{d}t}{f^2(t)}\right)^{-1/2}. \tag{A.12}$$

The Jacobian of the transformation (A.4), which was a Volterra equation, is obtained in terms of its kernel (Gel'fand and Yaglom 1960):

$$J(T) = \exp\left(\tfrac{1}{2}\int_0^T(\dot{f}/f)\mathrm{d}t\right) = \exp\{\tfrac{1}{2}\ln[f(T)/f(0)]\}$$

$$= [f(T)/f(0)]^{1/2}. \tag{A.13}$$

Inserting this into (A.12), we obtain

$$\Delta_1 = \left(\frac{-\mathrm{i}}{2\pi\hbar}\right)^{1/2}\left(f(T)f(0)\int_0^T\frac{\mathrm{d}t}{f^2(t)}\right)^{-1/2}. \tag{A.14}$$

This quantity is simply related to the parent periodic orbit $q_{\mathrm{cl}}(t)$. First note that

$$f(t) = \frac{\mathrm{d}}{\mathrm{d}t}(q_{\mathrm{cl}}(t)) \tag{A.15}$$

is a solution of (A.5). Since the orbit $q_{\mathrm{cl}}(t)$ is periodic,

$$f(T) = f(0) = \dot{q}_{\mathrm{cl}}(0). \tag{A.16}$$

Also

$$\int_0^T\frac{\mathrm{d}t}{f^2(t)} = \int_0^T\frac{\mathrm{d}t}{(\dot{q}_{\mathrm{cl}}(t))^2} = 2n\int_{q_1}^{q_2}\frac{\mathrm{d}q_{\mathrm{cl}}}{(\dot{q}_{\mathrm{cl}}(t))^3}$$

$$= 2n\int_{q_1}^{q_2}\frac{\mathrm{d}q}{(2(E-V))^{3/2}}, \tag{A.17}$$

where the classical orbit may have completed n traverses of a basic periodic path in time T, and q_1, q_2 are the turning points. In the last equality we have used

$$\tfrac{1}{2}\dot{q}^2 = E - V(q)$$

and dropped the subscript indicating that everything corresponds to the classical orbit. Now,

$$T = 2n \int_{q_1}^{q_2} \frac{\mathrm{d}q}{[2(E - V(q))]^{1/2}}. \tag{A.18}$$

Hence,

$$2n \int_{q_1}^{q_2} \frac{\mathrm{d}q}{[2(E-V)]^{3/2}} = -\frac{\mathrm{d}T}{\mathrm{d}E}. \tag{A.19}$$

Using (A.16)–(A.19) in (A.14), we obtain

$$\Delta_1 = \left(\frac{-\mathrm{i}}{2\pi\hbar}\right)^{1/2} \frac{1}{\dot{q}(0)} \left(-\frac{\mathrm{d}E}{\mathrm{d}T}\right)^{1/2}. \tag{A.20}$$

Remember that $q(0)$ is the end point of both ends of the path. For the periodic orbit, $q(0)$ is just one point on the orbit. When Δ_1 is inserted into the trace integral in (6.59), we obtain

$$G(T) = \sum_n \int \mathrm{d}q(0) \exp\left(\frac{\mathrm{i}}{\hbar} S_{\mathrm{cl}}^{(n)}(\mathrm{T})\right) \frac{1}{\dot{q}(0)} \left(-\frac{\mathrm{d}E}{\mathrm{d}T}\right)^{1/2} \left(\frac{-\mathrm{i}}{2\pi\hbar}\right)^{1/2}. \tag{A.21}$$

But over a closed orbit, S_{cl} and $\mathrm{d}E/\mathrm{d}T$ are independent of the starting point $q(0)$; also, every orbit is counted twice to include its time-reversed partner. Then

$$2\int_{q_1}^{q_2} \frac{\mathrm{d}q(0)}{\dot{q}(0)} = \oint \frac{\mathrm{d}q(0)}{\dot{q}(0)} = \oint \mathrm{d}t = T/n. \tag{A.22}$$

is the period per cycle.

We thus have the desired result (6.62)–(6.63) except for the phase factor $\mathrm{e}^{\mathrm{i}\pi n}$. This comes from turning points, whose singular nature we had not worried about so far. Near a turning point, say $q = q_2$, the velocity $\dot{q}$ has a zero. Thus, the integral $\int \mathrm{d}t/\dot{q}^2 = \int \mathrm{d}q/\dot{q}^3$ is singular as either limit approaches a turning point, although the entire object of the form

$$\dot{q}_a \dot{q}_b \int_{q_a}^{q_b} \frac{dq}{\dot{q}^3}$$

that occurs in (A.14) is not divergent as either q_a or q_b approaches a turning point. This can be checked by expanding $\dot{q} = [2(E - V(q))]^{1/2}$ about the turning point where it has a square root zero. But although Δ_1 is finite it acquires additional phases due to the cut structure produced by the singularity of $\dot{q}$ everytime the path goes through a turning point. The reader is referred to Keller (1958) and Levit and Smilansky (1977) where a discussion is given of why this phase is $e^{i\pi/2}$ per turning point. A similar phase in the wavefunction also arises in the conventional treatment of the WKB method (Landau and Lifshitz 1958).

For a path which traverses the same one-dimensional orbit n times (i.e. $2n$ turning points), the resultant phase factor is $e^{in\pi}$. On inserting this into (A.21), we obtain the results in (6.62)–(6.63).

APPENDIX B

We have seen in chapter 8 that when we compute quantum corrections to the contribution of classical solutions zero-modes arise due to continuous symmetries, and have to be replaced by the corresponding collective coordinates. We shall now show how this replacement may be done for the case of a one-parameter translational symmetry, in the functional-integral formalism. This discussion complements to some small extent that of sections 8.3 and 8.4 where canonical operator methods were used. We shall illustrate the procedure using the simple example of the one-instanton contribution in one-dimensional quantum mechanics. In particular, our discussion will apply to the periodic potential problem (PPP) of chapter 10. We shall derive eq. (10.13). The notation in section 10.1 will be employed.

Let us begin with the functional integral in eq. (10.10). We are of course interested in the limit where τ, the Euclidean transition time (i.e. the range of τ') tends to infinity. In this limit the Euclidean action of PPP is translationally invariant in τ'. $q_{\rm cl}(\tau'-\tau_0)$ is an instanton solution for all real τ_0. We expand $q(\tau')$ about $q_{\rm cl}(\tau'-\tau_0)$, for any given τ_0; then,

$$q(\tau') \equiv q_{\rm cl}(\tau'-\tau_0)+\eta(\tau') \tag{B.1}$$

$$S_{\rm E}[q(\tau')] = S_{\rm E}[q_{\rm cl}(\tau'-\tau_0)]+\tfrac{1}{2}\int \eta(\tau')\hat{O}\eta(\tau')\,{\rm d}\tau'+{\rm O}(\eta^3) \tag{B.2}$$

where

$$\hat{O} \equiv -\frac{\partial^2}{\partial\tau'^2}+\left.\frac{{\rm d}^2 V}{{\rm d}q^2}\right|_{q_{\rm cl}(\tau'-\tau_0)} \tag{B.3}$$

and

$$S_{\rm E}[q_{\rm cl}(\tau'-\tau_0)] = \int_{-\infty}^{\infty} {\rm d}\tau'\left(\frac{1}{2}\left(\frac{{\rm d}q_{\rm cl}}{{\rm d}\tau'}\right)^2+V[q_{\rm cl}]\right)$$

$$= \int_{-\infty}^{\infty} \mathrm{d}\tau' \left(\frac{\mathrm{d}q_{\mathrm{cl}}}{\mathrm{d}\tau'}\right)^2, \qquad \text{(by 2.20)}$$

$$\equiv S_0. \tag{B.4}$$

S_0 is independent of τ_0 by translation invariance. The normal modes of fluctuations are eigenfunctions $\eta_n(\tau' - \tau_0)$ of the operator $\hat{O}$, obeying

$$\hat{O}\eta_n(\tau' - \tau_0) = \omega_n^2 \eta_n(\tau' - \tau_0). \tag{B.5}$$

In particular the normalised function

$$\eta_0(\tau' - \tau_0) = \frac{1}{\sqrt{S_0}} \left(\frac{\mathrm{d}q_{\mathrm{cl}}(\tau' - \tau_0)}{\mathrm{d}\tau'}\right) \tag{B.6}$$

carries a zero eigenvalue, thanks to translation invariance. We would like to replace integration over this zero-mode by an integration over the location τ_0 of the instanton. This is done by the following trick (Gildener and Patrasciou 1977) which is an adaptation of the method of Faddeev and Popov (1967). Consider the identity

$$\int_{-\infty}^{\infty} \mathrm{d}\tau_0 \delta[(\int_{-\infty}^{\infty} \mathrm{d}\tau' \eta_0(\tau' - \tau_0) q(\tau')) - R]\, \Delta[q] = 1 \tag{B.7}$$

where $\delta[\ldots]$ is the Dirac δ-function of its argument,

$$\Delta[q] \equiv \int_{-\infty}^{\infty} \frac{\mathrm{d}}{\mathrm{d}\tau_0} (\eta_0(\tau' - \tau_0))\, q(\tau')\mathrm{d}\tau'$$

$$= \int_{-\infty}^{\infty} \eta_0(\tau' - \tau_0) \frac{\mathrm{d}q}{\mathrm{d}\tau'} \mathrm{d}\tau', \tag{B.8}$$

and

$$R \equiv \int_{-\infty}^{\infty} \mathrm{d}\tau' \eta_0(\tau' - \tau_0) q_{\mathrm{cl}}(\tau' - \tau_0). \tag{B.9}$$

Note that, by virtue of (B.6),

$$R = \frac{1}{2\sqrt{S_0}} \left[q_{\mathrm{cl}}^2(\tau' - \tau_0) \right]_{\tau' = -\infty}^{\tau' = +\infty}$$

and is independent of τ_0. For the particular $V(q)$ used in (PPP), R is $2\pi^2/\sqrt{S_0}$. In the identity (B.7) we have tacitly assumed that the argument of the δ-function vanishes only for one value of τ_0 for a given $q(\tau')$. The possibility of several roots and its consequences have been discussed by Hietarinta (1980).

Consider the Euclidean transition amplitude in (10.10).

$$G \equiv \lim_{\tau \to \infty} \langle 2\pi | e^{-H\tau/\hbar} | 0 \rangle = \int \mathscr{D}[q(\tau')] \exp\left(-\frac{1}{\hbar} S_E[q(\tau')] \right). \quad \text{(B.10)}$$

Insert the identity (B.7) into (B.10) and exchange the order of integration. Then

$$G = \lim_{\tau \to \infty} \int_{-\tau/2}^{\tau/2} d\tau_0 \int \mathscr{D}[q(\tau')] \exp\left(-\frac{1}{\hbar} S_E[q(\tau')] \right)$$
$$\times \delta[(\int d\tau' \eta_0(\tau' - \tau_0)\, q(\tau')) - R]\, \Delta[q]. \quad \text{(B.11)}$$

Now, for any given τ_0, expand $q(\tau')$ in the functional integral about $q_{cl}(\tau' - \tau_0)$, using (B.1)–(B.4). In the leading Gaussian approximation

$$G = \lim_{\tau \to \infty} \int_{-\tau/2}^{\tau/2} d\tau_0 e^{-S_0/\hbar} \Delta[q_{cl}(\tau' - \tau_0)] \int \mathscr{D}[\eta(\tau')]$$
$$\times \exp\left(-\frac{1}{2\hbar} \int d\tau' \eta \hat{O} \eta \right) \delta\left[\int \eta_0(\tau' - \tau_0) \eta(\tau') d\tau' \right]. \quad \text{(B.12)}$$

In this approximation, we have neglected $O(\eta^3)$ terms in the exponent and replaced $q(\tau')$ by $q_{cl}(\tau' - \tau_0)$ in $\Delta[q]$, precisely as required by the stationary-point approximation first introduced in eq. (6.13). In the argument of the δ-function, we have used (B.9) to cancel the term R. Now, all the terms in the integral over τ_0 in (B.12) are independent of τ_0. S_0 is independent of τ_0; so is

$$J \equiv \Delta[q_{cl}(\tau' - \tau_0)] = \int \eta_0(\tau' - \tau_0) \frac{dq_{cl}(\tau' - \tau_0)}{d\tau'} d\tau'.$$
$$= \sqrt{S_0}, \quad \text{(B.13)}$$

using (B.6) and (B.4). Finally, although τ_0 appears in the functional integral over $\eta(\tau')$, the value of this functional integral will not depend on τ_0. This will be evident if we decompose $\eta(\tau')$ in terms of the normal modes

$\eta_n(\tau' - \tau_0)$ in (B.5), and write the functional integral as a multiple integral over the mode coefficients. It is just a Gaussian integral over all functions $\eta(\tau')$ orthogonal to the zero mode, and can be written as $B_E(\tau)\{\mathrm{Det}'[\hat{O}]\}^{-1/2}$, where $B_E(\tau)$ is just the usual measure factor in the functional integral and $\mathrm{Det}'[\hat{O}] = \prod_{n>0}\omega_n^2$ stands for the determinant of $\hat{O}$ in the space orthogonal to the zero-mode. Putting this together, (B.12) reduces to

$$G = \lim_{\tau\to\infty} \tau e^{-S_0/\hbar} J\, B_E(\tau)\{\mathrm{Det}'[\hat{O}]\}^{-1/2}. \tag{B.14}$$

This is the result given in (10.13). If we compare (B.14) with the naive Gaussian approximation (10.12) (where the zero-mode was not carefully treated) we see that the zero-mode has been replaced by an integration over the corresponding collective coordinate τ_0, with an associated Jacobian $J = \sqrt{S_0}$.

Incidentally, as pointed out in chapter 6, the action in most cases of interest may be written in the form $S[q] = (1/g^2)\bar{S}[\bar{q}]$, where g is the coupling constant, which is required to be weak, and $\bar{S}[\bar{q}]$, in terms of a suitably rescaled variable $\bar{q}$, is independent of g. Hence the action, S_0, of the instanton will have the form $S_0 = \bar{S}_0/g^2$, and the Jacobian J in (B.13) is proportional to $1/g$. We have demonstrated this only for the example of a single zero-mode due to translation symmetry, but the result is more general (see for instance Coleman 1979). For *each* zero-mode which is replaced by the corresponding collective coordinate, the resulting Jacobian will pick up a factor of $1/g$.

REFERENCES

Abbott, L. F. (1978), Nucl. Phys. **B139**, 159.
Abers, E. S. and Lee, B. W. (1973), Phys. Rep. **9C**, 1.
Ablowitz, M. J., Kaup, D. J., Newell, A. C. and Segur, H. (1973), Phys. Rev. Lett. **31**, 125
Actor, R. (1979), Rev. Mod. Phys. **51**, 461.
Adler, S L. (1969), Phys. Rev. **177**, 2426.
Adler, S. L. (1979), Phys. Rev. **D19**, 2997.
Adler, S. L. and Dashen, R. F. (1968), 'Current Algebras and Applications to Particle Physics', W. A. Benjamin Inc. New York and Amsterdam.
Affleck, I. (1980) Nucl. Phys. **B162**, 461.
Appelquist, T. and Shankar, R. (1978), Phys. Rev. **D18**, 2952.
Arafune, J., Freund, P. G. O. and Goebel, C. J. (1975), J. Math. Phys. (N. Y.) **16**, 433.
Arefieva, I. Ya. and Korepin, V. E. (1974), Pizma v. Zh. E. T. F. **20**, 680. (JETP. Lett. **20**, 312, (1974)).
Atiyah, M. and Singer, I. (1968), Ann. Math. **87**, 484.
Atiyah, M., Patodi, V. and Singer, I. (1976), Math. Proc. Camb. Phil. Soc., **79**, 71.
Atiyah, M. and Ward, A. (1977), Comm. Math. Phys. **55**, 117.
Atiyah, M. F., Hitchin, N. J., Drinfeld, V. G. and Manin Yu. I. (1978), Phys. Lett. **65A**, 185.
Baacke, J. and Rothe, H. J. (1977), Nucl. Phys. **B118**, 371.
Babelon, O. (1977), Nucl. Phys. **B131**, 519.
Bais, F. A. and Weldon, H. A. (1978), Phys. Rev. Lett. **41**, 601.
Balachandran, A. P., Stern, A. and Trahern, G. (1979), Phys. Rev. **D19**, 2416.
Baluni, V. (1978), Phys. Rev. Lett. **40**, 1358.
Bardakci, K. and Samuel, S. (1978), Phys. Rev. **D18**, 2849.
Bardakci, K. and Samuel, S. (1979), Phys. Rev. **D19**, 2357.
Bardeen, W. A. (1974), Nucl. Phys. **B75**, 246.
Bardeen, W. A., Chanowitz, M. S., Drell, S. D., Weinstein, M. and Yan, T. M. (1975), Phys. Rev. **D11**, 1094.
Barnard, T. (1973), Phys. Rev. **A7**, 373.
Barone, A., Esposito, F., Magee, C. J. and Scott, A. C. (1971), Riv. Nuovo Cim., **1**, 227.
Baseyan, G. Z. and Matinyan, S. G. (1980), JETP Lett. **31**, 70.
Bechis, D. J., Dombeck, T. W., Ellsworth, R. W., Sager, E. V., Steinberg, P. H., Tieg, L. J., Yoh, J. K. and Weitz, R. L. (1979), Phys. Rev. Lett. **42**, 1511.
Belavin, A. A., Polyakov, A. M., Schwartz, A. S. and Tyupkin, Yu. S. (1975), Phys. Lett. **59B**, 85.
Belavin, A. A. and Polyakov, A. M. (1975), JETP Lett. **22**, 245.
Belavin, A. A. and Polyakov, A. M. (1977), Nucl. Phys. **B123**, 429.
Bell, J. S. and Jackiw, R. (1969), Nuovo Cim. **60A**, 47.

Belotti, E., Fiorini, E. and Zanotti, L. (1978), Phys. Lett. **76B**, 223.
Bender, C. M. and Wu, T. T. (1973), Phys. Rev. **D7**, 1620.
Bender, C. M. and Wu, T. T. (1976), Phys. Rev. Lett. **37**, 117.
Bender, I. and Rothe, H. J. (1978), Nucl. Phys. **B142**, 177.
Berezin, F. A. (1966), 'The Method of Second Quantisation', Academic Press, New York and London.
Berg, B. and Luscher, M. (1979a), Comm. Math. Phys. **69**, 57.
Berg, B. and Luscher, M. (1979b), Nucl. Phys. **B160**, 281.
Bernard, C. W. and Weinberg, E. J. (1977), Phys. Rev. **D15**, 3656.
Bernard, C. (1979), Phys. Rev. **D19**, 3013.
Bernstein, J. (1974), Rev. Mod. Phys. **46**, 7.
Bernstein, J., Fubini, S., Gell-Mann, M. and Thirring, W. (1960), Nuovo Cim., **17**, 757.
Berry, M. V. and Mount, K. E. (1972), Rep. Prog. Phys. **35**, 315.
Berry, M. V. and Tabor, M. (1977), J. Phys. A: Math. Gen. **10**, 371.
Bishop, A. R., Krumhansl, J. A. and Trullinger, S. E. (1980), Physica **1D**, 1.
Bitar, K. and Chang, S. J. (1978a), Phys. Rev. **D17**, 486.
Bitar, K. and Chang, S. J. (1978b), Phys. Rev. **D18**, 435.
Bjorken, J. D. and Drell, S. D. (1964), 'Relativistic Quantum Mechanics', McGraw-Hill Book Co., New York.
Bjorken, J. D. and Drell, S. D. (1965), 'Relativistic Quantum Fields', McGraw-Hill Book Co., New York.
Bloch, F. (1928), Z. Phys. **52**, 155.
Bogoliubov, N. N. and Tyablikov, S. (1949), Zh.E.T.F. (U.S.S.R.) **19**, 256.
Bogoliubov, N. N. and Shirkov, D. V. (1959), 'Introduction to the Theory of Quantised Fields', Interscience Publishers Inc. (New York).
Bogomol'nyi, E. B. (1976), Sov. J. Nucl. Phys. **24**, 449.
Bogomol'nyi, E. B. (1980), Phys. Lett. **91B**, 431.
Bott, R. (1956), Bull. Soc. Math. France **84**, 251.
Brezin, E., Le Guillou, J. C. and Zinn-Justin, J. (1977a), Phys. Rev. **D15**, 1544.
Brezin, E., Le Guillou, J. C. and Zinn-Justin, J. (1977b), Phys. Rev. **D15**, 1558.
Brezin, E., Mikami, S. and Zinn-Justin, J. (1980), Nucl. Phys. **B165**, 528.
Brown, L. S. and Creamer, D. B. (1978), Phys. Rev. **D18**, 3695.
Brown, L., Carlitz, R. and Lee, C. (1977), Phys. Rev. **D16**, 417.
Cahill, K. (1974), Phys. Lett. **53B**, 174.
Caldi, D. C. (1977), Phys. Rev. Lett. **39**, 121.
Callan, C. G. Jr., Dashen, R. F. and Gross, D. J. (1976), Phys. Lett. **63B**, 334.
Callan, C. G. Jr., Dashen, R. F. and Gross, D. J. (1977a), Phys. Lett. **66B**, 375.
Callan, C. G. Jr., Dashen, R. F. and Gross, D. J. (1977b), Phys. Rev. **D16**, 2526.
Callan, C. G. Jr., Dashen, R. F. and Gross, D. J. (1978a), Phys. Rev. **D17**, 2717.
Callan, C. G. Jr., Dashen, R. F. and Gross, D. J. (1979a), Phys. Rev. **D19**, 1826.
Callan, C. G. Jr., Dashen, R. F. and Gross, D. J. (1979b), Phys. Rev. **D20**, 3279.
Callan, C. G. Jr., Dashen, R. F. and Gross, D. J. (1980), Phys. Rev. Lett. **44**, 435.
Callan, C. G. Jr., Dashen, R. F., Gross, D. J., Wilczek, F. and Zee, A. (1978b), Phys. Rev **D18**, 4684.
Campbell, D. K. and Liao, Y-T (1976), Phys. Rev. **D14**, 2093.
Cant, R. J. (1980), Phys. Lett. **96B**, 380.
Carlitz, R. D. and Creamer, D. B. (1979), Ann. Phys. (N. Y.) **118**, 429.
Castillejo, L., Dalitz, R. H., and Dyson, F. J. (1956), Phys. Rev. **101**, 453.

Chadha, S., D'Adda, A., diVecchia, P. and Nichodemi, F. (1977), Phys. Lett. **72B**, 103.
Chakrabarti, A. (1975), Nucl. Phys. **B101**, 159.
Chang, S. J., Lee, B. W. and Ellis, S. D. (1975), Phys. Rev. **D11**, 3572.
Chang, S. S. (1978), Phys. Rev. **D17**, 2595.
Chew, G. F. (1965), 'High Energy Physics', Les Houches, Ed. C. de Witt and M. Jacob, Gordon and Breach, London, p. 189.
Christ, N. H. and Lee, T. D. (1975), Phys. Rev. **D12**, 1606.
Christ, N. H., Weinberg, E. J. and Stanton, N. K. (1978), Phys. Rev. **D18**, 2013.
Christ, N. H. and Jackiw, R. (1980), Phys. Lett. **91B**, 228.
Coleman, S. (1973), Comm. Math. Phys. **31**, 259.
Coleman, S. (1975a) in 'Laws of Hadronic Matter', Ed. A. Zichichi, Academic Press, New York and London.
Coleman, S. (1975b), Phys. Rev. **D11**, 2088.
Coleman, S. (1977a) 'Classical lumps and their Quantum Descendants', 1975 Erice Lectures published in 'New Phenomena in Sub-Nuclear Physics', Ed. A. Zichichi, Plenum Press, New York.
Coleman, S. (1977b), Comm. Math. Phys. **55**, 113.
Coleman, S. (1979), 'Uses of Instantons', in: 'The Whys of Subnuclear Physics', Ed. A. Zichichi, Plenum Press, New York.
Coleman, S. and Mandula, J. (1969), Phys. Rev. **159**, 1251.
Corrigan, E. and Fairlie, D. B. (1977), Phys. Lett. **67B**, 69.
Corrigan, E., Olive, D. I., Fairlie, D. B. and Nuyts, J. (1976), Nucl. Phys. **B106**, 475.
Cremmer, E., Schaposnik, F. and Scherk, J. (1976), Phys. Lett. **65B**, 78.
Cremmer, E. and Scherk, J. (1978), Phys. Lett. **74B**, 341.
Creutz, M. (1975), Phys. Rev. **D12**, 3126.
Creutz, M. (1980), Phys. Rev. **D21**, 2308.
Crewther, R. (1979), Riv. Nuovo Cim. **2**, 63.
Crutchfield II, W. Y. (1979), Phys. Rev. **D19**, 2370.
D'Adda, A., Luscher, M., and DiVecchia, P. (1978), Nucl. Phys. **B146**, 63.
Dashen, R. F., Hasslacher, B. and Neveu, A. (1974a), Phys. Rev. **D10**, 4114.
Dashen, R. F., Hasslacher, B. and Neveu, A. (1974b), Phys. Rev. **D10**, 4130.
Dashen, R. F., Hasslacher. B. and Neveu, A., (1975a), Phys. Rev. **D11**, 3424.
Dashen, R. F., Hasslacher, B. and Neveu, A. (1975b), Phys. Rev. **D12**, 2443.
Dashen, R. F., Ma, S. K. and Rajaraman, R. (1975c), Phys. Rev. **D11**, 1499.
De Alfaro, V., Fubini, S., Furlan, G. (1977), Phys. Lett. **65B**, 1631.
Derrick, G. H. (1964), J. Math. Phys. **5**, 1252.
Deser. S. (1976), Phys. Lett. **64B**, 463.
de Vega, H. J. (1976), Nucl. Phys. **B115**, 411.
de Vega, H. J., Gervais, J. L. and Sakita, B. (1978), Nucl. Phys. **B143**, 125.
Din, A. J. and Zakrewski, W. J. (1980), Nucl. Phys. **B168**, 173.
Dirac, P. A. M. (1931), Proc. R. Soc. (London) **A133**, 60.
Dirac. P. A. M. (1948), Phys. Rev. **74**, 817.
Dirac, P. A. M. (1950), Can. J. Math. **2**, 129.
Dobrushin, R. L. and Minlos, R. A. (1973), Func. Anal. Appl. **7**, 324.
Donnelly, T. W., Freedman, S. J., Lytel, R. S., Peccei, R. D. and Schwartz, M. (1978), Phys. Rev. **D18**, 1607.
Duncan, A. (1978), Phys. Rev. **D18**, 1988.

Dyson, F. J. (1952), Phys. Rev. **85**, 631.
Edwards, S. F. and Gulyaev, Y. V. (1964), Proc. R. Soc. **A279**, 229.
Eguchi, T. and Hanson, A. J. (1978), Phys. Lett. **74B**, 249.
Eichenherr, H. (1978), Nucl. Phys. **B146**, 215.
Englert, F. and Brout, R. (1964), Phys. Lett. **13**, 321.
Ezawa, Z. F. (1978), Phys. Rev. **D18**, 2091.
Ezawa, T. (1979), Phys. Lett. **81B**, 325.
Faddeev, L D. (1974), Leningrad Preprint MPI – PAE/16.
Faddeev, L. D. (1975), Pisma v. Zh.E.T.F., **21**, 141.
Faddeev, L. D. and Popov, V. N. (1967), Phys. Lett. **25B**, 29.
Faddeev. L. D. and Takhtajan, L. A. (1974), Teor, Mat. Fiz. **21**, 160 (see also communications of JINR, E2–7998, Dubna (1974)).
Faddeev, L. D., Kulish, P. E. and Korepin, V. E. (1975), Pisma v. Zh.E.T.F., **21**, 302 (JETP Lett. **21**, 138, (1975)).
Faddeev, L. D. and Korepin, V. E. (1975), Teor. Mat. Fiz., **25**, 147.
Faddeev, L. D. and Korepin, V. E. (1976), Phys. Lett. **63B**, 435.
Faddeev, L. D. and Korepin, V. E. (1978), Phys. Rep. **42C**, 1.
Faddeev, L. D. and Slavnov, A. A. (1980), 'Gauge Fields—Introduction to Quantum Theory' Benjamin Cummings Publishing Co., Reading, Mass. U.S.A.
Fainberg, V. Ya. and Iofa, M. Z. (1980), Nucl. Phys. **B168**, 495.
Fateev, V. A., Frolov, I. V., and Schwarz, A. S. (1979a), Nucl. Phys. **B154**, 1.
Fateev, V. A., Frolov, I. V. and Schwarz, A. S. (1979b), Yad. Phys. **8**.
Feynman, R. P. and Hibbs, A. R. (1965), 'Quantum Mechanics and Path Integrals', McGraw-Hill Book Co., New York.
Finkelstein, D. (1966), J. Math. Phys. **7**, 1218.
Finkelstein, D. and Misner, C. (1959), Ann. Phys. (N.Y.) **6**, 230.
Floquet, G. (1883), Ann. de l'Ecole Norm. Sup. (2), XII.
Frampton, P. (1976), Phys. Rev. **D14**, 528.
Frautschi, S. (1963), 'Regge Poles and *S*-Matrix Theory', W. A. Benjamin Inc., New York.
Friedberg, R., Lee, T. D. and Sirlin, A. (1976a), Phys. Rev. **D13**, 2739.
Friedberg, R., Lee, T. D. and Sirlin, A. (1976b), Nucl. Phys. **B115**, 1.
Friedberg, R., Lee, T. D. and Sirlin, A. (1976c), Nucl. Phys. **B115**, 32.
Friedberg, R. and Lee, T. D. (1977), Phys. Rev. **D15**, 1694.
Fritzsch, H. and Gell-Mann, M. (1972), Proceedings of the 16th International Conference on High Energy Physics held in Chicago–Batavia, Vol. 2, p. 135.
Fritzsch, H., Gell-Mann, M. and Leutwyler, H. (1973), Phys. Lett. **47B**, 365.
Frohlich, J. (1976), Comm. Math. Phys. **47**, 269.
Gardner, C. S. Greene, J. M., Kruskal, M. D. and Miura, R. M. (1967). Phys. Rev. Lett. **19**, 1095.
Gel'fand, J. M. and Yaglom, A. M., (1960), J. Math. Phys. **1**, 48.
Georgi, M. and Glashow, S. L. (1972), Phys. Rev. Lett. **28**, 1494.
Gervais, J-L. and Sakita, B., (1975), Phys. Rev. **D11**, 2943.
Gervais, J-L., Jevicki, A. and Sakita, B. (1975), Phys. Rev. **D12**, 1038.
Gervais, J-L. and Jevicki, A. (1976a), Nucl. Phys. **B110**, 93.
Gervais, J-L. and Jevicki, A. (1976b), Nucl. Phys. **B110**, 113.
Gervais, J-L. and Sakita, B. (1977), Phys. Rev. **D16**, 3507.
Geshkenbein, B. V. and Ioffe, B. L. (1980), Nucl. Phys. **B166**, 340.

Giambiaggi, J. J. and Rothe, K. D. (1977), Nucl. Phys. **B129**, 111.
Gibbons, G. W. and Perry, M. J. (1978), Nucl. Phys. **B146**, 90.
Gibbons, G. W. and Hawking, S. W. (1979), Comm. Math. Phys. **66**, 291.
Gildener, E. and Patrascioiu, A. (1977), Phys. Rev. **D16**, 423.
Glaser, V. (1958), Nuovo Cim. **9**, 990.
Glimm, J., Jaffe, A. and Spencer, T. (1975), Comm. Math. Phys. **45**, 203.
Glimm, J. and Jaffe, A. (1978a), Phys. Rev. Lett. **40**, 277.
Glimm, J. and Jaffe, A. (1978b), Phys. Lett. **73B**, 167.
Goddard, P., Nuyts, J., and Olive, D. (1977), Nucl. Phys. **B125**, 1.
Goddard, P., Olive, D. (1978), Rep. Prog. Phys. **41**, 1357.
Goldhaber, A. S. (1976), Phys. Rev. Lett. **36**, 1122.
Goldman, T. and Hoffmann, C. M. (1978), Phys. Rev. Lett. **40**, 220.
Goldstein, H. (1950), 'Classical Mechanics', Addison–Wesley Publishing Co., Reading, Mass., U.S.A.
Goldstone, J. (1961), Nuovo Cim. **19**, 154.
Goldstone, J., Salam, A. and Weinberg, S. (1962), Phys. Rev. **127**, 965.
Goldstone, J. and Jackiw, R. (1975), Phys. Rev. **D11**, 1486.
Golo, V. and Perelemov, A. M. (1978), Phys. Lett. **79B**, 112.
Gradshteyn, I. S. and Ryzhik, I. M. (1965), 'Tables of Integrals, Series and Products', Academic Press, New York.
Gross, D. J. and Wilczek, F. (1973), Phys. Rev. Lett. **30**, 1343.
Gross, D. J. and Neveu, A. (1974), Phys. Rev. **D10**, 3235.
Guralnik, G. S., Hagen, C. R. and Kibble, T. W. B. (1964), Phys. Rev. Lett. **13**, 585.
Guth, A. and Weinberg, E. J. (1976), Phys. Rev. **D14**, 1660.
Gutzwiller, M. (1967), J. Math. Phys. **8**, 1979.
Gutzwiller, M. (1970), J. Math. Phys. **11**, 1791.
Gutzwiller, M. (1971), J. Math. Phys. **12**, 343.
Hasenfratz, P. and 'tHooft, G. (1976), Phys. Rev. Lett. **36**, 1119.
Hawking, S. W. (1977), Phys. Lett. **60A**, 81.
Hietarinta, J. (1980), Nucl. Phys. **B164**, 343.
Higgs, P. W. (1964), Phys. Lett. **12**, 132.
Higgs, P. W. (1966), Phys. Rev. **145**, 1156.
Hobart, R., (1963), Proc. Phys. Soc. **82**, 201.
Horvath, Z. and Palla, L. (1976), Nucl. Phys. **B116**, 500.
Jackiw, R. (1972) in 'Lectures in Current Algebra and its Applications', Ed. Treiman, S. B., Jackiw, R. and Gross, D. J., Princeton University Press, Princeton, N. J., U.S.A.
Jackiw, R. (1976), in 'Gauge Theories and Modern Field Theories', Ed. R. Arnowitt and P. Nath, M. I. T. Press, Cambridge, Mass., U.S.A.
Jackiw, R. (1977), Rev. Mod. Phys. **49**, 681.
Jackiw, R. (1980), Rev. Mod. Phys. **52**, 661.
Jackiw, R. and Woo, G. (1975), Phys. Rev. **D12**, 1643.
Jackiw, R. and Rebbi, C. (1976a), Phys. Rev. Lett. **36**, 1116.
Jackiw, R. and Rebbi, C. (1976b), Phys. Rev. **D14**, 517.
Jackiw, R. and Rebbi, C. (1976c), Phys. Rev. **D13**, 3398.
Jackiw, R. and Rebbi, C. (1976d), Phys. Rev. Lett. **37**, 172.
Jackiw, R., Nohl, C. and Rebbi, C. (1977), Phys. Rev. **D15**, 1642.
Jackiw, R., Nohl, C. and Rebbi, C. (1978), in 'Particles and Fields', Ed. David Boal and A. N. Kamal, Plenum Publishing Co., New York, p. 199.

Jacobs, L. (1976), Phys. Rev. **D14**, 2739.
Jersak, J., Kiera, M. and Magg, M. (1977), Nuovo Cim. **40A**, 269.
Jevicki, A. (1976), Nucl. Phys. **B117**, 365.
Jevicki, A. (1979), Phys. Rev. **D20**, 3331.
Johnson, K. (1961), Nuovo Cim. **20**, 773.
Julia, B. and Zee, A. (1975), Phys. Rev. **D11**, 2227.
Kadanoff, L. and Ceva, H. (1971), Phys. Rev. **B3**, 3918.
Kandaswamy, J., Salomonson, P. and Schechter, J. (1978), Phys. Rev. **D17**, 3051.
Karowski, M. (1979), Phys. Rep. **49c**, 229.
Karowski, M., Thun, H.-J., Truong, T. T. and Weisz, P. (1977), Phys. Lett. **67B**, 321.
Karowski, M., and Thun, H-J., (1977), Nucl. Phys. **B130**, 295.
Keller, J. (1958), Ann. Phys. (N.Y.) **4**, 180.
Kerman, A. and Klein, A. (1963), Phys. Rev. **132**, 1326.
Khare, A. (1980), J. Phys. A: Math. Gen. **13**, 2253.
Khrustalev, D. A., Razumov, A. Y. and Taranov, A. Yu. (1980), Nucl. Phys. **B172**, 44.
Kilpatrick, J. and Kilpatrick, M. (1948), J. Chem. Phys. **16**, 781.
Kim, S. T. and Woo, C. H. (1979), Nucl. Phys. **B155**, 357.
Kiskis, J. (1977), Phys. Rev. **D15**, 2329.
Kittel, C. (1959) 'Introduction to Solid State Physics', Second Edition, John Wiley and Sons, New York.
Klaiber, B. (1968), 'Lectures in Theoretical Physics', **XA**, Ed. A. Barut and W. Britten, Gordon and Breach, London.
Kogut, J. and Susskind, L. (1975), Phys. Rev. **D11**, 3594.
Krumhansl, J. A. and Schrieffer, J. R. (1975), Phys. Rev. **B11**, 3535.
Kulish, P. P. and Nisimov (1976), JETP. Lett., **24**, 220.
Lamb, Jr. G. L. (1971), Rev. Mod. Phys. **43**, 99.
Landau, L. D. and Lifshitz, E. M. (1958), 'Quantum Mechanics', Pergamon Press Ltd., London and Paris.
Langer, R. E. (1937), Phys. Rev. **51**, 669.
La Rue, G., Fairbank, W. M. and Hebard, A. F. (1977), Phys. Rev. Lett. **38**, 1011.
Lee, T. D. (1976), 'Non-Topological Solitons', Lectures given at the Symposium on Frontier Problems in High Energy Physics, Pisa, Columbia University Preprint Co. 2271–76.
Lee, T. D. and Wick, G. C. (1974), Phys. Rev. **D9**, 2291.
Levine, H. (1979), Nucl. Phys. **B157**, 237.
Levine, H. and Yaffe, L. G. (1979), Phys. Rev. **D19**, 1225.
Levit, S. (1978), Ph.D. Thesis submitted to the Weitzmann Institute of Science, Israel (1978).
Levit, S. and Smilansky, U., (1977), Ann. Phys. (N.Y.) **103**, 198.
Lipatov, L. N. (1976). Pisma v. Zh. E .T .F., **24**, 179 (JETP Letters **24**, 157).
Lipatov, L. N. (1977) Pisma v. Zh. E. T. F., **25**, 116 (JETP Letters **25**, 104).
Luscher, M. (1978), Nucl. Phys. **B135**, 1.
Luscher, M. and Pohlmeyer, K. (1978), Nucl. Phys. **B137**, 46.
Ma, S-K. and Rajaraman, R. (1975), Phys. Rev. **D11**, 1701.
Makhankov, V.G. (1978), Phys. Rep. **35C**, 1.
Mandelstam, S. (1975), Phys. Rev. **D11**, 3026.
Mandelstam, S. (1979), Phys. Rev. **D19**, 2391.
Manton, N. S. (1977), Nucl. Phys. **B126**, 525.
Manton, N. S. (1979), Nucl. Phys. **B150**, 397.
Marciano, W. and Pagels, H. (1975), Phys. Rev. **D12**, 1093.

Marciano, W., Parsa, Z. and Pagels, H. (1977), Phys. Rev. **D15**, 1044.
Marciano, W. and Pagels, H. (1978), Phys. Rep. **36C**, No. 3, 137.
Maslov, V. (1970), Teor. Mat. Fiz., **2**, 21.
Matsumoto, H., Sodano, P. and Umezawa, H. (1979), Phys. Rev. **D19**, 511.
Matveev, V. A. (1977), Nucl. Phys. **B121**, 403.
McDougall, N. A. (1980), Phys. Lett. **89B**, 397.
McGuire, J. B. (1964), J. Math. Phys. **5**, 522.
Mermin, N. D. (1979), Rev. Mod. Phys. **51**, 591.
Michel, L., O'Raifeartaigh, L., and Wali, K. C. (1977), Phys. Lett. **67B**, 198.
Mitra, P. (1977), Phys. Lett. **72B**, 62.
Montonen, C. (1976), Nucl. Phys. **B112**, 349.
Montonen, C. and Olive, D. (1977), Phys. Lett. **72B**, 117.
Morse, P. and Feshbach, H. (1953), 'Methods of Mathematical Physics', McGraw-Hill Book Co., (New York).
Mottola, E. (1978), Phys. Lett. **79B**, 242.
Mukunda, N. and Sudarshan, E. C. G. (1974), 'Classical Dynamics: A Modern Perspective', John Wiley and Sons, New York.
Nambu, Y. (1960), Phys. Rev. Lett. **4**, 386.
Neuberger, H. (1980), Phys. Lett. **94B**, 199.
Nielsen, H. B. and Olesen, P. (1973), Nucl. Phys. **B61**, 45.
Nielsen, N. K. and Schroer, B. (1977), Nucl. Phys. **B127**, 493.
Novikov, V. A., Okun, L. B., Shifman, M. A., Vainshtein, A. I., Voloshin, M. B. and Zakharov, V. I. (1978), Phys. Rep. **41C**, 1.
Olive, D. (1976), Nucl. Phys. **B113**, 413.
Olive, D. (1979), Riv. Nuovo Cim. **2**, 2.
O'Raifaertaigh, L. (1977), Lett. Nuovo Cim. **18**, 205.
O'Raifaertaigh, L., Park, S. Y. and Wali, K. C. (1979), Phys. Rev. **D20**, 1941.
Ore, F. (1977), Phys. Rev. **D16**, 2577.
Osborn, H. and Moody, G. P. (1980), Nucl. Phys. **B173**, 422.
Pagels, H. (1976), Phys. Rev. **D13**, 343.
Pagels, H. (1977), Phys. Lett. **68B**, 466.
Parisi, G. (1977a), Phys. Lett. **66B**, 167.
Parisi, G. (1977b), Phys. Lett. **66B**, 382.
Parke, S. (1980), Nucl. Phys. **B174**, 166.
Pati, J. C. and Salam, A. (1973), Phys. Rev. **D8**, 1240.
Patrasciou, A., (1975), Phys. Rev. **D12**, 523.
Peccei, R. D. and Quinn, H. R. (1977a), Phys. Rev. Lett. **38**, 1440.
Peccei, R. D. and Quinn, H. R. (1977b), Phys. Rev. **D16**, 1791.
Perring, J. K. and Skyrme, T. H. R. (1962), Nucl. Phys. **31**, 550.
Pohlmeyer, K. (1976), Comm. Math. Phys. **46**, 207.
Politzer, H. D. (1973), Phys. Rev. Lett., **30**, 1346.
Polyakov, A. M. (1974), Pisma v. Zh. E.T.F., **20**, 430, (JETP Lett. **20**, 194 (1974).
Polyakov, A. M. (1975a), Pisma v. Zh. E.T.F., **20**, 430.
Polyakov, A M. (1975b), Phys. Lett. **59B**, 82.
Polyakov, A. M. (1977a), Phys. Lett. **72B**, 224.
Polyakov, A. M. (1977b), Nucl. Phys. **B120**, 429.
Prasad, M. K. and Sommerfield, C. H. (1975), Phys. Rev. Lett. **35**, 760.
Rajaraman, R. (1975), Phys. Rep. **21C**, 227.
Rajaraman, R. (1977), Phys. Rev. **D15**, 2866.

Rajaraman, R. (1979), Phys. Rev. Lett. **42**, 200.
Rajaraman, R. and Weinberg, E. J. (1975), Phys. Rev. **D11**, 2950.
Raj Lakshmi, M. (1979), Pramana **12**, 447.
Rossi, P. (1979), Nucl. Phys. **B149**, 170.
Rothe, K. D. and Swieca, J. A. (1978), Nucl. Phys. **B138**, 26.
Rothe, K. D. and Swieca, J. A. (1979), Nucl. Phys. **B149**, 237.
Rubinstein, J. (1970), J. Math. Phys. **11**, 258.
Ruck, H. M. (1980), Nucl. Phys. **B167**, 320.
Saha, M. N. (1936), Indian J. Phys. **10**, 145.
Salam, S. and Strathdee, J. (1976), Phys. Lett. **61B**, 375.
Salomonson, P. (1977), Nucl. Phys. **B121**, 433.
Schaposnik, F. A. (1978), Phys. Rev. **D18**, 1154.
Schrieffer, J. R. (1964), 'Theory of Superconductivity', W. A. Benjamin Inc., New York.
Schroer, B. and Swieca, J. (1977), Nucl. Phys. **B121**, 505.
Schwarz, A. S. (1977), Phys. Lett. **67B**, 172.
Schwinger, J. (1951), Phys. Rev. **82**, 664.
Schwinger, J. (1966), Phys. Rev. **144**, 1087.
Sciuto, S. (1979), Riv. Nuovo Cim. **2**, 16.
Scott, A. C., Chiu, F. Y. F. and Mclaughlin, D. W. (1973), Proc. I.E.E.E. **61**, 1443.
Seeger, A., Donth, H. and Kochendorfer, A. (1953), Z. Phys. **134**, 173.
Shankar, R. (1976), Phys. Rev. **D14**, 1107.
Shankar, R. (1978), Invited talk given at the American Physical Society Meeting, Washington D.C., Yale University Preprint C00-3075-199.
Shankar, R. and Witten, E. (1978a), Phys. Rev. **D17**, 2134.
Shankar, R. and Witten, E. (1978b), Nucl. Phys. **B141**, 349.
Shaw, R. (1955) Ph.D. Thesis, Cambridge University, U.K.
Sinha, A. (1976), Phys. Rev. **D14**, 2016.
Skryme, T.H.R. (1958), Proc. R. Soc. **A247**, 260.
Skryme, T.H.R. (1961), Proc. R. Soc., **A262**, 233.
Skyrme, T.H.R. (1962), Nucl. Phys. **31**, 556.
Sommerfield, C. (1963), Ann. Phys. (N.Y.), **26**, 1.
Steenrod, N. (1951), 'The Topology of Fibre Bundles', Princeton University Press, Princeton, N. J., U.S.A.
Steimann, O. (1978), Nucl. Phys. **B145**, 141.
Subbaswamy, K. R. and Trullinger, S. E. (1980), Phys. Rev. **D22**, 1495.
Suzuki, M. (1978), Phys. Lett. **78B**, 466.
Taubes, C. M. (1980), Comm. Math. Phys. **72**, 277.
Taylor, J. C. (1976), 'Gauge Theories of Weak Interactions', Cambridge University Press, U.K.
Taylor, J.G. (1978), Ann. Phys. (N.Y.) **115**, 153.
Thirring, W. (1958), Ann. Phys. (N.Y.), **3**, 91.
'tHooft, G. (1974a), Nucl. Phys. **B75**, 461.
'tHooft, G. (1974b), Nucl. Phys. **B79**, 276.
'tHooft, G. (1976a), Phys. Rev. Lett., **37**, 8.
'tHooft, G. (1976b), Phys. Rev. **D14**, 3432.
'tHooft, G. (1978), Nucl. Phys. **B138**, 1.
'tHooft, G. (1979), in 'The Whys of Subnuclear Physics', Ed. A. Zichichi, Plenum Press, New York.

Tomboulis, E. (1975), Phys. Rev. **D12**, 1678.

Tomboulis, E. and Woo, G. (1976), Nucl. Phys. **B107**, 221.

Toulouse, G. and Kleman, M. (1976), J. Physique Lett. **37**, 149.

Treiman, S. B. and Wilczek, F. (1978), Phys. Lett. **74B**, 381.

Tyupkin, Y., Fateev, V., and Shvartz, A. (1976), Teor. Mat. Fiz., **26**, 270.

Verwaest, J. (1977), Nucl. Phys. **B123**, 100.

Volovik, G. E. and Mineev, V. P. (1977), Sov. Phys. (J.E.T.P.) **45**, 1186.

Wadia, S. and Yoneya, T. (1977), Phys. Lett. **66B**, 341.

Weinberg, E. J. (1980), Nucl. Phys. **B167**, 500.

Weinberg, S. (1975), Phys. Rev. **D11**, 3583.

Weinberg, S. (1978), Phys. Rev. Lett. **40**, 223.

Weisz, P. (1977), Nucl. Phys. **B112**, 1.

Whitham, G. B. (1974), 'Linear and Non-linear Waves', John Wiley and Sons, New York.

Whittaker, E. T. and Watson, G. N. (1920), 'A Course on Modern Analysis' (3rd Edition), Cambridge University Press, U.K.

Wilczek, F. (1976), Phys. Lett. **65B**, 160.

Wilczek, F. (1977) in 'Quark confinement and field theory', Ed. D. Stump and D. Weingarten, John Wiley and Sons, New York.

Wilczek, F. (1978), Phys. Rev. Lett. **40**, 279.

Wilczek, F. and Zee, A. (1978), Phys. Rev. Lett. **40**, 83.

Wilson, K. G. (1974), Phys. Rev. **D10**, 2445.

Witten, E. (1977), Phys. Rev. Lett. **38**, 121.

Witten, E. (1978), Nucl. Phys. **B142**, 285.

Witten, E. (1979a), Nucl. Phys. **B149**, 285.

Witten, E. (1979b), Lectures at Cargese Summer School (1979), Harvard Preprint HUTP-79/A078.

Wu, T. T. and Yang, C. N. (1975), Phys. Rev. **D12**, 3845.

Yang, C. N. and Mills, R. (1954), Phys. Rev. **96**, 191.

Yang, C. N. and Wu, T. T. (1968), in 'Properties of Matter under Unusual Conditions', Ed. H. Mark and S. Fernbach, Interscience, New York.

Yang, T. C. (1978), Phys. Rev. Lett. **41**, 523.

Zakharov, V. E., Takhtajan, L. A. and Faddeev, L. D. (1974), Dokl. Akad. Sci. USSR, **219**, 1334.

Zamolodchikov, A. B. (1977a), Comm. Math. Phys. **55**, 183.

Zamolodchikov, A. B. (1977b), I.T.E.P. Preprint 112 (U.S.S.R.).

Zamolodchikov, A. B. and Zamolodchikov, Al. B. (1978a), Nucl. Phys. **B133**, 525.

Zamolodchikov, A. B. and Zamolodchikov, Al. B. (1978b), Phys. Lett. **72B**, 481.

Zamolodchikov, A. B. and Zamolodchikov, Al. B. (1979), Ann. Phys. (N.Y.) **120**, 253.

Zwanziger, D. (1968), Phys. Rev. **176**, 1480 and 1489.

INDEX